Urban Microclimate Modelling for Comfort and Energy Studies

Massimo Palme • Agnese Salvati
Editors

Urban Microclimate Modelling for Comfort and Energy Studies

Springer

Editors
Massimo Palme
Catholic University of the North
Antofagasta, Chile

Agnese Salvati
Brunel University London
London, UK

ISBN 978-3-030-65420-7 ISBN 978-3-030-65421-4 (eBook)
https://doi.org/10.1007/978-3-030-65421-4

This Springer imprint is published by the registered company Springer Nature Switzerland AG
The registered company address is: Gewerbestrasse 11, 6330 Cham, Switzerland

Foreword

Cities are both the primary source of energy, carbon, pollution and waste problems facing the world as well as potential wellsprings of solutions in our Anthropocene age. The urban climate anomaly exemplifies this duality of unintended climate, comfort and energetic consequences of urbanisation on the one hand, and the potential offered by its mitigation to act as a framework for the wider societal, environmental and well-being benefits on the other. Yet, the study and, more importantly, the application of urban climate knowledge still remains weak, compared to other socio-environmental conditions associated with cities. Part of the reason is the often narrow focus of the limited studies on the subject, as well as the limited metrics, models and toolkits available to researchers and practitioners. Equally important are the incomplete and erroneous assumptions about urban microclimate by practitioners and policy makes and a weak understanding of what this knowledge could and could not deliver in specific contexts.

I am therefore pleased to be asked to write this foreword to a book that attempts to bridge the gap between the 'big picture' as well as tools and metrics that provide the wherewithal of applicable knowledge. The editors are to be congratulated for bringing together such a variety of voices from across the climatic, socio-economic and developmental contexts of world's cities as well as modellers and practitioners from diverse professional and research backgrounds.

The organisation of the book in three parts is highly useful in addressing the challenges in applying the limited urban climate knowledge to policy and practice of holistic sustainability in cities. With its overview of the energetics, thermodynamic and thermal comfort aspects of urban climate and linking these to the wider sustainability issues in cities, Part I provides a good overview of the wider picture. The exploration of energy and comfort issues in specific climates (high latitude, Mediterranean, tropical and arid climates) adds value to this overview by situating the key issues in specific geographical contexts.

While it is true that urban climate studies are relatively few in number, the situation is changing rapidly in recent years and Part II provides an excellent overview of popular models and approaches. Both the overview of the tools and the mechanics of their utilisation should be of interest to new and experienced researchers aiming to push the urban climate frontier further towards practical application.

Ultimately, the mitigation of the negative consequences of urban climate boils down to three broad strategies: manipulation of urban form, green infrastructure and materials. These are well captured by Part III in different climatic contexts. The inclusion of case studies further enriches the discussions in this section.

We thus have an urban climate compendium of high value in our hands. By placing the modelling of urban climate in the wider context of sustainability as well as showcasing applications and exemplars from different contexts, the present volume could add to our limited but growing understanding of the phenomenon of urban climate and, more importantly, integrate such knowledge within actions that address the wider challenges facing the twenty-first century cities.

The Built Environment Asset Management (BEAM) Centre Rohinton Emmanuel
Glasgow Caledonian University
Glasgow, UK

October 2020

Preface

The idea of this book has roots at the Barcelona School of Architecture, UPC, where we both achieved the PhD degree in Architecture, Energy and Environment and happened to meet in 2014. Since then, we established a very fruitful scientific collaboration based on the common interest for the environmental performance of cities and urban scale energy and climate phenomena. We both believe that new multi-scale approaches have to be envisioned to assess the reciprocal interactions between urban climate and building performance and so to improve indoor and outdoor thermal comfort and decrease the energy consumption of buildings in urban environments.

In October 2018, Springer approached us to submit a book project and we accepted the challenge without hesitation. We wanted this book to be useful to both academics and practitioners and we started to contact both established and emerging experts working on the many aspects related to urban climate modelling: physicists and climatologists, geographers, architects and urban planners, building and mechanical engineers. Since from the beginning, we wanted to provide a multidisciplinary and holistic approach to this topic and we reflected this intention in the structure of the book: one introductory section describing the importance and complexity of the energy and comfort issues in different urban contexts and climates; one more technical part addressing the description and development of models and tools for urban climate modelling; and one last section reporting examples of practical applications of urban climate modelling to real planning problems.

We also committed to maintain a good balance among the countries, institutions and gender of the authors that contributed to the book, to benefit from the diversity of the points of view. We received 36 abstract proposals, which turned out in the 24 chapters of this book after our screening and two rounds of peer review. It was a long process and we want to acknowledge and thank all the people that took part in the peer review, editing and design of the final product. Miss Amanda Quinn was the first Springer person that contacted us and this book would have not existed without her intuition. Miss Cynthya Pushparaj was our contact during the editing process, attending all ours questions and resolving any issue we had. Many thanks also to Mr. Michel McCabe, Miss Zoe Kennedy and Mr. Brian Halm that ware involved in the editorial process. A warm thanks to our institutions, the Catholic University of the North and the Brunel University London, who supported the project and allowed us to spend the necessary time on the book preparation. A special thanks to Professor Maria Kolokotroni for totally supporting and encouraging Agnese in this independent editorship experience.

Ultimately, this book was only possible thanks to the efforts of the extraordinary group of authors that responded to our call, including some real masters in this field that taught us most of things we know about the topic but also young and brilliant colleagues and friends that enthusiastically contributed to make real our idea. All of them showed an extraordinary engagement to the improvement of science, contributing to pushing forward the world we live in by collaborating and sharing knowledge. We are very grateful to each one of them and we hope that this book will meet their expectations.

Antofagasta, Chile Massimo Palme
London, UK Agnese Salvati

October 2020

Contents

Editors and Contributors

About the Editors

Massimo Palme is materials engineer and holds a PhD in Architecture, Energy and Environment and a professional degree in Data Science. Currently, he is associate professor and director of the Urban and Built Environment Research Lab, Catholic University of the North in Antofagasta, Chile. Since 2006 develops research on the topics of building simulation, sustainability of urban environment, and climate change mitigation and adaptation. President of the Chilean Chapter of the International Building Performance Simulation Association and Secretary of the Chilean Chapter of the International Landscape Ecology Association, Massimo is author of about 100 publications in Journals, Books and Conference Proceedings. He has been founded by Senescyt in Ecuador and Fondecyt in Chile to develop research projects as principal investigator; he has been visiting researcher at several universities, and recently he has been invitational fellow of the Japanese Society for the Promotion of Science.

Agnese Salvati is research fellow in Resource Efficient Future Cities at the Institute of Energy Futures at Brunel University London. She has a double PhD degree in "Architecture, Energy and Environment" and "Engineering-based Architecture and Urban Planning" from UPC and Sapienza University of Rome. Her research focuses on urban climate modelling for building energy performance simulation and outdoor thermal comfort. She is currently involved in international and UK-based projects: the IEA EBC Annex 80 on Resilient Cooling, the EPSRC project Urban Albedo, the EU projects ReCO2St and Prelude. She formerly worked at the Low Carbon Building group at Oxford Brookes University and at the School of Architecture of the Catholic University of the North in Chile.

About the Contributors

Noelia Liliana Alchapar architect, PhD in Science—area renewable energy—(2015). Doctoral Thesis Subject: "Materials of the Urban Envelope. Assessment of its Aptitude to Mitigate the Heat Island in Cities of Arid Zones". Assistant researcher in the National Council of Scientific and Technological Research (CONICET) of Argentina, with a place of work at the Institute of Environment, Habitat and Energy (INAHE) in Mendoza, Argentina. The research activity is focused on the energy efficiency of building materials and in the urban cooling for the sustainable planning of cities. The main objective is to generate strategies for mitigation and adaptation to global climate change. On this theme, research projects have been financed by national and international agencies. The results have been published in national and international scientific journals and in proceedings on the subject.

Marie-Hélène Azam is PhD student at Cerema and at GeM laboratory (Nantes University—France); she works on the application of model reduction methods (i.e. Proper Generalized Decomposition and Proper Orthogonal Decomposition) to heat transfers in urban simulation tools.

Julieta Balter architect and master in Renewable Energies at the International University of Andalusia, Spain. PhD in Architecture at the National University of La Plata, Argentina. Assistant researcher at the Institute of Environment, Habitat and Energy (INAHE) of the National Council for Scientific and Technical Research (CONICET) Mendoza, Argentina. Teacher in the Architecture Career at the National University of Cuyo. Expertise in Bioclimatic Architecture, Energy Efficiency in Building, Building Simulation. She participates in research projects with distinguished national researchers. She is a peer reviewer of international journals. The results have been published in national and international scientific journals and in proceedings of congresses on the subject.

Gustavo Javier Barea Paci is graduate in Architecture and Urbanism of Universidad de Mendoza, Argentina. Specialist in Design Update of Universidad de Buenos Aires, PhD in Sciences—Renewable Energy from the National University of Salta. Assistant researcher at the Institute of Environment, Habitat and Energy (INAHE) of the National Council for Scientific and Technical Research (CONICET) Mendoza, Argentina. Professor at Universidad Nacional de San Luis, at the chair of Solar Architecture, as well as for post-graduate courses on energy simulation. His research field is Building Bioclimatology. He participates in research projects with distinguished national researchers. He is a peer reviewer of international journals, such as Energy and Building, Solar Energy, and Renewable Energy. He has participated in numerous national and international architecture and urbanism competitions, obtaining first and second prizes.

Umberto Berardi is professor and Canada Research Chair in Building Science at Ryerson University in Toronto (ON, Canada). His main research interests are related to the study of innovative materials for improving the performance of the built environment. He has published over 120 peer-reviewed journals, 150 international conference papers, and 5 books. Dr. Berardi has current research funds of $2M. He is the founding director of the BeTOP group, which counts ten researchers. He is the vice-president of IAQVEC and seats on the board of several associations, including CAA and IBPSA-Canada.

Federico M. Butera professor emeritus, Politecnico di Milano. For more than 40 years he has been active in the fields of solar energy applications, low energy architecture and sustainable urban development. He is author of more than 200 scientific publications and 12 books. He was awarded: "Pioneer for contributions in renewable Energy" by the World Renewable Energy Network in 1998; by Eurosolar in 2004 for his "outstanding service to the utilization and promotion of Renewable Energies and Sustainable Architecture", by PLEA in 2015 as Pioneer in Passive and Low Energy Architecture.

Marco Cacciani Italian "Laurea" in Physics (score 110/110 "cum laude"). Dissertation title: "Measurements of ozone absorption coefficient in the Huggins bands". 1990 Permanent Position as Research Scientist at the Physics Department of the University of Rome "La Sapienza". From 1985 he has been working on several projects regarding the measurements of ozone and stratospheric aerosols. In particular, he has been involved in the construction and deployment of lidars at the South Pole, at Thule and Seychelles. He has also been involved in the construction and in the operations of an airborne lidar installed in the Russian stratospheric aircraft M55 "Geophysica". His main fields of interest have been ozone spectroscopy; ozone, water vapor, clouds and aerosols measurements and modelling in the polar, tropical and urban regions.

M. Alicia Canton architect. DAE in Architecture. University Professor in Mendoza University (UM) and Researcher INAHE-CONICET, Argentina. The research activity is focused on the study of the open space as regulator of environmental conditions in arid zones urban contexts. The main objective is to conduct the urban and building development towards sustainability.

Helena Coch Roura Dr. in Architecture, full professor in the department of Architectural Technology, Polytechnic University of Catalunya (UPC). She is coordinator of the PhD programme "Architecture Energy and Environment" (AiEM) and the corresponding research line in the MBArch, advanced master course in Architectural studies. She leads various teaching modules in the graduate and masters' programmes of the School of Architecture of Barcelona (ETSAB) and the School of Architecture of Valles (ETSAV). Her research has focused on the energy performance of architecture and urban spaces and thermal comfort of indoor and outdoor spaces. She participated and coordinated various national and international research projects and attended prestigious international conferences. Her work is published in numerous books and scientific journals.

Thibaut Colinart holds a PhD in Civil Engineering and is assistant professor in building physics at Université de Bretagne Sud, France. He is researcher on the hygrothermal behavior of building envelope and on the interaction with build environment.

Erica Norma Correa Cantaloube chemical engineer, PhD in Science—Renewable Energy Area—(2006). Researcher at INAHE-CONICET and Professor at National Technological University (UTN), Argentina. Their research activity is focused on the energy and environmental sustainability of cities, in particular, the assessment of the efficiency of diverse heat island mitigation strategies in arid zones urban contexts. Their main objective is to contribute to the sustainable urban planning and growth.

Vincenzo Costanzo MEng in Architectural Engineering and PhD in Energetics. Currently researcher at the University of Catania (Italy), his main research interests are building energy efficiency and simulation, indoor and outdoor thermal and visual comfort, renewable energy production and daylighting. He is author of several international publications on such topics and editor of various prestigious indexed scientific journals.

Francisco De la Barrera ecologist, Dr. in Geography. Associated Professor at the Universidad de Concepción and Associated Researcher at the Center for Sustainable Urban Development (CEDEUS). His research interests focus on urban and landscape ecology analyzing the effects of large and rapid landscape changes on ecosystem services at different scales (e.g. megafires, urbanization processes) and generating data for planning the sustainability of landscapes.

Annalisa Di Bernardino has been a researcher at the Physics Department of the University of Rome "La Sapienza" since 2019. She obtained her PhD in Hydraulics and Environmental Engineering in 2016 at the University of Rome "La Sapienza". Her research activity is mainly focused on remote sensing investigation of planetary boundary layer meteorology and aerosols, micrometeorology, laboratory modelling of urban canopies and dispersion models.

Gianpiero Evola has a degree in Mechanical Engineering and a PhD in Building Physics, and is currently a tenure-track researcher at the University of Catania. His research activity has mainly focused on dynamic energy simulation of buildings, indoor and outdoor thermal comfort, solutions for passive cooling in warm climates and the performance of renewable energy systems in buildings. On these topics, he wrote 50 papers published on peer-review journals, and more than 70 communications to international and national conferences.

Claudia Fabiani graduated in Building Engineering and Architecture at University of Perugia, Italy, in 2015. She received her PhD in Energy and Sustainable Development from University of Perugia in 2019. She was visiting research collaborator at Princeton University (NJ, USA) in 2017. She has been post-doc fellow of Applied Physics since 2019, and she is the holder of a research grant in Applied Physics at University of Perugia, Italy. She is author of more than 30 papers, most of which are published in international refereed journals. She has been working in several European projects under the framework of Horizon 2020 program. She is mentoring several PhD students and has been co-advisor of more than ten master thesis in Building/Civil and Mechanical Engineering. Her research interests cover innovative smart materials for building envelope applications, and indoor-outdoor environmental comfort analysis and energy saving.

Dominik Fröhlich was research associate at the Research Centre Human Biometeorology (ZMMF) of the German Meteorological Service (DWD) and at the Chair of Environmental Meteorology of the University of Freiburg. His research focuses on the development of a microscale model for the thermal environment in complex areas (SkyHelios). Formerly he was involved in the research programs KLIMOPASS and MOSAIK.

Julie Futcher is a chartered Architect and a built environment consultant with interests in climate responsive urbanism, and the influence of building form on modifying the climates of neighbouring buildings and the outdoors. Her research focuses on the dynamic and interdependent influence of built form (particularly tall buildings) on the urban setting, and how these can be optimised to promote comfortable healthy environments that encourage more sustainable urban practice. Disseminating activities form an important aspect of this work. To address this, she has developed novel approaches to link urban design with the experiences of pedestrians and urban dwellers through a series of workshops which include a series of "urban climate" walking tours.

Marcel Gangwisch is research associate at the Research Centre Human Biometeorology (ZMMF) in Freiburg of the German Meteorological Service (DWD). His research focuses on the development of numeric urban microscale models to assess thermal impact of urban green on the atmosphere and human thermal perception based on computational fluid dynamics. He is currently involved in the GrüneLunge project.

Carolina Ganem Karlen is senior lecturer in the School of Design at the National University of Cuyo. (FAD-UNCuyo), Mendoza, Argentina, and senior researcher at the National Research Council of Argentina (INAHE-CONICET); architect from the University of Mendoza, Argentina, and post-graduate from the ABITA Centre at the Universitá degli Studi di Firenze (UNIFI), Florence, Italy. She holds a PhD from the Universitat Politécnica de Catalunya (ETSAB-UPC), Barcelona, Spain. Her research focuses are on energy efficiency; indoor environment and well-being;

micro urban environments, occupants' thermal comfort and adaptive behavior, and energy assessment and certification of buildings. She has completed numerous research projects by a wide variety of resources and has authored or co-authored many book chapters and more than 150 papers and research reports. She is a peer reviewer of prestigious national and international journals

Mahendra Gooroochurn BEng (Hons), MSc, PhD (Loughborough, UK), CEng, LEED AP BD+C, MIET, works as senior lecturer in the Mechanical and Production Engineering Department of the University of Mauritius. He has over 3 years industry experience as research manager and head of sustainability in building services consultancy. He is the Ellen McArthur Foundation Circular Economy Pioneer for Mauritius as part of a network of 10 countries in Sub-Saharan Africa. He is specifically interested in researching passive and low energy solutions to promote energy efficiency and thermal comfort in buildings. He has secured several grants during his time in industry and in academia to research into the design of green buildings for the local context and worked in UNDP-funded national projects with international partners, including BRE, UK, to devise an energy audit management scheme for Mauritius and with Sarah Wigglesworth Architects, UK, to develop sustainability principles for residential and non-residential buildings in Mauritius.

Sihem Guernouti holds a PhD in Civil Engineering and is a senior researcher at Cerema (Nantes, France) in the team BPE. Her research focusses mainly on heat and moisture modelling for buildings and the application of model reduction techniques. She also studies the interaction between buildings and the urban environment at both building and urban scales.

Zahra Jandaghian is a research associate in the Performance of Roofing Systems and Insulation (PRSI) group at the National Research Council Canada (NRC). Her research revolves around the fields of building energy performance, residential and commercial roofing system durability performance and resilience to climate change. She has a PhD in Building engineering from Concordia University on the effects of increasing surface albedo on urban climate, air quality, and human health. She then joined Ryerson University working with Prof. Umberto Berardi on the effects of urbanization and greenery enhancement on urban climate. She has authored and co-authored more than 30 journal and conference papers.

Maria Kolokotroni is professor of Energy in the Built Environment at Brunel University London. She is a Chartered Engineer (CEng) and Fellow of CIBSE (Chartered Institute of Building Services Engineers) in the UK. She is a member of IEA (International Energy Agency) AIVC (Air Infiltration and Ventilation Centre) Board and currently participates to the international project "Annex 80—Resilient Cooling for Buildings" by the IEA EBC (Energy in Buildings and Communities). She has co-ordinated and participated to a number of UK, European and International projects on Energy Efficient Technologies for Buildings, Nearly Zero Energy Building (NZEB) Retrofit, Low Energy Cooling and Ventilation, Urban Heat Island (quantification for London and its impact on energy demand by buildings) and Mitigation Strategies such as cool materials and microclimatic interventions as well as the impact of climate change on the energy performance of buildings and urban areas.

Daniele La Rosa (PhD in Urban and Regional Planning) is associate professor in Urban and Environmental Planning at the Department Civil Engineering and Architecture of the University of Catania (Italy). He teaches spatial planning and urban design in Building Engineering MSc course at the University of Catania. His research topics include sustainable planning, Ecosystem Services, GIS applications for urban and landscape planning, environmental indicators, Environmental Strategic Assessment, Land Use science and landscape studies.

Nicolas Lauzet holds a PhD since 2019 from University of Lorient (France). In his PhD, realised from the consultancy agency TRIBU, he studied the climate change effect and urban heating phenomenon for future buildings' frugal design by focusing on occupants comfort.

Giovanni Leuzzi is associate professor at the University of Rome "La Sapienza". Obtained his PhD in Applied Mechanics at the University of Rome "La Sapienza" in 1993. He teaches "Fluid Mechanics" and "Numerical Modeling of Pollutant Dispersion". Research topics: environmental fluid mechanics, Lagrangian dispersion models and marine turbulence.

Jiachen Mao is a research associate at the Laboratory for Information and Decision Systems (LIDS) within Massachusetts Institute of Technology (MIT). He is broadly interested in physics-based mathematical modeling and data-driven algorithm development for sustainable design and operation of energy and environmental systems, particularly at the building and urban scale. He is a co-author of many peer-reviewed articles in international journals and conferences, and he has worked on research projects supported by the National Natural Science Foundation of China, Renewable Energy and Energy Efficiency Partnership, MIT Energy Initiative, etc.

Irene Marincic is full professor-researcher, and researcher of the National Researchers System SNI (Mexico). Her expertise areas are thermal building behavior, thermal comfort, energy efficiency, and daylighting. Some developed activities are: monitoring of outdoor and indoor thermal conditions (evaluation post-occupancy), data processing of measurements, thermal comfort surveys, and monitoring of lighting conditions.

Andreas Matzarakis is leading the Research Centre Human Biometeorology of the German Meteorological Service in Freiburg since 2015. Appointed as extraordinary professor at the University of Freiburg since October 2006. Developer of several models and tools in applied climatology and biometeorology, i.e. RayMan Model, SkyHelios Model, Climate Mapping Tool and CTIS (Climate-Tourism-Information-Scheme).

Dasaraden Mauree holds a PhD from University of Strasbourg (France). He has worked as a senior scientist and lecturer at EPFL on climate change adaptation of cities and efficient energy neighbourhoods and works now as a consultant.

Gerald Mills is a climatologist whose research focusses on the development of an urban climate science. His work has focused on modelling and measuring the urban effect at the micro-scale and, more recently, on the development of global urban databases to support climate studies. He is a former President of the International Association for Urban Climate and co-author of Urban Climates, which is a modern textbook on the field.

Paolo Monti is associate professor at the Engineering Faculty of the University of Rome "La Sapienza". Obtained his PhD in Environmental Engineering at the University of Rome "La Sapienza" in 1994. He teaches "Urban Climatology" and "Environmental Fluid Mechanics". His research interests include turbulence and mixing in natural flows, Lagrangian stochastic dispersion models and urban fluid mechanics.

Michele Morganti architectural engineer by education, is research fellow at Politecnico di Milano. In 2013, he obtained a joint PhD Cum Laude from the Barcelona School of Architecture of the UPC Barcelona Tech and the Engineering Faculty of Sapienza University of Rome. Since 2014, Michele is Adjunct professor in Building Design for Sustainable Architecture at Sapienza. He is member of the Sustainable Energy Efficient Design (Politecnico), the Architecture, Energy and Environment (UPC Barcelona) and the SOS Urban Lab (Sapienza) research teams and his research focuses on the relation among climate change, urban microclimate, building design and energy performance, with special interests in housing. Current research includes sustainable building and regenerative design, urban morphometrics and energy performance, building and urban nature-based solutions.

Benjamin Morille is an expert in the urban microclimate simulation field at district scale. He has worked as researcher during 8 years investigating the Urban Heat Island (UHI) issues and especially the interaction between vegetation and climate. In 2017, he created the company SOLENEOS to propose to cities to evaluate their UHI mitigation strategies using the simulation tool SOLENE-microclimate. He was involved in various research projects (ANR VegDUD, EVA, ANR ECOVILLE) and continues to develop SOLENE-microclimate through other projects like CoolParks supported by ADEME (French Agency of Environment and Energy Management).

Marjorie Musy is doctor in Civil Engineering and has been qualified to direct research. She is scientific director at Cerema and leads the "Buildings' Performance in their Environment" team. She is the scientific director of the H2020 project Nature4Cities. Much of her research is conducted at the interface between building physics (buildings) and urban physics (city), including nature (soil, water, vegetation). This research is conducted at multiple scales, from the performance of individual buildings in their urban context through the performance of a district to the analysis and modeling of urban heat island. It mainly focusses on the impact of climate adaptation strategies.

Matías Nieto-Tolosa born and raised in a semi-arid Mediterranean landscape at the southeastern Iberian Peninsula in the 1980s, educated in architecture and regional planning in public institutions in Spain (1999–2010), The Netherlands (TUe, 2003–2004) and Brazil (UNICAMP, USP, UFABC, 2010–2020), worked for private offices in Madrid, Rotterdam and São Paulo. Founder of Artemátyika (2004, www.artematyika.com), personal trademark and architectural practice doing applied research on cybernetics, visual programming and compassionate automation. Authorized Rhino Trainer since 2010 (McNeel Barcelona). Researcher and author of scientific papers on computational design, architecture, landscape and spatial planning.

Marialena Nikolopoulou is professor of Sustainable Architecture and Deputy Head of School at the Kent School of Architecture & Planning, University of Kent. She has extensive experience in monitoring of environmental conditions and post-occupancy surveys in different operational contexts, and is on the Steering Committee of the CIBSE Guide A on Environmental Design. Her work on outdoor comfort has received awards from diverse bodies (RIBA, International Society of Biometeorology) and best paper prizes, and has influenced understanding of the topic across the world. She is currently principal investigator of the EPSRC-funded project "Urban Albedo Computation in high latitude locations: An experimental approach" and the UK lead for the European H2020 Marie Sklodowska-Curie Industrial Doctorate "Solutions for Outdoor Climate Adaptation".

Leslie K. Norford is professor of Building Technology in the Department of Architecture at Massachusetts Institute of Technology (MIT). His research and teaching focus on reducing building energy use and associated resource consumption and carbon emissions. He has developed fault detection and optimal control strategies and low-energy latent cooling options for HVAC (heating, ventilation, and air conditioning) systems. He has also explored the interaction of buildings with both the electricity grid and the urban microclimate and the impact of local and transboundary pollutant emissions on indoor and outdoor air quality.

Andre Santos Nouri is an architect and urban designer who specialises in improving/maturing in situ outdoor/indoor interdisciplinary bottom-up adaptation approaches that address both: (1) existing human biometeorological risk factors and (2) those to be expected as a result of climate change impacts. After working in the UK within the private and public sectors, he obtained his PhD in Urbanism at the Faculty of Architecture at the University of Lisbon. In addition to his publications, he serves as reviewer for many international journals, and is on the topic editor board for "Atmosphere". Furthermore, he is a collaborator of numerous international research groups, and currently coordinates a funded research project attributed by The-

Scientific-and-Technological-Research-Council-for-Turkey (TÜBİTAK) through the European-Union's-Horizon 2020/Marie-Skłodowska-Curie grant agreement. He has international teaching experience in different countries, including in institutions located in Portugal, Romania, and New Zealand. Presently, he is positioned in Ankara, as an Assistant Professor at Bilkent University.

José Manuel Ochoa is full researcher and head of the Laboratory of Energy, Environment and Architecture and general director of the National Laboratory of Housing and Sustainable Communities of Mexico. His research is on the development of environmental technology applied to sustainable habitability in urban and architectural design. In the area of education, he teaches courses in the field of sustainable design, building technology and architectural photography, both undergraduate, and also to graduate students and professionals.

Martina Pacifici is an architectural engineer, licensed in Italy (Sapienza University of Rome) in 2013. Over the last years, she specialised in climate and energy modelling applied to urban and building domains. From 2014 to 2019, she used to live in Latin-America (Brazil), where she performed a PhD degree at the Department of Civil Engineering of the University of São Paulo (held in May, 2019). Main research activities: microclimate modelling, data-driven environmental design, energy/climatic data analysis and interpretation. Currently, Martina works as Building Performance Engineer in the Sustainability Team of Allford Hall Monaghan Morris practice in UK (London), carrying on energy performance-oriented modelling and analytical procedures applied to architectural projects. Current focus: natural ventilation modelling, free running buildings and adaptive thermal comfort, IES, Grasshopper.

Olga Palusci is a consultant and PhD candidate at Sapienza University of Rome and Eindhoven University of Technology. During her studies, Olga developed a strong interest in urban regeneration, energy consumption reduction, and the implementation of sustainability and resilience in the built environment. She graduated "cum laude" in Building Engineering—Architecture at the Faculty of Civil and Industrial Engineering at Sapienza. In the same university, Olga obtained a second level master degree in sustainability and energy savings and started working as a researcher. In 2012, she started her PhD working on the microclimate modelling in real urban areas at Sapienza. In 2014, Olga started her joint PhD program between Sapienza and the Eindhoven University of Technology.

Armando Pelliccioni obtained his PhD in Hydraulic Engineering at the University of Rome "La Sapienza" in 2009. From 1994, he is a researcher at the Ispesl (currently Inail). His research interests include neural networks, turbulence and mixing in stratified flows, atmospheric flows and Lagrangian dispersion models. In the last years, he is working on indoor air particulate matter models coupled with turbulence measurements in open and confined environments.

Agnese Pini received her master degree in Environmental Engineering from the University of Rome "La Sapienza" in 2014. In 2018, she earned her PhD in Environmental and Hydraulic Engineering at the University of Rome "La Sapienza". Her research interests include geophysical fluid mechanics, marine turbulence, Lagrangian dispersion models, ocean models, and computational fluid dynamics applied to particle dispersion in the atmospheric environment.

Anna Laura Pisello graduated cum laude in Building Engineering at Polytechnic University of Milan, Italy, in 2009. She received her PhD in Energy Engineering from University of Perugia in 2013. She was visiting scholar at Columbia University, Virginia Tech and City University of New York in 2010–2012. She is currently assistant professor (tenure track) of Applied Physics at University of Perugia, Italy, and visiting research associate at Princeton University (NJ, USA). She got the national professorship qualification as associate professor. She is author of more than 150 papers (more than 90 in international refereed journals). She won seven international academic awards in 2013–2018 and several Horizon 2020 European projects. Member of the teaching board of the Doctorate school of Energy and Sustainable Development, she is mentoring several PhD students. She is lecturer of Environmental Applied Physics and has co-advised more than 30 master thesis in Building/Civil and Mechanical Engineering.

Riccardo Privitera is research fellow in Urban and Spatial Planning at the Department of Civil Engineering and Architecture, University of Catania (Italy). He holds a PhD in Urban Planning and teaches "Urban Design" as a lecturer. His scientific interests include non-urbanised areas planning, urban green infrastructures, climate change adaptation and mitigation strategies, ecosystem services, urban morphology analysis, real estate development processes, and transfer of development rights.

Sonia Reyes-Paecke biologist, PhD, assistant professor at the Pontificia Universidad Catolica de Chile and associated researcher at the Center for Sustainable Urban Development (CEDEUS). Her research interests are on the interactions between ecological and social systems in human-dominated environments, focused on ecosystem services, land use planning, green infrastructure and urban sustainability. She is president of the Chilean Association of Landscape Ecology (IALE-Chile).

Giridharan Renganathan BArch(IIT); MUr Dgn (HKU); PhD (HKU); AIA (SL), is a senior lecturer at University of Kent. He has held research appointments at University of Hong Kong (2006), Brunel University (2007) and Loughborough University (2009). Prior to taking up the academic career, he had worked for organisations such as State Engineering Corporation, Urban Development Authority and Avant Garde Urban Design Partnership in Sri Lanka. His research focus is on Urban Morphology and Climatology (environmental design), with specific interest in Urban Heat Island effect and indoor overheating. He has secured large research grants from RGC (Hong Kong SAR) and EPSRC (UK), European Commission, and small grants from European Union Regional Development Fund (EU), Heritage lottery fund/Kent County Council (UK) and University of Moratuwa (Sri Lanka). He has over 25 peer reviewed publications and 4 book chapters.

Auline Rodler holds a PhD in Civil Engineering and is a researcher at Cerema (Nantes, France), in the team "Buildings' Performance in their Environment (BPE)". Her research focusses on the building physics and its interaction with urban microclimate, addressed at a building and district scale.

M. Angélica Ruiz is an agricultural engineer graduated from the Universidad Nacional de Cuyo (2007) and a Doctor of Science—Renewable Energy Area from the Universidad Nacional de Salta (2013). Her research has been developed around the study of the impact of vegetation on the urban microclimate, on the habitability of open spaces and on associated energy consumption.

M. Belén Sosa Architect graduated from the Universidad Mendoza—Argentina (2009), Master in Sustainable Architecture and Energy Efficiency from the Universidad Ramón Llull—Spain (2010) and PhD in Science—Renewable Energy area from the Universidad Nacional de Salta—Argentina (2018). Her main research scope has been developed around the analysis of urban form impact's over the microclimate, the outdoor thermal comfort and the energy consumption.

Iain Stewart is a research fellow at the Global Cities Institute (University of Toronto) where he contributes to the development of new ISO performance standards for sustainable and resilient cities. His research and educational interests focus on the field of urban climatology. He recently developed an urban climatology course for students of engineering, architecture, and planning. The course was launched at the Universidad Nacional de Colombia in 2019. He also co-developed the Local Climate Zone scheme that is widely used in urban studies to describe neighbourhood-scale landscapes and their climate impacts.

Chapter 1
Introduction: Anthropocene or Urbanocene?

Massimo Palme and Agnese Salvati

1.1 Anthropocene or Urbanocene?

We live in times of great changes. Global warming is altering the planetary thermal equilibrium, causing deep changes in our lifestyles. Economic transformations are also pushing the society toward new organisational structures, able to respond to the characters of a deeply globalised world. The human activity has an impact on the ecological system that had never been seen before. As a result, we are exposed to new risks, including extreme climatological events, pandemic diseases and massive migrations. For this reason, many specialists in different fields, geologists in particular, are proposing the term "Anthropocene" for the geological epoch in which we are living (Crutzen and Stoemer 2000; Zalasiewics et al. 2018). This epoch could have started with the great acceleration in economic growth of the 1950s, or with the industrial revolution or with the trip of Columbus to America and its ecological consequences; it could even coincide with the entire Holocene. One of the fundamental characteristics of the "Anthropocene", especially starting from the industrial revolution, is the urban living. Cities have expanded into large metropolitan areas and megacities and are still expanding at a very fast pace, especially in Asia and Africa. The urban population passed from 33.6% to 55.7% of the total population between 1960 and 2019 (United Nations Population Division 2018) and it is expected to be 70% by the middle of the century. The London School of Economics and the Alfred Herrhausen Gesellschaft have extensively analysed the deep economical, societal and environmental consequences of these urban transformations, defining this epoch as the "Urban Age" (Burdett and Rode 2018; Burdett

M. Palme (✉)
Escuela de Arquitectura and CEITSAZA, Universidad Católica del Norte, Antofagasta, Chile
e-mail: mpalme@ucn.cl

A. Salvati
Institute of Energy Futures, Brunel University London, London, UK
e-mail: Agnese.salvati@brunel.ac.uk

M. Palme, A. Salvati (eds.), *Urban Microclimate Modelling for Comfort and Energy Studies*, https://doi.org/10.1007/978-3-030-65421-4_1

and Sudjic 2007, 2011). Some scientists proposed the term "Urbanocene" (West 2017) to describe this new epoch, arguing that even the stratigraphic signal of a "golden spike" could be found in urban geomorphology (Certini and Scalenghe 2011; Elias 2018).

As a matter of fact, the world is increasingly populated and urbanised. Cities are the places where we live and use almost 80% of the final energy, with associated dissipation of heat to the environment. For this reason, many scientists from different fields highlighted the need of studying the urban metabolism, to understand the fluxes of materials, energy, food, water and information across our cities. The urban metabolism is directly associated to entropy production, whose most visible effect is the generation of local microclimates in cities (Palme et al. 2017). The existence of these microclimates is often neglected when studying global warming. For a matter of scale and computational efficiency, the IPCC scenarios and the global climate models do not explicitly consider the presence of cities and their different thermal behaviours compared to non-urbanised areas in their predictions. However, even though cities occupy a negligible percentage of the earth surface, they concentrate 75% of global CO_2 emissions (Burdett and Sudjic 2007) and are warmer than their surroundings (Stone 2012). This could lead to underestimate of the pace of warming of cities that is higher than that of rural environment.

1.2 Cities' Metabolism and Urban Climate

The concept of urban metabolism was firstly proposed by Wolman (1965) and today it is an increasing field of study (Kennedy et al. 2007; Princetl et al. 2012; Zhang et al. 2015; Cui 2018; Dijst et al. 2018), whose purpose is to understand the physical processes taking place in a city, by analogy with living structures. It grounds on seminal contributions by Schrodinger (1944) regarding the quintessence of life, Prigogine and Stengers (1984) regarding far-from-equilibrium thermodynamics and Odum (1988) and Allen (1994) regarding complex systems dynamics. Allen (1998) and Pumain (1998) were among the first to apply these concepts to urban systems.

The study of urban climate should be considered as one of the most important areas in this field of research today. Starting from the pioneering study by Howard on the climate of London (1888), urban climatology has turned into an established academic discipline. The energetic basis of the urban heat island effect and other urban microclimatic phenomena has been deeply studied and described in influential books such as "Boundary Layer Climates" (Oke 1987) and "Urban Climates" (Oke et al. 2017). Urban climate can be seen as the resulting modification introduced in the environment by urban structure and activities taking place in cities. Anthropogenic heat generation, changes in land cover and land use and building density contribute to the generation of new climatic patterns that modify the environment in which cities are placed. The modified urban climate has a deep impact on both building energy performance and comfort of inhabitants. Santamouris (2014) showed that the increase in cooling needs due to urban warming could rise up to 500% by the middle of the century. Many studies carried out in different cities of the world confirmed that the urban heat island causes an energy penalty on urban buildings, increasing their annual energy demand (Santamouris 2020).

1.3 Integrating Urban Climate Knowledge in Urban Policies, Planning and Design: What Went Wrong?

The growing concern for the risks posed by climate change to the liveability of cities has certainly also contributed to increase in the awareness on the impact of the built environment on urban microclimate. Nevertheless, the attempts to integrate climate principles in the planning, design and renovation of cities and buildings are still scarce, or not effective.

Oke advocated the need for a greater consideration of climate principles in urban planning yet in the 1980s (Oke 1984), encouraging urban climatologists to translate their knowledge into practical applications and guidelines for the design of street canyon geometry (Arnfield 1990; Oke 1988). Since then, the body of knowledge in urban climatology has grown, as well as the computation capabilities to carry out more advanced modelling studies (Chen et al. 2011; Grimmond et al. 2011). Other attempts to disseminate this knowledge to a wider audience have been done (Grimmond et al. 2010; Oke et al. 2017). Today, a wide and multidisciplinary research community recognises the importance of interactions among buildings in the urban environment to assess urban sustainability and energy performance (Emmanuel and Steemers 2018). However, it is still rare to find good applications of climate principles in urban planning and architectural practice. The basic concepts of urban climatology are often unknown also to building energy modellers that normally carry out energy performance simulations just neglecting any urban effect.

The poor importance given to urban climate in planning and design is partially due to the absence of microclimate regulations. As a matter of fact, the introduction of regulations on the energy efficiency has improved the thermal performance of buildings, making architects and engineers accountable for the energy consequences of their choices. However, as shown by Futcher et al. (2017), the energy impact of new buildings, especially tall ones, can go far beyond their envelope, modifying the solar access, wind speed and air quality of surrounding areas and consequently the energy use of other buildings. As Futcher and Mills rightly point out (Futcher and Mills 2015), no one is accountable for the microclimate and energy impact of new buildings beyond their envelope and this should be better regulated to achieve sustainable urban development.

Other motivations behind this persistent gap between urban climatology and planning are still those enumerated by Oke: “the inherent complexity of the subject, its interdisciplinary nature and lack of meaningful dialogue between planners and the climatological research community” (Oke 1984).

The communication gap between climatologists and designers, including planners, architects and building engineers, is basically due to the lack of a common ground in terms of vocabulary, educational background and scale of analysis. The designers do not have any background on climatology and urban physics, while climatologists know very little about architectural design and planning processes. Architects and planners are more prone to think about buildings and cities in relation to the users' practical needs, while climatologists tend to abstract the urban context into surface properties (i.e. roughness, albedo) and heat fluxes. Designers and climatologists also associate different meanings to key properties of the built environment such as urban morphology, compactness or density. Ultimately, they are interested in the same things, buildings and cities, but they use different languages and, as a result, they struggle to understand each other's.

This communication issue has also led to some misinterpretations of the concept of urban heat island (UHI) by architects, planners and engineers. The most common misunderstanding is the consideration of the UHI intensity as a constant phenomenon in time and space. Therefore, the first myth to debunk is that there is no single UHI effect in a city, but different types of urban climate modifications that influence thermal comfort and building energy performance (Salvati et al. 2020).

Another difficulty is given by the multiple scale of urban climate phenomena and the interconnections between meso-climate, local climate and microclimate. For instance, to assess the impact of urban microclimate on the energy performance of buildings in a certain street, a multiscale approach including at least three steps is needed: (1) an analysis of the geographic and topographic features of the city and its surroundings, (2) an analysis of the urban fabric characteristics in terms of local climate zones (Palme et al. 2018; Salvati et al. 2019; Stewart and Oke 2012) and (3) an analysis of the three-dimensional shape and arrangement of buildings in the street canyon, the thermal and optical properties of urban and building materials and the thermal performance and function of buildings (Futcher et al. 2018; Salvati and Kolokotroni 2019). Due to this complexity, it is not possible to draw easy-to-apply and universally valid climate guidelines, because every city is different, and so are the districts, streets and buildings across a city.

Another aspect to consider is that urban climate modifications may have favourable or negative impact depending on the season and the objective of the analysis, which can be thermal comfort and air quality in the outdoors or energy efficiency and environmental quality indoors. For this reason, it is crucial to understand the net impact of architectural and planning choices on the environmental performance of cities, including urban microclimate, building energy performance, air quality, noise and daylight access at least.

In light of these issues, more has to be done to bridge urban climatology to planning, architectural design and building energy simulations. This entails an effort from both disciplines. Architects, planners and engineers need to expand their understanding of the microclimate impact of design choices and the impact of urban context on the building performance. Urban climatologists and geographers need to understand the complexity of the planning and design processes and work with architects and planners to develop more design-oriented tools.

1.4 The Book Structure

This book provides the state of the art of applied modelling and simulation of cities' thermodynamics, including contributions by some of the most influencing experts and emergent researchers in the field. Information is organised in three parts. In the first part, the general topic of the city as a complex adaptive system is presented, putting in evidence of the relation between the city metabolism and the resulting microclimate, as well as the implications for comfort and energy consumption. The second part includes a range of studies on modelling methodologies and available tools for urban climate simulation. The third part is a compendium of planning and design applications applying the described methodologies to improve cities' resilience and sustainability.

1.4.1 Urban Climate and Sustainability: Energy Performance and Thermal Comfort in Cities

In the first part, Butera and Palme (Chap. 2) introduce the city as a complex thermodynamic system. Mills, Futcher and Stewart (Chap. 3) discuss the energetic basis of the urban heat island and Nicolopoulou (Chap. 4) introduces the concept of outdoor thermal comfort.

The chapters from 5 to 8 are focused on specific issues in different climate regions. Kolokotroni and Salvati (Chap. 5) explore the energy implications of urban microclimates in high-latitude cities, Coch and Salvati (Chap. 6) focus on comfort and energy performance of Mediterranean cities, Gooroochurn and Renganathan (Chap. 7) describe how to enhance energy performance in tropical climates, while Marincic and Ochoa (Chap. 8) explore thermal comfort in arid climates. The first section is concluded by a position paper by Mills and Futcher (Chap. 9), making an argument for a redefinition of the "urban canopy layer" to include both the indoor and outdoor space in a single domain of investigation, namely the zone of human occupation. Perhaps, this could be an effective attempt to overcome the current lack of integration between complementary disciplines that study this space in a fragmented way and a step forward for the management of urban climate and energy issues under the same context, scale and regulation framework.

1.4.2 Urban Climate Modelling and Simulation: Physics and Tools

The second part of the book is dedicated to modelling studies. Di Bernardino et al. (Chap. 10) review the topic of air circulation in urban environments. Jandaghian and Berardi (Chap. 11) show how to couple weather forecasting models and urban canopy models to obtain data for building performance simulation. Mao and Norford (Chap. 12) present the last-released version of the Urban Weather Generator tool, while Musy et al. (Chap. 13) present the SOLENE-microclimat model. Pacifici and Nieto-Tolosa (Chap. 14) compare ENVI-met and Grasshopper modelling strategies to assess local climate and urban heat island. Rodler et al. (Chap. 15) review the most significant urban microclimate and building energy simulation coupling techniques. Matzarakis et al. (Chap. 16) present RayMan and SkyHelios models, while Ganem and Barea (Chap. 17) present a methodology to assess the impact of climate change on energy performance of buildings.

1.4.3 Applying Urban Climate Modelling in Policy, Planning and Design: Case Studies

In the third part, Morganti (Chap. 18) presents spatial metrics to investigate the impact of urban morphology on microclimate and energy performance of buildings. De la Barrera and Reyes (Chap. 19) explore the possibility to mitigate extreme temperatures by using green infrastructure, while Privitera et al. (Chap. 23) present the capability of green infrastructure to reduce energy needs of a city. Correa et al. (Chap. 20) discuss urban morphology as a mitigation strategy for urban warming, Santos-Nouri and Matzarakis (Chap. 21) present human biometeorological studies in Lisbon and Ganem et al. discuss the impact of urban climate on building energy performance in the city of Mendoza (Chap. 22). Finally, Fabiani and Pisello (Chap. 24) review the importance of cool materials for passive cooling of buildings.

1.5 Conclusion

The Anthropocene-Urbanocene epoch is pushing humanity beyond the planetary limits. Many processes involved in the maintenance of our economic system and social structure have been found already out of control, while others are facing the limits of the planet resources (Rockstrom et al. 2009; Steffen et al. 2015). At least two of these processes are directly related to urban living (land use and climate change), and all the others are strictly connected (nitrogen and phosphorus cycles, biodiversity loss, ocean acidifications, etcetera). To maintain our living standards (and to permit the same rights to all the world population) and to stay in the so-called Safe Operating Space, we should deeply change our vision of prosperity and society (O'Neil et al. 2018). Urban climate and building operation are key factors in both defining new standards of acceptable prosperity and achieving the challenge of staying in the aforementioned limits.

As a final note, this book has been mostly written and edited over the 2020 COVID-19 pandemic, which has definitely made clear to all of us how it is important to have healthy, liveable and comfortable houses and cities for our physical and mental well-being. We hope this further lesson will foster all the stakeholders involved with the study and management of cities and buildings, including architects, engineers, planners, physicists, geographers, climatologists and economists, to effectively collaborate to achieve more resilient, sustainable and liveable cities.

References

Allen, P. (1994). Evolutionary complex systems: Models of technology change. In *Evolutionary Economics and Chaos Theory*. London: Pinter Ed.

Allen, P. (1998). Cities as self-organizing complex systems. In *The city and its sciences*. Heidelberg: Physica-Verlag.

Arnfield, A. J. (1990). Street design and urban canyon solar access. *Energy and Buildings, 14*, 117–131. https://doi.org/10.1016/0378-7788(90)90031-D.

Burdett, R., & Rode, P. (Eds.). (2018). *Shaping cities in an urban age*. London: Phaidon Press.

Burdett, R., & Sudjic, D. (2007). *The endless city—Urban age project*. London: Phaidon Press.

Burdett, R., & Sudjic, D. (Eds.). (2011). *Living in the endless city*. London: Phaidon Press.

Certini, G., & Scalenghe, R. (2011). Anthropogenic soils are the golden spikes for the Anthropocene. *The Holocene, 21*(8), 1269–1274.

Chen, F., Kusaka, H., Bornstein, R., Ching, J., Grimmond, C. S. B., Grossman-Clarke, S., et al. (2011). The integrated WRF/urban modelling system: Development, evaluation, and applications to urban environmental problems. *International Journal of Climatology, 31*(2), 273–288. https://doi.org/10.1002/joc.2158.

Crutzen, P. J., & Stoermer, E. (2000). The Anthropocene. *Global Change Newsletter, 41*, 17.

Cui, X. (2018). How can cities support sustainability: A bibliometric analysis of urban metabolism. *Ecological Indicators, 93*, 704–717.

Dijst, M., Worrel, E., Bocker, L., Brunner, P., Davoudi, S., Geertman, S., Narmsen, R., Helbich, M., Holtslag, A. A. M., Kwan, M. P., et al. (2018). Exploring urban metabolism—Towards an interdisciplinary perspective. *Resources, Conservation and Recycling, 132*, 190–203.

Elias, S. A. (2018). *Finding a "golden spike" to mark the Anthropocene. Encyclopedia of the Anthropocene*. Amsterdam: Elsevier.

Emmanuel, R., & Steemers, K. (2018). Connecting the realms of urban form, density and microclimate. *Building Research and Information, 46*, 804–808. https://doi.org/10.1080/09613218.2018.1507078.

Futcher, J., & Mills, G. (2015, April). Pushing the envelope. *CIBSE Journal*, 13–27. Retrieved from http://portfolio.cpl.co.uk/portfolio/CIBSE/201504/opinion-futcher/.

Futcher, J., Mills, G., Emmanuel, R., & Korolija, I. (2017). Creating sustainable cities one building at a time: Towards an integrated urban design framework. *Cities, 66*, 63–71. https://doi.org/10.1016/j.cities.2017.03.009.

Futcher, J., Mills, G., & Emmanuel, R. (2018). Interdependent energy relationships between buildings at the street scale. *Building Research and Information, 46*, 829–844. https://doi.org/10.1080/09613218.2018.1499995.

Grimmond, C. S. B., Roth, M., Oke, T. R., Au, Y. C., Best, M., Betts, R., & Carmichael, G. (2010). Climate and more sustainable cities: Climate information for improved planning and management of cities (producers/capabilities perspective). *Procedia Environmental Sciences, 1*, 247–274. https://doi.org/10.1016/j.proenv.2010.09.016.

Grimmond, C. S. B., Blackett, M., Best, M. J., Baik, J.-J., Belcher, S. E., et al. (2011). Initial results from Phase 2 of the international urban energy balance model comparison. *International Journal of Climatology, 31*, 244–272. https://doi.org/10.1002/joc.2227.

Kennedy, C., Cuddiby, J., & Engel-Yan, J. (2007). The changing metabolism of cities. *Journal of Industrial Ecology, 11*, 43–59.

O'Neil, D., Fanning, A., Lamb, W., & Steinberger, J. (2018). A good life for all within planetary boundaries. *Nature Sustainability, 1*, 88–95.

Odum, H. T. (1988). Self-organization, transformity and information. *Science, 242*, 1132–1139.

Oke, T. R. (1984). Towards a prescription for the greater use of climatic principles in settlement planning. *Energy and Buildings, 7*(1), 1–10. https://doi.org/10.1016/0378-7788(84)90040-9.

Oke, T. R. (1987). *Boundary layer climates (2nd edition)*. London: Taylor & Francis.

Oke, T. R. (1988). Street design and urban canopy layer climate. *Energy and Buildings, 11*, 103–113. https://doi.org/10.1016/0378-7788(88)90026-6.

Oke, T. R., Mills, G., Christen, A., & Voogt, J. A. (2017). *Urban climates*. Cambridge: Cambridge University Press. https://doi.org/10.1017/9781139016476.

Palme, M., Inostroza, L., Villacreses, G., Lobato, A., & Carrasco, C. (2017). From urban climate to energy consumption. Enhancing building performance simulation by including the urban heat island effect. *Energy and Buildings, 145*, 107–120.

Palme, M., Inostroza, L., & Salvati, A. (2018). Technomass and cooling demand in South America: A superlinear relationship? *Building Research and Information, 46*(8), 864–880. https://doi.org/10.1080/09613218.2018.1483868.

Prigogine, I., & Stegers, I. (1984). *Order out of chaos*. New York: Bantam Ed.

Princetl, S., Bunie, P., & Holmes, T. (2012). An expanded urban metabolism method: Toward a system approach for assessing urban energy processes and causes. *Landscape and Urban Planning, 107*, 193–202.

Pumain, D. (1998). Urban research and complexity. In *The city and its sciences*. Berlin: Springer Verlag ed.

Rockstrom, J., Steffen, W., Noone, K., et al. (2009). Planetary boundaries: Exploring the safe operating space for humanity. *Ecology and Society, 14*(2), 32.

Salvati, A., & Kolokotroni, M. (2019). Microclimate data for building energy modelling: Study on ENVI-met forcing data. In V. Corrado & A. Gasparella (Eds.), *Proceedings of the 16th IBPSA conference* (pp. 3361–3368). Rome, Italy: IBPSA.

Salvati, A., Monti, P., Coch Roura, H., & Cecere, C. (2019). Climatic performance of urban textures: Analysis tools for a Mediterranean urban context. *Energy and Buildings, 185*, 162–179. https://doi.org/10.1016/j.enbuild.2018.12.024.

Salvati, A., Palme, M., Chiesa, G., & Kolokotroni, M. (2020). Built form, urban climate and building energy modelling: case-studies in Rome and Antofagasta. *Journal of Building Performance Simulation, 13*(2), 209–225. https://doi.org/10.1080/19401493.2019.1707876.

Santamouris, M. (2014). On the energy impact of urban heat island and global warming on buildings. *Energy and Buildings, 82*, 100–113.

Santamouris, M. (2020). Recent progress on urban overheating and heat island research. Integrated assessment of the energy, environmental, vulnerability and health impact. Synergies with the global climate change. *Energy and Buildings, 207*, https://doi.org/10.1016/j.enbuild.2019.109482

Schrodinger, E. (1944). *What is life?* Cambridge: Cambridge University Press.

Steffen, W., Richardson, K., & Rockstrom, J. (2015). Planetary boundaries: Guiding human development on a changing planet. *Science, 347*(6223), 1259855.

Stewart, I. D., & Oke, T. R. (2012). Local climate zones for urban temperature studies. *Bulletin of the American Meteorological Society, 93*(12), 1879–1900. https://doi.org/10.1175/BAMS-D-11-00019.1.

Stone, B. (2012). *The city and the coming climate. Climate change in the places we live*. Cambridge: Cambridge University Press.

United Nations Population Division. (2018). World urbanization prospects: 2018 Revision. Retrieved 26 August, 2020, from https://data.worldbank.org/indicator/SP.URB.TOTL.IN.ZS?end=2019&start=1960&view=chart.

West, G. (2017). *Scale*. London: Penguin Books.

Wolman, A. (1965). The metabolism of cities. *Scientific American, 213*(3), 178–190.

Zalasiewics, J., Waters, C., Summerhayes, C., & Williams, M. (2018). The Anthropocene. *Geology Today, 34*(5), 162–200.

Zhang, Y., Yang, Z., Yu, X. (2015). Urban metabolism: A review of current knowledge and directions for future study. *Environmental Science and Technology, 49*(19). https://doi.org/10.1021/acs.est.5b03060

Part I
Urban Climate and Sustainability: Energy Performance and Thermal Comfort in Cities

Chapter 2
The City as a Complex Thermodynamic System

Federico Maria Butera and Massimo Palme

2.1 Introduction: Non-equilibrium Thermodynamics and Complex Systems

Cities are complex adaptive systems, which means, under a thermodynamic point of view, open systems far from equilibrium, continuously importing energy, matter and information and dissipating heat as a result of energy transformations taking place inside of system boundaries (Filchakova et al. 2007). The process results in a "metabolism" of the city, in analogy with natural living systems (Wolman, 1965; Kennedy et al. 2007; Zhang et al. 2015). In physics, a system is defined as "isolated" when it does not exchange matter and energy with the environment. A system that exchanges only energy is defined as "closed" and a system that exchanges both matter and energy is defined as "open". Open systems depend on the fluxes of matter and energy that move through them. The branch of physics that studies the theory of this kind of systems is the non-equilibrium thermodynamics, which considers the system as time dependent. Simple open systems are often near the equilibrium, or in a quasi-steady state. On the other hand, complex systems—which are alive and continuously evolving—move far from equilibrium, where new behaviours emerge. A most important contributor to the development of far-from-equilibrium thermodynamics was Ilya Prigogine (1984), who introduced the concept of dissipative systems to explain the emergence of order out of chaos. However, the first idea with respect to such systems' fundamentals came from the researches on the concept of living systems done by Erwin Schrodinger (1944). In this chapter we follow the

F. M. Butera
Politecnico di Milano, Milan, Italy
e-mail: federico.butera@polimi.it

M. Palme (✉)
Escuela de Arquitectura and CEITSAZA, Universidad Católica del Norte, Antofagasta, Chile
e-mail: mpalme@ucn.cl

M. Palme, A. Salvati (eds.), *Urban Microclimate Modelling for Comfort and Energy Studies*, https://doi.org/10.1007/978-3-030-65421-4_2

entropy approach to explain the dynamics followed by cities in their expansion and evolution as complex adaptive systems. Entropy has been found to be quite useful in the study of urban systems, because of the complementary means of the concept, coming, respectively, from thermodynamics, statistical mechanics and information theory (Purvis et al. 2019). Firstly, we will discuss the application of the dissipative systems concept to cities. Then, we will focus on the evolutionary dynamics of urban systems. Finally, we will study the adaptation and learning processes of these kinds of structures, concluding that cities are the result of a balance of diversity and efficiency, fluctuating between an excess of the first or of the second depending on resource availability in the environment.

2.2 Cities as Dissipative Systems

Prigogine states that a dissipative structure is a structure that produces order by self-adaptive processes. Such a structure is emergent, in the sense that new laws have to be formulated to study its behaviour, and hierarchical, in the sense that it organizes itself on different levels, responding to different functions. Living beings and ecological systems are dissipative systems. Social systems, like cities, are surely complex systems, but are they also dissipative systems? Some attempts to answer this question have been carried out in the recent past.

Lai et al. (2013) developed a model to simulate urban development and used it to predict the entropy evolution of urban structures. Their conclusion is that cities should be considered as dissipative systems, moving to low internal entropy values according to Prigogine definition. However, the author also notes that cities are not just free-running and self-organizing systems. Planning is always present to guide the adaptive process.

Portugali (2000) states that cities evolve under the pressure of different agents like urban planners and firms that design buildings and public equipment, but with results that are always different from original planning ideas and drivers. The process appears different when timescale is changed: if postmodern cities seem to evolve under planner guidelines—most of the time based on pattern recognitions (Alexander et al. 1977; Linch 1960)—the shift from rural villages to middle-age cities and then to modern urban settlements presents complex behaviour similar to an adaptive self-organizing process.

Rees (2012) evidences that cities are inserted in a more complex dynamic, which can be studied by applying the SOHO (self-organizing holarchic open) complex system theory. This model can explain the negative entropy creation of a subsystem in terms of degradation of the immediate hosting system. Rees puts also in evidence that there exist many differences between non-human and human-driven complex systems, with probably the most important fact being that the human ecosystems present an excess of catabolic processes, which are destructive instead of regenerative, like anabolic processes.

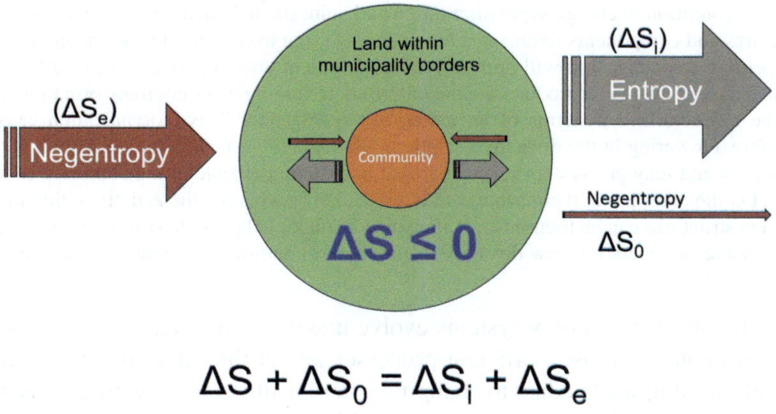

Fig. 2.1 The city as an open thermodynamic system. ΔS is the system's entropy variation per unit of time; ΔS_i is the system's production of entropy (always positive); ΔS_e is the flow of entropy coming from the environment (always negative); and ΔS_0 is the possible negentropy coming out of the system. (Authors' drawing)

One of the visible effects of such kind of destructive dissipation is the urban climate degradation. If the climate of the biosphere is able to regulate itself and adapt to changes on different scales of time, urban climate seems to be driven by humans in a no-return way towards constantly increasing heat dissipation. Urban heat island is one of the most immediate consequences of human-driven evolution of cities (Santamouris 2014; Palme et al. 2017).

Figure 2.1 presents the concept of a city as a negative entropy processor. The system imports information, matter and energy; then proceeds to internal processing of these; and is able to increase its own organization and even to export more information to the environment. The associated entropy dissipated is the price that has to be paid for the order production that takes place inside of the system.

2.3 Evolution: Between Specialization and Diversification

Complex systems evolve in a circular manner, moved by two very different principles—so different that it could be regarded as apparently opposite. The first is the principle of the minimum entropy, proposed by Nicolis and Prigogine (1977): an evolving system constantly tries to put itself in a local equilibrium point where entropy production is minimized. This principle relates to the efficiency of energy transformations, as noted for example by Labanca (2017). The second principle is the maximum power production (Odum and Pinkerton 1955), expressed also in terms of maximum exergy dissipation (Morowitz 1979): an evolving system should develop itself generating a variety of different solutions or capacities to respond to external solicitations. In the aforementioned study, Labanca states:

> […] in a condition of energy supply limitation and quite stable boundary conditions, system structures and components requiring a lower energy input to produce a given output have a competitive advantage and will prevail over less efficient ones (i.e. over system structures requiring more energy to produce a same output) determining a system transformation that can be characterized in terms of an increased organization. This reorganization causes therefore a lowering in the diversity of options available to perform a same function in the short term and may put system survival at risk in case of a change in the boundary conditions. On the other hand, it contributes to liberate energy whereby the activity within more efficient structures can be focused and intensified so making the whole systems more robust and capable of generating new diversity in case a new condition of energy abundance will be achieved.

It seems finally that complex systems evolve under both principles with the objective to eliminate the most inefficient processes and at the same time to generate a variety of possibilities to adapt to changing external situation. The predominance of one principle or another is set by the availability of external resources: if the environment is rich of resources the evolution will privilege the creation of diversity, while if resources start to be scarce, then evolution will privilege the efficiency of a hierarchical order. So, ecosystems will constantly move in this oscillation between biodiversity generation and species specialization (Fig. 2.2). Glansdorff and Prigogine (1971) called that "thermodynamic fluctuations" and suggested that the probability of occurrence of such events is directly linked to the distance from equilibrium of the systems, which should be assumed as an indicator of the state of far-from-equilibrium systems, just like temperature is assumed as an indicator of the state of a system in equilibrium. Thermodynamic fluctuations should have caused, following this interpretation, the numerous massive extinctions and the periods of species growth occurred on the planet Earth. A similar interpretation was also proposed by Cavallaro (1998), who identified a loop of growth and decline of urban systems through urbanization, suburbanization, deurbanization and re-urbanization phases.

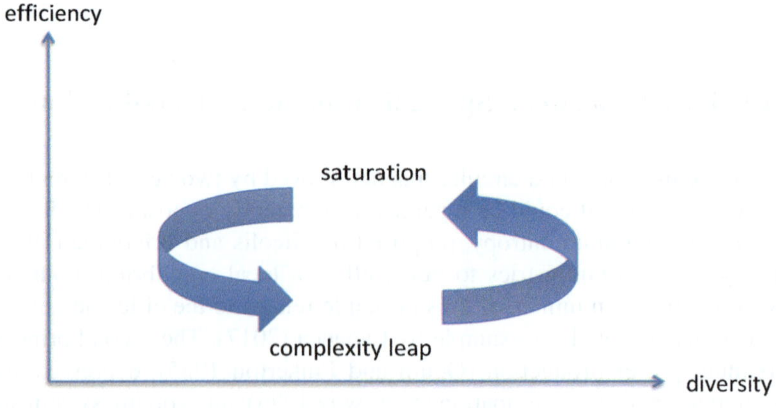

Fig. 2.2 Complex system dynamics from growth to saturation and complexity leap (a change in boundary conditions that enables new diversity explosion). (Authors' drawing)

2.4 Adaptation and Autopoiesis

Dissipative structures are adaptive systems. Such kind of systems depend on incoming matter and energy flows and cannot avoid outgoing residuals. A dissipative structure is not a stand-alone one: on the contrary, it is strongly dependent on the environment. However, as order is extracted out of chaos, a dissipative structure has to be considered as autopoietic, in the sense that it generates its own rules to function (Allen 1998). Labella (1998) proposed different modelling approaches to describe such kind of systems, the most important being the perturbation theory approach, the fractal approach and the catastrophe theory approach. Each of these approaches is useful to describe some characteristics of complex systems like cities. The perturbation theory shows that small changes in boundary conditions can lead to bifurcations and increasing consequences. Fractal approach puts in evidence of the existence of scaling laws that should be studied in topological sense. Evidences of scaling effects in cities are numerous (Isalgué et al. 2007; Bettencourt et al. 2007; Li et al. 2017). The catastrophe theory approach underlines the occurrence of external and unpredictable events pushing the system into new states. As a consequence, it is fundamental to understand the importance of resilience in urban studies.

Even if it may seem that they are similar, there is, instead, a large difference between ecological systems and technological or social systems. Adaptation process of the latter ones is not really a free self-adaptation process. They evolve using guidelines, normally by imitating other structures that are functioning somewhere.

Butera (1998) proposed the concept of "guided self-organization process" to describe the way in which cities evolve. This kind of process takes place as a learning process on the basis of the communication between the subsystem and the hosting system, to use the Reese conceptualization, or between the system and the environment, to use a more common way to express the same.

It can be appropriate to use, for cities, information theory methods, particularly in the formulation done by Jantsch (1980), who focused his work on the pragmatic information. Pragmatic information is a kind of information that generates changes in the receiver. In other words, it is a newly discovered negentropy, recognized and metabolized because of structural changes in the system. If we conceive the city as a negentropy processor, like exposed in Fig. 2.1, we can conclude that the evolution of such a system is only possible by recognizing new things as resources. Changes in the system are structural changes when something that was only noise before the change is now appreciated as a useful resource for the system metabolism. Pragmatic information can be regarded indeed as a balance between novelty and confirmation. The ability to recognize new resources, or to assign value to resources that was not considered before, comes out of the noise through a communication process that uses pattern recognition as a driver, according to the hypothesis developed by Haken and Portugali (2003). The new information is interpreted as a message, depending on the ability of the system's actors to analyse other structures and recognize relevant patterns. This is exactly the guide-learning process that permits self-organization of urban systems.

During a transition between states of the system, the novelty component overpasses the confirmation and the entropy released by the system tends to a maximum (A). When novelty is absorbed by the system and recognized as useful information, the balance of novelty and confirmation tends to be 50% and the entropy reaches a minimum (B). This is the local quasi-stationary point in a phase space where the system will stand for a while. Kauffmann (1993) called this phenomenon "structural stability", referring to a far-from-equilibrium point that found the energy flows needed to maintain the system organization (Fig. 2.3).

To visualize the learning process, the description by Haken (1988) can be used. Patterns can be assumed as constituting a landscape, with hills and valleys. The system leads to a valley, that is, a local quasi-steady-state condition attractor. When a message comes to the system, two situations could happen. The message can be recognized as information, so the system is pushed to new attractors, or the message is not recognized as information and the system does not move from the actual state. When the message is recognized as information, the evolution to new system's quasi-equilibrium can be reached by different ways: the new attractor can be uniquely determined, or the message can give rise to different attractors. Finally, different messages can give rise to the same attractor. Not all the messages or the attractors have the same importance, of course. The redundancy of the system is determined by the quantity of messages pushing the system to one specific configuration. The evolution is a learning process that is possible only in a combination of novelty and confirmation, in terms adopted by the communication theory. Redundancy is an aspect that increases in mature systems, where efficiency is privileged with respect to power maximization and diversity explosion. However, mature systems arrive to saturation and collapse, moving the dynamic forward to new

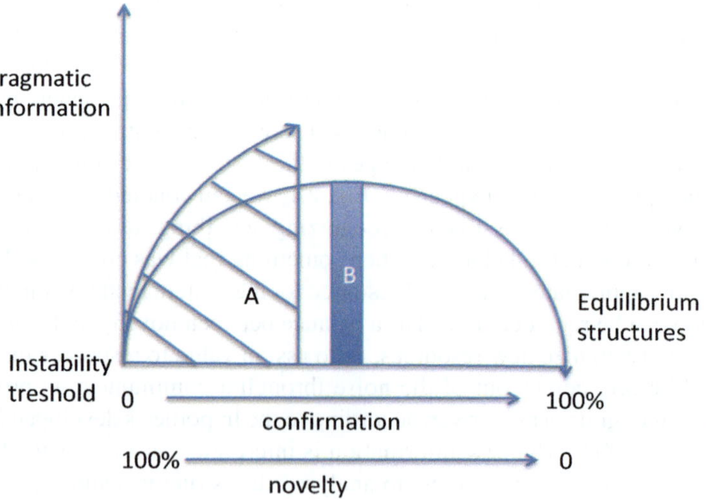

Fig. 2.3 Fluctuation novelty-confirmation and pragmatic information according to Jantsch. (Author redrawing from Butera 1998)

configurations. In this sense, the novelty-redundancy loop expresses the same dynamics of the efficiency-diversity principles exposed by Labanca.

2.5 Conclusion: The Future of Cities as a Balance of Efficiency and Diversity

Future cities should close as much as possible the circles of production and consumptions in order to lower the local entropy production but conserve sufficient diversity to assure resilience. It was timely observed by some authors (Ho 2015; Pelorosso et al. 2017), who proposed the "low-entropy city" as a structure able to pack together many metabolic processes respecting the emergency of possible adaptation strategies. It has to be observed that this vision refers to the entropy processed in the system at a certain time.

Ensuring diversity and resilience of a system implies that its efficiency is not maximized and, thus, the entropy production is not minimized. Otherwise, in systems whose efficiency is maximized entropy production is minimized. This is true in principle but not in the real systems created by humans, because of the so-called rebound effect or Jevons paradox (Wallenborn 2018; Polimeni et al. 2008), according to which each efficiency improvement gives rise to an increase in the amount of resources metabolized by the system and an increase of consumption—thus increase of entropy production.

The dynamic explored in previous sections shows that all complex systems move in a circular way, evolving through different phases from chaos to order and then to chaos again. This movement always permits the complex system to be resilient, which means to be able to reconfigure itself to adapt to changed boundary conditions.

However, we should take care of the concept of resilience we use. The dynamic described by Labanca suggests that the energy dissipated in the whole system (environment) makes it (the environment) able to generate new configurations once energy abundance would be achieved. However, such resilience concept implies the destruction of the subsystem that was critically exposed to resource scarcity. Is that the type of resilience we are looking for? It appears quite obvious that the planet Earth will be able to reconfigure itself after the eventual urban civilization collapses. But what about us? We should construct a very different type of resilience: a resilience internal to the system, that is, something able to generate the capacity of the urban structures to constantly renew themselves to respond to environmental pressure.

So, the idea could be to generate many local subsystems with high levels of energy efficiency and order, maintaining certain degree of entropy production inside the system representing the city as a whole (not just in the external environment). One of the problems is then what is defined as "external" to the system. Probably, these considerations should also lead us to a reflection on the dichotomies we still use in urban science—such as rural-urban just to name one of these. Recently, different researches expressed new vision as an attempt to overpass this dichotomy and to reach a more dialectical dialogue with nature (Inostroza et al. 2019).

Urban microclimate probably evolves under the same dynamics of the city as a whole complex system. In times of resource availability, different configurations of urban subsystems generate different microclimates. Then, the city evolves with efficiency and specialization, generating a diffuse urban microclimate. Probably, in a short future, rural-urban distinction will not be used anymore and we will live during a time in a complex urbanized world, with consequences on the macroclimate, which will evolve to a general urban macroclimate in a dissipative world. The study of the conditions that are generating today the urban climates, as well as the solutions for comfort and energy savings in such a situation, is a growing field that involves the efforts of both academia and practitioners to arrange countermeasures and build resilience in the way described before as internal adaptation capacity (Fig. 2.4).

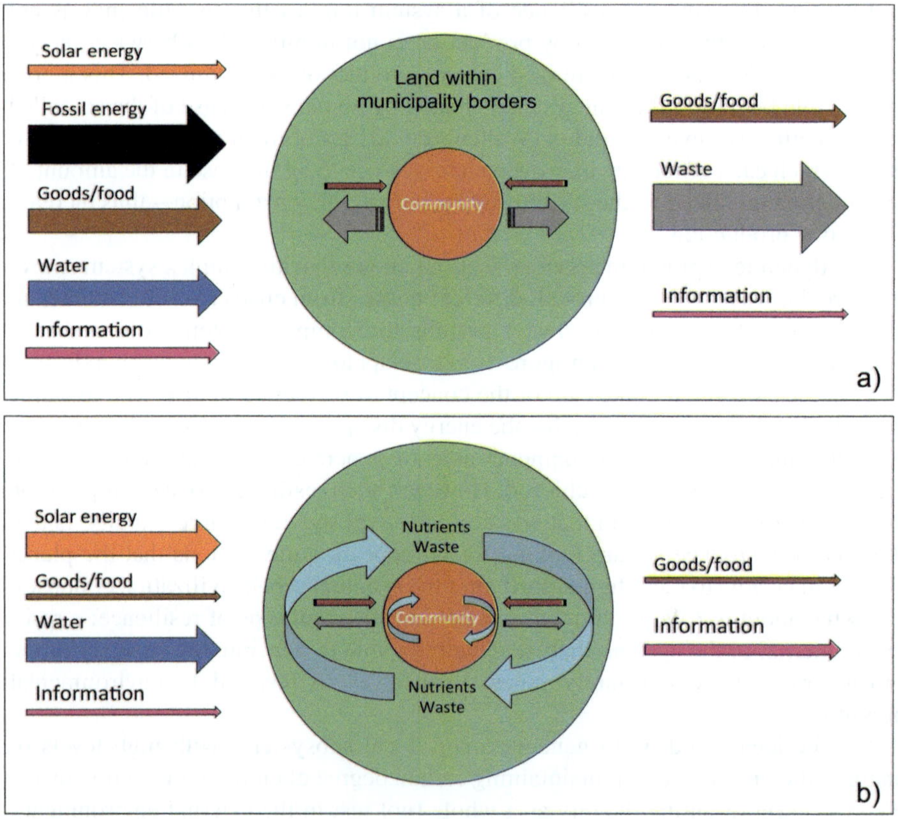

Fig. 2.4 High (**a**) and low (**b**) entropy cities. In a high-entropy city, flows are linear, required inputs are high as well as the waste is released to the environment. In a low-entropy city, the inputs' need is lowered (and fossil energy disappears), recycling and reusing contribute to close loops of materials and energy inside of the system, and waste release is reduced to almost zero. (Authors' drawing)

A combination of two objectives, in terms of metabolic change of future urban settlements, should be achieved. The first objective is to reduce the n-entropy flows but maintaining the services we need. It implies a more efficient use of natural resources and a reduction of waste. The second objective is to develop solutions responding to local stimuli that implies an increase in diversity and redundancy of the system. Sample actions to be taken are implicit in the concept of circular economy (Leone et al. 2018) and can be listed as follows:

- To decentralize energy production mainly by renewable sources (coupled with an increased energy efficiency of building and industry sector)
- To improve the efficiency of transport system, substituting private-car mobility with public transportation, car sharing, walking and biking (supported by new mixed-use planning)
- To incentivize local food consumption and a proper return of organic waste in the soil
- To optimize water cycles introducing new uses for waste water
- To reduce material flows through maintenance, repair and reuse of any kind of goods

All these actions imply not only an increase in the efficiency of resource use, but also a change of lifestyle, focusing on local diversity and avoiding the alienated cycle of production and consumption of goods at a global scale. A new economical thinking is, of course, the base for the new cities' metabolism development. A deeper insight into these issues is required, involving all disciplines: from physics to ecology, from economics to political sciences, from sociology to information science. Otherwise, without any tool for planning sustainable and resilient cities, there is a high risk of total collapse of their present structure and rise of new degraded ones.

References

Alexander, C., et al. (1977). *A pattern language*. New York: Oxford University Press.
Allen, P. (1998). Cities as self-organizing complex systems. In *The city and its sciences*. London: Physica-Verlag Ed.
Bettencourt, L., Lobo, J., Helbing, D., Kuhnert, C., & West, G. (2007). Growth, innovation, scaling and the pace of life in cities. *Proceedings of the National Academy of Science, 104*(17), 7301–7306.
Butera, F. (1998). Urban development as a guided self-organization process. In *The city and its sciences*. London: Physica-Verlag Ed.
Cavallaro, V. (1998). Strange loops, tangled hierarchies and urban self-regulation. In *The city and its sciences*. London: Physica-Verlag.
Flichakova, N., Robinson, D., & Scartezzini, J.-L. (2007). Quo vadis thermodynamics and the city: A critical review of applications of thermodynamic methods to urban systems. *International Journal of Ecodynamics, 2*(4), 222–230.
Glansdorff, P., & Prigogine, I. (1971). *Thermodynamic theory of structure, stability and fluctuations*. New York: John Wiley and Sons.
Haken, H. (1988). *Information and self-organization*. Berlin: Springer-Verlag.

Haken, H., & Portugali, J. (2003). The face of the city is its information. *Journal of Environmental Psychology, 23*, 385–408.

Ho, M. W. (2015). Sustainable cities as organisms, a circular thermodynamics perspective. *International Journal of Design & Nature and Ecodynamics, 10*(2), 127–139.

Inostroza, L., Hamstead, Z., Spyra, M., & Qureshi, S. (2019). Beyond urban-rural dichotomies: Measuring urbanization degrees in central European landscapes using the technomass as an explicit indicator. *Ecological Indicators, 96*, 466–476.

Isalgué, A., Coch, H., & Serra, R. (2007). Scaling laws and the modern city. *Physica A, 382*, 643–649.

Jantsch, E. (1980). *The self-organizing universe*. Oxford: Pergamon Press.

Kaufman, S. (1993). *The origin of order*. New York: Oxford University Press.

Kennedy, C., Cuddihy, J., & Engel-Yan, J. (2007). The changing metabolism of cities. *Journal of Industrial Ecology, 11*(2), 43–59.

Labanca, N. (2017). Energy and complex systems dynamics. In N. Labanca (Ed.), *Complex systems and social practices in energy transitions*. Cham: Springer Nature.

Labella, A. (1998). A short discussion of alternative approaches to modelling complex self-organizing systems. In *The city and its sciences*. London: Physica-Verlag Ed.

Lai, S. K., Han, H., & Ko, P. (2013). Are cities dissipative structures? *International Journal of Urban Sciences, 17*, 46–55.

Leone, A., Gobattoni, F., & Pelorosso, R. (2018). *Pianificazione e incertezza. Una bussola e alcune mappe per navigare nel mondo liquido*. Milan: Franco Angeli Ed.

Li, R., Zhang, J., Wang, X., Wang, W., Di, Z., & Stanley, D. E. (2017). Simple spatial scaling rules behind complex cities. *Nature Communications, 8*, 1841.

Linch, K. (1960). *The image of the city*. Boston: MIT Press.

Morowitz, H. J. (1979). *Energy flow in biology*. Woodbridge, CT: Ox Bow Press.

Nicolis, G., & Progogine, I. (1977). *Self-organization in nonequilibrium systems*. New York: Wiley-Interscience.

Odum, H. T., & Pinkerton, R. C. (1955). Time's speed regulator: The optimum efficiency for maximum power output in physical and biological systems. *American Scientist, 43*, 331–343.

Palme, M., Inostroza, L., Villacreses, G., Lobato, A., & Carrasco, C. (2017). From urban climate to energy consumption. Enhancing building performance simulation by including the urban heat island effect. *Energy and Buildings, 145*, 107–120.

Pelorosso, R., Gobattoni, F., & Leone, A. (2017). The low-entropy city: A thermodynamic approach to reconnect urban systems with nature. *Landscape and Urban Planning, 168*, 22–30.

Polimeni, J., Mayumi, K., Gianpietro, M., & Alcott, B. (2008). *The Jevons paradox and the myth of resource efficiency improvements*. London: Earthscan by Rutledge.

Portugali, J. (2000). Self-organizing city. In *Self-organization and the city*. Berlin: Springer Series in Synergetics.

Prigogine, I., & Stegers, I. (1984). *Order out of chaos*. New York: Bantam Ed.

Purvis, B., Mao, Y., & Robinson, D. (2019). Entropy and its application to urban systems. *Entropy, 21*, 56.

Rees, W. (2012). Cities as dissipative structures: Global change and the vulnerability of urban civilization. In *Sustainability science*. New York: Springer Nature.

Santamouris, M. (2014). On the energy impact of urban heat island and global warming on buildings. *Energy and Buildings, 82*, 100–113.

Schrodinger, E. (1944). *What is life?* Cambridge: Cambridge University Press.

Wallenborn, G. (2018). Rebounds are structural effects of infrastructures and markets. *Frontiers in Energy Research, 6*, 99.

Wolman, A. (1965). The metabolism of cities. *Scientific American, 213*(3), 179–190.

Zhang, Y., Yang, Z., & Yu, X. (2015). Urban metabolism: A review of current knowledge and direction for future studies. *Environmental Science and Technology, 49*, 19.

Chapter 3
The Urban Heat Island: Its Energetic Basis and Management

Gerald Mills, Julie Futcher, and Iain D. Stewart

3.1 Introduction

> The Mean Temperature of the Climate [...] is strictly about 48.50° Fahr.: but in the denser parts of the metropolis, the heat is raised, by the effect of the population and fires, to 50.50°; and it must be proportionately affected in the suburban parts. The excess of the Temperature of the city varies through the year, being least in spring, and greatest in winter; and it belongs, in strictness, to the nights; which average three degrees and seven-tenths warmer than in the country; while the heat of the day, [...] falls, on a mean of years, about a third of a degree short of that in the open plain (Howard, The Climate of London 1818).

The urban heat island (UHI) has been a subject of study for a very long time. Luke Howard was the first to measure the magnitude of the urban effect on near-surface air temperature when he compared measurements that he was making at his home(s) outside London with the observations being made at the Royal Society in central London, that is, the temperature difference (ΔT_{u-r}) between the countryside (rural) and the city (urban) observation stations (Mills 2008). He concluded that the temperature in the city was elevated owing to a number of causes, including the heat added by humans, the lack of vegetation (reduced evaporation), the reduced airflow (diminished mixing) and the geometry of urban surfaces that 'traps' radiation and obstructs 'free radiation to the sky'. Modern research has confirmed these causes and included one more, the thermal and radiative properties of construction materials that influence the storage and release of heat energy by the urban fabric. Howard's work on the UHI provided an auspicious start to the observation and understanding

G. Mills (✉)
School of Geography, UCD, Dublin, Ireland
e-mail: gerald.mills@ucd.ie

J. Futcher
Urban Generation, London, UK

I. D. Stewart
Global Cities Institute, University of Toronto, Toronto, ON, Canada

M. Palme, A. Salvati (eds.), *Urban Microclimate Modelling for Comfort and Energy Studies*, https://doi.org/10.1007/978-3-030-65421-4_3

of the UHI phenomenon but it would take almost 200 years for some of his ideas to be formalised and tested. To understand the UHI phenomenon (and to manage it) requires that it be placed within the framework of the surface-atmosphere exchanges of energy (expressed formally as an energy budget) and in the context of atmospheric boundary-layer formation. In the remainder of the introduction, we will outline the formation of the planetary boundary layer and the urban boundary layer, especially.

3.1.1 The Planetary Boundary Layer

The interaction of the atmosphere with the Earth's surface results in the formation of a planetary boundary layer (PBL) that has properties (temperature, humidity and wind profiles) that are acquired by exchanges with the underlying landscape. The PBL grows in depth during the daytime as solar radiation energy is absorbed at the surface and is transferred to the overlying air. Under clear conditions, the effects of the Earth's surface will extend vertically to 1–2 km by mid-afternoon. The distinct impact of different landcover types is clearest in the lowest 10% of the PBL where the surface layer (SL) is found. Above this layer lies the mixed layer where the contributions of the underlying surface are mixed with the contributions of upwind surfaces and these effects are transferred downwind. The SL is itself comprised of a roughness sublayer (RSL) that forms in and around the surface roughness elements (grass, crops, trees, etc.) that churn up the atmosphere close to the ground and an overlying inertial sublayer (ISL) where the effects of the individual roughness elements have been blended. The properties of the PBL are formed by energy (and momentum) exchanges at the Earth's surface that are transferred to the depth of the overlying air through the surface layer (via the RSL and ISL) and into the mixed layer.

Energy flows are expressed as the magnitude of energy (Joules, J) standardised by time (seconds, s)—referred to as Watts (W)—and surface area (square metres, m^2); this common set of units is termed a flux density (Wm^{-2}). The energy exchanges are categorised by type (radiation, sensible and latent heat) and by mode of exchange (radiation, convection and conduction).

Radiation describes both a form of energy and a mode of transfer; radiation is transferred as waves and the intensity of this energy is inversely proportional to wavelength. Radiation transfer is classified by source into solar (Sun) and terrestrial (Earth-atmosphere system). Solar energy is described as short wave as the energy it emits has wavelengths between 0.1 and 3 μm with a peak between 0.4 and 0.7 μm (visible radiation). Terrestrial radiation is emitted by the components of the Earth-atmosphere system (e.g. clouds, air and surface) at wavelengths between 3 and 100 μm (with a peak at 10–12 μm) and is categorised as long wave (or thermal infrared). Net radiation (Q^*) is the sum of all incoming (↓) and outgoing (↑) short- (K) and long-wave radiation (L) at a surface:

$$Q^* = \left(K\downarrow - K - + L\downarrow - L -\right) \tag{3.1a}$$

On a horizontal surface, received short-wave radiation ($K\downarrow$) is a function of the solar altitude (the angle of the Sun above the horizon) and the transmissivity of the atmosphere; a proportion (α) of $K\downarrow$ is reflected (albedo) from the surface as $K\uparrow$. Received long-wave radiation ($L\downarrow$) is emitted by the atmosphere toward the surface at a rate based on its air temperature (T_a) and atmospheric emissivity (ε_a); $L\uparrow$ is emitted from the surface based on surface temperature (T_s) and emissivity (ε_s). For long-wave radiation exchanges, the big difference between the Earth's surface and the atmosphere is their respective emissivities; whereas ε_s is generally around 0.9, ε_a is lower and much more variable. When $\varepsilon_s \approx \varepsilon_a$, cloud cover is low and extensive and surface cooling by long-wave radiation exchange is weak. Typically, over the course of a clear day the magnitude of Q^* is negative overnight ($K\downarrow$ is zero and $L\uparrow > L\downarrow$) and positive during the day ($K\downarrow > 0$) peaking at noon when solar altitude is highest. The cycle of radiation gain and loss drives the diurnal pattern of surface warming (energy gain) and cooling (energy loss). This basic pattern is the same throughout the year but the magnitudes change mostly as a result of the seasonal change in the position of the Sun in relation to the Earth, which controls the length of day and the intensity of $K\downarrow$. Net radiation can be treated as available energy that is transformed at the surface into non-radiative energy exchange into sensible and latent heat.

Temperature is a measure of the sensible heat content of a medium; the connection is based on the heat capacity of the medium, that is, the energy required to change the temperature of a volume ($Jm^{-3}\ K^{-1}$). At the surface-air interface, sensible heat is transferred by conduction into the substrate and by convection into the atmosphere based on the temperature difference. The different modes of transfer are the result of the medium itself; whereas energy is transferred slowly by molecular action in the solid substrate, it is transferred quickly by mixing in the fluid atmosphere. The rate of transfer (Wm^{-2}) into the substrate (Q_G) is regulated by the difference in temperature (expressed as Kelvins, K) over a distance (Km^{-1}) and the thermal conductivity (k) of the material, that is, the Watts per metre per Kelvin ($Wm^{-1}\ K^{-1}$). The rate of transfer into the atmosphere (Q_H) is related to the temperature difference and the mixing ability of the near-surface atmosphere. By analogy with Q_G, mixing ability is often expressed as the eddy conductivity, the magnitude of which is related to both atmospheric and surface properties such as windspeed, roughness of the surface and stability. Wind and roughness together govern the forced mixing of the air which is greater where the surface is rougher and the wind stronger. Stability is a measure of the buoyancy of the air and the tendency to mix by free convection. When the atmosphere is unstable, Q_H is greater for the same surface-air temperature difference but when stable, Q_H is suppressed.

Water has extraordinary energy properties related to its physical state and changing from a solid to a liquid to a gas requires a considerable input of energy that is not revealed by temperature. The transformation of liquid to gas via evaporation is a major driver of the near-surface climate; this requires about 2.5 MJ kg^{-1}. Humidity is a measure of the latent heat content of air (measured as vapour density, ρ_v) and the surface-air transfer of latent heat (Q_E) is regulated by the difference in humidity and the mixing ability of the atmosphere.

Figure 3.1 and Table 3.1 present the variables that regulate the magnitude of Q^* ($K\downarrow$, $K\uparrow$, $L\downarrow$, $L\uparrow$), Q_H, Q_E and Q_G at a surface. At this interface, the sum of the terms must balance

$$Q^* = Q_H + Q_E + Q_G \tag{3.1b}$$

These terms are dynamic and change over the course of a day and year in response to solar inputs and atmospheric and soil conditions. Following sunrise, as Q^* increases and becomes positive, the majority of the available energy is expended as Q_G as the temperature difference is large (following night-time cooling) and

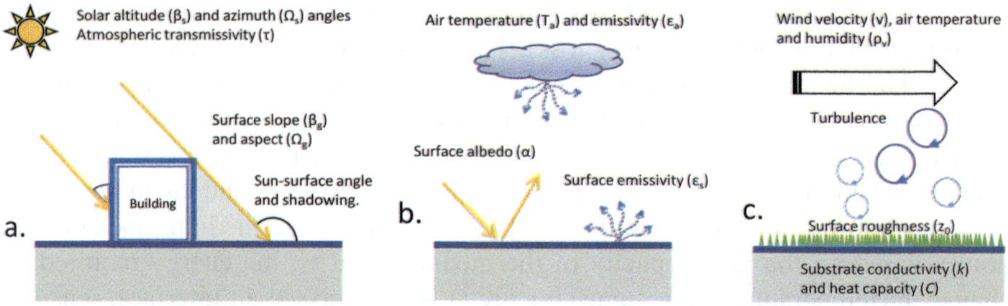

Fig. 3.1 The controls on short-wave (**a**) and long-wave (**b**) radiation exchanges at the surface and on the surface-air and substrate exchanges (**c**)

Table 3.1 Energy budget terms and controls on the magnitude of fluxes (Wm^{-2})

Energy budget term	Controls
Received short-wave radiation ($K\downarrow$)	Solar geometry: azimuth (Ω) and altitude (Z)
	Atmospheric transmissivity (τ) and cloud (k)
	Surface geometry: slope (β_s) and aspect (Ω_s)
Reflected short-wave radiation ($K\downarrow$)	Surface reflectivity or albedo (α)
Received long-wave radiation ($L\downarrow$)	Atmospheric temperature (T_a)
	Atmospheric emissivity (ε_a)
Emitted long-wave radiation ($L\uparrow$)	Surface temperature (T_s)
	Surface emissivity (ε_s)
Turbulent sensible heat flux (Q_H)	Surface-air temperature difference (ΔT_a)
	Atmospheric stability (Φ)
	Wind (v) and surface roughness (z_o)
Turbulent latent heat flux (Q_E)	Surface-air humidity difference ($\Delta\rho_a$)
	Atmospheric stability (Φ)
	Wind (v) and surface roughness (z_o)
Conductive sensible heat flux (Q_G)	Surface-substrate temperature difference (ΔT_s)
	Thermal conductivity (k_s) and heat capacity (C_s)

atmospheric mixing is weak. By mid-morning, the surface temperature has risen and the air begins to respond by becoming more turbulent. As a result, Q_H and Q_E start to increase and more of Q^* is expended on transfer by convection (also known as eddy transfer). By mid-afternoon, these eddy terms dominate the energy balance as Q_G declines, becoming negative before sunset even as ($Q_H + Q_E$) remain positive. Once the Sun sets, the convective exchanges decline and Q_H becomes negative. Overnight, Q^* is negative ($L\uparrow > L\downarrow$) and balance is achieved by heat transfer from the substrate ($Q_G < 0$) and from the overlying air ($Q_H < 0$); in general, Q_E will approximate zero unless condensation occurs. Figure 3.2a shows the typical diurnal pattern of these fluxes for a mid-latitude grass surface on a clear day. The ratio of Q_G/Q^* is a good indicator of the relative role of convective versus conductive exchanges of available energy at a surface and the Bowen ratio ($\beta = Q_H/Q_E$) is a measure of the dominant type of convective exchange. For a moist surface, $\beta < 1$ and surface-air exchanges are dominated by evaporation (increasing atmospheric humidity) but where $\beta > 1$, it is direct warming of the atmosphere that dominates.

The terms of the surface energy balance cannot be measured at the surface-air interface directly; instead, the radiative (Q^*) and turbulent ($Q_H + Q_E$) fluxes are observed above the ground, within the inertial sublayer (ISL) and above the surface roughness elements (e.g. grass or trees). Technically then, the instruments in the ISL measure the vertical fluxes at a plane representing the top of an imaginary control volume that extends to a depth in the substrate and has sides. The sum of all the exchanges across the facets of this volume describes an energy *budget* that allows for storage (Fig. 3.3a):

$$Q^* = Q_H + Q_E + \Delta Q_S + \Delta Q_A \tag{3.2}$$

Here, the new terms are heat storage (ΔQ_S) and horizontal energy transfer by wind or advection (ΔQ_A). As the purpose of these measurements is to record the surface-air exchanges over different land-cover types, good practice is to place the observation site over an extensive and flat homogenous surface so that ΔQ_A is negligible. The storage term (ΔQ_S) accounts for the energy stored in the layer of atmosphere and substrate; the response of the system to energy storage is change of temperature (air and substrate) and of humidity (air). The dominant store of energy is in the substrate, which acts as the thermal 'memory' in the system.

Conventionally, the instruments placed in the ISL consist of radiometers (Q^*) and eddy correlation (EC) sensors (Q_H and Q_E). The EC system is comprised of very-fast-responding instruments that record the 3D wind, air temperature and humidity (Fig. 3.4). Closure of the budget is achieved by assigning the residual term to storage (ΔQ_S) although it can also include significant error associated with the EC measurements.

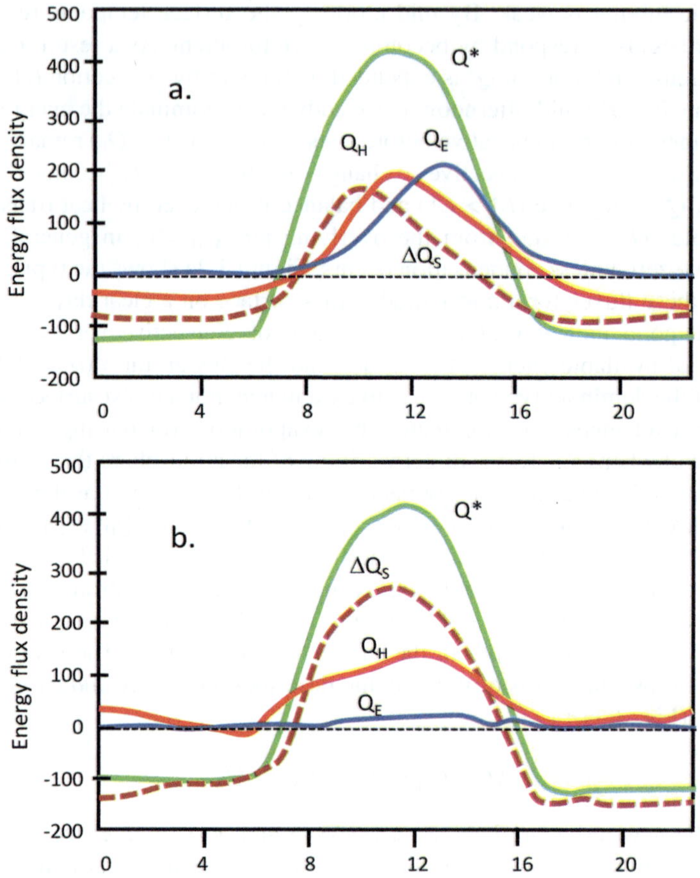

Fig. 3.2 Typical energy exchanges over an extensive grass-covered site in the mid-latitudes (**a**) and ensemble average fluxes for a site in central Mexico City for six almost cloudless days in December 1993 (**b**, redrawn from Oke et al. 1999)

3.1.2 The Urban Boundary Layer (UBL)

When air passes the upwind edge of an urbanised landscape a distinct urban boundary layer (UBL) begins to form and its depth grows with distance downwind (at a slope of about 1 unit vertically for every 100 m of horizontal travel) as the effects of the city below are mixed upwards. The UBL is distinguished by its contents; it is generally warmer and drier, and more turbulent than the PBL over a natural surface. The depth of the UBL reaches a maximum (1–2 km) near the city centre and does not grow significantly beyond this point. A new boundary layer forms over the non-urban landscape at the downwind edge and as it grows vertically from the ground upwards the UBL remains as an elevated plume that extends well beyond the urban footprint (Fig. 3.5). The vertical structure of the UBL is the same as that described

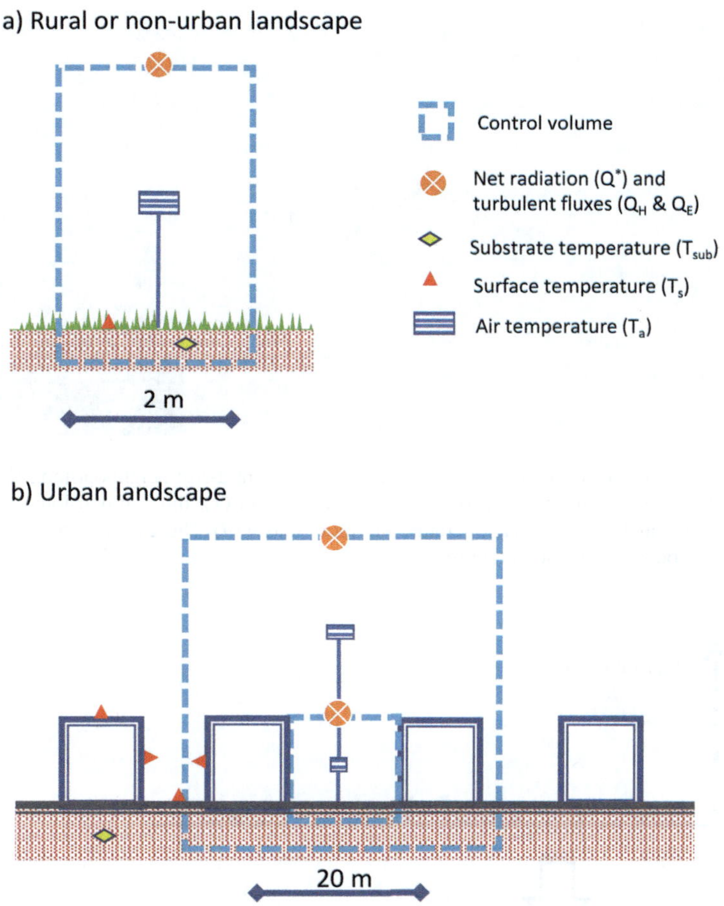

Fig. 3.3 Observations over a natural (grass) surface and over an urbanised landscape. The surface temperature (T_s) is an instantaneous response to the energy exchanges at the interface between the air and the substrate. The air (T_a) and substrate (T_{sub}) temperatures represent a response to energy storage in a volume. The energy fluxes (Q^*, Q_H and Q_E) are typically measured at the top of a 'control volume' that encloses a layer of atmosphere and underlying substrate to a depth unaffected by the surface exchanges

in Sect. 3.1.1; however, we can identify the urban canopy layer (UCL) as a distinct feature within the roughness sublayer (RSL) and surface layer. The term 'canopy' here is analogous to that of a forest canopy and forms an interface between the ground and the top of the canopy. However, unlike the trees in a forest, the canopy elements (buildings) are solid and exhibit considerable spatial variation in layout and dimensions which results in a deep turbulent layer (RSL) extending to 2–4 times the heights of buildings (Fig. 3.5).

The challenge in studying the urban energy balance is to identify the relevant *surface* at which the fluxes should be measured and/or modelled (Fig. 3.3b). Unlike

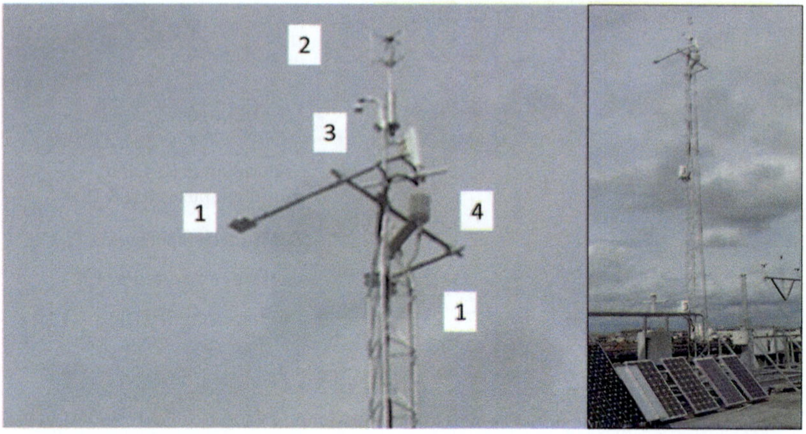

Fig. 3.4 Eddy covariance instrument platform located in a mid-rise neighbourhood (LCZ 2) in Dublin (Ireland). The instruments shown are net radiometers (*1*), three-dimensional sonic anemometer (*2*), gas analyser (*3*) and temperature/humidity sensor (*4*). These instruments are mounted at height well above the surrounding urban landscape

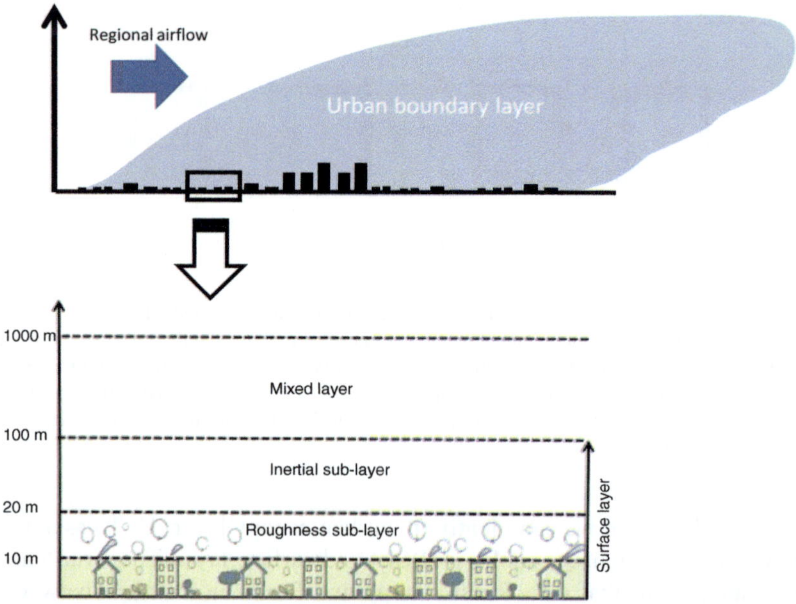

Fig. 3.5 The development of the urban boundary layer (UBL) as airflow crosses the upwind urban edge (top diagram). The UBL can be divided into a surface layer, which occupies the lower 10%, and a mixed layer. The surface layer is comprised of an inertial sublayer (ISL) where flux densities are nearly constant with height and a roughness sublayer (RSL) that is characterised by enhanced turbulence caused by underlying urban surface. The urban canopy layer (UCL) is immersed within the RSL and extends from the ground to mean building height

natural surfaces, where interest is usually in the layer of atmosphere above the canopy layer, in cities, we are most interested in the canopy layer (below roof level) where people live. This means that urban observations are made at two levels, each of which is exposed to different signals: within the UCL (and RSL) and above the UCL (and RSL) within the inertial sublayer. This can cause some confusion on the meaning of the term 'surface', which is used to describe both exchanges at the solid-air interface (ground, walls, roof, etc.) and exchanges at the top of the surface layer (the base of the mixed layer), located well above the underlying urbanised landscape. Clarity on these two surface types is important for structuring and interpreting urban observational and modelling projects.

In this chapter, we will outline the development of the field of urban climate science and place the urban temperature effect (that is, the heat island) within the framework of boundary-layer meteorology and energy exchanges. The structure of this chapter is as follows: first, the different types of UHI and the evolution of their study; second, placing the UHI within the framework of the surface energy balance; third, the current 'state of the art' with regard to measuring and simulating the UHI; fourth, UHI mitigation and adaption policies; and finally, the future of UHI studies.

3.2 The Types of UHI

The UHI is a consequence of the differences in the energy balance of the urban surface (and the energy budget of the urban atmosphere/substrate) when compared to that of the surrounding non-urban landscape. The layering of the near-surface atmosphere described in Sect. 3.1.2 results in four types of UHIs (Fig. 3.6), each of which has distinct temporal and spatial patterns associated with different dominant processes:

- The canopy-level UHI (CUHI)
- The boundary-layer UHI (BUHI)
- The surface UHI (SUHI)
- The substrate UHI (GUHI)

Of these types, the CUHI and SUHI have been studied most and are often confused in popular understanding of the phenomenon. The SUHI represents an instantaneous temperature response to the exchanges of energy at the solid-air interface; this is where the surface energy balance (Eqs. 3.1a and 3.1b) is appropriate. Measurements of air temperature used to study the CUHI and BUHI represent the change in energy storage of an air volume (ΔQ_S) that is the outcome of the convergence of energy in a layer of atmosphere (Eq. 3.2, Fig. 3.3b). The subsurface UHI is a result of ΔQ_S in the layer of substrate below the ground. Recognising the difference between these UHI types is critical to identifying and understanding the processes responsible for each and guiding any adaptation and/or mitigation efforts in response.

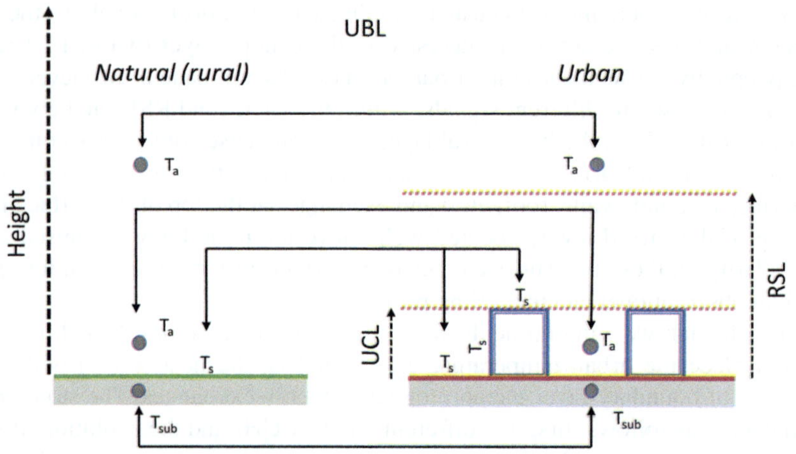

Fig. 3.6 The vertical structure of the atmosphere over rural (short grass) and urban surfaces. In the rural environment, the grass blades form a very shallow canopy and roughness layer close to the ground. In the urban environment the vertical geometry results in deep layers associated with the UCL and RSL. The locations of instruments to measure substrate (T_{sub}), surface (T_s) and air (T_a) temperatures are used to evaluate the types of urban heat islands as a difference between urban and rural values

3.2.1 A Brief History of UHI Studies

The great majority of UHI observational studies conducted until the 1980s focussed on near-surface air temperatures (that is, the CUHI). Most followed Howard's urban-rural temperature difference approach (ΔT_{u-r}) but provided more details on the spatial and temporal patterns of its magnitude. This was accomplished by using thermistor sensors and storage devices that could be placed on mobile platforms, such as cars (e.g. Sundborg 1951). Much of this work was completed for mid-latitude cities and shows that the CUHI:

- Is strongest under clear skies and calm conditions during a dry spell (weather)
- Increases in magnitude from sunset and is largest about 4 h after sunset
- Spatially increases from the low-density edge of the city toward its higher density core(s)
- Includes cooler areas which coincide with areas of green/water cover
- Has implications for agriculture, settlement design, weather forecasting and human health

These results capture the effects of the UCL (the space below roof level) on air temperature and the microscale processes in the surrounding walls, street, vegetation, etc. Figure 3.7 shows the CUHI for London during a warm summer; note the increasing warmth toward the city centre and the cooler areas associated with large green parks.

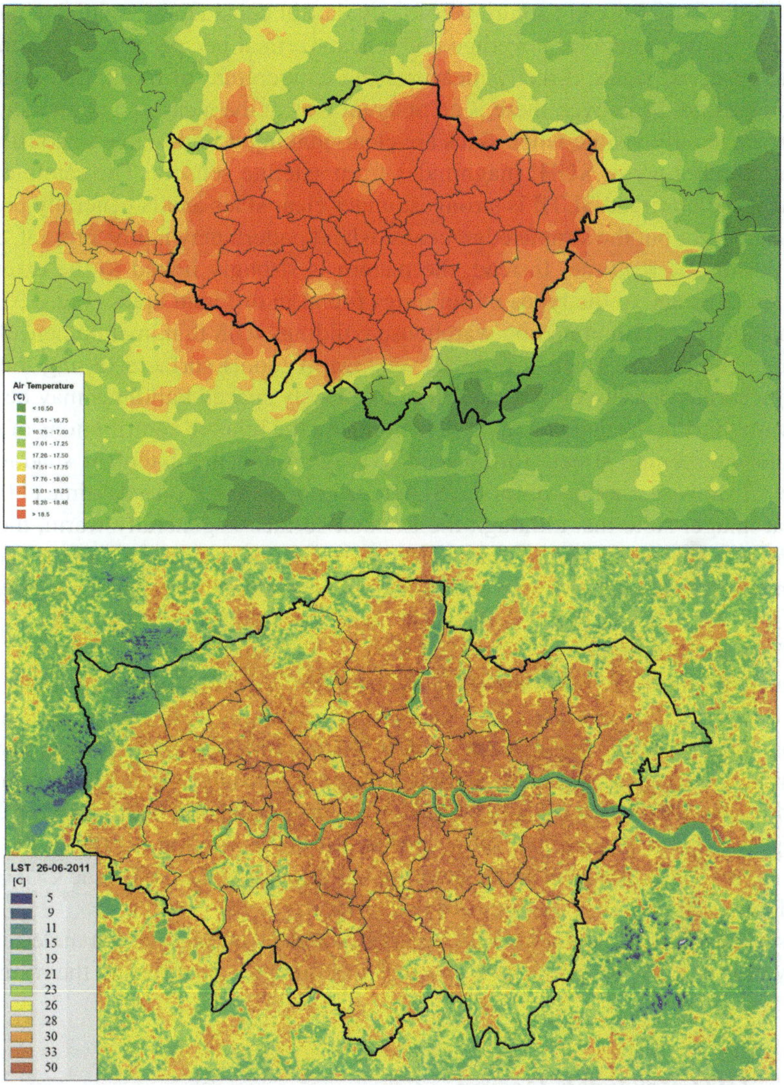

Fig. 3.7 Summertime CUHI (top) and SUHI (bottom) for the London area. The Greater London boundary is shown in black. The SUHI is measured using daytime land surface temperatures obtained from the Landsat satellite for 26th June 2011 (credit: Arup/UK Space Agency). The CUHI is modelled by average air temperature for May–July 2006 (credit: Arup/University College London). (Source: Arup (2014) with permission)

From the 1980s onwards the availability of thermal sensors that could measure the surface temperature has added greatly to the body of UHI knowledge; until this time, the properties of the SUHI were largely imputed from the canopy level UHI (CUHI). The number of SUHI studies has substantially increased since satellite-based thermal infrared (TIR) observations became widely available. These observational systems can capture whole-city data at considerable spatial detail but they have limited temporal coverage and a near-vertical perspective, which means that the observed signal directly measures the temperature of roofs, tree tops and exposed ground. Moreover, there is an inherent bias in the measured temperature as the surface is only visible when skies are clear, making it impossible to construct a general climatology of SUHI magnitudes from satellite sensors. The satellite data show that the observed SUHI is strongest under clear skies and is generally present at day and night but the simple edge-centre island pattern may not be present, especially during the daytime. Instead, the pattern is strongly linked to the extent of paved surfaces and to the types of buildings found in different areas (Fig. 3.7). For example, districts dominated by warehouse buildings with thin, metallic envelopes and large roof surfaces are visible as very warm (cool) surfaces during the day (night) whereas densely built areas with tall, closely spaced buildings may be seen cool/warm during the day and night depending on the relative contributions of rooftops and intervening streets to the received signal and the extent of shadowing. Figure 3.7 shows a satellite-observed SUHI for London based on a daytime (1100 h) Landsat image. Note that although the London surface is generally warm, there is considerable spatial variability that reflects the variation in surface properties across the urbanised landscape. Moreover, while there is evident correspondence between the CUHI and SUHI in Fig. 3.7, the scales (and magnitudes) of the urban temperature effect are different.

The BUHI has received comparatively little scientific attention (despite its relevance for air quality) and there have been few observations in the mixed layer where the contributions from the urban SL are blended with contributions from upwind non-urban landscapes. Similarly, there have been few studies of the GUHI which is caused by the conductive transfer of heat from the ground and buildings above and heating by piping systems and infrastructure buried at varying depths in the substrate. In the UHI mitigation and adaptation literature, it is the SUHI and CUHI that have dominated and form the focus of this chapter.

It is worth outlining assumptions, techniques and developments that underpin the history of much UHI research as it provides a context for interpreting the results.

3.2.2 UHI Research Design

The majority of UHI studies have adopted Howard's simple framework to measure urban effects on air temperature (ΔT_{u-r}), often without a complete consideration for its underlying implicit assumption. While the urban station is taken to represent the character and climate of the urbanised landscape, the non-urban (rural) station is taken to represent the climate that would exist in the absence of urbanisation (Fig. 3.6). In other words, the non-urban station represents an unchanging 'baseline' against which the urban effect is measured. Lowry (1977) articulated the problematic nature of the ΔT_{u-r} methodology and identified a number of confounding effects that limit this as a measure of the CUHI. First, the natural landscape experiences change just as the urban landscape does, so that considerable care needs to be taken in identifying both urban and rural stations for a UHI study. Ideally, any landscape changes that have taken place in the vicinity of all stations are recorded and form part of the station metadata. Second, the stations are each influenced by background weather conditions but for ΔT_{u-r} to be attributed to only human effects on the landscape, both stations must have experienced the same sequence of weather and must be located in areas of similar relief and elevation. Third, the influence of the city extends outside the urban boundary and the extent and intensity of this influence depend on background weather. In other words, observations at a nearby rural station may contain an unknown urban effect that is variable in size and timing. To minimise these issues, Lowry stressed the need to select urban/rural comparison stations carefully and control for weather conditions especially in analysing any differences that emerge. He also indicated that modelling may provide the only means of separating out different processes so that the urban effect on atmospheric variables could be isolated.

3.2.3 Describing the Urban Landscape

Measuring the UHI as a difference between two categories—urban and rural (non-urban)—pre-supposes that these terms are meaningful descriptions of the landscape. CUHI studies have repeatedly shown intra-urban variations in near-surface air temperature associated with land-cover changes. Overall, the pattern shows an increased magnitude of CUHI from the urban ‘edge’ to the city centre, which is generally where the built density is highest, road width is narrowest and there is little vegetation. However, within this ‘island’ there are pools of cooler air located over well-vegetated urban areas, parks and rivers/lakes (Fig. 3.7). These variations can be usefully categorised into temperature effects at the urban, local (≥1 km^2) and microscales (100–500 m).

To cope with the variety of microscale environments found in cities, climatologists adopted the urban canyon as a model of the city street, a ubiquitous urban feature (Fig. 3.3b). A canyon is formed by a long row of terraced buildings of a given height (H) separated by a street of given width (W). This model encapsulates the properties of the construction materials and geometry that distinguish urban microscale environments and it has provided an effective frame for analysing and modelling the climate effects within the UCL. At a local scale, UHI researchers have used a variety of urban land-use and land-cover categories in the past to interpret UHI observations; however, the lack of consistent terminology and the absence of details on vegetative cover, construction materials, etc. greatly limited their diagnostic value. A significant recent development has been the development of the local climate zone typology as a description of landscape types suited for CUHI studies (Stewart and Oke 2012); there are 17 LCZ types, 10 of which are built up and 7 of which are natural (Fig. 3.8).

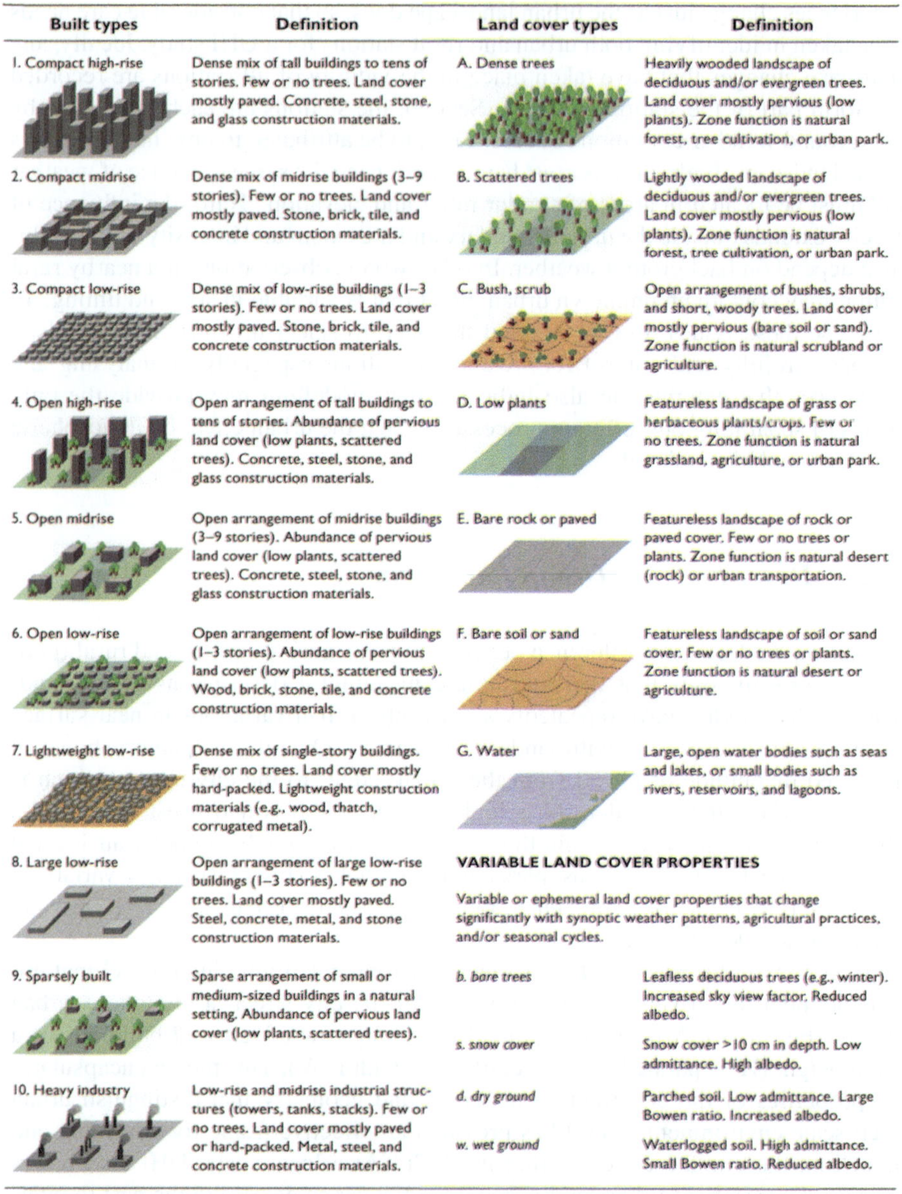

Built types	Definition	Land cover types	Definition
1. Compact high-rise	Dense mix of tall buildings to tens of stories. Few or no trees. Land cover mostly paved. Concrete, steel, stone, and glass construction materials.	A. Dense trees	Heavily wooded landscape of deciduous and/or evergreen trees. Land cover mostly pervious (low plants). Zone function is natural forest, tree cultivation, or urban park.
2. Compact midrise	Dense mix of midrise buildings (3–9 stories). Few or no trees. Land cover mostly paved. Stone, brick, tile, and concrete construction materials.	B. Scattered trees	Lightly wooded landscape of deciduous and/or evergreen trees. Land cover mostly pervious (low plants). Zone function is natural forest, tree cultivation, or urban park.
3. Compact low-rise	Dense mix of low-rise buildings (1–3 stories). Few or no trees. Land cover mostly paved. Stone, brick, tile, and concrete construction materials.	C. Bush, scrub	Open arrangement of bushes, shrubs, and short, woody trees. Land cover mostly pervious (bare soil or sand). Zone function is natural scrubland or agriculture.
4. Open high-rise	Open arrangement of tall buildings to tens of stories. Abundance of pervious land cover (low plants, scattered trees). Concrete, steel, stone, and glass construction materials.	D. Low plants	Featureless landscape of grass or herbaceous plants/crops. Few or no trees. Zone function is natural grassland, agriculture, or urban park.
5. Open midrise	Open arrangement of midrise buildings (3–9 stories). Abundance of pervious land cover (low plants, scattered trees). Concrete, steel, stone, and glass construction materials.	E. Bare rock or paved	Featureless landscape of rock or paved cover. Few or no trees or plants. Zone function is natural desert (rock) or urban transportation.
6. Open low-rise	Open arrangement of low-rise buildings (1–3 stories). Abundance of pervious land cover (low plants, scattered trees). Wood, brick, stone, tile, and concrete construction materials.	F. Bare soil or sand	Featureless landscape of soil or sand cover. Few or no trees or plants. Zone function is natural desert or agriculture.
7. Lightweight low-rise	Dense mix of single-story buildings. Few or no trees. Land cover mostly hard-packed. Lightweight construction materials (e.g., wood, thatch, corrugated metal).	G. Water	Large, open water bodies such as seas and lakes, or small bodies such as rivers, reservoirs, and lagoons.
8. Large low-rise	Open arrangement of large low-rise buildings (1–3 stories). Few or no trees. Land cover mostly paved. Steel, concrete, metal, and stone construction materials.	**VARIABLE LAND COVER PROPERTIES**	Variable or ephemeral land cover properties that change significantly with synoptic weather patterns, agricultural practices, and/or seasonal cycles.
9. Sparsely built	Sparse arrangement of small or medium-sized buildings in a natural setting. Abundance of pervious land cover (low plants, scattered trees).	*b. bare trees*	Leafless deciduous trees (e.g., winter). Increased sky view factor. Reduced albedo.
		s. snow cover	Snow cover >10 cm in depth. Low admittance. High albedo.
10. Heavy industry	Low-rise and midrise industrial structures (towers, tanks, stacks). Few or no trees. Land cover mostly paved or hard-packed. Metal, steel, and concrete construction materials.	*d. dry ground*	Parched soil. Low admittance. Large Bowen ratio. Increased albedo.
		w. wet ground	Waterlogged soil. High admittance. Small Bowen ratio. Reduced albedo.

Fig. 3.8 Local climate zone (LCZ) types (Stewart and Oke 2012). (© American Meteorological Society. Used with permission)

This scheme provides a framework for CUHI studies by decomposing the city (and its surroundings) into neighbourhoods, each of which is comprised of a characteristic mix of microclimatic environments that exerts an effect on the near-surface atmosphere. A temperature sensor that makes a traverse across an urban landscape in ideal CUHI weather reveals variations in T_a in response to the microscale influence that characterise that neighbourhood type. The LCZ map of a city provides a spatial context for sampling and analysing the CUHI and reframes the CUHI as a difference in air temperature between types of built and natural landscapes, rather than simply urban and rural. Critically, the LCZ types are linked to numerical descriptions of the variables that regulate the surface energy exchanges and T_a (Table 3.2).

Table 3.2 Local climate zone types and associated ranges of associated parameter values (Stewart and Oke 2012)

LCZ type	Mean building height (m)	Building surface fraction (%)	Impervious surface fraction (%)	Pervious surface fraction (%)	Sky view factor	Q_F Wm^{-2}
Compact high-rise	>25	40–60	40–60	<10	0.2–0.4	50–300
Compact mid-rise	10–25	40–70	30–50	<20	0.3–0.6	<75
Compact low-rise	3–10	40–70	20–50	<30	0.2–0.6	<75
Open high-rise	>25	20–40	30–40	30–40	0.5–0.7	<50
Open mid-rise	10–25	20–40	30–50	20–40	0.5–0.8	<25
Open low-rise	3–10	20–40	20–50	30–60	0.6–0.9	<25
Lightweight low-rise	2–4	60–90	<20	<30	0.2–0.5	<35
Large low-rise	3–10	30–50	40–50	<20	>0.7	<50
Sparsely built	3–10	10–20	<20	60–80	>0.8	<10
Heavy industry	5–15	20–30	20–40	40–50	0.6–0.9	>300
Dense trees	3–30	<10	<10	>90	<0.4	0
Scattered trees	3–15	<10	<10	>90	0.5–0.8	0
Bush/scrub	<2	<10	<10	>90	0.7–0.9	0
Low plants	1	<10	<10	>90	>0.9	0
Bare rock or paved	<0.25	<10	>90	<10	>0.9	0
Bare soil or sand	<0.25	<10	<10	>90	>0.9	0
Water	N/A	<10	<10	>90	>0.9	0

3.2.4 Process Vs. Responses

The majority of UHI studies up to the 1970s were descriptive analyses of temperature observations that were accompanied by hypothesised causes. Statistical tests were used to link the observed CUHI magnitudes to the weather and landscape controls on heating and cooling, especially the impact of cloud cover and wind speed. Research in the field however was gravitating toward a process-based understanding of the phenomenon that was founded on the surface energy balance. Initially, progress was based on theoretical models but as the instruments to measure the radiative, turbulent and conductive heat fluxes became available, UHI studies developed into a more sophisticated science in which observations were used to test theories and models used observations to evaluate simulations.

Myrup (1969) was the first to model the UHI by treating the city as a rough and dry paved surface ($Q_E = 0$) and using inputs from a nearby weather station to represent the ambient atmosphere (that is, the boundary conditions). The model cleverly specified all of the terms in the surface energy balance as a function of an equilibrium surface temperature (T_s), such that

$$Q^* - Q_H - Q_G = 0 \tag{3.3}$$

Figure 3.9 shows the simulated diurnal T_s for a paved area (city) and grass (park).

This model simulates a SUHI (not the CUHI) at the top of the UCL. The absence of geometry means that it does not account for radiation effects (such as shading and multiple radiation exchanges) or sensible heat exchanges at the walls and street surfaces. These processes are critical to the cooling and heating of the building

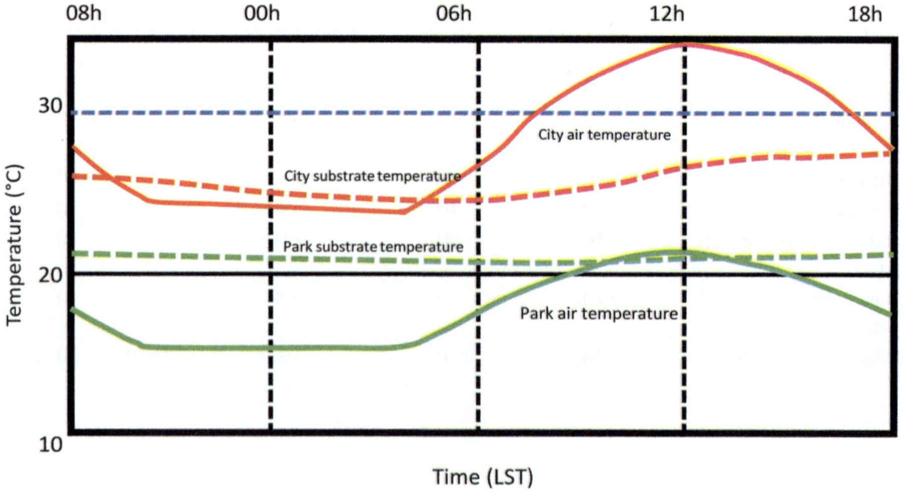

Fig. 3.9 Simulated near-surface and substrate temperatures for an urban park and a city centre at 38.5° N, 121.7° W on June 21st. Redrawn from Myrup (1969)

walls and street and to the formation of the surface and air temperature heat island within the canopy. This approach was used for early urban-scale models that simulated the effect of differential surface heating and roughness on the atmospheric boundary layer (e.g. Bornstein 1975). Other modelling efforts focussed on the UCL and used simplified forms such as street canyons to include the effects of geometry on radiation exchanges especially (e.g. Terjung and Louie 1974; Arnfield 1982; Mills 1997).

Observations of the urban energy balance started in the 1980s using methods that had already been developed over extensive and homogenous natural surfaces, such as grassland or forest biomes. As such, energy flux measurements were usually made above the UCL (within the urban ISL) and represented the local-scale landscape (Fig. 3.4). In contrast to the natural surfaces, the urban energy budget must account for the energy emitted by human activities (that is, the anthropogenic heat flux, Q_F):

$$Q^* + Q_F = Q_H + Q_E + \Delta Q_S + \Delta Q_A \tag{3.4}$$

Applying this to an urban landscape requires finding a 'near'-homogenous neighbourhood so that the advective term (ΔQ_A) is zero. In practice, neither Q_F nor ΔQ_S are observed directly; Q_F is embedded within the measured terms and may be estimated and ΔQ_S is obtained as a residual. Figure 3.2b shows the diurnal energy budget measured using this approach over a compact mid-rise neighbourhood in Mexico City. The selected site has little vegetation and measurements were made during a dry period. The lack of water availability is clear in observations of Q_E which is negligible; instead the majority of the available energy (Q^*) is channelled into sensible heat exchanges with the substrate (ΔQ_S) and the atmosphere (Q_H). This site is similar (albeit with geometry) to the urban surface simulated by Myrup.

Very few observational studies have been completed within the UCL. In a groundbreaking study, Nunez and Oke (1977) observed the energy balances of the walls and floor of a small street and the energy budget of the street volume that they enclosed. The measurements detected the distinct pattern of net radiation at each facet and the reflection of solar radiation from one to another (multiple reflection); however, when the exchanges at all facets were added and expressed as exchanges at roof level (that is, ignoring the microscale variations associated with geometry) the diurnal pattern looked remarkably similar to that of a paved flat surface. Unfortunately, few studies since have repeated this scale of work (Offerle et al. 2007).

3.3 Current Knowledge

Each UHI type is a response to changes to the energy balance/budget associated with aspects of urban form and function; however, the relative role of each in regulating UHI magnitude will vary.

3.3.1 Urban Form

This describes the physical components of the city which are in place and generate an urban temperature effect without the direct addition of new heat sources (Q_F). The relevant variables of form are:

- Surface cover: the proportions that are paved and impervious and vegetated
- Fabric: the thermal and radiative properties of construction materials
- Geometry: the three-dimensional organisation of buildings to create the urban canopy

Surface cover: The replacement of natural cover by paving has the effect of reducing the latent heat flux (Q_E) over urban areas; as a consequence net radiation at a surface is preferentially channelled into sensible heat flux with the substrate (Q_G) and atmosphere (Q_H). The response of the surface and overlying air will be to increase temperature. A simple way to reduce surface (air) temperature is to increase the vegetative cover so that more of the available energy is used to evaporate water.

Fabric: Construction materials have very diverse properties that are suited to their purpose. Paving materials are usually impermeable and dense with high thermal conductivity (k) and heat capacity (C) when compared to natural materials (Tables 3.3 and 3.4), which means that they can transfer and store heat more effectively. Most fabrics will have high emissivity values but can be distinguished by their ability to reflect solar radiation (albedo). Urban paving varies from dark (asphalt) to light (concrete) compared to natural surfaces. The chief characteristic of all manufactured materials is their consistency, while their thermal and radiative

Table 3.3 Thermal properties of natural materials and manufactured fabrics (Oke et al. 2017)

Materials	Heat capacity (C) $MJ\ m^{-3}\ K^{-1}$	Thermal conductivity (k) $W\ m^{-1}\ K^{-1}$
Clay soil		
• Dry	1.42	0.25
• Saturated	3.10	1.58
Sandy soil		
• Dry	1.28	0.3
• Saturated	2.96	2.2
Water (4 °C)	4.18	0.57
Air (10 °C)		
• Still	0.0012	0.025 (still)
• Turbulent	0.0012	125 (turbulent)
Asphalt	1.94	0.75
Brick	1.37	0.83
Clay tiles	1.77	0.84
Glass	1.66	0.74
Concrete		
• Aerated	0.28	0.08
• Dense	2.11	1.51

Table 3.4 Radiative properties of natural and manufactured materials

Surfaces	Albedo (α)	Emissivity (ε)
Grass	0.16–0.26	0.90–0.98
Forest	0.13–0.20	0.90–0.99
Water	0.03–0.10	0.92–0.97
Desert sand	0.20–0.45	0.84–0.92
Snow	0.50–0.90	0.82–0.99
Asphalt	0.05–0.27	0.89–0.96
Brick	0.2–0.6	0.90–0.92
Clear glass	0.08	0.87–0.95
Concrete	0.10–0.35	0.85–0.97
Tile	0.01–0.35	0.90
Tar and gravel	0.08–0.18	0.92
White paint	0.50–0.90	0.85–0.95

The values for water and glass are dependent on the angle of the solar beam at the surface; those for natural cover are dependent on seasonal growth and those for snow depend on condition. The properties of manufactured materials change with age and become darker (lighter) with time (Oke et al. 2017)

properties may change over long periods of time; they generally do not respond to weather or seasons. By contrast, the thermal properties of natural materials can change considerably depending on their water content, which increases k and C. In addition the radiative properties can change with seasons and vegetative growth cycles. Hence, the **relative** responses of urban and natural fabrics depend greatly on a number of factors. For example, while the urban paving is designed to shed water quickly and return to its dry state, the natural cover will retain rainfall and meltwater for a period of time. For this reason, the differential thermal response of the urban and non-urban surfaces is greatest during dry weather when soils have dried; in these circumstances the natural surface will heat up quickly during the daytime but cool quickly at night. Building materials share the property of imperviousness with paving materials but have far more diverse thermal and radiative properties as suits their purpose. The outer envelope of a building is designed primarily to provide shelter and to manage energy exchanges for benefit of the indoor space; as a result the materials can have extraordinary radiative properties especially.

Geometry: The urban surface is crenulated so that individual facets (roofs, walls, streets) have unique slopes and azimuths which affects the receipt and loss of short-wave (K) and long-wave (L) radiation. Apart from the impact of shadowing and the loss of direct (beam) short-wave radiation, the geometrical attributes mean that facets share diffuse radiation exchanges with each other and the sky. A simple measure of the canopy effect on the latter is the sky view factor (SVF), which indicates the proportion of radiation exiting a surface in the direction of the sky vault; where SVF equals unity the surface has an uninterrupted view of the vault. SVF is especially relevant to the formation of the night-time CUHI, which is a result of surface cooling. Oke (1981) illustrated the significance of SVF (represented as the H/W ratio of streets) in a well known equation that summarized the emprical evidence from observations of the maximum CUHI made in different cities around the world:

$$\Delta T_{\mathrm{u-r(mx)}} = 7.54 + 3.97\ln\left(H/W\right) \tag{3.5}$$

3.3.2 Urban Function

Urban functions describe the range of activities that results in the anthropogenic heat flux (Q_{F}), which has three sources:

$$Q_{\mathrm{F}} = Q_{\mathrm{Fb}} + Q_{\mathrm{Fv}} + Q_{\mathrm{Fm}} \tag{3.6}$$

where Q_{Fb} is heat from electricity use and combustion of domestic and industrial fuels in buildings; Q_{Fv} is heat from fuel combustion in ground vehicles; and Q_{Fm} is heat from human and animal metabolism. Table 3.2 provides general values for Q_F for LCZ types; note the association between the intensity of occupation (or industrial activity) and the magnitude of the flux. The highest values (up to 300 Wm^{-2}) are to be found in areas of heavy industry and in compact high-rise neighbourhoods. Of the three sources, Q_{Fm} is generally small except for high-rise, densely occupied neighbourhoods (Stewart and Kennedy 2017). The other terms have distinct diurnal (and weekday/weekend) patterns associated with commuting patterns and time spent at work and home. In addition, Q_{Fb} depends on the seasonal need for heating and cooling. Each of these sources adds heat to the urban atmosphere in different ways: Q_{Fb} will enter through building outlets (heating/cooling systems) and via the envelope; Q_{Fv} injects heat directly into the UCL mostly through exhausts; and Q_{Fm} warms the atmosphere directly (outdoors) or indirectly (indoors) where energy systems are used to offset the heat generated. The role of Q_{F} in the CUHI is clear but its contribution (compared to aspects of urban form) to the urban-scale phenomenon is highly variable in time and space and will be greatest where natural energy sources are small (high latitudes in winter) and heating demand is large. Elsewhere it will make a significant contribution at local and microscales that are densely occupied and used. However, one should note that the intensity of the CUHI is greatest at night when Q_{F} is lowest.

3.3.3 Progress in Urban Climate Science

Progress has come about through both observations and modelling. Increasingly, these are integrated such that the models need observations to be evaluated and observational strategy is guided by simulations.

Observations: Observations of the canopy-level UHI have continued using the same basic historic approach (that is, ΔT_{u-r}) but the geographical (and climatological) range of studies has increased and the observation methods have changed as inexpensive instruments that can be Web-enabled have proliferated. There are now more studies of cities in very hot tropical and very cold polar climates to add to the canon. In addition, the availability of privately owned weather stations that report directly to websites means that detailed spatial and temporal data is available, often in sufficient amounts to create a climatology. However, systemic problems often make it difficult to incorporate the findings into the corpus of existing knowledge. Stewart (2011) identified two areas of weakness in the body of UHI work from 1950 to 2007: first, the lack of controlled measurement for isolating the confounding effects of weather and surface relief on UHI magnitudes, and second, the failure to communicate essential metadata on instrumentation and field site characteristics. Hopefully, highlighting these issues will have an effect on reporting UHI studies. This is especially relevant as UHI studies are increasingly carried out in non-climatology fields and are being used to inform policy.

The biggest change that has taken place in direct observations of the UHI over the last two decades has been the proliferation of satellite-derived surface UHI assessments. There are a number of satellite platforms that have sensors that measure in the thermal infrared (TIR) radiation band. These instruments are tuned to wavelengths between 8 and 14 μm so as to acquire radiation emitted by the Earth's surface and transmitted through the atmosphere. TIR observations are available across a range of scales but probably the best known and used are those obtained by the Landsat programme, which has been in place for over 40 years. The great advantage of TIR data is that the means of acquisition is consistent and the coverage is global; this permits the study of multiple cities over lengthy periods. The disadvantages of these observation systems' data include the timing of the satellite passage and the need for clear skies. In the case of the Landsat satellite the publicly available data is acquired at about 1100 h local time every 16 days; however, the duration of this mission has allowed the compilation of data over many years. These data are processed to generate land surface temperature (LST) at a spatial resolution of 120 m (Fig. 3.7).

Satellite data is ideally suited to the development and assessment of UHI mitigation strategies as it can provide timely information on the response of surface temperature to actions, such as white roof programmes or enhanced greening. However, it is important to understand that these data represent an aggregate surface temperature (T_s) based on a near-vertical perspective of the urban surface. As a result the signal is dominated by the roofs of buildings, the canopy top of trees and the exposed ground. The effects of geometry on T_s are included indirectly through their effects on the surfaces that can be seen.

The results of satellite TIR studies show a very high correlation between surface temperature and fraction of the surface that is vegetated. The temperature of plant leaves will be close to air temperature when they are freely transpiring and their small surface/volume ratio means that they have very small capacity for heat storage. By comparison, manufactured surfaces are usually dry so that evaporation does not take place. Where the material is ground cover (such as asphalt or concrete paving), the daytime surface temperature is a response to the absorption of short-wave radiation and the relative ease of transfer to the underlying substrate and overlying atmosphere. Typically, surface temperature will increase from sunrise to mid-afternoon and then decrease. The rate of cooling is modulated by the ability to withdraw heat stored in the substrate during the daytime. From a satellite perspective, roofs are prominent and can have unusual properties as many have dark surfaces and are thin and consequently have the potential to heat and cool quickly. The ratio between the turbulent sensible and latent heat (Q_H/Q_E or the Bowen ratio, β) is an indicator of the type of surface-air exchanges. Where $\beta < 1$ it is evaporation/transpiration that dominates and T_s and T_a variations will be moderated. Conversely, where $\beta > 1$, T_s and T_a variations will be large. The importance of the vegetative surface fraction (λ_V) at a neighbourhood scale is evident in Fig. 3.10, which compiles data from a number of field studies and shows a linear relationship with the Bowen ratio. Simply put, increasing λ_V decreases β and moderates warming of the overlying air. In terms of the UHI magnitude, these exchanges have to be compared against the non-urban background context; if a city is located in an arid environment, then the urban-rural temperature difference will be relatively small and may even be reversed.

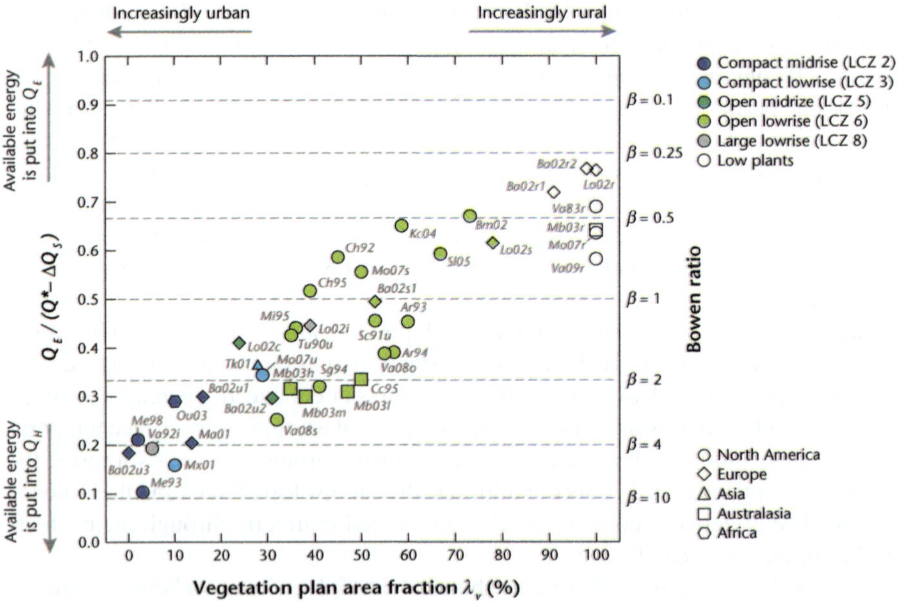

Fig. 3.10 The proportion of available energy ($Q^*-\Delta Q_S$) that is expended on the latent heat flux (Q_E) as a function of vegetative plan area fraction. This information has been compiled from energy flux studies conducted at a neighbourhood level at sites (indicated by the label) in cities across the world (source: Oke et al. 2017, with permission)

Modelling: A number of urban energy budget models have been developed that simulate the processes responsible for the UHI (Fig. 3.7). These can be categorised into two types: those developed as 'stand-alone' models that employ meteorological observations to simulate energy exchanges at a detailed scale and those developed as land-surface schemes to be incorporated into mesoscale models, such as the Weather Research Forecasting (WRF) model. In general, the stand-alone models incorporate more details on the characteristics of the UCL and were developed for detailed microscale studies (e.g. Krayenhoff and Voogt 2007). The land-surface schemes are used to capture the fundamental UCL processes within models that have a much larger extent (domain) and devote the majority of the available computer resources to the study of the boundary layer (e.g. Masson 2000; Kusaka et al. 2001). Typically, the stand-alone models are initiated using standard meteorological data and used to simulate energy exchanges at a high spatial and temporal resolution and many are designed to be used by non-experts in evaluating urban impacts and potential mitigation measures (e.g. Broadbent et al. 2019). By comparison, the land-surface schemes are part of a larger modelling environment that generates the urban meteorology internally based on boundary conditions at the outer edges that have been modified by exchanges with the underlying surface. A comparison of the performance of a large number of UEB models (Grimmond et al. 2010) concluded that Q_E simulations were poorest and that simpler models generally performed best.

One of the better known 'stand-alone' models is the Surface Urban Energy and Water Balance Scheme (SUEWS) which simulates radiation, energy and water balances using only standard meteorological variables and information about the surface cover (Sun et al. 2019). It is designed for neighbourhood-scale simulations and generates fluxes at the ISL level (equivalent to the level of observations). It describes the underlying urban surface in terms of seven surface types: paved, buildings, evergreen trees/shrubs, deciduous trees/shrubs, grass, bare soil and water. The scheme cleverly uses the diurnal relationship between heat storage (ΔQ_S) and net radiation (Q^*) to estimate the turbulent flux exchanges and a well-established meteorological scheme to simulate Q_E. Q_H is obtained as a residual. Typically, SUEWS is applied to a grid cell that is comprised of fractions of surface types, each of which exerts a distinct influence on the overlying atmosphere. It can also be applied to a domain that is divided into grid cells; however, it does not account for advection ($Q_A = 0$). The model has been evaluated and applied in a number of different circumstances and can generate surface and near-surface air temperatures for use in land-use planning (e.g. Alexander et al. 2015).

3.4 UHI Management

Until the 1980s studies of the UHI were driven by consideration (and anticipation) of urban effects on frost formation, survival of plants, weather forecasts, atmospheric pollution and human health (Chandler 1965). However, these studies were generally of esoteric interest. This has changed since the 1980s for a number of reasons. First, in assessing the instrumental evidence for climate change, the impact of the UHI on the temperature records of stations in and around settlements had to be removed as it confounded the regional and global temperature signal (Kataoka et al. 2009). Second, the public health consequences of heatwave events (periods of exceptional high temperatures) in cities became more apparent. The urban contribution to elevated heat risk in these circumstances is complex (Klinenberg 2015) but the UHI is a contributory factor. Third, the projections of climate change include global warming, which will add further to the heatwave hazard in cities (Li and Bou-Zeid 2013). As a consequence, managing the UHI is now considered part of urban policies that seek to create more sustainable and resilient cities that are able to cope with climate change at global, regional and urban scales.

The policies for UHI management are typically divided into those designed to reduce its magnitude (mitigate) or adjust (adapt) people's response to its effect, although most policies will address both simultaneously. For example, replacing conventional urban paving in part of a city with a vegetative surface cover reduces local heating (mitigation) but it may also become a place for people to go to seek relief from heat elsewhere (adaption) (Fig. 3.11). The key questions for any climate policy are which UHI is being managed and, for what purpose(s)? Answering these at the outset can set limits on the types of, and the scale of, actions (micro- or local or urban scale) needed. Most policies are designed primarily to reduce heat stress in outdoor and indoor environments; in the case of the former, this is often evaluated as the deviation from human thermal comfort while, for the latter, it is the building energy needed to achieve indoor comfort. The focus of these actions is the urban canopy layer where the population is directly exposed to the urban effect. Understanding of the processes that drive the SUHI and CUHI is needed to design appropriate policies.

To understand the contribution of the canopy and surface UHIs to discomfort and heat strain in humans, it must be placed in the context of the suite of meteorological variables that create environmental stress; these include radiation (Q^*), wind velocity (v), air temperature (T_a) and humidity (RH). The radiation inputs to the body include receipt of direct (beam) and diffuse solar radiation and of diffuse long-wave radiation. The body reflects short-wave radiation from the clothed skin and emits radiation to the surrounding environment. The radiation environment is often expressed through the mean radiant temperature (T_{MRT}), that is, the surface temperature of a uniform enclosure that generates the same radiation receipt as that received in the actual setting. The SUHI raises the T_{MRT} while the CUHI raises T_a; in addition, the RH and v are lowered in the urban canopy. The net impact of these changes is generally to increase heat stress during the daytime (SUHI and reduced wind) and night-time (CUHI and reduced wind). However, the magnitude (and even the direction) of these effects is not uniformly distributed across the urban landscape and the impact of these changes on heat stress depends on the background climate, which is often based on measurements made outside the city in a 'natural' environment. The heat stress may be expressed in terms of an equivalent temperature, that is, the actual air temperature under standard conditions that would be needed to produce the same level of stress (e.g. Matzarakis et al. 1999).

Fig. 3.11 Water and vegetation to offset urban warming in Nîmes, France. This can be viewed as both a mitigation strategy (lowering surface and air temperature in its vicinity) and an adaptation strategy (citizens availing of the microclimate during hot weather)

3.4.1 Mitigation

Reducing the magnitude of each UHI requires moderating surface temperature (T_s) at relevant space and timescales. Fundamentally, this means managing the exchange terms in the surface energy balance to reduce energy gain (Q^*) and/or channel available energy into evaporation (Q_E) over heating ($Q_H + Q_G$).

The diurnal driver of Q^* is received short-wave radiation ($K\downarrow$) which is regulated by time of year and time of day (Earth-Sun geometry) and atmospheric transmission, neither of which are controllable. At the surface, receipt can be managed through organising the 3D elements (trees and buildings mainly) to create areas of shade; while this does not change the total $K\downarrow$ at the urban surface it does manage the intensity at individual facets. Creating areas of shade to correspond with the hottest time of day is an effective way of managing the daytime SUHI at a microscale. Changing surface reflectivity (albedo) to manage ($K\uparrow$) is a simple way of lowering absorption at a surface (Tables 3.3 and 3.4) but it is important to consider the fate of the displaced energy. For example, raising the albedo of the ground surface by applying a reflective coating will transfer the reflected energy to other nearby objects, increasing their heat load. Reflective paving will reduce T_s of the ground surface but will increase radiation receipt on pedestrians; if the radiative properties of the reflected radiation match those of solar energy generally, this can also cause visual glare. By comparison, reflective roofs can be an effective means of reducing the energy gain of a building while not imposing a burden on other outdoor buildings. Applying this technique at neighbourhood and urban scales can reduce cooling energy demand and anthropogenic heat emissions at the building scale. However, altering roof properties will have a limited impact on the climate at street level.

Over the day, the net exchange of long-wave radiation ($L{\downarrow}$ and $L{\uparrow}$) is negative but typically shows little variation (by comparison with K^*); however, at night when K^* is zero, long-wave radiation loss is the driver for surface cooling. The receipt of long-wave radiation ($L{\downarrow}$) on a flat surface originates from the sky hemisphere and is a function of its temperature (T_a) and emissivity (ε_a); the latter varies with the water content of the atmosphere and typically lies between 0.6 and 0.9. Most fabrics have a surface emissivity of >0.90 (Tables 3.3 and 3.4) so the proportion of $L{\downarrow}$ that is reflected is small and most $L{\uparrow}$ is a result of emission regulated by surface temperature (T_s). The sky hemisphere represents a radiative sink from the perspective of the surface such that limiting the portion of the sky that is visible from the surface (the sky view factor) reduces the net radiative loss and the rate of surface cooling via this exchange. Within the UCL, horizon screening by nearby 3D elements means that SVF < 1 and a portion of $L{\uparrow}$ is recycled within the canopy. As there is a limited capacity to increase ε_s, the emissivity, the only route to increase the cooling rate and lower the SUHI at night within the UCL is to increase the SVF by widening streets or lowering buildings, neither of which are viable options in most cities.

The capacity of a material to store sensible heat is related to its heat capacity (C) and thermal conductivity (k). Higher values for both allow energy to be transferred over a deeper volume (Q_G) with a smaller change in its temperature (ΔQ_S). Manufactured materials have consistent C and k values, while the values for soils change radically with their water content (Tables 3.3 and 3.4). During ideal weather conditions for UHI formation, the natural soil is dry. As a consequence, low k values means that heat is stored during the daytime close to the surface and low C values result in high temperature; as a result these materials will respond quickly to energy gain and loss at the surface. Urban fabrics then are usually a better store of daytime energy that can sustain surface temperatures overnight (note, however, that if the natural soils are wet, the urban-rural difference may be reversed). Urban fabrics that cover the ground are designed to be strong and to shed water and are usually dense and impermeable as a result. Modifying the urban fabric to be less efficient reservoirs of heat means making these fabrics less dense and permeable. In general, this type of paving has been used in areas of light traffic or to create car parks and not as a replacement for road surfaces. The modified fabric will behave thermally more like the natural soils and its thermal properties will respond to wetting in a similar fashion.

The turbulent transfers of sensible (Q_H) and latent (Q_E) heat are driven by the surface-air differences in temperature and humidity, respectively, and the magnitude of each is regulated by atmospheric mixing. The absence of water and the paucity of vegetation on the urban surface means that the great majority of available energy

($Q^{*}-Q_{G}$) is channelled into Q_H (that is, $\beta \gg 1$) which raises near-surface air temperature. To increase Q_E and reduce T_s means increasing water availability at the surface by adding vegetation or simply wetting. While the latter can be used in particular circumstances where water is plentiful, increasing the vegetative cover is a more effective long-term solution. Within the UCL, vegetation can replace paved surfaces on the ground or provide cover for walls. The capacity of the ground-level vegetation to take up moisture depends on the volume of soil and its moisture status. Green walls typically consist of climbing plants that cover the wall surface. The leaves of plants have temperatures close to T_a so that the radiant temperature (T_{MRT}) for people outdoors is reduced. On roofs, green infrastructure provides the same function but the impact is felt mostly by the building itself and where buildings are tall, the impact of this facet on climate in the UCL is small.

The anthropogenic heat flux (Q_F) is an additional source of heating in cities and enters the atmosphere either via the building envelope (Q_{Fb}) or through direct injection (Q_{Fv}). Mitigating this contribution to the UHI is obviously a matter of reducing this input by either reducing the demand or making energy use more efficient. Managing energy exchanges at the building envelope with green walls and cool roofs, for example, will affect Q_{Fb}. Even decisions on where to place the exhausts of buildings (on the roof or along walls) will influence the microclimate. Other strategies to reduce energy consumption at various scales include use of energy-efficient household appliances, passive heating and cooling systems and building designs, public transportation networks and mixed land-use planning.

3.4.2 *Trees and the Urban Forest*

Trees are the most versatile of urban design tools because they can modify the properties of both the air and surface (in)directly. The ecosystem service of trees is related to species, age and health which together control the architecture of the leafy canopy and exchanges at its leaf surfaces and roots, both of which may vary with seasons. Below the tree canopy, trees provide shade (lowering daytime T_s) but reduce the sky view factor and impede radiative loss and night-time surface cooling. The temperature of the canopy leaves is close to that of the air throughout the day owing to evaporative cooling; as a result air that passes through the tree canopy will be slowed, and have its temperature reduced and its humidity raised. In addition to these effects, trees provide a host of environmental services (Nowak 2006) and are often presented as a panacea for many undesirable urban effects.

The impact of trees on the microclimate depends on scale and perspective. The aggregate impact will depend on the layout of trees, that is, their placement in relation to each other to create large areas of shade or shelterbelts, for example. Within the UCL, trees are found in private gardens/grounds, in public parks and on streets/plazas forming what may be described as the urban forest. In terms of UHI mitigation/adaptation, each component of this forest can serve a different purpose. For example, in public green spaces trees can be part of an adaptation strategy to provide cool daytime spaces, while along streets and next to buildings they may be used as shading devices to mitigate the UHI within the UCL. Trees will lower the daytime SUHI below the leafy canopy by shading surfaces mainly; this will reduce the mean radiant temperature (T_{MRT}) and lower the radiative load on outdoor pedestrians. Shade on buildings reduces the external energy load and may offset the need for internal cooling; in this respect trees can help reduce Q_F. At night, trees will slow the rate of surface cooling below the leafy canopy so that the net effect is to dampen the diurnal temperature range. If trees are planted closely so that they form a closed leaf canopy and limit air movement, the T_s of the surfaces beneath and the T_a of the enclosed air will experience a small diurnal range; by comparison to a grass-surface 'rural' site, the temperature differences may be negative (or positive) during daytime (or night-time) under ideal UHI weather conditions. The impact of these changes on people outdoors must also account for the potential loss of ventilation in these circumstances which will affect heat loss and dilution of emissions at street level.

The top of the tree canopy is cool by day and night and tree planting is an effective strategy for managing the boundary-layer UHI. The impact of trees at this scale will depend on the nature of the urban forest including the species composition and extent of the leafy canopy. However, it is also important to acknowledge that trees also contribute to the chemical make-up of the overlying air and can aid in the formation of ozone, which affects public health.

3.4.3 Adaption

Adaptation in its strictest sense means adjustment to the reality of the UHI rather than attempting to manage its magnitude. Practically, however, it is difficult to separate UHI mitigation from adaption strategies and it is better to distinguish strategies based on emphasis; while mitigation focusses on aspects of the physical infrastructure of the city (land cover), adaptation focusses on human behaviour, land use and anthropogenic heat flux (Q_F). A critical part of adaptation is the creation of climate 'refuges' that are accessible to the exposed population and provide more comfortable microclimates.

At a building scale, the UHI increases the cooling energy demand. In many cities, much of this demand is met through mechanical air conditioners that refrigerate the indoor space and expel warm and humid air through vents into the outdoor space, thereby exacerbating the UHI. There are alternative approaches that rely on passive methods of energy management; these could include architecture designs such as the use of shutters and shading devices to limit solar gain and maximise airflow. Most often however it falls to the occupants to make decisions to open/close windows and relocate within the building to manage their heat load; for example, in hot and arid climates, it is common to use flat roof space as a living (and sleeping) space that is cooler at night owing to its exposure to the sky above. At a neighbourhood scale, adaptation requires moving to a more comfortable setting (Fig. 3.11); in ideal UHI conditions, this would mean locating in shaded spaces during the day and open spaces after sunset; each of these environments is associated with lowered T_s and T_a. Evaporative cooling in either setting (vegetative or water cover) will have the same effect. Green parks will cool the overlying air and some of this air will 'leak' out into nearby streets and neighbourhoods (Eliasson and Upmanis 2000) and contribute to mitigation.

At an urban scale, adaptation strategies can be embedded into the land-use planning by taking into account the topographic setting. A great many cities are located close to coasts where the calm and clear conditions that produce the UHI also create the daytime sea breeze, which draws cooler and more humid air off the ocean. Similarly, places with complex orography will experience gravity-based flows at night. Settlements designed to maximise the benefits of these local circulations will manage its land use/land cover to ensure that these winds penetrate far into the urbanised landscape by removing obstacles along routes (Hebbert and Webb 2012; Wong et al. 2012). The UHI itself generates a weak circulation that draws cooler 'country breeze' from the natural areas outside of the city; however, allowing this airflow into the heat island core is likely to diminish its magnitude and slow the circulation. A different approach is to limit population exposure to excessive heating through mobility, moving to different parts of the city that are designed for distinct uses. In fact, in many Western cities, the movement of much of the working population from open low- and mid-rise suburbs (LCZ types 5 and 6) to a compact city centre (LCZ types 1 and 2) in the mornings and vice versa in the evenings means that many do not experience the strongest canopy-layer UHI. This is not of course the situation for densely built and occupied cities or for those that live in impoverished environments where there is limited green space and ability to afford air conditioning.

3.5 Conclusion

The UHI phenomenon has been the subject of enquiry for a long time and there is a very large literature on the topic, which has grown exponentially since the 1970s. This interest reflects concerns for environmental quality in cities and concerns for synergies between local warming (driven by urban landscape change mainly) and regional and global warming (driven by human activities that are focussed in cities). Addressing the causes of the UHI can have wider benefits for the urban environment and climate change but it is only in the last few decades that the causative processes have been examined in any detail. Although the UHI is an apparently simple outcome of urbanisation, it belies easy explanation. This can be attributed to a number of factors including the existence of different UHIs which respond to different dominant processes at a hierarchy of scales; the complexity of the urbanised landscape; and the difficulty in obtaining representative air and surface temperatures in this context. Proper evaluation (and management) requires that the UHI is placed within an energy budget framework at relevant scales. In the current literature the UHI focus is on the canopy level and surface UHI types, which are clearly linked but are not equivalent.

Managing the UHI relies on three interlinked components: observations, understanding and response. Observations of the UHI can now be done on a routine basis; the CUHI can be assessed using well-positioned air temperature sensors (using the local climate zone framework) and the SUHI can be (partially) evaluated using satellite observations to obtain surface temperature. Ideally, these data are complemented by high-quality meteorological data that records the climate drivers, and urban land-cover/land-use data to analyse and interpret the UHI information. Understanding the phenomenon and its temporal and spatial expressions is needed to interpret the observational data. Ideally, this is done with the aid of surface energy balance models that place observations in a process-based context and allow simulations of the UHI in response to interventions. Finally, policy interventions should be linked to measurable and verifiable outcomes at identified space and time scales.

References

Alexander, P. J., Mills, G., & Fealy, R. (2015). Using LCZ data to run an urban energy balance model. *Urban Climate, 13*, 14–37.

Arnfield, A. J. (1982). An approach to the estimation of the surface radiative properties and radiation budgets of cities. *Physical Geography, 3*(2), 97–122.

Bornstein, R. D. (1975). The two-dimensional URBMET urban boundary layer model. *Journal of Applied Meteorology, 14*(8), 1459–1477.

Broadbent, A. M., Coutts, A. M., Nice, K. A., Demuzere, M., Krayenhoff, E. S., Tapper, N. J., & Wouters, H. (2019). The Air-temperature Response to Green/blue-infrastructure Evaluation Tool (TARGET v1. 0): an efficient and user-friendly model of city cooling. *Geoscientific Model Development, 12*(2), 785–803.

Chandler, T.J. (1965). *The climate of London. Hutchinson*. Retrieved from urban-climate.org.

Eliasson, I., & Upmanis, H. (2000). Nocturnal airflow from urban parks-implications for city ventilation. *Theoretical and Applied Climatology, 66*(1-2), 95–107.

Grimmond, C. S. B., Blackett, M., Best, M. J., Barlow, J., Baik, J. J., Belcher, S. E., Bohnenstengel, S. I., Calmet, I., Chen, F., Dandou, A., & Fortuniak, K. (2010). The international urban energy balance models comparison project: first results from phase 1. *Journal of Applied Meteorology and Climatology, 49*(6), 1268–1292.

Hebbert, M., & Webb, B. (2012). Towards a liveable urban climate: Lessons from Stuttgart. Liveable Cities: Urbanising World, pp. 132–150.

Howard, L. (1818). *The climate of London: Deduced from meteorological observations, made at different places in the neighbourhood of the metropolis (Vol. 1)*. W. Phillips, George Yard, Lombard Street, sold also by J. and A. Arch, Cornhill; Baldwin, Cradock, and Joy, and W. Bent, Paternoster Row; and J. Hatchard, Piccadilly.

Kataoka, K., Matsumoto, F., Ichinose, T., & Taniguchi, M. (2009). Urban warming trends in several large Asian cities over the last 100 years. *Science of the Total Environment, 407*(9), 3112–3119.

Klinenberg, E. (2015). *Heat wave: A social autopsy of disaster in Chicago*. Chicago: University of Chicago Press.

Krayenhoff, E. S., & Voogt, J. A. (2007). A microscale three-dimensional urban energy balance model for studying surface temperatures. *Boundary-Layer Meteorology, 123*(3), 433–461.

Kusaka, H., Kondo, H., Kikegawa, Y., & Kimura, F. (2001). A simple single-layer urban canopy model for atmospheric models: Comparison with multi-layer and slab models. *Boundary-Layer Meteorology, 101*(3), 329–358.

Li, D., & Bou-Zeid, E. (2013). Synergistic interactions between urban heat islands and heat waves: The impact in cities is larger than the sum of its parts. *Journal of Applied Meteorology and Climatology, 52*(9), 2051–2064.

Lowry, W. P. (1977). Empirical estimation of urban effects on climate: a problem analysis. *Journal of Applied Meteorology, 16*(2), 129–135.

Masson, V. (2000). A physically-based scheme for the urban energy budget in atmospheric models. *Boundary-Layer Meteorology, 94*(3), 357–397.

Matzarakis, A., Mayer, H., & Iziomon, M. G. (1999). Applications of a universal thermal index: physiological equivalent temperature. *International Journal of Biometeorology, 43*(2), 76–84.

Mills, G. (1997). An urban canopy-layer climate model. *Theoretical and Applied Climatology, 57*, 229–244.

Mills, G. (2008). Luke Howard and the climate of London. *Weather, 63*(6), 153–157.

Myrup, L. O. (1969). A numerical model of the urban heat island. *Journal of Applied Meteorology, 8*(6), 908–918.

Nowak, D. J. (2006). Institutionalizing urban forestry as a "biotechnology" to improve environmental quality. *Urban Forestry & Urban Greening, 5*(2), 93–100.

Nunez, M., & Oke, T. R. (1977). The energy balance of an urban canyon. *Journal of Applied Meteorology, 16*(1), 11–19.

Offerle, B., Eliasson, I., Grimmond, C. S. B., & Holmer, B. (2007). Surface heating in relation to air temperature, wind and turbulence in an urban street canyon. *Boundary-Layer Meteorology, 122*(2), 273–292.

Oke, T. R. (1981). Canyon geometry and the nocturnal urban heat island: comparison of scale model and field observations. *Journal of Climatology, 1*(3), 237–254.

Oke, T. R., Mills, G., Christen, A., & Voogt, J. A. (2017). *Urban climates*. Cambridge: Cambridge University Press.

Oke, T. R., Spronken-Smith, R. A., Jáuregui, E., & Grimmond, C. S. (1999). The energy balance of central Mexico City during the dry season. *Atmospheric Environment, 33*(24-25), 3919–3930.

Stewart, I. D., & Oke, T. R. (2012). Local climate zones for urban temperature studies. *Bulletin of the American Meteorological Society, 93*(12), 1879–1900.

Stewart, I. D. (2011). A systematic review and scientific critique of methodology in modern urban heat island literature. *International Journal of Climatology, 31*(2), 200–217.

Stewart, I. D., & Kennedy, C. A. (2017). Metabolic heat production by human and animal populations in cities. *International Journal of Biometeorology, 61*(7), 1159–1171.

Sun T., Järvi L., Omidvar H., Theeuwes N., Lindberg F., Li Z., & Grimmond S. (2019). *Urban-Meteorology-Reading/SUEWS: 2019a Release (Version 2019a)*. https://doi.org/10.5281/zenodo.3533450.

Sundborg, Å. (1951). Climatological studies in Uppsala with special regard to the temperature conditions in the urban area. *Geographical, 22*, 79–89.

Terjung, W. H., & Louie, S. S. F. (1974). A climatic model of urban energy budgets. *Geographical Analysis, 6*(4), 341–367.

Wong, K., Ng, E., & Yau, R. (2012). Urban ventilation as a countermeasure for heat islands toward quality and sustainable city planning in Hong Kong. *Journal of Heat Island Institute International, 7*(2), 11–17.

Chapter 4
Thermal Comfort in Urban Spaces

Marialena Nikolopoulou

4.1 Introduction

At a time when climate change presents the most critical threat to humanity (IPCC 2018), UN predictions of nine billion people living in cities by 2050 and the wake of a global COVID-19 pandemic request for an urgent rethink of how we design cities for all. It is now more important than ever to evaluate the need for density, access to open space as well as fresh air, intertwined with social justice, support liveability and resilience. The abstract concept of *thermal comfort* thus becomes a focal point affecting all the relevant professional bodies.

A 'comfortable and stimulating public realm that encourages social interaction' (Llewelyn-Davies 2007: 99) has long been considered an important criterion for a thriving public realm. However, the 'comfortable' component is solely assessed around the type of surfaces used—hard vs. soft—security and lighting. Thermal comfort, an integral state which cannot be disassociated from the experience of any space, is glaringly absent from professional urban design guidance in the twenty-first century.

This chapter focuses on the concept of thermal comfort and how it can be achieved in open urban spaces. Encompassing the approaches beyond thermal physiology, it aims to provide a more comprehensive framework for urban designers and planners. The enhanced adaptive capacity, critical for densely populated regions, is capable of improving thermal comfort and resilience highlighting the need for open space, an important resource for climate regulation, social cohesion as well as health and well-being.

M. Nikolopoulou (✉)
Kent School of Architecture and Planning, University of Kent, Canterbury, UK
e-mail: M.Nikolopoulou@kent.ac.uk

M. Palme, A. Salvati (eds.), *Urban Microclimate Modelling for Comfort and Energy Studies*, https://doi.org/10.1007/978-3-030-65421-4_4

4.2 What Is Thermal Comfort?

ASHRAE (2013), the professional association for the American Society of Heating, Refrigerating and Air-Conditioning Engineers, defines comfort as the 'condition of mind in which satisfaction is expressed with the environment'. Such an unusually vague definition for a building service professional association highlights the complexity and justifies the extended body of research on the topic conduced over the years.

In typical conditions, heat transfer from the body to the environment takes place predominantly by radiation (45%), convection (30%) and evaporation (25%), although these proportions vary over a wide range. The key parameters influencing thermal comfort are environmental and personal. The environmental parameters include (a) air temperature; (b) radiant environment, which is influenced by both short-wave and long-wave radiation from the sun and from the surrounding materials; (c) humidity; and (d) air movement. The personal parameters refer to clothing insulation and one's metabolic rate, which is affected by their activity. To achieve thermal balance, the heat losses must equal the heat gains between the individual and their environment to maintain a deep core body temperature of 37 °C. This thermal equilibrium provides the basis for thermal comfort according to the standards.

In an outdoor environment, this heat exchange mechanism is encountered at a greater temporal and spatial complexity and a range of indicators have been developed. From Siple and Passel's (1945) wind chill index developed for cold climates to Thom's (1959) discomfort index for warm climates, we have seen a wide range of models and indices developed to assess outdoor thermal comfort. Graphic representations of comfort conditions, such as Olgyay's (1963) bioclimatic chart and Penwarden's (1973) nomograms, enabled greater engagement with designers, particularly professionals such as architects, engineers and planners, to incorporate comfort into building proposal developments. Such concerns became particularly important in the rapid developments of city centres with predominantly commercial activities, where the space around buildings was being ignored.

The last 20 years have witnessed development of additional models for outdoor comfort in the field of biometeorology, aided by the rapid progress of computing power and sophisticated simulation tools, where different aspects of urban microclimate and thermal comfort can be assessed. The most extensively used indicator, also embedded in various microclimatic simulation software tools, is PET, the physiological equivalent temperature (Höppe 1999). Using °C for the units, it is easily comprehended by the various disciplines and has been integrated in simulation programmes such as RayMan (Matzarakis 2000) and ENVI-met (Bruse and Fleer 1998). Other indicators, such as the PMV-PPD model, originally developed for indoors predominantly for use in mechanically heated and cooled buildings (ISO 7730 2005) have been worrying when extensively used as an indicator, although they have been proven to be inadequate for outdoor use (Nikolopoulou et al. 2001). The most recent indicator is the Universal Thermal Comfort Index (UTCI), which enables calculations of the thermal state of different parts of the human body (Jendritzky et al. 2008).

There is however an important limitation to the widely accepted thermoregulatory models for outdoor thermal comfort. The dynamic nature of the external conditions, whether a breeze, gust or sunshine hiding behind the clouds, means that thermal balance is often unachievable. This has been the basis for extensive outdoor thermal comfort surveys across the world in different climatic contexts. In parallel to developments with indicators, the last 20 years have also seen an extensive amount of field surveys conducted in different geographic and climatic locations.

Field surveys have cast doubt on the extensive use of heat balance models, as they have revealed the limitations of the thermoregulatory approach. Aiming to understand the human parameter, findings from field surveys have enabled understanding of some key priorities and issues that can inform the design process.

4.3 The Role of Field Surveys

Following the seminal studies of Nikolopoulou in Cambridge (Nikolopoulou et al. 2001; Nikolopoulou and Steemers 2003) and the subsequent large-scale RUROS project, still the largest single study and database available for meta-analysis with nearly 10,000 interviews (CRES 2004), investigating comfort studies in Greece, Italy, Switzerland, Germany and the UK (Nikolopoulou and Lykoudis 2007), various reviews (Nikolopoulou 2012; Mahgoub and Hamza 2019; Lai et al. 2020) have enabled a systematic and comprehensive understanding of outdoor thermal comfort.

Field surveys have taken different forms. They are normally associated with extensive collection of microclimatic data—such as air temperature, globe temperature as a proxy for solar radiation where pyranometers are not available, wind speeds and humidity with portable weather stations—and are supplemented by structured questionnaire-guided interviews with participants in open spaces. The most frequent type is *transverse* surveys, where the sampling methods aim to include representation from large groups of participants to enable statistical analysis, in contrast to *longitudinal* surveys where the same participants are recruited for different days (Nicol et al. 2012). The advantage of longitudinal surveys is that they can capture detailed information on the participants, which is not normally possible with transverse surveys, and as a result can gain great insight on individual differences. However, they require extensive commitment from the participants, and they may lead to sampling bias, as they rely on a small number of people, so they would not be recommended for the development of empirical models.

Outdoor comfort surreys have been conducted in different parts of the world and different climatic zones in all continents, with the exception of Antarctica. Even a recent special issue on Subjective Approaches to Thermal Perception (Lenzholzer and Nikolopoulou 2020) included surveys from different cities in Europe (from Porto to Amsterdam and Kassel) to Harbin, Adelaide, New Delhi, São Paulo, Dhaka, Phoenix and Marrakech. Field surveys have frequently been accompanied with extensive observations of the population in open spaces, which has informed usage and activity patterns.

Due to the multitude of methods employed for the surveys and the survey protocols, comparison is not always feasible. This is one of the reasons that Johansson et al. (2014) have called for standardisation in outdoor comfort studies. Apart from differences in the equipment used, there have been great differences in the questionnaires employed. For example, with the use of Likert scales, there is a range of comfort, thermal sensation or satisfaction and preference scales. Even for thermal sensation scales, these vary from the standard ASHRAE 7-point scale developed for indoors (ISO 7730 2005 and ASHRAE 2013, which is the most extensively used in hundreds of surveys across the world; see Lai et al. 2020) to simplified 5-point scales (e.g. Nikolopoulou et al. 2001; Nikolopoulou and Lykoudis 2006; Fontes et al. 2008) or even 9-point scales (e.g. Kántor et al. 2012), with more recent attempts to rescale categories to enable consolidation of large databases (Nikolopoulou et al. 2018; Pantavou et al. 2019).

Additionally, there are challenges with the language as scales are translated in different languages and how nuances are understood by the local population (Pitts 2006; Al-Khatri and Gadi 2019; Pantavou et al. 2020).

Despite such intricacies, which make direct comparison of the developed empirical models more cumbersome, there are some consistent messages that are coming out and strengthen not only the need for detailed surveys, but more importantly understanding of the findings for cities enhancing resilience of urban environments.

A more specialised type of field survey includes thermal walks. Investigating the urban continuum in transit, they normally focus on the dynamic interaction between different spaces and the impact of the thermo-spatial variation on comfort (Vasilikou and Nikolopoulou 2013, 2015, 2020; Futcher and Mills 2015; Lemonsu et al. 2020).

Further developments around the Internet of Things and social media provide additional platforms for outdoor comfort, reaching out to wider audiences, as the recent work of Larsson and Chapman (2020) and Pantavou et al. (2020) demonstrated, using Web-based platforms, which could be forwarded via Facebook and other applications, or as shown by Giuffrida et al. (2020) who used Twitter data feeds.

4.4 Use of Space and Activities

Various urban planning and urban design theorists have highlighted the role of microclimate and comfort as a prerequisite for successful places (Carmona et al. 2003; Gehl 1987; Westerberg 1994). One of the consistent findings of outdoor comfort surveys refers to such importance for use of space. In fact, the range of activities and use greatly depend on the resulting comfort conditions, irrespective of season.

Although the majority of the work has focused on temperate (Nikolopoulou et al. 2001; Thorsson et al. 2004; Eliasson et al. 2007; Fontes et al. 2008) and warm (Nikolopoulou and Lykoudis 2007; Lin 2009; Aljawabra and Nikolopoulou 2010;

Lin et al. 2013; Chen et al. 2015) climates, even the limited work on cold climates has linked attendance and use of public space to microclimatic conditions (Zacharias et al. 2001; Larsson and Chapman 2020; Leng et al. 2020).

Such understanding is critical for urban planning, as even if responses to the microclimate are unconscious, they result in different patterns of use. Depending on the climatic context, access to sunlight and protection from winds can be key strategies for planning urban developments, or the emphasis may be on providing shade and exposure to breezes.

One of the interesting issues from studies on thermal comfort and attendance is the different temporal scales and how they affect attendance. This is clearly demonstrated in a study on an urban square in southern Athens (Fig. 4.1). In the cooler seasons, in this case, autumn and winter, attendance is highest during the warmest parts of the day, from mid-morning to lunchtime, and drops quite sharply afterwards.

Beyond the daily profile, however, there is a strong seasonal effect with the attendance profile reversing in the summer. The number of people found outdoors is quite small in the morning, decreasing further after midday, which coincides with the warmest part of the day and intense solar radiation. In the evening, however, attendance increases exponentially, particularly after sunset, which coincides with dropping air temperature. At the time when the conditions inside the blocks of flats are the most uncomfortable with the concrete radiating heat, the open square provides a much cooler place to escape from the hot indoor environment, highlighting the essential role of public open space in a warming climate.

The changing needs and seasonality are important for northern climates as well, as a recent study in Umeå, northern Sweden, highlighted (Costamagna et al. 2019).

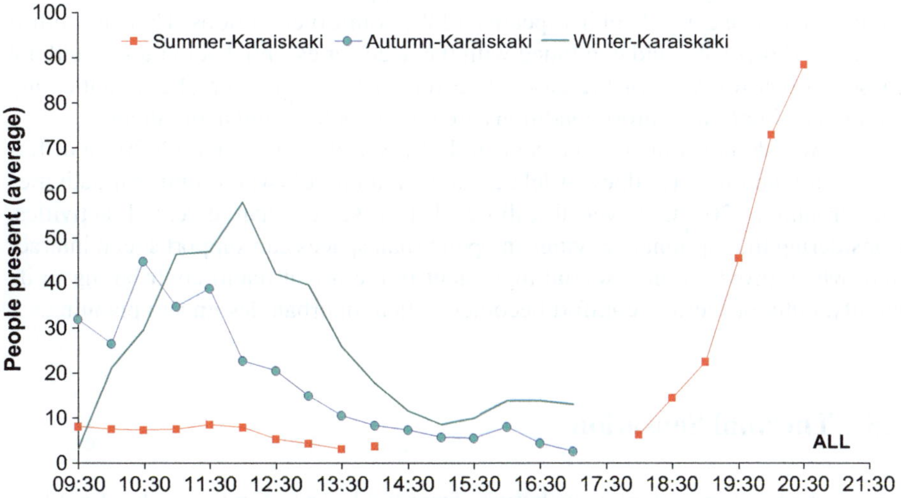

Fig. 4.1 Seasonal variation of mean diurnal distribution of the number of people in an urban square in southern Athens (source: Nikolopoulou and Lykoudis 2007)

Despite the city being located in the subarctic climate, it also experiences warm summers for a short period of time, and as a result it is important to address the different seasons. Extensive interviews with planners and users of public spaces, supplemented by observations, demonstrated that public spaces that are designed for both winter and summer are preferred, important aspects for liveable cities. As de Vet (2017) highlights, weather is an integral part rather than a backdrop to people's daily life.

Outdoor comfort conditions do influence not only activities and use of open spaces, but also the duration spent there. A study of the hot arid climate in Marrakesh (Morocco) and Phoenix (Arizona, USA) highlighted the strong influence of solar radiation, wherein as the latter increased, the time people spent outdoors decreased (Aljawabra and Nikolopoulou, 2010). Similar findings have been presented by Faustini et al. (2020) in the climate of São Paulo, Brazil, where the space with the highest degree of shading and most favourable microclimatic conditions led to longest duration time compared with other spaces exposed to the sun. Even in the temperate climate of Cambridge, people spent significantly longer periods of time in areas with a variety of spaces in sun and shade, 50 min on average in the summer, while in areas where no shade was available, the average was 16 min only (Nikolopoulou and Steemers 2003).

Moving the argument to the influence of outdoor comfort on the type of activity outdoors, favourable conditions support optional activities, which require a degree of choice and are an important indicator for liveability (Gehl 1987; Carr et al. 1992). With this in mind, Nikolopoulou et al. (2001) had differentiated open spaces in routes and resting places, as people choose to sit somewhere, whereas they do not necessarily choose a particular route, to avoid discomfort. Hence crossing through a space on the way to work or for essential shopping is an essential activity which would need to be carried out irrespective of the comfort conditions. This discomfort would be short-lived, and combined with the need for essential activities would not cause serious distress. In the case of resting places, however, the activities are optional and unless comfort conditions are met, people would avoid them.

This was demonstrated in a recent study by Sharifi and Boland (2020) in different Australian cities (Sydney, Adelaide and Melbourne), where optional activities had around a 20 °C lower threshold (UTCI scale) than essential activities. Considering that optional activities in open urban spaces can support social interaction, which provides an essential ingredient in the urban realm contributing to its vitality, outdoor thermal comfort becomes critical in urban design and planning.

4.5 Thermal Sensation

One of the most important contributions from field surveys in the study of outdoor comfort is an in-depth understanding of thermal sensation. The research in the past 20 years has significantly enriched a more holistic approach of how thermal comfort is perceived in different latitudes, different cultures and even socio-economic backgrounds. With the contribution from different disciplines, including environmental psychology, the importance of behavioural and other cognitive and psychological factors on thermal comfort became apparent.

One of the main issues highlighted is the discrepancy between theoretical thermoregulatory models and actual conditions of thermal comfort with environmental parameters accounting for only about half of the variation in thermal sensation (Nikolopoulou et al. 2001; Nikolopoulou 2004; Eliasson et al. 2007; Nikolopoulou 2012; Lai et al. 2020). Conducting outdoor comfort surveys in different parts of the world enabled the development of empirical comfort models as a function of the different microclimatic conditions experienced, the weighting of which varied depending on the climatic context. This was possible through the extensive data collected from objective and subjective data, i.e. microclimatic monitoring and questionnaires, from a variety of open spaces. Using advanced statistical modelling techniques, different types of regression and/or probit analysis provided the means to fully explore different aspects of thermal comfort, such as thermal sensation, acceptance, satisfaction and preference, to fully understand outdoor comfort. These large data sets also present an extensive potential resource available worldwide, which if combined in the future can provide additional meta-analysis to inform future debates on thermal comfort further.

Field surveys highlighted the wide range of thermal comfort zones, which also strengthened the argument for adaptation. For example, in Cambridge in the UK, thermal comfort was always around 10% in different seasons (Nikolopoulou et al. 2001), in line with acceptable range of dissatisfaction rates even in tightly controlled indoor environments (ISO 7730 2005; ASHRAE 2013). Subsequent studies in different European cities as part of the RUROS project (CRES 2004), with mean air temperature varying from 30.1 °C in Athens, Greece, to 5.4 °C in Kassel, Germany, demonstrated very high levels of overall comfort, over 75% for all cities (Fig. 4.2). The seasonal variation was lowest in Kassel in winter (43%) and highest in Cambridge (91%).

Aiming to quantify and identify comfort temperatures, the concept of neutral temperature was introduced by Humphreys (1975), to define the temperature where people feel neither warm nor cool and are in a state of thermal neutrality, corresponding to the central category of the ASHRAE thermal sensation scale. This was calculated for the cities involved in the RUROS project and as demonstrated in Fig. 4.3, there is a large variation of over 10 K in neutral temperatures across the different cities, as well as within the different seasons for the same city.

Different studies have used different indicators to calculate neutral temperature, and demonstrated equally big seasonal differences. For example, Aljawabra and Nikolopoulou (2018) found a winter and summer neutral globe temperature of 19.5 °C and 25.5 °C for Marrakech, while for Phoenix the respective temperature was 23.5 °C and 26.5 °C. In the hot and humid climate of Belo Horizonte, Brazil, using PET temperature, Hirashima et al. (2016) demonstrated neutral temperatures of 27.7 °C, in summer, and 15.9 °C, in winter.

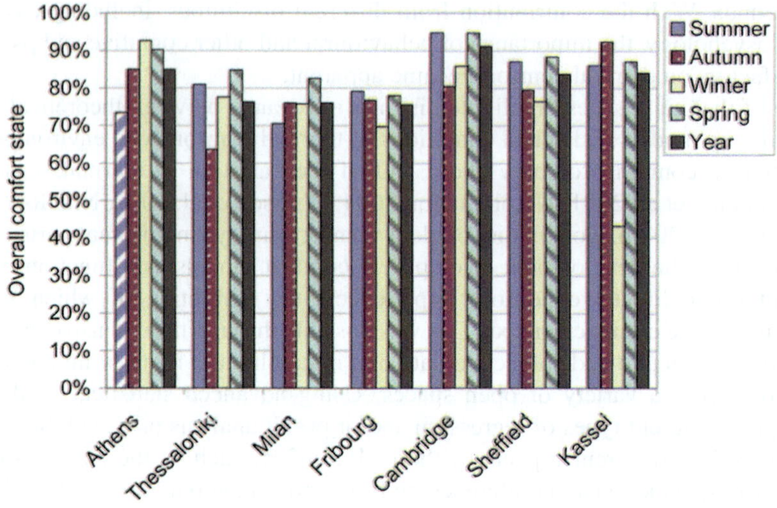

Fig. 4.2 Percentage distribution of overall comfort conditions, for different European cities at different seasons (source: Nikolopoulou and Lykoudis 2006)

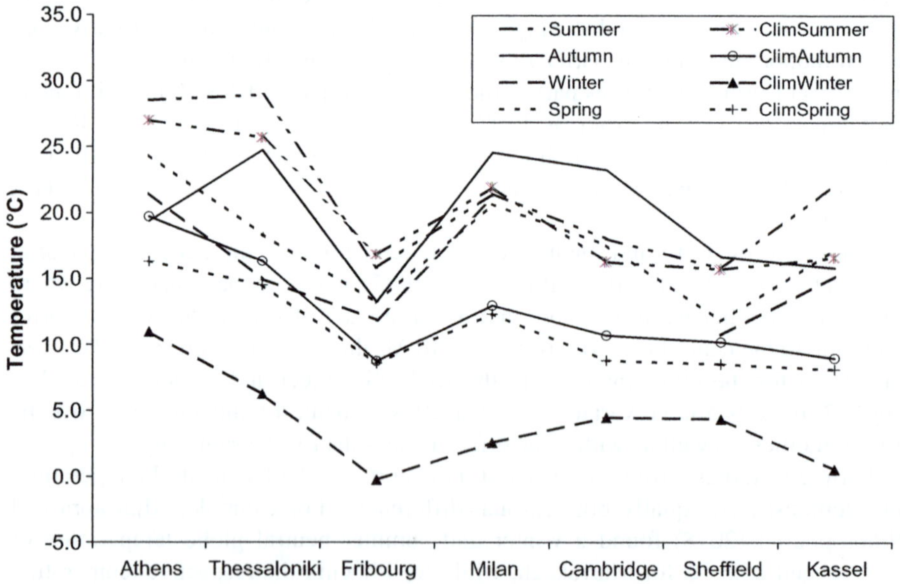

Fig. 4.3 Neutral temperatures compared with the relevant climatic air temperature for different seasons, for the different cities (source: Nikolopoulou and Lykoudis 2006)

All these highlight that thermal physiology alone is not sufficient to justify the big differences encountered with thermal comfort and adaptation mechanics enable people to adapt in a wide range of thermal environments.

4.6 Adaptation

In the context of thermal comfort, we define adaptation as the gradual decrease of the organism's response to repeated exposure to a stimulus, involving all the actions that make them better suited to survive in the thermal environment. Nikolopoulou and Steemers (2003) and Nikolopoulou (2004) discussed extensively the theoretical framework for adaptation in this context, identifying three key categories: *physical*, *physiological* and *psychological*.

Physiological is not a scale relevant to urban design, as it implies changes in the physiological responses. *Physical* adaptation refers to predominately conscious changes to one's behaviour, e.g. changing of clothing in different conditions. For example, research has highlighted the seasonal variation of clothing (Aljawabra and Nikolopoulou 2010, 2018; de Vet 2017; Leng et al. 2020), and as a direct function of temperature (Nikolopoulou et al. 2001; Nikolopoulou 2004; Nikolopoulou and Lykoudis 2006; Yahia and Johansson 2013; Aljawabra and Nikolopoulou 2018; Lai et al. 2020).

Other adaptive actions have included consumption of hot or cold drinks to increase or decrease one's metabolic heat according to the weather (Nikolopoulou 2004; Nikolopoulou and Lykoudis 2006; de Vet 2017).

Seeking shade is one of the most common options individuals can take to improve thermal comfort. Rarely seen as a portable option (Fig. 4.4), this is most commonly available as permanent fixtures in the urban environment, such as built projections, bus-stop shelters or trees (Fig. 4.4), and occasionally through temporary installations for special occasions such as religious festivals in warm climates (Fig. 4.4).

Fig. 4.4 Different forms of shade: (left) personal umbrella used as a parasol to protect individuals from the sun; (middle) large-scale shading installed in the centre of Seville, preparations for a religious ceremony; (right) dense vegetation providing shaded footpath

Fig. 4.5 (Left) Variation of percentage of people in shade as a function of air temperature (Nikolopoulou et al. 1999); (right) people carefully avoiding the shade on a spring day

Research has demonstrated the direct link between shaded areas being used and increasing temperatures, with nearly 50% of the people found in shade attributed to air temperature alone in a study in Cambridge (Fig. 4.5) (Nikolopoulou et al. 1999). This explains the more intense use of space where such options are available, as mentioned earlier (Faustini et al. 2020) and being selected as the most popular adaptive action to improve thermal comfort in Taiwan (Lin 2009). A study in New Delhi also showed similar patterns, with declining attendance as sunlight increased, as evaluated by the proxy (Tglobe–Tair) indicator.

Interestingly, not all shade has the same effect. Shade from trees is particularly effective in improving comfort conditions, with the difference in PET values between shaded and sunlit areas being 12–22 °C lower, even if the difference in air temperature is as low as 1 K (Klok et al. 2020). Considering the synergistic health effects of greenery from the improved microclimate and in reducing stress and facilitating social cohesion, such design interventions should be encouraged (de Vries et al. 2013).

4.6.1 Psychological Adaptation

Addressing the last part of the adaptive theory, *psychological adaptation*, it is worth remembering that different people perceive the environment in different ways and environmental psychologists have extensively informed this debate (Cabanac 1996; Knez and Thorsson 2006) suggesting 'the psychology of outdoor place and weather assessment' (Knez et al. 2009).

Considering Hopkinson's stimulus and sensation models (Hopkinson 1963), in outdoor comfort the response to physical stimulus is not simply a function of its magnitude, but depends on wider factors, such as personal, cognitive and psychological. Nikolopoulou and Steemers (2003) provided a comprehensive framework for psychological adaptation, identifying the following parameters: the naturalness of a space, experience and expectations, time of exposure, perceived control and environmental stimulation.

This section focuses predominantly on individual *experiences* and the resulting *expectations* which influence outdoor comfort, along with the aspect of *personal control*, contextualising with evidence from different studies around the world to highlight the key messages.

Figure 4.3 showed the large variation in neutral temperatures between different cities in Europe as well as between the different seasons. Examining the data closer, it becomes apparent that for all cities, neutral temperatures follow the profile of their climatic temperature in all seasons. This highlights the importance of recent experiences and expectations, influencing the seasonal comfort temperatures, with warmer neutral temperatures expected in autumn following the summer, and the cooler spring neutral temperatures following the colder winter conditions.

Recent meta-analysis of the original data sets attempted to quantify the contribution of climate. Using the PET and UTCI index they established different thresholds for the different scales of indices, which were influenced by the climatic air temperatures, demonstrating a 0.6 °C average increase of the index threshold for each 1 K increase of the annual air temperature (Pantavou et al. 2020). In fact, there have been attempts to calibrate scales of outdoor thermal indices to take account of climatic backgrounds, so they may become more widely used in planning (Tseliou et al. 2010; Hirashima et al. 2016; Yang et al. 2017).

Published comfort zones from different cities around the world further highlight the influence of climate and long-term experiences, particularly interesting when comparing comfort zones from central and western Europe (18–23 °C PET (Matzarakis et al. 1999)) to more extreme climatic contexts. For example, in Taiwan, the thermal comfort zone for PET neutral is 21.3–28.5 °C (Lin 2009), while in Belo Horizonte it is 19–27 °C (Hirashima et al. 2016) with the wide zones also encompassing the seasonal variation. In subarctic climate zones, we see a dramatic reduction in the comfort zone, albeit still quite wide, with people adapting at the 10–17 °C PET comfort zone in Umeå (Yang et al. 2017), with a similar lower threshold of 10.2 °C PET for the winter city of Harbin (Lai et al. 2020).

The influence of *expectations* becomes more apparent in a study which looked at the time spent outdoors evaluated against people's weather expectations. In both Phoenix and Marrakech, the time people spent outdoors was 10% for those who considered the conditions to be typical for the year (Aljawabra and Nikolopoulou 2010, 2018). As de Vet (2017) argues, understanding the local weather and context facilitates its integration into everyday life.

All the studies mentioned so far have focused on the native population. The strength of the influence of climatic background was demonstrated in a recent longitudinal series of surveys, which involved two distinct groups of a diverse student population: native Israelis or those from similar climatic context and others from colder environments (Brychkov et al. 2018). The participants' individual thermal history and long-term experiences were consistently found to influence their expectations and thermal perception under the same conditions. More specifically, the locals were better adapted to the hot summer conditions, perceiving the conditions as less hot and with far less discomfort. The reverse was evident in winter, with those from northern climates being better adapted to the cold conditions, and closer to thermal neutrality. Ultimately, their *long-term thermal experiences* lead to divergent *expectations* and thermal comfort, even under identical conditions.

Shifting the discussion to *short-term experience*, a study in Israel identified interesting patterns in the variation of thermal sensation and thermal neutrality (Pearlmutter et al. 2014), using the index of thermal stress (ITS). ITS represents the rate of sweat required for the body to maintain thermal equilibrium with its surroundings through evaporative cooling, which is suitable for the hot arid climate of the Negev region. Focusing on the daily variation of ITS, the authors demonstrated a different rate of increase of thermal sensation as a function of ITS at different times of day, i.e. morning, midday and evening, with a steeper slope in the morning (Fig. 4.6). The results indicate a desensitisation in comfort over the course of the day, as a result of short-term adaptation, from the morning following the cooler part of the day to the evening following the hottest part of the day late afternoon.

Another interesting point arises when comparing the comfort range and conditions in Phoenix and Marrakech. The two cities have very similar climate; however, the comfort range and sensitivity of the comfort conditions differ between them (Aljawabra and Nikolopoulou 2018). The comfort zone is wider in Marrakech, 19.5–25.5 °C, while for Phoenix it is 23.5–26.5 °C. Additionally, the thermal sensitivity is steeper for Phoenix, suggesting that people are more sensitive to changes in the external conditions. The authors speculated that this could be due to the extensive prevalence of air-conditioning cooling systems in Phoenix. In fact,

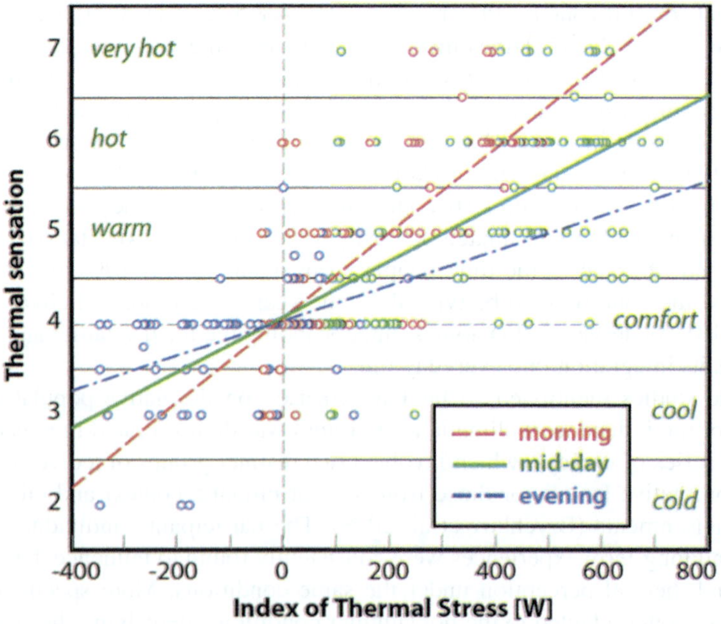

Fig. 4.6 Variation of the scatter plot of thermal sensation with ITS for three different periods during the day, morning, midday and evening (source: Pearlmutter et al. 2014)

approximately 75% of the interviewees in Phoenix in the summer period had been in an air-conditioned space within the last hour, while for Marrakech this was only 25%, where most indoor spaces are naturally ventilated (Aljawabra and Nikolopoulou 2010). The loss of acclimatisation for people spending large amounts of time in air-conditioned spaces was also identified in a similar context in Damascus, Syria, where people without air conditioning could accept higher temperatures (Yahia and Johansson 2013).

Another important factor in the contextual framework of psychological adaptation is *perceived control*. A term originally used by Paciuc (1990) for the indoor environment, it implies the availability of control which is not necessarily exercised. As environmental psychology highlights, a high degree of control over a source of discomfort leads to a reduction in the negative emotional responses (Evans 1982). Comfort surveys have provided evidence of the important role of personal control, which Lin (2009) identifies as the most important parameter, and how it leads to increased tolerance for conditions that would otherwise be expected to lead to great discomfort.

Design of urban spaces has an important role to play in providing people with choice and as already discussed, areas with a variety of microclimates and choices to sit in sun or shade or avoid a microclimatic stressor altogether benefit from more extensive use.

Perceived choice is also identified with attendance in a space as an optional activity. Various studies in different climatic contexts have identified that when someone is in a space not because they wanted to be there but for other reasons, e.g. to meet someone, for working or simply passing through, discomfort was increased for the same conditions (Nikolopoulou and Steemers 2003; Nikolopoulou 2004; Nikolopoulou and Lykoudis 2006; Lin 2009).

The important role of personal choice in increasing thermal tolerance was demonstrated in what could be described as harsh conditions, in a recent study in New Delhi (Manavvi and Rajasekar 2020). The study focused on religious squares, integral for Indian cities, used for worship purposes as well as shopping activities and meeting of people. Although the calculated neutral PET was 24.7 °C in line with other studies in hotter climates, the average PET encountered varied in the different subspaces between 36.8 °C and 52.4 °C, with 80% acceptability at 39.4 °C. Furthermore, at 42 °C, corresponding to 'strong heat stress', only 44% of the respondents desired for cooler conditions!

The intended purpose of the visit, clearly, had a significant impact on thermal comfort. The neutral PET for those who were visiting for non-religious purposes was 2.7 °C lower than that for worship. However, with rising PET, the decline of thermal acceptability was steeper for those who were visiting for non-religious purposes than that for worship. In fact, the group for whom the predominant purpose was non-worship demonstrated very high thermal tolerance, which can only be explained through personal choice.

It is important to remember that cultural norms also influence outdoor thermal comfort conditions, manifesting in different ways, e.g. from purposes such as congregation and worship as discussed above to appropriate clothing or social distancing and need for privacy—particularly for women. It is customary in Marrakech, for example, for clothing to cover most of the body, which results in a higher clothing insulation value than in Phoenix, where T-shirts and shorts are regularly worn in the summer (Aljawabra and Nikolopoulou 2010, 2018).

4.7 Social (In)equalities?

So far, with the exception of small-scale surveys focusing on the student population, the studies have broadly assumed that the population groups participating in the surveys are representative of the local demographics. However, is there any indication that different groups may be exposed differently?

A recent article in the New York Times discussed shade as a mark of privilege in Los Angeles, sitting 'at the intersection of two crises: climate change and income inequality' with officials aiming to incorporate shading to approximately 750 bus stops (Arango 2019). Social inequality in relation to outdoor comfort becomes more apparent when viewing shade as a public resource, or combined with the need for public transport; in the poorest areas of Los Angeles, tree canopy approximates 10% whereas in wealthier areas it increases to around 35%, and in Bel Air it is 53% (Bloch 2019).

The research conducted in Phoenix and Marrakech revealed some pertinent issues around outdoor comfort and social inequalities, which still remains the most detailed study in this aspect. The participants' socio-economic status was evaluated based on education, job type and financial self-assessment (Platt 2006) and was grouped into three categories. On the low end, people identified with basic education were jobless or with low income and with a difficult financial situation. The upper end consisted of people with higher education, highly skilled jobs and comfortable financial living. It was a subjective evaluation, rather than using objective indicators such as GDP, but more informative to better understand people's socio-economic inequalities (Aljawabra 2014).

Evaluating comfort conditions against socio-economic background with both cities combined provides interesting insight. The neutral temperatures (using globe temperature as the indicator) for the three groups, i.e. high, medium and low, were 23.5 °C, 21.5 °C and 19.5 °C, respectively, while additionally, the highest group also had the steepest slope (Aljawabra and Nikolopoulou 2018). This is probably indicative of the fact that the highest group spends a lot of time in air-conditioned places and is more sensitive to changes in temperature. The lower socio-economic group, however, mentioned that they cannot afford to install air conditioning at their homes. To escape the indoor heat, they take refuge outdoors and become less sensitive to changes in temperatures, better tolerating the thermal conditions.

When examining the two cities separately, comparing the use of space and time spent outdoors in the summer, a similar picture emerges. People from the lower socio-economic background spent more time outdoors (Fig. 4.7). This was particularly noticeable in Phoenix, where the mean time spent outdoors was 140 min,

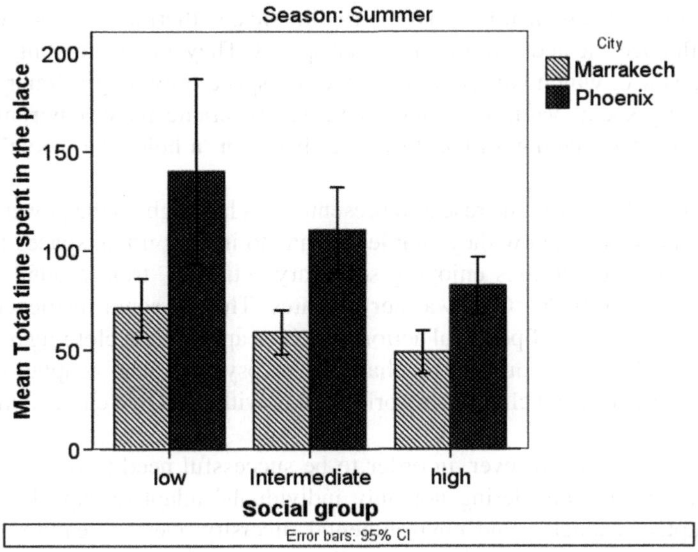

Fig. 4.7 Mean time spent in outdoors for the different socio-economic groups in Marrakech and Phoenix (source: Aljawabra and Nikolopoulou 2009)

whereas for the group from the higher background it was 80 min (Aljawabra and Nikolopoulou 2010).

Adding to the social segregation, research from UC Berkeley also exposed racial inequalities and recommended an environmental justice perspective for urban comfort. Their work identified that densely populated and segregated neighbourhoods of racial/ethnic groups were disproportionately exposed to health risks from increased urban heat, resulting from increased impervious surfaces such as asphalt and concrete and lack of trees (Jesdale et al. 2013). All these highlight the importance of microclimatic interventions to provide comfortable public open space, to support different groups of the population with limited access to resources such as air conditioning. Given the social implications, one can easily appreciate how in a warming climate outdoor comfort becomes an important commodity, where open spaces play a critical role in public health.

4.8 Conclusions

So what does that mean for the field of outdoor thermal comfort? How does it impact attempts to address resilience and climate change?

It becomes clear across the chapter that current physiological approaches to quantify thermal comfort in urban spaces are very limiting. A new paradigm is required taking into account the wider context with lasting implications for urban design and planning, energy conservation, liveability and public health.

It has also been demonstrated that microclimatic and thermal comfort conditions directly influence use and activities in urban spaces. They are an inherent characteristic of the space, which can either enhance the space providing a 'microclimatic island' adding social value (Roesler 2019: 12) or create harsh environments to remain abandoned making all the difference between 'a hole or a whole' (Zucker 1959: 3).

At the individual level, the research presented has highlighted the powerful adaptation mechanisms and how they enable humans to inhabit urban spaces in a wide range of comfort conditions enjoying sedentary activities, from around 10 °C in cold climates to over 30 °C in warmer climates. This personal resilience is supported through a range of personal actions, such as appropriate clothing and access to shade, and the more complex mechanisms of psychological adaptation, where personal experiences and climatic histories, along with available choice, inform and influence thermal comfort.

Such mechanisms, however, in order to be successful need to be supported by appropriate design, considering not only individuals' adaptive capacity, but also spatial adaptive capacity, and environmental diversity. Design features enabling protection and exposure to different conditions depending on seasonality are an important consideration for open space and can support a range of activities for longer periods of time. For example, as mentioned before, in Cambridge people spent nearly four times as long in the area with ample shade (Nikolopoulou and Steemers 2003). In the surveys in Belo Horizonte in two nearby squares with similar thermal conditions, individuals demonstrated higher levels of tolerance in the square which had vegetation, water features and interesting design, in contrast to the one with impervious surfaces and heavy vehicular traffic (Hiroshima et al. 2016). Even among the religious squares in New Delhi, where great tolerance was demonstrated, the square with predominantly hard surfaces and no trees had a lower neutral PET (22.1 °C rather than 26.4 °C), lower tolerance and almost half the acceptability rate (18.5% as opposed to 35% in the square with available shade) (Manavvi and Rajasekar 2020).

The requirement for environmental diversity can even be formulated and evaluated through advanced computational analysis. Recent work investigated thermal diversity in open spaces in London and Paris, in relation to density and coverage values as indicators. Such knowledge can inform debates around densification of different neighbourhoods and how the built form can provide open space with varying degrees of sun and wind exposure, for spatial thermal diversity as appropriate for different seasons (Chatzipoulka et al. 2016).

Moving forward, it is thus conceivable that open spaces support various functions currently addressed by indoor spaces. Most of the work on outdoor comfort has predominantly focused on urban squares, parks or streets, open spaces around transport hubs, or areas of commercial interest, such as open-air markets. More recently there is a shift of focus to areas of rich heritage, such as archaeological sites (e.g. Fabbri et al. 2020), and even outdoor workspaces and business parks in hotter climates (e.g. Atwa et al. 2020). Is there more scope?

Fig. 4.8 Open-air school in the freezing cold, the Netherlands, 1918 (source: Flickr—Nationaal Archief, no known copyright restrictions)

About a 100 years ago, as a response to a different health crisis, the rapid spread of tuberculosis saw the development of the open-air school movement, where teaching was conducted outdoors to ensure fresh air and avoid contamination (Fig. 4.8). At the wake of the global COVID-19 pandemic, after extended period of lockdown, as the economies reopen, the need for social distancing makes indoor activities challenging. Within such a framework, open-air schools are being revisited providing partial solutions, with Denmark being the first to aim for more lessons to be taught outside (Orange 2020), while we were already seeing architectural proposals for open-air classrooms (O'DonnellBrown 2019).

All the above cases highlight that the need for comfortable open spaces has become more relevant than ever. From vernacular solutions adapted for urban environments (Nikolopoulou 2012) to technological advances in materials with special coatings and pavements to combat excess heat (Kolokotroni et al. 2013; Santamouris and Kolokotsa 2016) microclimatic design for urban climate mitigation is progressing rapidly. Additionally, there is cross-fertilisation between the different disciplines with the knowledge from urban climatology (Oke 1987; Erell et al. 2016) being translated to landscape architecture (Brown 2010; Lenzholzer 2015), and design research being employed as a new methodology to address adaptive capacity to climate change and urban comfort (e.g. projects REALCOOL, SOLOCLIM). Even professional bodies are embracing the new challenges with the most recent CIBSE Guide A (2015) incorporating advice for outdoor comfort.

Advances with software and computational power enable microclimatic simulation for proposed developments with a range of software tools available, e.g. ENVI-met, SOLWEIG, RayMan and SkyHelios, as well as the increasingly available open-source resources, such as OpenFoam CFD. These have been used for numerous projects of varying complexity, focusing on the urban realm with some recent projects modelling large areas in cities to evaluate the range of comfort conditions resulting from changes in urban morphology (Chatzipoulka et al. 2016; Chatzipoulka and Nikolopoulou 2018; Tsichritzis and Nikolopoulou 2019; Chatzipoulka et al. 2020).

Careful design for thermal diversity is thus more possible than ever. What remains is to ensure that the human-centric approach is prioritised. Increasing our adaptive capacity becomes relevant for all the different scales, from the individual person to the street space and public square, and to the urban scale. It then becomes possible to adopt solutions synergistically, which has been proven to improve both climatic and socio-economic conditions (Foshag et al. 2020).

Under a warming climate, it is not only sensitive groups of the population, outdoor workers or the homeless that are at risk (Russell et al. 2020). The British Medical Journal recently recommended person-centred design rather than infrastructure-led plan, for places to support healthy communities (McKinnon et al. 2020). As highlighted, urban comfort has ramifications for climate change, liveability and social cohesion. It is thus imperative that—paraphrasing Bloch's (2019) manifesto—it becomes 'a requirement for public health, and a mandate for urban planners and designers'.

References

Software

ENVI-met. Retrieved from https://www.envi-met.com/.

OpenFOAM. Retrieved from https://openfoam.org/.

RayMan. Retrieved from https://www.urbanclimate.net/rayman/.

SkyHelios. Retrieved from https://www.urbanclimate.net/skyhelios/.

SOLWEIG. Retrieved from https://umep-docs.readthedocs.io/en/latest/OtherManuals/SOLWEIG.html.

Twitter platform. Retrieved from https://twitter.com/.

Project Websites

REALCOOL-Really Cooling Water Bodies in Cities. Retrieved from https://www.wur.nl/en/project/Really-Cooling-Water-Bodies-in-Cities-REALCOOL.htm.

RUROS-Rediscovering the Urban Realm and Open Spaces. Retrieved from http://alpha.cres.gr/ruros.

SOLOCLIM-Solutions for Outdoor Climate Adaptation. Retrieved from https://soloclim.eu/.

Publications

Aljawabra, F. (2014). *Thermal Comfort in Outdoor Urban Spaces: the Hot Arid Climate*, University of Bath, PhD thesis (unpublished).

Aljawabra, F., & Nikolopoulou, M. (2009). Outdoor thermal comfort in the hot arid climate: The effect of socio-economic background and cultural differences. *Proceedings: PLEA 2009-26th Conference on passive and low energy architecture, Canada.*

Aljawabra, F., & Nikolopoulou, M. (2010). The influence of hot arid climate on the use of outdoor urban spaces and thermal comfort: Do cultural and social backgrounds matter? *Intelligent Buildings International, 2*(3), 2010. https://doi.org/10.3763/inbi.2010.0046.

Aljawabra, F., & Nikolopoulou, M. (2018). Thermal comfort in urban spaces: A cross-cultural study in the hot arid climate. *International Journal of Biometeorology, 62*, 1901–1909.

Al-Khatri, H., & Gadi, M. B. (2019). Investigating the behaviour of ASHRAE, Bedford, and Nicol thermal scales when translated into the Arabic language. *Building and Environment, 151*, 348–355.

Arango, T. (2019). 'Turn off the sunshine': Why shade is a mark of privilege in Los Angeles, *The New York Times*, 1st December 2019. Retrieved from https://www.nytimes.com/2019/12/01/us/los-angeles-shade-climate-change.html.

ASHRAE. (2013). *Standard 55:2013—Thermal environmental conditions for human occupancy, American Society of Heating, Refrigerating and Air-Conditioning Engineers*.

Atwa, S., Ibrahim, M. G., & Murata, R. (2020). Evaluation of plantation design methodology to improve the human thermal comfort in hot-arid climatic responsive open spaces. *Sustainable Cities and Society, 59*, 102198. https://doi.org/10.1016/j.scs.2020.102198.

Bloch, S. (2019). Shade. *Places Journal*. https://doi.org/10.22269/190423.

Brown, R. D. (2010). *Design with microclimate: The secret to comfortable outdoor space*. Washington: Island Press.

Bruse, M., & Fleer, H. (1998). Simulating surface-plant-air interactions inside urban environments with a three dimensional numerical model. *Environmental Modelling and Software, 13*(3–4), 373–384.

Brychkov, D., Garb, Y., & Pearlmutter, D. (2018). The influence of climatocultural background on outdoor thermal perception. *International Journal of Biometeorology, 62*(10), 1873–1886. https://doi.org/10.1007/s00484-018-1590-7.

Cabanac, M. (1996). The place of behavior in physiology. In M. J. Fregly & C. M. Blatteis (Eds.), *Handbook of Physiology: A critical, comprehensive presentation of physiological knowledge and concepts, Sect 4 Environmental Physiology* (pp. 1523–1536). New York: American Physiological Society, Oxford University Press.

Carmona, M., Heath, T., Oc, T., & Tiesdell, S. (2003). *Public places, urban spaces*. Oxford: Architectural Press.

Carr, S., Francis, M., Rivlin, L. G., & Stone, A. M. (1992). *Public space*. Cambridge: Cambridge University Press.

Chatzipoulka, C., & Nikolopoulou, M. (2018). Urban geometry, SVF and insolation of open spaces: London and Paris. *Building Research and Information, 46*, 881–889. https://doi.org/10.1080/09613218.2018.1463015.

Chatzipoulka, C., Compagnon, R., & Nikolopoulou, M. (2016). Urban geometry and solar availability on façades and ground of real urban forms: using London as a case study. *Solar Energy, 138*, 53–66. https://doi.org/10.1016/j.solener.2016.09.005.

Chatzipoulka, C., Steemers, K., & Nikolopoulou, M. (2020). Density and coverage values as indicators of thermal diversity in open spaces: Comparative analysis of London and Paris based on sun and wind shadow maps. *Cities, 100*, 102645. https://doi.org/10.1016/j.cities.2020.102645.

Chen, L., Wen, Y., Zhang, L., & Xiang, W.-N. (2015). Studies of thermal comfort and space use in an urban park square in cool and cold seasons in Shanghai. *Building and Environment, 94*, 644–653.

CIBSE. (2015). *Guide A: Environmental design*. London: Chartered Institute of Building Services Engineers.

Costamagna, F., Lind, R., & Stjernström, O. (2019). Livability of urban public spaces in Northern Swedish cities: The case of Umeå. *Planning Practice and Research, 34*(2), 131–148. https://doi.org/10.1080/02697459.2018.1548215.

CRES. (2004). *Centre for renewable energy sources*. RUROS project & database. Retrieved from http://alpha.cres.gr/ruros.

de Vet, E. (2017). Experiencing and responding to everyday weather in Darwin, Australia: The important role of tolerance, Weather. *Climate and Society, 9*, 141–154. https://doi.org/10.1175/WCAS-D-15-0069.1.

de Vries, S., van Dillen, S. M. E., Groenewegen, P. P., & Spreeuwenberg, P. (2013). Streetscape greenery and health: Stress, social cohesion and physical activity as mediators. *Social Science & Medicine, 94*, 26–33. https://doi.org/10.1016/j.socscimed.2013.06.030.

Eliasson, I., Knez, I., Westerberg, U., Thorsson, S., & Lindberg, F. (2007). Climate and behaviour in a Nordic city. *Landscape and Urban Planning, 82*, 72–84.

Erell, E., Pearlmutter, D., & Williamson, T. (2016). *Urban microclimate: Designing the spaces between buildings*. Routledge: Taylor & Francis Group.

Evans, G. W. (1982). *Environmental stress*. Cambridge: Cambridge University Press.

Fabbri, K., Ugolini, A., Iacovella, A., & Bianchi, A. P. (2020). The effect of vegetation in outdoor thermal comfort in archaeological area in urban context. *Building and Environment, 175*, 106816. https://doi.org/10.1016/j.buildenv.2020.106816.

Faustini, F. B., de Faria, J. R. G., & Fontes, M. G. D. (2020). The influence of thermal comfort conditions on user's exposure time in open spaces. *International Journal of Biometeorology, 64*, 243–252. https://doi.org/10.1007/s00484-019-01749-3.

Fontes, M., Aljawabra, F., & Nikolopoulou, M. (2008). Open urban spaces quality: a study in a historical square in Bath-UK, *Proceedings 25th International Conference on Passive and Low Energy Architecture, Dublin.*

Foshag, K., Aeschbacha, N., Höfled, B., Winkler, R., Siegmund, A., & Aeschbach, W. (2020). Viability of public spaces in cities under increasing heat: A transdisciplinary approach. *Sustainable Cities and Society, 59*, 102215. https://doi.org/10.1016/j.scs.2020.102215.

Futcher, J., & Mills, G. (2015). Walking among giants, *CIBSE Journal*, February 2015. Retrieved from https://www.cibsejournal.com/uncategorized/walking-among-giants/.

Gehl, J. (1987). *Life between buildings*. New York: Van Nostrand-Reinhold.

Giuffrida, L., Lokys, H., & Klemm, O. (2020). Assessing the effect of weather on human outdoor perception using Twitter. *International Journal of Biometeorology, 64*, 205–216. https://doi.org/10.1007/s00484-018-1574-7.

Hiroshima, S., Assis, E., & Nikolopoulou, M. (2016). Daytime thermal comfort in urban spaces: A field study in Brazil. *Building and Environment, 103*, 2016. https://doi.org/10.1016/j.buildenv.2016.08.006.

Hopkinson, R. G. (1963). *Architectural physics: Lighting*. London: Building Research Station HMSO.

Höppe, P. (1999). The physiological equivalent temperature—a universal index for the biometeorological assessment of the thermal environment. *International Journal of Biometeorology, 43*, 71–75.

Humphreys, M.A. (1975). *Field studies of thermal comfort compared and applied, building research establishment, current paper 76/75, Watford, UK.*

IPCC. (2018). Summary for policymakers. In V. Masson-Delmotte, P. Zhai, H. O. Pörtner, D. Roberts, J. Skea, P. R. Shukla, A. Pirani, W. Moufouma-Okia, C. Pan, R. Pidcock, S. Connors, J. B. R. Matthews, Y. Chen, X. Zhou, M. I. Gomis, E. Lonnoy, T. Maycock, M. Tignor, & T. Waterfield (Eds.), *Global warming of 1.5°C. An IPCC Special Report on the impacts of global warming of 1.5°C above pre-industrial levels and related global greenhouse gas emission pathways, in the context of strengthening the global response to the threat of climate change, sustainable development, and efforts to eradicate poverty*. Geneva: World Meteorological Organization.

ISO 7730. (2005). *Ergonomics of the thermal environment—Analytical determination and interpretation of thermal comfort using calculation of the PMV and PPD indices and local thermal comfort criteria*. Geneva: International Standards Organisation.

Jendritzky, G., Havenith, G., Weihs, P., Batchvarova, E., & de Dear, R. (2008). The universal thermal climate index UTCI-goal and state of COST action 730 and ISB commission 6. In: *Proceedings 18th International Congress of Biometeorology, Tokyo, Japan.*

Jesdale, B. M., Morello-Frosch, R., & Cushing, L. (2013). The racial/ethnic distribution of heat risk–related land cover in relation to residential segregation. *Environmental Health Perspectives, 121*, 811–817. https://doi.org/10.1289/ehp.1205919.

Johansson, E., Thorsson, S., Emmanuel, R., & Krüger, E. (2014). Instruments and methods in outdoor thermal comfort studies—The need for standardization. *Urban Climate, 10*, 346–366.

Kántor, N., Égerházi, L., & Unger, J. (2012). Subjective estimation of thermal environment in recreational urban spaces—Part 1: investigations in Szeged, Hungary. *International Journal of Biometeorology, 56*, 1075–1088.

Klok, L., Rood, N., Kluck, J., & Kleerekoper, L. (2020). Assessment of thermally comfortable urban spaces in Amsterdam during hot summer days. *International Journal of Biometeorology, 64*(2), 303. https://doi.org/10.1007/s00484-019-01818-7.

Knez, I., & Thorsson, S. (2006). Influences of culture and environmental attitude on thermal, emotional and perceptual evaluations of a public square. *International Journal of Biometeorology, 50*, 258–268. https://doi.org/10.1007/s00484-006-0024-0.

Knez, I., Thorsson, S., Eliasson, I., & Lindberg, F. (2009). Psychological mechanisms in outdoor place and weather assessment: Towards a conceptual model. *International Journal of Biometeorology, 53*, 101–111. https://doi.org/10.1007/s00484-008-0194-z.

Kolokotroni, M., Gowreesunker, B. L., & Giridharan, R. (2013). Cool roof technology in London: An experimental and modelling study. *Energy and Buildings, 67*, 658–667.

Lai, D., Lian, Z., Liu, W., Guo, C., Liu, W., Liu, K., & Chen, Q. (2020). A comprehensive review of thermal comfort studies in urban open spaces. *Science of the Total Environment, 742*, 140092. https://doi.org/10.1016/j.scitotenv.2020.140092.

Larsson, A., & Chapman, D. (2020). Perceived impact of meteorological conditions on the use of public space in winter settlements. *International Journal of Biometeorology, 64*, 631–642. https://doi.org/10.1007/s00484-019-01852-5.

Lemonsu, A., Amossé, A., Chouillou, D., Gaudio, N., Haouès-Jouve, S., Hidalgo, J., Le Bras, J., Marchandise, S., & Tudoux, B. (2020). Comparison of microclimate measurements and perceptions as part of a global evaluation of environmental quality at neighbourhood scale. *International Journal of Biometeorology, 64*, 265–276. https://doi.org/10.1007/s00484-019-01686-1.

Leng, H., Liang, S., & Yuan, Q. (2020). Outdoor thermal comfort and adaptive behaviors in the residential public open spaces of winter cities during the marginal season. *International Journal of Biometeorology, 64*, 217–229. https://doi.org/10.1007/s00484-019-01709-x.

Lenzholzer, S. (2015). *Weather in the city-how design shapes the urban climate*. Rotterdam: Nai 010 Uitgevers/Publishers.

Lenzholzer, S., & Nikolopoulou, M. (2020). Foreword to the Special Issue on Subjective approaches to thermal perception. *International Journal of Biometeorology, 64*, 167–171. https://doi.org/10.1007/s00484-019-01857-0.

Lin, T.-P. (2009). Thermal perception, adaptation and attendance in a public square in hot and humid regions. *Building and Environment, 44*, 2017–2026. https://doi.org/10.1016/j.buildenv.2009.02.004.

Lin, T.-P., Tsai, K.-T., Liao, C.-C., & Huang, Y.-C. (2013). Effects of thermal comfort and adaptation on park attendance regarding different shading levels and activity types. *Building and Environment, 59*, 599–611.

Llewelyn-Davies. (2007). *Urban design compendium English partnerships*. London: Housing Corporation.

Mahgoub, M. H., & Hamza, N. (2019). Behavioural perspectives of outdoor thermal comfort in urban areas: A critical review. *Atmosphere, 11*(1), 51. https://doi.org/10.3390/atmos11010051.

Manavvi, S., & Rajasekar, E. (2020). Semantics of outdoor thermal comfort in religious squares of composite climate: New Delhi, India. *International Journal of Biometeorology, **64***, 253–264. https://doi.org/10.1007/s00484-019-01708-y.

Matzarakis, A. (2000). Estimation and calculation of the mean radiant temperature within urban structures. In *Manual to RayMan*. Germany: University of Freiburg.

Matzarakis, A., Mayer, H., & Izimon, M. G. (1999). Applications of a universal thermal index: physiological equivalent temperature. *International Journal of Biometeorology, 43*, 76–84.

McKinnon, G., Pineo, H., Chang, M., Taylor-Green, L., Johns, A., & Toms, R. (2020). Strengthening the links between planning and health in England. *BMJ, 369*, m795. https://doi.org/10.1136/bmj.m795.

Nicol, F., Humphreys, M. A., & Roaf, S. (2012). *Adaptive thermal comfort: Principles and practice*. London: Routledge.

Nikolopoulou, M. (2004). *Designing open spaces in the urban environment: A bioclimatic approach, centre for renewable energy sources, EESD, FP5*.

Nikolopoulou, M. (2012). Urban open spaces and adaptation to climate change. In M. Richter & U. Weiland (Eds.), *Applied urban ecology* (pp. 106–122). Chichester: Wiley-Blackwell.

Nikolopoulou, M., & Lykoudis, S. (2006). Thermal comfort in outdoor urban spaces: Analysis across different European countries. *Building and Environment, 41*, 1455–1470.

Nikolopoulou, M., & Lykoudis, S. (2007). Use of outdoor spaces and microclimate in a Mediterranean urban area. *Building and Environment, 42*, 3691–3707.

Nikolopoulou, M., & Steemers, K. (2003). Thermal comfort and psychological adaptation as a guide for designing urban spaces. *Energy and Buildings, 35*, 95–101.

Nikolopoulou, M., Baker, N., & Steemers, K. (1999). *Thermal comfort in outdoor urban spaces: Different forms of adaptation, Proc. Rebuild 1999*. Barcelona: The Cities of Tomorrow.

Nikolopoulou, M., Baker, N., & Steemers, K. (2001). Thermal comfort in outdoor urban spaces understanding the human parameter. *Solar Energy, 70*, 227–235.

Nikolopoulou, M., Kotopouleas, A., Lykoudis, S. (2018). From indoors to outdoors and in-transition; thermal comfort across different operation contexts. In *Proceedings of Windsor Conference. Rethinking Comfort* (pp. 747–759).

O'DonnellBrown (2019) *The community classroom, ArchDaily*. Retrieved from https://www.archdaily.com/936.845/the-community-classroom-odonnellbrown.

Oke, T. R. (1987). *Boundary layer climates* (2nd ed.). London: Methuen.

Olgyay, V. (1963). *Design with climate: Bioclimatic approach to architectural regionalism*. Princeton: Princeton University Press.

Orange, R. (2020). *Split classes, outdoor lessons: what Denmark can teach England about reopening schools after Covid-19, The Observer, Sunday 17th May*. Retrieved from https://www.theguardian.com/education/2020/may/17/denmark-can-teach-england-safe-reopening-of-schools-covid-19.

Paciuc, M. (1990). The role of personal control of the environment in thermal comfort and satisfaction at the workplace. In R. I. Selby, K. H. Anthony, J. Choi, & B. Orland (Eds.), *Coming of age*. Syracuse: Environment Design Research Association.

Pantavou, K., Lykoudis, S., Delibasis, K., Tseliou, A., Koletsi, I., Nikolopoulou, M. and Tsiros, I.X. (2019). The integration of three field survey datasets in Athens, Greece: Transformation of five-point to seven-point thermal sensation scale. In: *16th International Conference on Environmental Science and Technology, Rhodes, Greece*.

Pantavou, K., Koletsis, I., Lykoudis, S., Melas, E., Nikolopoulou, M., & Tsiros, I. X. (2020). Native influences on the construction of thermal sensation scales. *International Journal of Biometeorology, 64*, 1497–1508. https://doi.org/10.1007/s00484-020-01927-8.

Pearlmutter, D., Jiao, D., & Garb, Y. (2014). The relationship between bioclimatic thermal stress and subjective thermal sensation in pedestrian spaces. *International Journal of Biometeorology, 58*, 2111–2127. https://doi.org/10.1007/s00484-014-0812-x[486].

Penwarden, A. D. (1973). Acceptable wind speeds in towns. *Building Science, 8*(3), 259–267.

Pitts, A. (2006). The languages and semantics of thermal comfort. In: *Proceedings: Network for Comfort and Energy Use in Buildings, Getting them Right. Cumberland Lodge, Windsor, UK*.

Platt, L. (2006). Poverty. In G. Payne (Ed.), *Social Divisions* (2nd ed.). Hampshire: Palgrave Macmillan.

Roesler, S. (2019). *On microclimatic islands; the garden as a place of intensified thermal experience, Les Cahiers de la Recherche Architecturale Urbaine et Paysagère*.

Russell, D., Gawthrop, E., Penney, V., Raj, A., & Hickey, B. (2020). Deadly heat is killing Americans: A decade of inaction on climate puts lives at risk, *The Guardian*, 16th June 2020. Retrieved from https://www.theguardian.com/us-news/2020/jun/16/climate-deaths-heat-cdc.

Santamouris, M., & Kolokotsa, D. (2016). *Urban climate mitigation techniques*. Abingdon: Routledge.

Sharifi, E., & Boland, J. (2020). Passive activity observation (PAO) method to estimate outdoor thermal adaptation in public space: case studies in Australian cities. *International Journal of Biometeorology, 64*, 231–242. https://doi.org/10.1007/s00484-018-1570-y.

Siple, P., & Passel, C. (1945). Measurements of dry atmospheric cooling in subfreezing temperatures. *Proceedings of the American Philosophical Society, 89*(1), 177–199.

Thom, E. C. (1959). The discomfort index. *Weatherwise, 12*, 57–60.

Thorsson, S., Lindqvist, M., & Lindqvist, S. (2004). Thermal bioclimatic conditions and patterns of behaviour in an urban park in Goteborg, Sweden. *International Journal of Biometeorology, 48*, 149–156.

Tseliou, A., Tsiors, I. X., Lykoudis, S., & Nikolopoulou, M. (2010). An evaluation of three biometeorological indices for human thermal comfort in urban outdoor areas under real climatic conditions. *Building and Environment, 45*, 1346–1352. https://doi.org/10.1016/j.buildenv.2009.11.009.

Tsichritzis, L., & Nikolopoulou, M. (2019). The effect of building height and façade area ratio on pedestrian wind comfort of London. *Journal of Wind Engineering and Industrial Aerodynamics, 191*, 63–75. https://doi.org/10.1016/j.jweia.2019.05.021.

Vasilikou, C. and Nikolopoulou, M. (2013). Thermal walks: Identifying pedestrian thermal comfort variations in the urban continuum of historic city centres. In *Proceedings of PLEA 2013: International conference on sustainable architecture for a renewable future, Munich.*

Vasilikou, C., & Nikolopoulou, M. (2015). Thermal perception of pedestrians moving in interconnected urban spaces: adaptive thermal comfort in irregular spatial sequences in Rome and London. In *Proceedings of PLEA 2015*. Bologna: Architecture in (R) Evolution.

Vasilikou, C., & Nikolopoulou, M. (2020). Outdoor thermal comfort for pedestrians in movement: thermal walks in complex urban morphology. *International Journal of Biometeorology, 64*, 277–291. https://doi.org/10.1007/s00484-019-01782-2.

Westerberg, U. (1994). Climatic planning—Physics or symbolism. *Architecture and Behaviour, 19*, 49–72.

Yahia, M. W., & Johansson, E. (2013). Evaluating the behaviour of different thermal indices by investigating various outdoor urban environments in the hot dry city of Damascus, Syria. *International Journal of Biometeorology, 57*, 615–630. https://doi.org/10.1007/s00484-012-0589-8.

Yang, B., Olofsson, T., Nair, G., & Kabanshi, A. (2017). Outdoor thermal comfort under subarctic climate of North Sweden—A pilot study in Umeå. *Sustainable Cities and Society, 28*, 387–397. https://doi.org/10.1016/j.scs.2016.10.011.

Zacharias, J., Stathopoulos, T., & Hanqing, W. (2001). Microclimate and downtown open space activity. *Environment and Behaviour, 33*(2), 296–315.

Zucker, P. (1959). *Town and square*. New York: Columbia University Press.

Chapter 5
Comfort and Energy Implications of Urban Microclimate in High Latitudes

Maria Kolokotroni and Agnese Salvati

Abbreviations

CFD	Computational fluid dynamics
DTM	Dynamic thermal modelling
EPBD	Energy Performance of Buildings Directive
H/*L*	Canyon aspect ratio
IPCC	Intergovernmental Panel on Climate Change
SUHI	Surface urban heat island
TRY	Typical reference year
UA	Urban albedo
UHI	Urban heat island

5.1 Introduction

The latitude of a location determines solar radiation received on horizontal and vertical building and urban surfaces; higher latitudes in both earth hemispheres receive less solar radiation than lower latitudes.

Figure 5.1 shows daily insolation for selected latitudes in the north hemisphere; higher latitudes receive most of the solar radiation during the summer months resulting in higher air temperatures in the summer compared with winter months. This results in seasonal variations in climatic conditions with winter months characterised by lower ambient temperatures compared to summer which in turn impact the energy demand by buildings to maintain thermal comfort.

In characterising the general climate of a location, geography studies typically use the Köppen-Geiger system. Figure 5.2 shows a world map of this classification. In simple terms, although latitude affects climatic conditions, the radiation energy

M. Kolokotroni (✉) · A. Salvati
Department of Mechanical and Aerospace Engineering, College of Engineering, Design and Physical Sciences, Brunel University London, London, UK
e-mail: maria.kolokotroni@brunel.ac.uk

M. Palme, A. Salvati (eds.), *Urban Microclimate Modelling for Comfort and Energy Studies*, https://doi.org/10.1007/978-3-030-65421-4_5

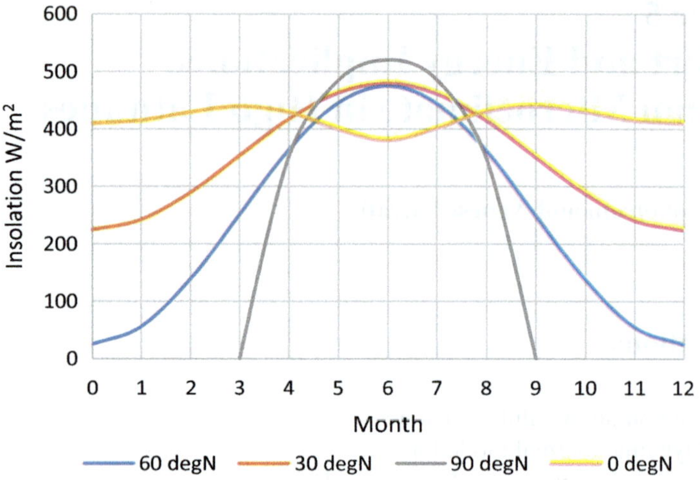

Fig. 5.1 Daily insolation through the year at four latitudes in the north hemisphere. (Redrawn from https://geography.name/insolation-over-the-globe/)

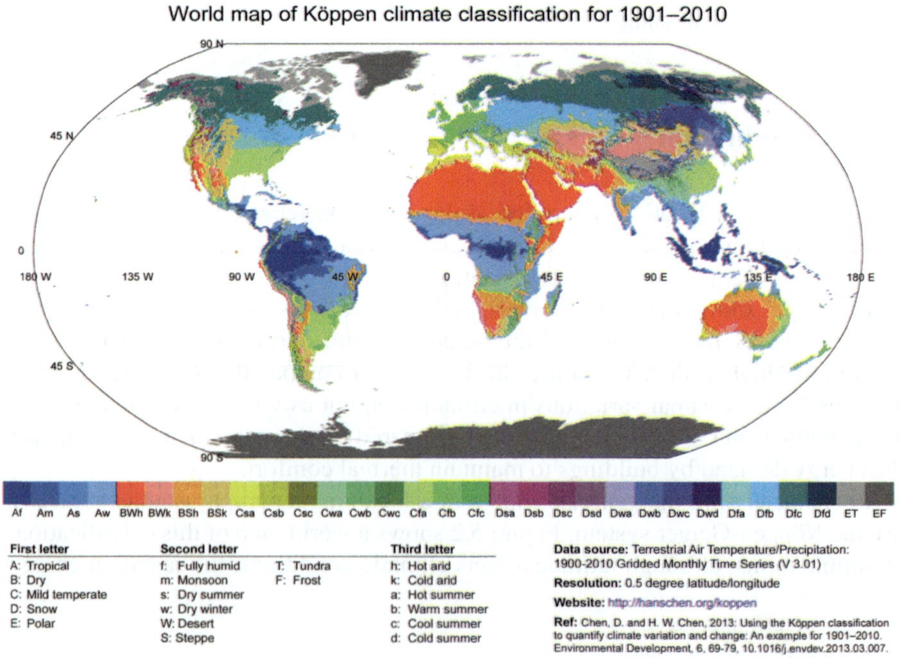

Fig. 5.2 World map of Köppen climate classification, (modified from Chen and Weiteng 2013)

balance, tilt of earth, altitude, proximity to water and prevailing winds form the general climatic conditions. There are five different types of Köppen-Geiger climate classification: (A) tropical, (B) dry, (C) temperate, (D) continental and (E) polar (World Atlas 2020), following latitude but with variations.

Tropical zones receive intense solar radiation throughout the year and seasonal variations start to exist as we extend to the Tropics of Cancer and Capricorn (23.5°). These areas have two peaks of solar radiation during the year. The dry zones lie between the tropics and around 35° and have a strong seasonal cycle and a large annual insolation. The mild temperate and continental zones between latitude 35° and 60°, north and south, are characterised by seasonal contrasts in insolation and experience a large range in annual surface temperature. The zones above 60° have an extremely large yearly variation in day length, with enormous contrasts in insolation over the year with some months' insolation approaching zero.

Within the mild temperate zones (type C), two main subgroups of climates exist, depending on the latitude and the amount of precipitation and its distribution over the year. Lower latitudes, between 35° and 45°, are characterised by dry summers and are known as "Mediterranean climate" (Csa and Csb) while higher latitudes, between 45° and 60°, have higher precipitation and lack of a dry season (Cfb and Cfc). This chapter focusses on these high-latitude temperate regions, with higher heating requirements for buildings compared to cooling requirements.

Section 5.2 presents a description of the urban microclimate from global to local scale. Studies on surface and air UHI in high-latitude cities are described as well as the relationship between UHI and building energy demand in high-latitude cities. It continues with a description of microclimate impacts on ambient temperature using London as a case study of high-latitude location.

Section 5.3 focusses on urban albedo. Its definition is given together with the contributing parameters to its value. Available literature on the impact of these parameters on the urban albedo is presented. Results of a recent study of urban albedo in London are presented.

Section 5.4 focusses on modelling tools enabling the study of microclimate impact on indoor thermal conditions. The application of such tools to a case study in London is presented as an example. Section 5.5 summarises the chapter.

5.2 Urban Microclimate: From Global to Local Scale

The thermal balance of cities is affected by the increased absorption of solar radiation, corresponding increase of sensible heat released by urban structures, higher anthropogenic heat, reduced urban vegetation and higher emission of infrared radiation (Akbari and Kolokotsa 2016).

The urban heat island (UHI), although more pronounced in high-radiation/high-ambient-temperature cities, has been measured and quantified in cities of high latitudes characterised by warm/cold summers. Studies can be divided into those based on surface temperature data and those measuring air temperature within cities. The first are useful in providing trends and mitigation measures at urban scales while the second can give insights into urban thermal comfort due to microclimatic variations and energy demand by urban buildings.

5.2.1 Surface and Air UHI in High-Latitude Cities

Peng et al. (2012) have reported on the surface UHI for 419 global big cities with population more than one million in 2007. The analysis was based on land surface temperature and considered the parameters of vegetation, urban albedo, climate and several socio-economic indexes. They showed that the average annual daytime is higher than the annual night surface UHI across big cities while no correlation was found between daytime and night. The distribution of night surface UHI correlates positively with the difference in albedo and night lights between urban and suburban areas indicating the impact of materials used, urban geometry and anthropogenic heat. They note that city structure such as building density, height and surface materials could also affect the night surface energy exchange, thus impacting the night surface UHI. No obvious effect of city size was found although it is noted that other studies found larger urban heat island intensity in bigger cities in one country or region. They concluded that because the 419 big cities are from different countries, different climatic zones and different economic development, the effects of city size could be masked by these factors. The distribution of daytime surface UHI (SUHI) correlates negatively with the difference of vegetation cover and activity between urban and suburban areas. The heat island intensity also varies seasonally. Daytime SUHI intensity has larger summer-to-winter amplitude than night SUHI intensity in the temperate regions over the northern hemisphere, and there are remarkable latitudinal variations in daytime SUHI intensity. Mid- to high-latitude cities (e.g. Beijing, Vancouver) have larger seasonal amplitude of daytime SUHI intensity than their low-latitude cities of similar size; for example, tropical cities where vegetation is active all year round show a flat seasonal profile of daytime SUHI intensity throughout the year. The evaporative cooling effects of the vegetation during the growing season may partly explain why the seasonal amplitude of heat islands differs between cities. The night SUHI intensity exhibits a small summer-to-winter amplitude of less than 1 °C for more than 86% of the big cities. The seasonal amplitude of night SUHI intensity is also more pronounced in cold regions (high latitudes) compared to warm regions (low latitudes). However, the seasonal amplitude of night SUHI intensity negatively correlates with both the mean temperature for the months when air temperature exceeds 0 °C and mean annual temperature. This result is consistent with the larger solar radiation energy transformed into SUHI intensity in summer over high-latitude regions. They note that the large SUHI intensity in summer not only increases the cooling power consumption in urban areas but also could increase the risk of heatwave extreme events over urban areas, as urban areas are affected by both global and local climate forcing.

The impact of latitude and climate on surface UHI was also reported by Zhou et al. (2014) for 32 cities in China. Results show that the SUHI differed substantially between day and night, with distinct climate-driven patterns, characterised by a higher SUHI in the south-eastern parts of China in the day and in the northern regions at night compared with other regions. In particular, the Northeast, with humid-cold climate, experienced the most intense SUHI across China. In addition, the SUHI intensity varied greatly with season, indicated by a relatively higher intensity in the day for the summer and at night for the winter across most cities. As a result, whether the daytime SUHI intensity was higher or lower than the night SUHI intensity depends strongly on the climate region and time period investigated. Overall, climate showed ultimate control on the SUHI spatial variability. In this study too the SUHI distribution in the day correlated significantly to vegetation activity and anthropogenic heat emissions in summer and to climate in winter, and at night was related closely to surface albedo, anthropogenic heat emissions, built-up intensity and climate in both seasons.

In studies of air temperature in urban areas, wind patterns are shown to have an impact. In Europe, UHI studies until 2007 are reported by Santamouris (2007); they are divided into the Mediterranean zone (Csa), central Europe and the UK (Cfb) and northern Europe (Dfc) as shown in Fig. 5.2. It was concluded that in cities in central Europe, the UK and northern Europe, maximum UHI values developed at night and similarly with Mediterranean cities do not show any relation to the city size. Greatest intensity was found in windless and cloudless conditions mostly in the summer similarly to Csa regions. Additional data for Europe are included in the work of Santamouris (2016).

At continental level Santamouris (2015) reported on UHI in Australian and Asian cities. Most cities reported are located in hot climatic zones but a few are cold/warm summer regions. The conclusions of this review introduce the impact of climatic conditions and wind on the formation of UHI which is not apparent in SUHI studies. The review concluded that the size of city correlates with magnitude but the method of data acquisition is important to quantify this. There is a seasonal variability as observed in SUHI studies; humid locations have the max UHI during dry periods rather than during warm periods. UHI is more pronounced at night consistent with SUHI studies but in some cities with hot climate and high density maximum UHI can be present during the day mainly due to anthropogenic heat. Also, cool islands have been observed due to shading, reduced anthropogenic heat and possible cooling effect of sea breeze. Calm and clear sky conditions help in the development of UHI but high wind speed can dramatically reduce it although wind speed thresholds to reduce UHI vary according to geometry and density while coastal cities mostly benefit from sea breeze.

The results reported at global level are more or less replicated in locations of high latitudes with cold/warm summers. An example is London. Figure 5.3 shows the UHI measured using fixed stations (Watkins et al. 2002a). The average annual UHI is 2.5 °C at night but there is wide variation including cool islands mainly during the day. Figure 5.4 shows the impact of solar irradiance which is mainly observed during days with meteorological wind speed more than 5 m/s (Kolokotroni and

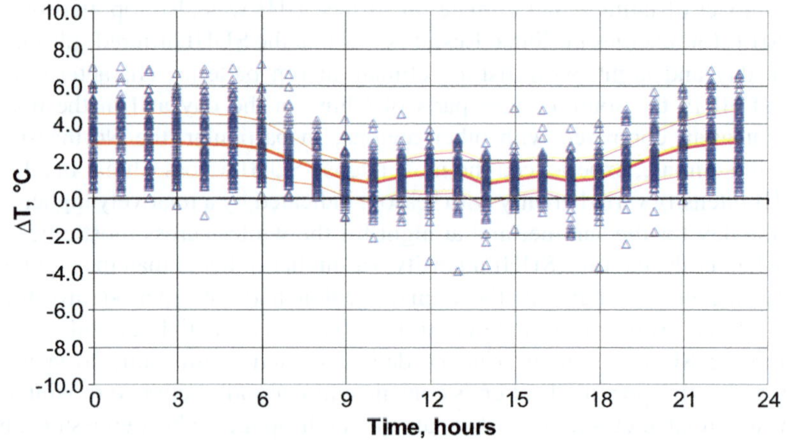

Fig. 5.3 Hourly air temperature difference over 24 h between urban and reference locations in London measured at fix stations (source: Watkins et al. 2002a)

Giridharan 2008) while Fig. 5.5 shows that night UHI is independent of daytime air temperature during the summer; during the night similar difference between urban and rural air temperature was measured during a typical and hot week (Kolokotroni et al. 2006).

5.2.2 UHI and Building Energy Demand in High-Latitude Cities

At global level Santamouris (2020) reported on recent progress of urban overheating and urban heat island. Air temperature experimental results from 400 cities show that the magnitude of the average temperature increase may exceed 4–5 °C while at the peak may exceed 10 °C. The paper reviews UHI impact on energy demand by buildings and impacts on outdoor air quality and health. In terms of energy demand, the analysis shows that the total increase in energy demand is close to 12% considering all locations and building types and the actual magnitude is strongly correlated with the magnitude of UHI. In most cases the reduction in heating energy demand is less than the increase in cooling energy demand resulting in a net increase. This net increase depends on ambient temperatures during the cooling season but is also related to building type and its energy efficiency strategies.

For locations with warm/cold summer conditions, low cooling demand for well-designed buildings should result in low increase in cooling demand despite the UHI; however cooling demand has increased during the last years in these locations too. There are two reasons for this: first increasing ambient temperatures due to climate change combined with the UHI, and second regulations for energy efficiency focused on winter conditions with the unintended consequence of overheating

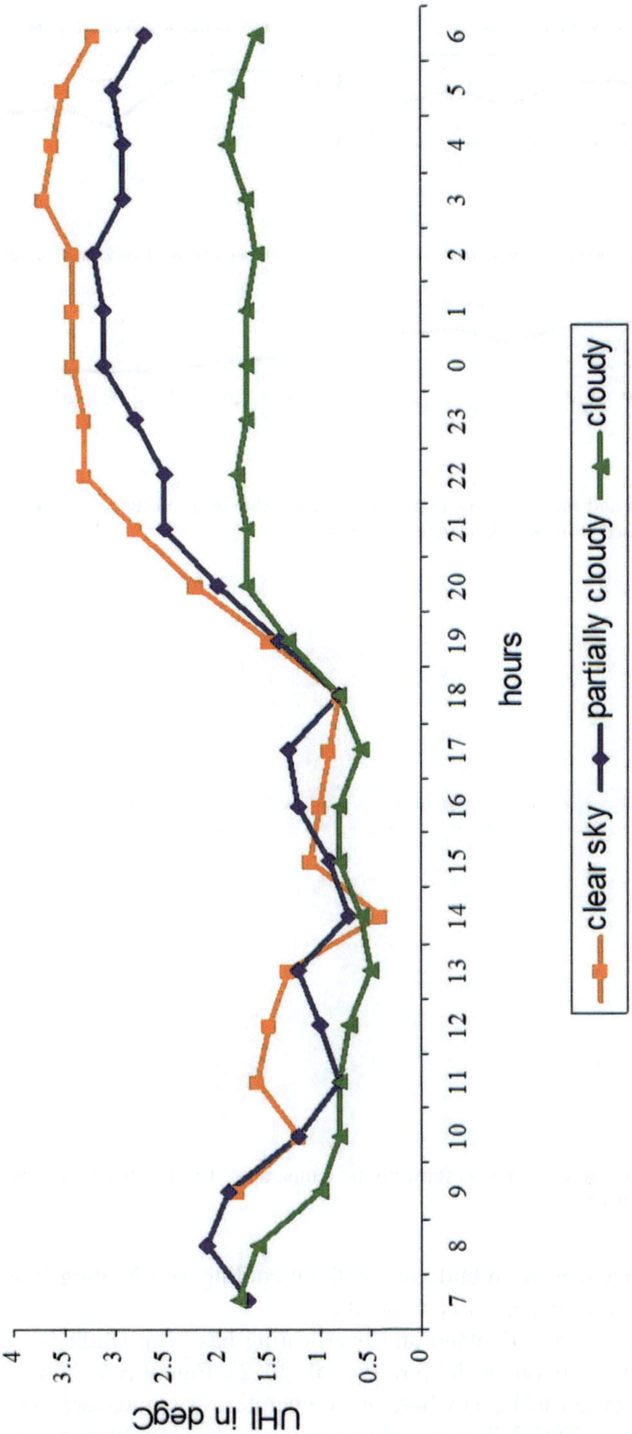

Fig. 5.4 Effect of solar irradiance on hourly air temperature difference over 24 h between urban and reference locations in London at fixed stations (adapted from Kolokotroni and Giridharan 2008)

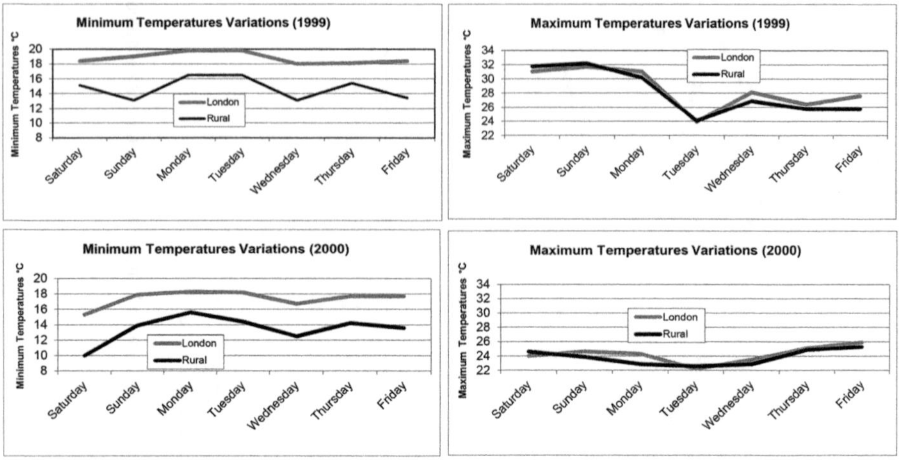

Fig. 5.5 Minimum and maximum air temperature in London and reference site during a typical and hot week (adapted from Kolokotroni et al. 2006)

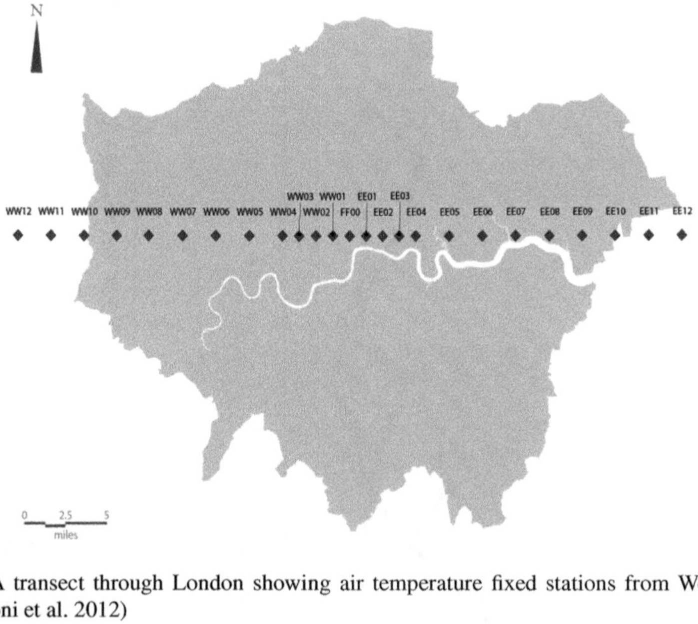

Fig. 5.6 A transect through London showing air temperature fixed stations from West to East (Kolokotroni et al. 2012)

during the summer season. In buildings without cooling, overheating is also a consequence which impacts comfort and health.

This impact on energy demand and overheating has been studied in detail for office buildings in London (Kolokotroni et al. 2012). Figure 5.6 shows a transect through London (West to East) where air temperature was measured at fixed stations (Watkins et al. 2002a). The model of a typical office building was created and thermally simulated using the city's Test Reference Year (TRY) weather file from the Heathrow Airport weather station and urban TRY files in which air temperature was changed to correspond with hourly measurements at each transect point. Figure 5.7 shows the increase in overheating hours in the free-floating office across the transect compared to the rural position. It also shows the impact of future climate change by presenting simulation results using the 2050 future weather file. This was created by first modifying the Heathrow TRY weather file using the climate projections of the Intergovernmental Panel on Climate Change (IPCC) for medium-high emission scenario. The air temperature was then changed to superimpose the measured air temperature in each of the transect locations assuming that the UHI will be the same in the future. Figure 5.8 shows the interurban and future variation of cooling energy demand assuming for an air-conditioned office with a cooling setpoint temperature of 24 °C.

The impact of UHI and climate change on energy demand is as expected and in the magnitude reported in literature (Santamouris 2016) for heating-dominated climates. What is interesting for building designers is the impact of energy efficiency measures and their performance in the future. Figure 5.8 shows that at present, offices with medium internal heat gains without night free cooling have similar cooling demand to offices with high heat gains and night free cooling. In the future, and because of higher night temperature due to the combined effect of climate change and UHI, offices with medium internal heat gains perform better than those with night free cooling.

Figure 5.9 shows that as expected heating demand is lower in offices within the UHI and even lower in the future. Night cooling is not shown as it is not used during the heating season.

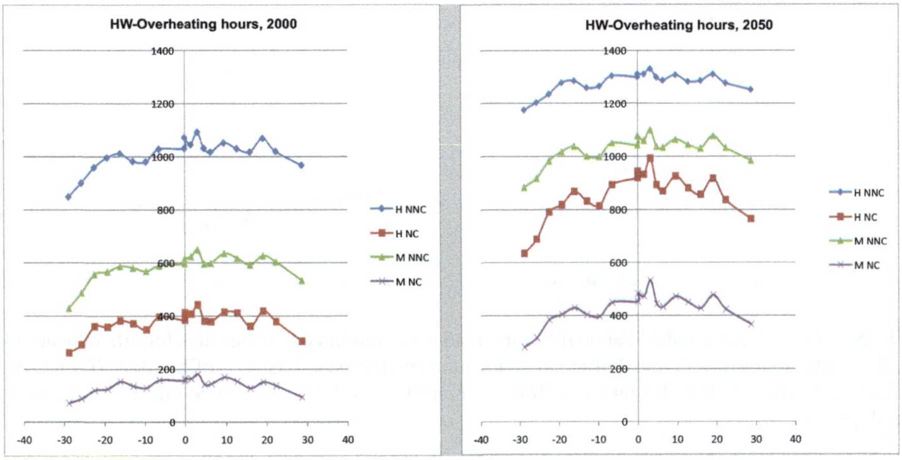

Fig. 5.7 Predicted overheating hours (>28 °C) during occupancy hours using modified weather files for urban location and climate change for a typical office in 20 locations on the West-East axis of London (*HW* heavy weight construction, *H* high heat gains, *M* medium heat gains, *NC* night cooling, *NNC* no night cooling) (Kolokotroni et al. 2012)

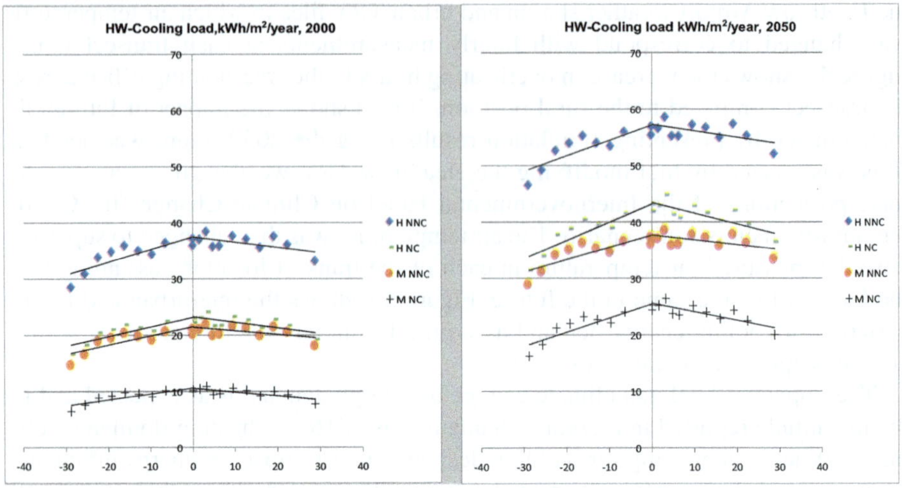

Fig. 5.8 Predicted normalised annual cooling load using modified weather files for urban location and climate change for a typical office in 20 locations on the West-East axis of London (*HW* heavy weight construction, *H* high heat gains, *M* medium heat gains, *NC* night cooling, *NNC* no night cooling) (Kolokotroni et al. 2012)

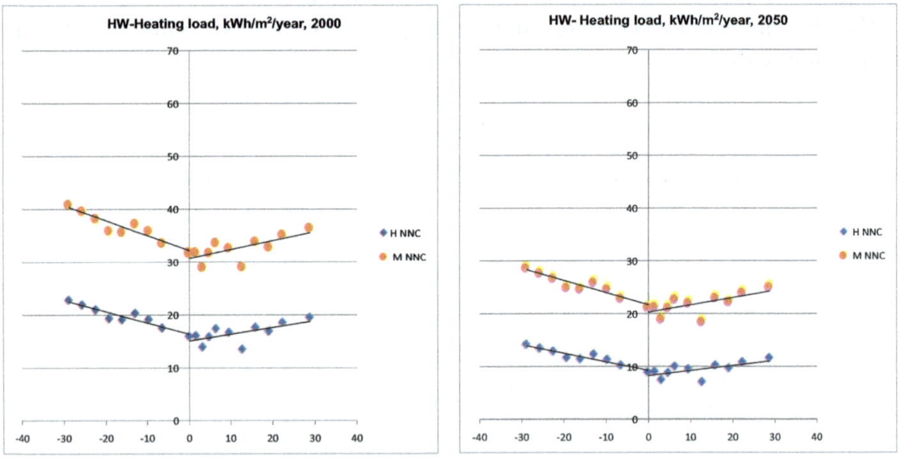

Fig. 5.9 Predicted normalised annual heating load using modified weather files for urban location and climate change for a typical office in 20 locations on the West-East axis of London (*HW* heavy weight construction, *H* high heat gains, *M* medium heat gains, *NNC* no night cooling) (Kolokotroni et al. 2012)

Therefore, in terms of energy demand by buildings, in high-latitude cities with warm/cold summers where most buildings are not air-conditioned, UHI helps reduce the building energy consumption, at present. However, building will be air-conditioned in the future because of higher temperatures and even higher in the city.

Once air conditioning is introduced, internal temperatures will be regulated at the range of 21–24 °C. At present, we tolerate higher. Estimations indicate a five-fold increase in carbon emissions by city buildings in 2050 (Kolokotroni et al. 2012).

There are various strategies to mitigate this; these include the following:

(a) Improve energy efficiency and passive cooling design of buildings in high-latitude cities, to avoid overheating risk due to urban and climate change effects. This is already regulated in many regions for example through the Energy Performance of Buildings Directive (EPBD) in Europe.
(b) Generate heat and electricity by building integrated renewables—nearly zero-carbon buildings—this is coupled to (a) and again regulated in many regions.
(c) Improve external thermal environment; this is not regulated yet but initiatives and studies indicate its importance. Mitigation strategies include the following:

    - Reduce anthropogenic heat from buildings and transport; this is related to improved energy efficiency in buildings and use of passive strategies where possible to reduce rejected heat from cooling systems. Transport is also regulated mainly due to interventions to reduce vehicle emissions with electric vehicles increasing in many cities.
    - Increase vegetation.
    - Increase albedo of surfaces.

The following section attempts to answer the question of 'what is the impact of vegetation and surface albedo in the microclimatic scale in high latitudes?' taking London as a case study.

5.2.3 Microclimate Impacts on Ambient Temperature: London Case Study

The impact of vegetation (parks and green spaces) on the ambient air temperatures has been studied in London. Figure 5.10 shows measurements inside and outside a park over 1 day (Watkins et al. 2002b). Air temperatures are lower inside the park. There is more pronounced variation outside indicating the impact of other microclimatic variables such as canyon geometry and prevailing wind. Figure 5.11 shows the air temperature difference between the park and outside during the daytime. Air temperature is lower in the park consistent with findings of the effect of vegetation at the scale of the city (Peng et al. 2012).

A study on the impact of small areas of vegetation was carried out in a suburban area of London during day and night-time (Maciel et al. 2013). Figure 5.12 shows that air temperature in green locations is lower during the night with the difference increasing as ambient temperature increases.

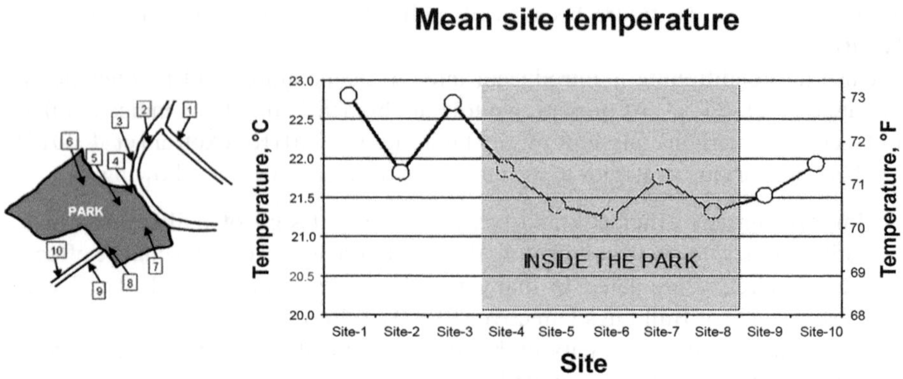

Fig. 5.10 Mean site air temperature over 1 day, showing the effect of vegetation (Watkins et al. 2002b)

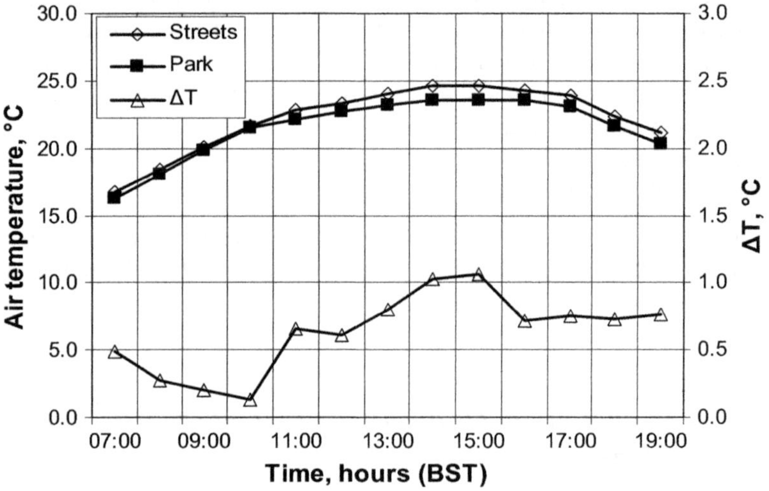

Fig. 5.11 Air temperature inside and outside the park over the day, indicating maximum difference during early afternoon (Watkins et al. 2002b)

The impact of roof reflectivity was measured on a building at the same site (Kolokotroni et al. 2013). Figure 5.13 shows the reduction of roof surface temperature with changed albedo from 0.1 to 0.6 (in situ measurements).

Road and pavement with high albedo can change the air temperature above these surfaces. This has been studied extensively as a method to mitigate UHI and increase external thermal comfort (Akbari and Kolokotsa 2016). In addition it can help to extend the effectiveness of passive cooling strategies such as natural ventilation (Maciel and Kolokotroni 2017) as shown diagrammatically in Fig. 5.14. Cool surface temperature could be 20–30 K less than asphalt and measurements have shown a 10 K air temperature difference at 75 cm above the surface.

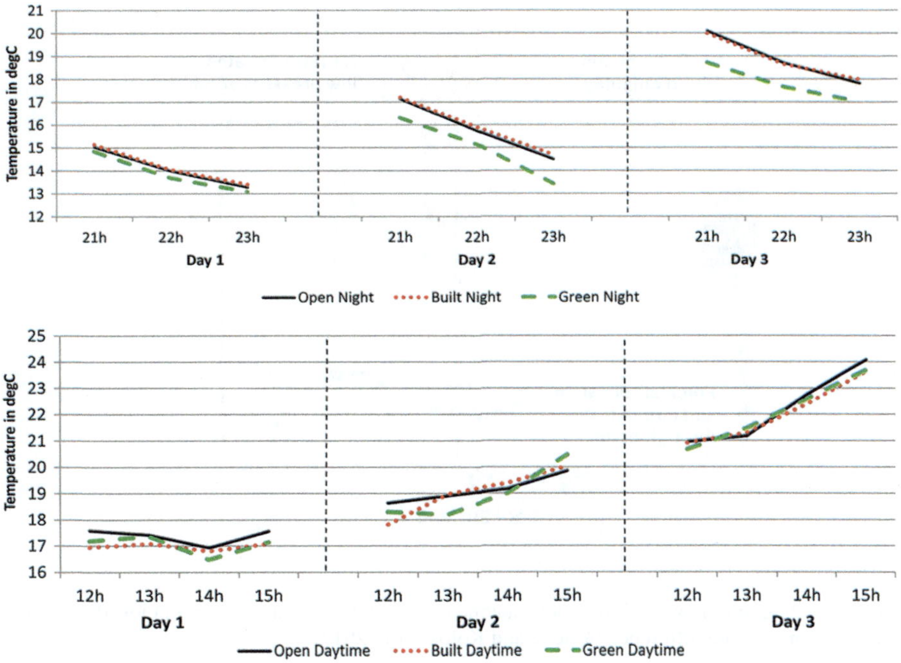

Fig. 5.12 Air temperature measured during day and night at locations with open aspect (mainly roads), built aspect (next to buildings) and green aspect (mainly grass) (Maciel et al. 2013)

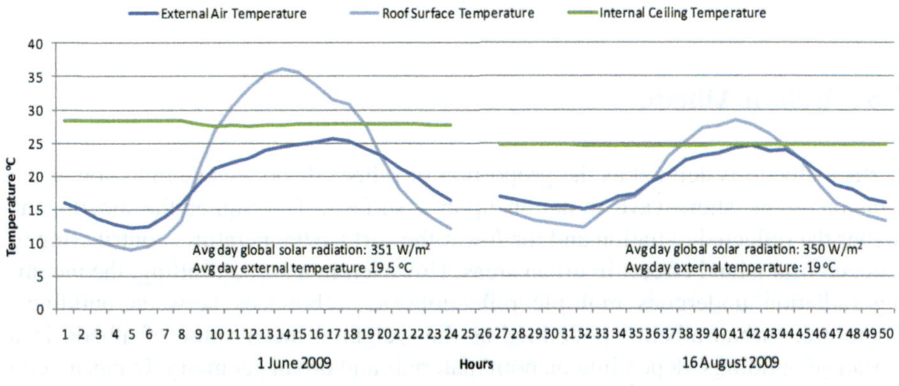

Fig. 5.13 Measured air and surface temperature of the case study building on 2 days, one in June (pre-paint period) and the other in August (after-paint period) during which external temperature and solar radiation are similar (Kolokotroni et al. 2013)

The effect of building facade albedo was studied in central London (Watkins et al. 2002b). Figure 5.15 shows the results of these measurements; it can be seen that dark-coloured facades result in higher surface temperatures but overshadowing effects were clear. The figure also shows that facade surface temperatures impact the

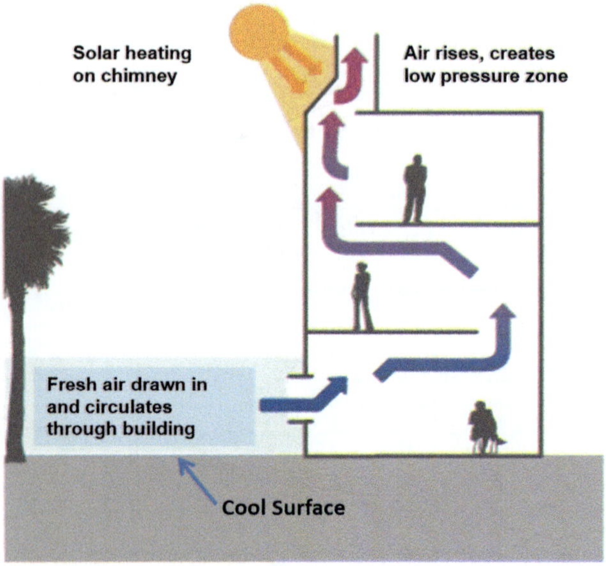

Fig. 5.14 Diagram showing how cool pavements can enhance passive ventilation due to the enhanced temperature difference (Maciel and Kolokotroni 2017)

canyon air temperature while climatic conditions and solar radiation intensity determine the magnitude. These relationships are explained in more detail in Sect. 5.3 where the urban albedo is discussed.

5.3 Urban Albedo

Surface albedo is defined as the proportion of reflected short-wave radiation to the total incoming short-wave radiation upon a surface. The higher the albedo, the higher the reflected radiation and the lower the surface temperature, contributing to reduced heat island effects in urban areas. However, in an urban setting, the incoming radiation undergoes multiple reflections as it bounces between buildings. Therefore, 'urban albedo' is defined as the weighted mean value of albedo in a cluster of buildings, depending on both materials and urban geometry. Urban albedo is a major contributor to changes in outdoor surface and air temperature as discussed in Sect. 5.2.1, intensifying the urban heat island phenomenon which in turn impacts the outside thermal comfort and energy demand by urban buildings.

The factors that contribute to the value of urban albedo can be grouped into the following interrelated parameters:

- *Geometry*—the geometry of individual buildings, the building group geometry and the geometry of buildings in relation to roads, usually referred to as urban canyon geometry

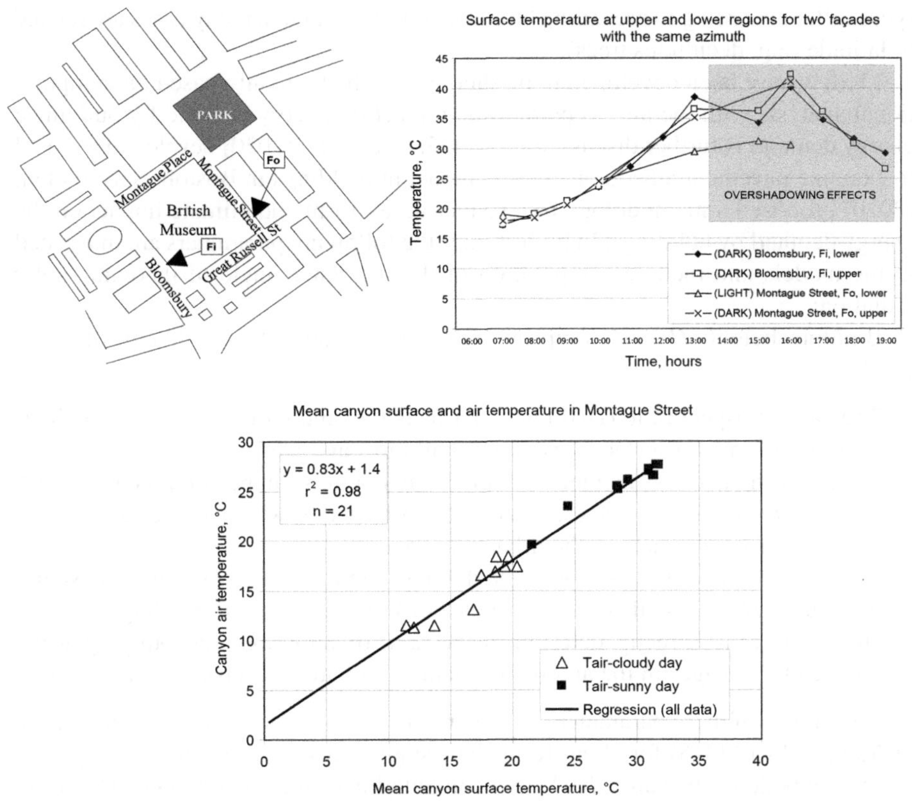

Fig. 5.15 Measured air and surface temperatures in two street canyons in central London (Watkins et al. 2002b)

- *Reflectance* of surfaces and their orientation—ground, facades and roofs

These two parameters are not related to latitude and climate, although climate impacts the urban development and materials used for buildings and there are indications that wetness of surfaces might affect reflectance.

- *Sun altitude and azimuth* determining the incidence angle of incoming solar radiation with strong seasonal contrasts in high latitudes

Geometry, surface reflectance and sun's position are in many cases used to calculate the urban albedo.

- *Moisture* of permeable surfaces (soil) and altering reflectivity of impermeable surfaces
- *Vegetation* acting as obstruction to the incoming radiation and reflection paths

These three parameters are closely related to climate. Moisture is connected to climatic conditions (air relative humidity) and precipitation. Similarly, vegetation types differ according to prevailing climatic conditions and in many cases are related to latitude (e.g. deciduous trees).

Models have been developed to predict urban albedo mainly based on computational analysis with the aim to predict the impact on energy demand by buildings; energy demand is further discussed in Sect. 5.4. Xu et al. (2020) developed a model to examine pavement albedo on energy impact in buildings in Boston. Salvati et al. (2020) proposed a methodology to consider urban form and climate impact on the energy demand by buildings. In both studies the following parameters are integrated in the models and include computation of the incidence angle of incoming solar radiation:

Building height (*H*) and footprint area, street width (*W*) and length of street (*L*) define:

- The canyon aspect ratio (*H*/*W*) which affects the amount of solar radiation incident on the canyon surfaces, as higher buildings and narrower canyons result in more shading and less incident radiation. This in turn can be computed as *building density* or site coverage ratio or plan area index which is the ratio of the buildings' footprint area to the urban site area.
- The ratio of the vertical surface area defines the *facade density* which is usually defined as $H \times P$ (with P the perimeter of buildings) to the building footprint area. This ratio directly determines the area of buildings and building walls affected by changes in the amount of incident radiation received and reflected.

The impact of these parameters on the urban albedo was studied computationally by Yang and Li (2015) for Hong Kong. The study gives a concise literature review of urban albedo studies until 2015 both computational and experimental. The study reports seasonal variations of urban albedo with low-rise medium density (0.44) having the greatest ability to absorb solar radiation. Figure 5.16 (for *H*/*L* ratio of 1) shows that as density increases or decreases, it plays a dominant role in the radiation reflection process leading to an increase in urban albedo. Urban albedo is minimum for building density of about 0.45; for lower densities, urban albedo is higher because the higher distance between buildings eases the ability of urban texture to

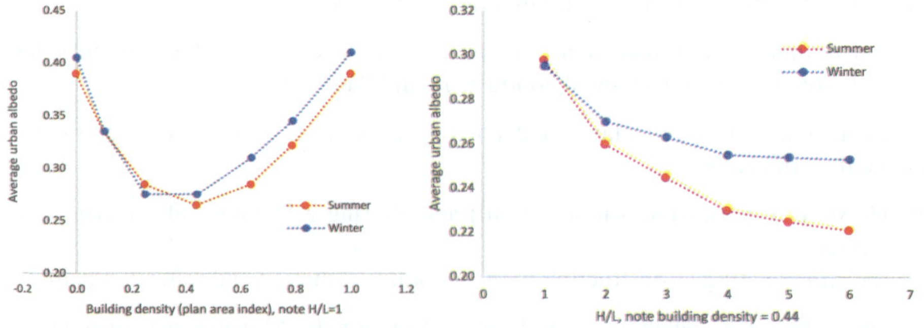

Fig. 5.16 Urban albedo relationship with season, building density (plan area) and canyon aspect ratio *H*/*L* (redrawn from Yang and Li 2015)

reflect solar radiation toward the sky. Urban albedo is higher for building density above 0.44 due to the increased contribution of roofs to reflecting radiation out of the urban fabric. Urban albedo was found to decrease with increasing building height for a fixed building density.

It should be noted that the same geometrical parameters influence wind patterns in the canyon. Although wind effects have no impact on urban albedo, they influence surface and air temperature in the canyon (Yang and Li 2015; Salvati et al. 2020).

The daily variation of the urban albedo was reported by Qin (2015) studied computationally with experimental validation of measurements at the latitude of ~51 in Poland for different canyon orientations and seasons (Pawlak and Fortuniak 2003). The parametric analysis carried out concluded that the canyon aspect ratio (*H*/*L*) dominates the urban albedo while other parameters play a secondary role.

Recently, the urban albedo has been studied in London as a case study of high latitude (Urban Albedo 2019). A typical residential area was identified based on the work of Kolokotroni and Giridharan (2008) and is being studied with the aim to identify how the urban albedo can be estimated. Methods include the construction of a 1:10 model of the urban area where extensive measurements are carried out, selective measurements in the actual case study area, CFD modelling using ENVI-met calibrated by the physical and scale model measurements, ray modelling using Radiance and laboratory characterisation of the surface materials of the urban case study.

Fig. 5.17 1:10 scale model constructed at the University of Kent and full-scale spot measurements at the case study area in London

Figure 5.17 shows photos of the scale model constructed at the University of Kent, UK, and full-scale measurements taken at the case study area. The scale model represents a real residential area in London as indicated in Fig. 5.18.

Results to date from the scale model indicate that consistently with previous studies, the observed albedo peaks occurred at lower sun angles during morning and late afternoon hours in summer months. Therefore, within the same geometry, albedo was measured to peak at low irradiance (50–150 W/m^2). As expected increase of paving reflectivity and wall reflectivity increased the urban albedo for the same geometry. This was observed mainly in dry conditions for radiation intensity up to 350 W/m^2 with marginal increase at higher irradiances. The observed increase of albedo due to the application of paving and facade colour disappears in rainy days, thus highlighting the dynamic nature of albedo. This is a significant observation for climates with frequent rainfall such as in London (Nikolopoulou et al. 2020).

Results from field spot measurements in the three urban canyons (as shown in Fig. 5.18) indicate no change in the urban albedo for different heights as shown in

Fig. 5.18 Residential area in London where measurements were carried out and the 1:10 scale model is based on. Study by Salvati and Kolokotroni (2020a), aerial views from Google Earth

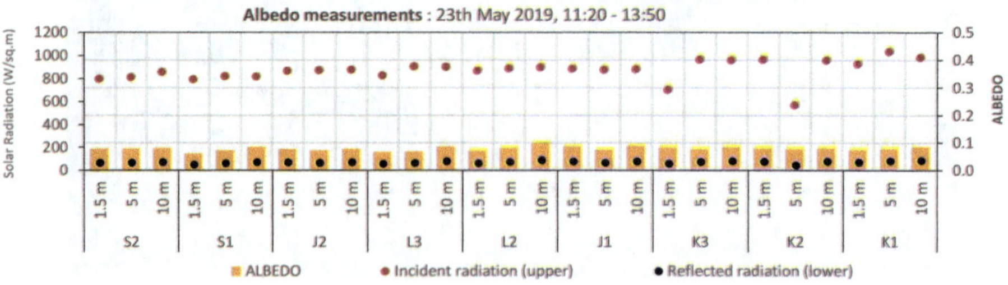

Fig. 5.19 Measured incoming (incident) and outgoing (reflected) radiation and urban albedo values within urban canyons in the case study area (Salvati and Kolokotroni 2020a)

Fig. 5.19 (Salvati and Kolokotroni 2020a). The three canyons have the same *H*/*W* ratio of approximately 0.75 but different orientation and material distribution. The radiation measurements showed that, for a datum geometry, the reflections are not influenced by street orientation and material distribution. Also, the measured reflections did not vary with measurement height, being approximately the same value at the street level (1.2 m height), second-floor height (5 m) and eaves level (10 m). Therefore, the measured urban albedo, namely the ratio of the measured outgoing to the incoming radiation, showed very small variation. This can be explained by the dominant role played by the canyon geometry in trapping solar radiation and the low albedo of the road material (asphalt) in comparison to the albedo and the orientation of the facades.

ENVI-met was calibrated using the scale model and field measurements (Salvati and Kolokotroni 2019) and was used to study the impact of urban albedo changes on microclimate and outdoor thermal comfort. These results were first reported by Salvati and Kolokotroni (2020a, b). The canyon aspect ratio in the case study area is approximately 0.75 (building height equal to 12 m and street width equal to 16 m). In this kind of geometry, at London's latitude, the simulation results show that increasing the wall albedo does not improve the outdoor thermal comfort while increasing the road albedo has a significant positive impact on the physiological equivalent temperature (PET) at the street level which is used for assessing outdoor thermal comfort (Hoppe 1999). This result is explained by the impact of the roads' and facades' albedo change on the air temperature and mean radiant temperature at street level, as reported in Fig. 5.20. The trends reported in the figure correspond to point S2 (see Fig. 5.18), but they are similar in all the canyons. In terms of air temperature, changing the albedo of facades does not influence air temperature. A

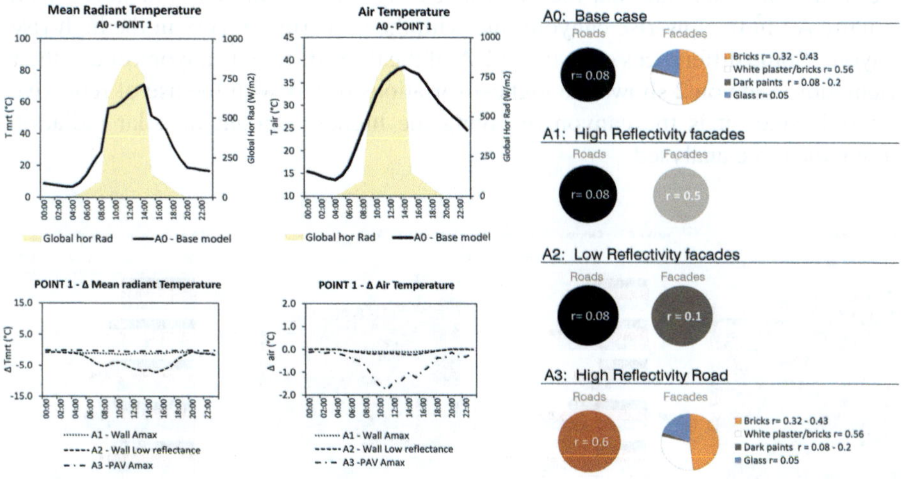

Fig. 5.20 Impact of roads and facades' material reflectivity on air temperature and mean radiant temperature at street level compared to the current scenario (base case) (source: Salvati and Kolokotroni 2020a)

significant decrease of air temperature is instead obtained with the increase in road albedo (scenario A3). On the other hand, for the studied geometry, increasing the reflectivity of facades (scenario A1) has a negligible impact also on the mean radiant temperature. Conversely, a substantial decrease in facade albedo (scenario A2) allows a reduction of the mean radiant temperature at the street level.

Similar results were also reported in other latitudes (Alchapar and Correa 2016; Erell et al. 2014) and can be explained by the impact of the walls' albedo on the multiple reflection of solar radiation within urban canyons. Similar results were found by Levinson (2019), who showed that the wall's solar availability in an urban context may increase up to 30% in deep urban canyons when the albedo of the neighbouring building is raised from 0.25 to 0.6.

The impact of canyon aspect ratio (*H*/*W*) and changes of reflective materials on UA was studied. The results are shown in Fig. 5.21 (Salvati and Kolokotroni 2020b). The reflective materials have a larger impact on low-rise aspect ratio canyons compared to high-rise ones. This is particularly evident for the scenario A2—reflective roads—which determines a much higher relative increase of UA in the canyons with aspect ratio of 0.75 compared to the ones with aspect ratio of 1.5. However, when the reflective materials are applied to the top half of the canyon's facades (A1 TOP), the two canyon geometries show a similar relative and absolute variation of the UA.

This might be explained by the fact that the top half of the canyon facades receives a similar amount of solar radiation despite the different canyon geometry. The results presented in Fig. 5.21 indicate that the two canyon geometries have less variation of the UA for the base case materials compared to variations for other cases examined. However, this similarity does not result in the same microclimate at the street level, because the increase of building height determines a significant reduction of the solar radiation reaching the street level in the high-rise urban area. The results also indicate that the highest relative increase of UA is achieved with scenario A2 in the low-rise canyons and with the scenario A1 TOP in the high-rise canyons. Some differences exist in the UA of the three different canyons due to their orientation. Canyon 2 shows the highest variations of UA with the use of reflective materials since it is the canyon receiving the highest amount of solar radiation among the three analysed.

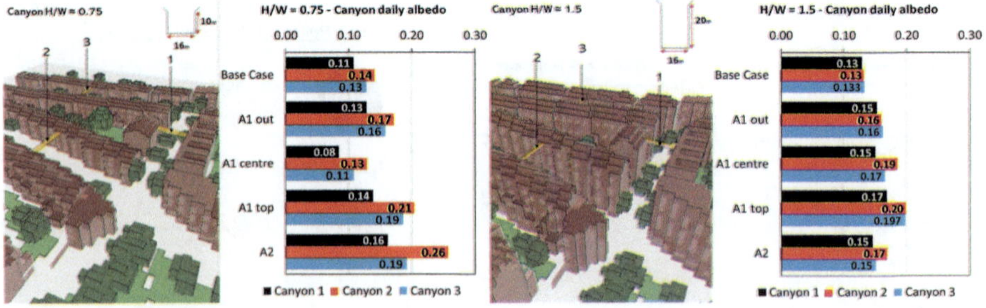

Fig. 5.21 Impact of the reflective material scenarios on the three canyons' daily urban albedo (source: Salvati and Kolokotroni 2020b)

5.4 Modelling Tools: Impact of Microclimate on Indoor Thermal Conditions

Approaches have been proposed to include local climate in the energy performance simulations of buildings in urban contexts. However, in most proposed approaches, only a subset of the urban climate modifications are included in the analysis. A recent study carried out by the authors in collaboration with two other authors (Salvati et al. 2020) showed that in a dense urban context, building energy performance is significantly affected by urban air temperature but also by urban wind speed, solar access and surface temperature of surrounding buildings. The methodology proposed by Salvati et al. (2020) is based on the use of a suite of climate tools, including the *Urban Weather Generator* (Bueno et al. 2013; Mao et al. 2017), to calculate urban air and surface temperatures and empirical models to estimate wind speed in urban canyons. The calculation of the canyon wind speed was based on the URBVENT project approach (Ghiaus et al. 2004), applying different empirical models for different urban situations, depending on the geometry and orientation of the canyon and the wind speed and direction at the meteorological station. This modelling approach is suitable for building in dense urban areas, characterised by urban canyons with high height-to-width ratio. In less dense suburban areas, the impact of atmospheric and surface heat island is lower. However, the distribution of wind speed at the microscale may affect the thermal performance of naturally ventilated buildings also in suburban locations.

At the microscale, CFD microclimate models such as ENVI-met (Huttner and Bruse 2009) allow to simulate the spatial and temporal distribution of the airflow around buildings, considering the actual three-dimensional geometry of the building and its surroundings. A case study in London is presented as an example of the potential impact of the microscale airflow on the indoor thermal conditions of buildings.

5.4.1 Case Study

The case study is a student accommodation building located at the campus of Brunel University London (Fig. 5.22). This case study is part of the H2020 project ReCO2ST (2020) and was used to study the effect of microclimatic conditions on energy simulations. ENVI-met V4.4.4 was used to run simulations in full forcing mode, namely using local hourly values of air temperature, relative humidity, wind speed and direction as input inflow boundary to the three-dimensional CFD atmospheric model, increasing the model accuracy significantly. ENVI-met simulation outputs include the spatial and temporal distribution of wind speed and directions at the building facades. These were used as boundary conditions for dynamic thermal simulations of the building using EnergyPlus, to investigate their impact on the building air change per hour (ACH) and the indoor thermal conditions.

The microclimate simulations confirmed that the wind speed and direction at the building facades are significantly modified by the presence of the nearby trees and buildings, for any wind direction (Fig. 5.23). The staggered shape of the building also plays an important role, determining significant variation in the wind speed and pressure on facades with same orientation.

These form-driven effects on airflow and consequent wind pressure on facades are normally neglected in building performance simulation, unless for tall or special buildings. The normal approach for the calculation of wind pressure on facades is based on Bernoulli's equation, considering the wind pressure coefficients and the wind speed at the building height, by applying the power law profile to the meteorological wind speed. The main weaknesses of this procedure consist of (1) the simplified building forms for which pressure coefficients are available and (2) the calculation of the local wind speed based on the power law profile, which is not valid in the presence of large obstacles such as buildings and trees. The inaccuracy of the ventilation potential due to this approximation can be relevant, since the wind pressure calculation is proportional to the second power of the wind speed.

To assess the impact of the microclimatic wind speed conditions in comparison to the meteorological weather file, the ENVI-met outputs were used to calculate the

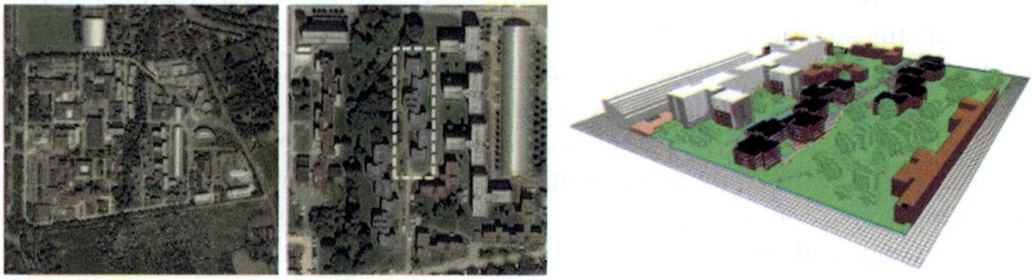

Fig. 5.22 Aerial view of the case study building and corresponding ENVI-met model

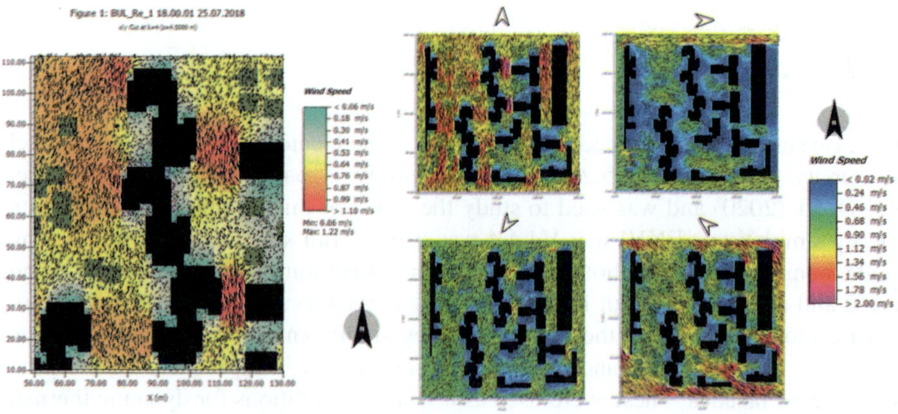

Fig. 5.23 ENVI-met results of the airflow at 4.5 m above ground level around the case study building for different dominant wind directions

wind pressure at each opening in the EnergyPlus simulations. The results showed significant reduction of the ventilation rates in some of the rooms (up to −35% in comparison to the standard pressure coefficient approach). In summer, the reduction of the ventilation rates has a clear negative impact on indoor thermal comfort, increasing the operative temperatures and overheating hours. The maximum increase of the indoor operative temperature due to reduced ventilation rates reached 3.7 °C at night, when the outdoor temperature decreased, and the natural ventilation had the maximum cooling potential. Larger discrepancies in the building ventilation rates were found for higher meteorological wind speeds.

These results confirm that the selection of a suitable weather file and microclimatic conditions is essential for more accurate predictions of internal thermal comfort and will assist in the sizing of passive and active systems to avoid overheating.

5.5 Chapter Summary

This chapter gave an outline of urban microclimate characteristics focusing on locations in high latitudes and its impact on comfort and energy demand. Throughout the chapter, London has been used as an example of a high-latitude location; results from various studies in London over the last two decades are presented including recent studies of urban albedo.

The relationship of surface and air UHI in high-latitude cities was described based on available literature at global scale. Energy demand is mostly affected by microclimatic air temperature while comfort is also affected by radiant temperature and wind patterns. In London, the average annual UHI is 2.5 °C at night but there is wide variation including cool islands mainly during the day. It was found that night UHI is independent of daytime air temperature during the summer; during the night similar difference between urban and rural air temperature was measured during a typical and hot week. A computational study was carried out on the impact of UHI using current and future weather files on energy demand and overheating for office buildings. As expected, energy demand for cooling is higher within the UHI. It was shown that offices with medium internal heat gains without night free cooling have similar cooling demand to offices with high heat gains and night free cooling. In the future, and because of higher night temperature due to the combined effect of climate change and UHI, offices with medium internal heat gains perform better than those with night free cooling. The study also concluded that as expected heating demand is lower in offices within the UHI and even lower in the future. Mitigation strategies are outlined with the suggestion that improving the external thermal environment is the main outstanding issue where further research is needed.

Of the parameters affecting external thermal environment vegetation and surface albedo have potentially a pronounced impact on the microclimate and were explored further. Measurements in London have shown that air temperatures are lower within the UHI in areas with vegetation. The effect of surface albedo was also measured indicating that reflective surfaces on the ground and on roofs can improve local microclimate by reducing surface temperatures. The effect of building facade albedo was studied in central London showing that dark-coloured facades result in higher surface temperatures, but overshadowing effects were clear.

Studies of urban albedo indicate that the contribution can be grouped into five interrelated parameters: (a) geometry—the geometry of individual buildings, the building group geometry and the geometry of buildings in relation to roads, usually referred to as urban canyon geometry; (b) reflectance of surfaces and their orientation—ground, facades and roofs; (c) sun altitude and azimuth determining the incidence angle of incoming solar radiation with strong seasonal contrasts in high latitudes; (d) moisture of permeable surfaces (soil) and altering reflectivity of impermeable surfaces; and (e) vegetation acting as obstruction to the incoming radiation and reflection paths. Of these (a) and (b) are not related to climate and latitude while (c), (d) and (e) are closely related to climate. Models have been developed to predict urban albedo mainly based on computational analysis with the aim to predict the impact on energy demand by buildings which have developed indices for these parameters such as the canyon aspect ratio which affects the amount of solar radiation incident on the canyon surfaces, building density which is the ratio of the buildings' footprint area to the urban site area and ratio of the vertical surface area that defines the facade density. Results of measurements and computational analysis for an area in London and their impact on urban albedo have been presented.

Approaches to include local microclimate in the energy performance simulations of buildings in urban contexts have been presented in Sect. 5.4 based on the use of a suite of climate tools such as Urban Weather Generator, calculation of the canyon wind speed by applying different empirical models and ENVI-met which allows to simulate the spatial and temporal distribution of the airflow around buildings, considering the actual three-dimensional geometry of the building and its surroundings. A case study in London was presented as an example of the potential impact of the microscale airflow on the indoor thermal conditions of buildings. The results showed significant reduction of the ventilation rates in some of the internal spaces resulting in a clear negative impact on indoor thermal comfort, increasing the operative temperatures and overheating hours.

Acknowledgement This work was funded by EPSRC, UK, under the project 'Urban albedo computation in high latitude locations: An experimental approach' (EP/P02517X/1).

References

Akbari, H., & Kolokotsa, D. (2016). Three decades of urban heat islands and mitigation technologies research. *Energy and Buildings, 133*, 834–842.

Alchapar, N. L., & Correa, E. N. (2016). The use of reflective materials as a strategy for urban cooling in an arid "OASIS" city. *Sustainable Cities and Society, 27*, 1–14. https://doi.org/10.1016/j.scs.2016.08.015.

Bueno, B., Norford, L., Hidalgo, J., & Pigeon, G. (2013). The urban weather generator. *Journal of Building Performance Simulation, 6*(4), 269–281. https://doi.org/10.1080/19401493.2012.718797.

Chen, D., & Weiteng, H. (2013). Using the Köppen classification to quantify climate variation and change: An example for 1901–2010. *Environmental Development, 6*, 69–79. https://doi.org/10.1016/j.envdev.2013.03.007].

Erell, E., Pearlmutter, D., Boneh, D., & Kutiel, P. B. (2014). Effect of high-albedo materials on pedestrian heat stress in urban street canyons. *Urban Climate, 10*, 367–386.

Ghiaus, C., Allard, F., Georgiakis, C., Santamouris, M., Roulet, C.-A., Germano, M., & Tillenkamp, F. (2004). URBVENT WP1 final report: Soft computing of natural ventilation potential. In *URBVENT: Natural ventilation in urban areas-potential assessment and optimal façade design. European commission*. London: Routledge. https://doi.org/10.4324/9781849772068.

Hoppe, P. (1999). The physiological equivalent temperature—A universal index for the biometeorological assessment of the thermal environment. *International Journal of Biometeorology, 43*, 71–75.

Huttner, S., & Bruse, M. (2009). Numerical modeling of the urban climate—A preview on ENVI-MET 4.0, in *The seventh International Conference on Urban Climate. Yokohama, Japan*. Retrieved from http://www.ide.titech.ac.jp/~icuc7/extended_abstracts/pdf/382880-2-090514010851-002.pdf.

Kolokotroni, M., & Giridharan, R. (2008). Urban Heat Island Intensity in London: An investigation of the impact of physical characteristics on changes in outdoor air temperature during summer. *Solar Energy, 82*, 986–998.

Kolokotroni, M., Giannitsaris, I., & Watkins, R. (2006). The effect of the London Urban Heat Island on building summer cooling demand and night ventilation strategies. *Solar Energy, 80*(4), 383–392.

Kolokotroni, M., Ren, X., Davies, M., & Mavrogianni, A. (2012). London's urban heat island: Impact on current and future energy consumption for heating and cooling. *Energy and Buildings, 47*, 302–311.

Kolokotroni, M., Gowreesunder, B., & Giridharan, R. (2013). Cool roof technology in London: An experimental and modelling study. *Energy and Buildings, 67*, 658–667. https://doi.org/10.1016/j.enbuild.2011.07.011.

Levinson, R. (2019). Using solar availability factors to adjust cool-wall energy savings for shading and reflection by neighboring buildings. *Solar Energy, 180*, 717–734. https://doi.org/10.1016/j.solener.2019.01.023.

Maciel, C. R., & Kolokotroni, M. (2017). Cool materials in the urban built environment to mitigate heat islands: Potential consequences for building ventilation. *38th AIVC International Conference, 13–14 Sep 2017, Nottingham, UK*.

Maciel, C. R., Kolokotroni, M., Nogueira, M. C. J. A., Giridharan, R., & Watkins, R. (2013). The impact of surface characteristics on ambient temperature at urban micro-scale: Comparative field study in two climates. *International Journal of Low Carbon Technologies*. https://doi.org/10.1093/ijlct/ctt016.

Mao, J., Yang, J. H., Afshari, A., & Norford, L. K. (2017). Global sensitivity analysis of an urban microclimate system under uncertainty: Design and case study. *Building and Environment, 124*, 153–170. https://doi.org/10.1016/j.buildenv.2017.08.011.

Nikolopoulou M, et al. (2020). Developing an Urban Albedo Calculator for London: The experimental campaign supporting the development of the tool. *CIBSE Technical Symposium. September 2020*.

Pawlak, W., & Fortuniak, K. (2003). Application of the physical model to study effective albedo of the urban canyon, in: *Fifth International Conference on Urban Climate, Lodz, Poland*.

Peng, S., Piao, S., Ciais, P., Friedlingstein, P., Ottle, C., Breon, F. M., Nan, H., Zhou, L., & Myneni, R. B. (2012). Surface urban heat island across 419 global big cities. *Environmental Science & Technology, 46*(2), 696–703.

Qin, Y. (2015). Urban canyon albedo and its implication on the use of reflective cool pavements. *Energy and Buildings, 96*, 86–94.

ReCO2ST. (2020). *Residential Retrofit assessment platform and demonstrations for near zero energy and CO_2 emissions with optimum cost, health, comfort and environmental quality.* https://reco2st.eu/

Salvati, A., & Kolokotroni, M. (2019). Microclimate data for building energy modelling: Study on ENVI-Met forcing data. In V. Corrado, E. Fabrizio, A. Gasparella, & F. Patuzzi (Eds.), *Proceedings of the 16th IBPSA Conference 2019, 2–4 September 2019, Rome, Italy* (pp. 3361–3368). Rome: IBPSA. https://doi.org/10.26868/25222708.2019.210544.

Salvati, A., & Kolokotroni, M. (2020a). Impact of urban albedo on microclimate and thermal comfort over a heat wave event in London. In S. Roaf, F. Nicol, & W. Finlayson (Eds.), *WINDSOR 2020: Resilient Comfort. Proceedings. Cumberland Lodge, 16–19 April 2020, Windsor UK* (pp. 566–578). London: WINDSOR.

Salvati, A. and Kolokotroni, M. (2020b). Impact of urban albedo on microclimate: Computational investigation in London. *35th PLEA Conference, sustainable architecture and urban design, planning post carbon cities, A Coruna, Spain, 1–3 September 2020.*

Salvati, A., Palme, M., Chiesa, G., & Kolokotroni, M. (2020). Built form, urban climate and building energy modelling: Case-studies in Rome and Antofagasta. *Journal of Building Performance Simulation, 13*, 209–225. https://doi.org/10.1080/19401493.2019.1707876.

Santamouris, M. (2007). Heat Island Research in Europe: The state of the art. *Advances in Building Energy Research, 1*(1), 123–150. https://doi.org/10.1080/17512549.2007.9687272.

Santamouris, M. (2015). Analyzing the heat island magnitude and characteristics in one hundred Asian and Australian cities and regions. *Science of the Total Environment, 512–513*, 582–598.

Santamouris, M. (2016). Innovating to zero the building sector in Europe: Minimising the energy consumption, eradication of the energy poverty and mitigating the local climate change. *Solar Energy, 128*, 61–94.

Santamouris, M. (2020). Recent progress on urban overheating and heat island research. Integrated assessment of the energy, environmental, vulnerability and health impact. Synergies with the global climate change. *Energy and Buildings, 207*, 109482.

Urban Albedo Project. (2019). https://research.kent.ac.uk/urbanalbedo/.

Watkins, R., Palmer, J., Kolokotroni, M., & Littlefair, P. (2002a). The London Heat Island—Results from summertime monitoring, Proc. Chartered Institution of Building Services Engineers, Series A. *Building Services Engineering Research and Technology, 23*(2), 97–106.

Watkins, R., Palmer, J., Kolokotroni, M., & Littlefair, P. (2002b). The London Heat Island—Surface and air temperature measurements in summer 2000. *ASHRAE Transactions, 108*(Pt1), 419–427.

World Atlas. (2020). *Köppen climate classification system.* https://www.worldatlas.com/articles/what-is-the-koppen-climate-classification-system.html.

Xu, X., Azari Jafari, H., Gregory, J., Norford, L., & Kirchain, E. (2020). An integrated model for quantifying the impacts of pavement albedo and urban morphology on building energy demand. *Energy and Buildings, 211*, 109759.

Yang, X., & Li, Y. (2015). The impact of building density and building height heterogeneity on average urban albedo and street surface temperature. *Building and Environment, 90*, 146–156.

Zhou, D., Zhao, S., Liu, S., Zhang, L., & Zhu, C. (2014). Surface urban heat island in China's 32 major cities: Spatial patterns and drivers. *Remote Sensing of Environment, 152*, 51–61.

Chapter 6
Urban Climate and Building Energy Performance in Compact Cities in Mediterranean Climate

Agnese Salvati and Helena Coch

6.1 Managing Urban Climate in Mediterranean Compact Cities

The shape, structure and colours of many Mediterranean villages, towns and cities reveal an old historical link between climate, architecture and built form in this region. The Mediterranean basin is gifted with one of the most hospitable climates on earth. Mild winters, sunny summers, good amount of precipitation and well-marked seasonal variability allow for a wide range of possibilities for people to thrive. In this context, historic city centres as well as old villages show recurrent characteristics such as high inertia constructions, use of local "cool" materials like light-colour stone and white paints, facades with low window-to-wall ratios and flexible shading systems to control solar radiation and ventilation in summer and winter. The urban fabric is very compact and continuous, and the most common building typology is the courtyard type (Fig. 6.1). The arrangement of buildings and urban blocks generates a dense network of public street canyons and inner courtyards for multipurpose outdoor activities.

All these are characteristic attributes of the "compact Mediterranean city", where compactness here refers to one specific physical characteristic, namely the ratio of building footprints to urban site area (known also as "plan area density" or "site

A. Salvati (✉)
Department of Mechanical and Aerospace Engineering, Institute of Energy Futures—Resource Efficient Future Cities, Brunel University London, London, UK
e-mail: agnese.salvati@brunel.ac.uk

H. Coch
AIEM Architecture Energy and Environment, Barcelona School of Architecture, Polytechnic University of Catalunya, Barcelona, Spain

M. Palme, A. Salvati (eds.), *Urban Microclimate Modelling for Comfort and Energy Studies*, https://doi.org/10.1007/978-3-030-65421-4_6

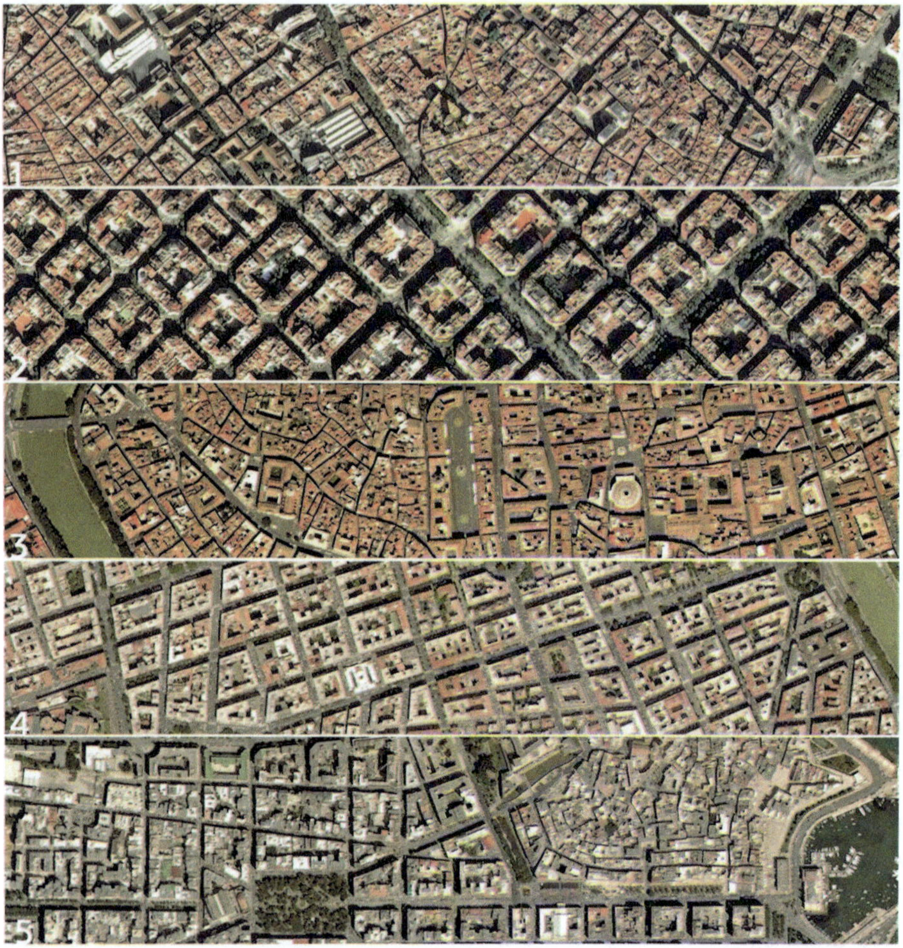

Fig. 6.1 Aerial views of typical urban textures of Mediterranean cities, showing the compactness of the fabric and the recurrent use of courtyard building typologies. (*1*) Barcelona Medieval city centre (Ciutat Vella). (*2*) Barcelona nineteenth-century Plan Cerda development (Eixample). (*3*) Rome Medieval city centre (Campo Marzio). (*4*) Rome twentieth-century development (Prati). (*5*) Medieval core and nineteenth-century development (Murat) in Bari, Italy. (Map data: Google Earth, Landsat/Copernicus)

coverage ratio[1]). Compactness indicates how close the buildings are in the urban fabric. In Mediterranean cities, the ratio of building footprint to the urban site area is normally higher than 0.4 but can be as high as 0.8, meaning that more than 40%

[1] Refer to the chapter "Spatial Metrics to Investigate the Impact of Urban Form on Microclimate and Building Energy Performance: An Essential Overview" by M. Morganti for a complete overview of density parameters (Chap. 18).

and up to 80% of the urban space is occupied by buildings (Morganti 2018). High compactness generates narrow street canyons that provide protection from the excess of solar radiation in summer and from strong winds in winter, improving outdoor and indoor thermal comfort throughout the year. Similarly, at the block and the building scale, compactness is achieved with courtyard building typologies that allow the best control of natural energy resources such as sun and wind to improve indoor thermal comfort throughout the year (Hsie 2008; Natanian et al. 2019; Rojas-Fernández et al. 2017).

In the Mediterranean context, as in any other climate region, the shape and morphology of vernacular architecture show that urban and building form can be designed to create an "artificial climate" that works better than the outdoor climate, protecting people from low and high temperatures, rain, strong solar radiation and winds without the use of active energy systems. Using Rafael Serra's words, a building is indeed "*an artificial refuge, as an island of calm in an un uncomfortable world*[2]" (Serra 1999). Ironically, today we refer to another kind of "island" in relation to climate, the "urban heat island" (Oke 1987; Oke et al. 2017) which is not a refuge, but instead a climate threat to the health and liveability of urban areas. The urban heat island (UHI), namely the increase of temperature in urban areas compared to the rural (or less urbanised) surroundings, is also a form of "artificial climate" resulting from the way buildings and cities are designed. However, this climate modification is not intentional, and, in most of the cases, it is not of help in creating sustainable, healthy and resilient cities.

The increase of temperature in urban areas is strictly linked to the nature of cities themselves, which aims at bringing together people and activities in one place. This needs a substantial modification of the environment, with a substitution of natural surfaces with impervious materials for the efficient run-off of water from precipitation and, above all, the concentration of buildings and functions and consequent energy consumption for climatisation and transportation. All these factors contribute to the increase of surface and air temperature in urban areas.[3]

In the Mediterranean context, the urban heat island intensity may have both positive and negative impact depending on the season. An increase of temperature is advantageous in winter, helping reducing the heating loads of buildings and improving thermal comfort both indoors and outdoors. In summer, the UHI is instead responsible for a rise in building energy demand for cooling and an increase of heat-related health diseases. Due to the intensity of solar radiation, the UHI is normally stronger in summer than in winter in this context, resulting in a net negative impact on the annual energy load of urban buildings (Salvati et al. 2015).

---

[2] "Los edificios son barreras a la lluvia, al viento y, a veces, filtros sutiles a la luz y al calor. Rodeados de entornos variables, donde cambian el día y la noche, el calor y el frio, el viento y la calma, la lluvia y el sol; se convierten en refugios de artificiales condiciones, como islas de tranquilidad en un mundo incómodo". "Arquitectura y climas" (Serra 1999) p. 7.

[3] A detail explanation of the causes and the characteristics urban heat islands in cities can be found in the chapter "The Energetic Basis of the Urban Heat Island" by G. Mills, J. Futcher and I.D. Stewart (Chap. 3).

In the Mediterranean region, the issue of heat in cities and buildings is becoming more and more alarming in light of global climate change, which is predicted to have strong impact in this region, with significant increase in temperature, decrease in precipitation in summer and increase in frequency, intensity and duration of extremely hot periods known as "heatwaves" (Bastin et al. 2019; IPCC 2014). Therefore, the UHI is a local climate modification that amplifies the negative impact of climate change in urban areas, increasing the exposure and vulnerability of urban population. Studies have shown that even small changes in the average air temperature above certain thresholds may determine an exponential increase in mortality, especially for vulnerable population groups like older people or people with chronic diseases. This threshold temperature depends on the background climate; in Rome, for instance, it is around 27.5 °C (Gasparrini et al. 2015), which is a typical mild daytime summer temperature for this context. In rural areas, the night-time summer temperature normally falls below this threshold even in the hottest days. Conversely, the air temperature may stay above this threshold for many consecutive days in urban areas, due to the night-time UHI intensity.

Considering the high level of urbanisation, the increasing ageing of population and the prediction of an increase in frequency and magnitude of heatwaves, the risk of premature heat-related deaths is considered one of the major threats to public health in the Mediterranean region in the next years (Linares et al. 2020). The risk is obviously higher in cities with strong UHI intensity, where mitigation and adaptation plans are imperative to protect the citizens' health and well-being and to improve the sustainability of urban areas. For these reasons, the need to effectively integrate urban climate knowledge and modelling into planning and design has never been so urgent to guide the policies and to assess the performance of existing and new buildings and urban spaces.

This chapter intends to provide an overview of the characteristics of urban climate in Mediterranean cities, discussing experimental and numerical studies investigating the UHI intensity in some representative cities in this region. Particular attention is posed on the strong relationships between urban form, urban climate modifications and impact on thermal comfort and building energy demand in typical urban fabric of compact Mediterranean cities. Case studies in Rome and Barcelona are presented in order to highlight the complexity of the phenomena involved at different scales and with different points of view (i.e. outdoor thermal comfort or building energy efficiency). In the last section, design strategies for "heat management" at the urban and building scale are discussed.

6.2 Designing with Climate in a Mediterranean Context

The most characterising feature of the Mediterranean climate is its complexity. The weather conditions change rapidly, over the year and over the day, and both *cold* and *heat* are a problem. Consequently, the design solutions that must be applied in architecture in this region are similarly complex (Coch 1998; Serra 1999).

The variable climatic conditions require the local architecture to have great flexibility in order to cope with periods of both excessive cold and oppressive heat. Moreover, since the changes occur rapidly and at any time of the year, the buildings need to be able to adapt quickly as well. In very short time, the external conditions can change from dry heat to humid heat, from calm to strong wind or from drought to downpours. Therefore, the protection from excessive heat is just one of the functions of buildings in this region, even though one of the more complex to achieve, especially in cities (Coch and Serra 1996; Salvati et al. 2017b).

In Mediterranean climates, the problem of overheating in buildings is a function of solar radiation, humidity and air temperature, in this order.

Intense solar radiation for long periods not only contributes to increased air temperature but, more importantly, also heats the masses of urban structures, which normally consist of heavy masonry buildings in this area. As a result, it is common to find worse environmental conditions indoors than outdoors during some periods of the year. In summer and autumn, for instance, the indoor environment can be very hot due to high internal mean radiant temperatures, even if the air temperature is not particularly high. These are the paradoxical cases in which the saying that "some buildings work worse than the climate" is unfortunately true (Coch 1998).

A similar phenomenon occurs with the humidity of the air. Frequently, although not always nor in all locations, conditions along the Mediterranean coasts are characterised by high humidity combined with relatively high temperatures. It is widely known that thermal comfort zones get narrower as humidity increases, in a way that sometimes one can go from feeling cold to feeling hot without achieving an intermediate comfortable perception. The control of humidity, which we consider to be underestimated in conventional comfort studies, is in fact an important factor for comfort in buildings in such climates. Even in this case, it is true to say that the outside climate is often more comfortable than the interior one, where in many cases there are elements that release humidity.

Finally, the temperature of the air is also important, along with the above parameters. In conditions of dry heat, daytime temperatures in the Mediterranean regions can exceed 35 °C or even 40 °C. In such conditions, the exterior becomes uninhabitable and the interiors are cool refuges because the building inertia enables the temperature of the air to be controlled, although in the night-time that same air temperature becomes uncomfortable.

Considering the above aspects of the Mediterranean climate, the design of buildings should be aimed at controlling these three variables all together to improve the indoor thermal conditions over hot periods. Concerning the issue of excess of heat, each climate variable should be controlled with specific strategies as follows (and in this order).

6.2.1 Control the Entry of Solar Radiation

- Totally blocking the entry of direct sunlight through openings, particularly on the east and west facades and through the roof.

- Controlling and optimising the entry of diffuse radiation through openings on any facade, not only coming from the sky but also from exterior reflecting surfaces: In fact, any excess of light indoors can be detrimental in this climate in summer due to its thermal effect.
- Avoiding the entry of long-wave radiation, both originating from surrounding surfaces heated by the sun and entering the building through its openings and by the energy stored in the walls of the building itself: Both kinds can have noticeable impact at night, long after the sun has set.

These three strategies against radiation should be used together, but we can say that (2) and (3) are of little help if (1) is not implemented.

6.2.2 Control the Humidity of the Indoor Air

- Efficient, controllable indoor ventilation to expel the excess of humidity in humid heat conditions.
- Ventilation in combination with humidification of the air (evaporative cooling) in dry heat conditions.
- Controlled ventilation using previously cooled air, by means of underground ducts and/or conveying air from cooler exterior zones.

The strategies (1) and (2) are conceptually contradictory; therefore, it is essential to find flexible solutions in those cases in which both climatic phenomena are likely to occur.

6.2.3 Regulate the Temperature of the Indoor Air

- High inertia in the interior and the outside walls, in climatic conditions with elevated temperature fluctuations, usually in dry heat climates.
- Reduced ventilation during the hottest part of the day, particularly important in the same cases as in the point above.
- Insulation on the exterior side of walls, thus decreasing the transmission of heat from the outdoors to the indoors over the hottest hours of the day.

It has to be noted that this last group, and the last strategy in particular, is of secondary importance in comparison to the two previous groups of strategies. In Mediterranean climates, the sole control of air temperature would result in just negligible improvement of indoor thermal comfort without first having dealt with the direct, diffuse and long-wave radiation and the humidity issues. As well known, all these climate variables are profoundly modified in urban areas, due to the UHI effect and other urban climate modifications determined by urban morphology, such as solar access, radiation trapping and wind obstruction. Therefore, preliminary study and prediction of the urban microclimate conditions around urban buildings are crucial to develop successful architectural designs, adopting the most suitable passive technologies to improve indoor thermal comfort based on each site-specific climate boundary conditions.

6.3 Measuring and Modelling Urban Climate in Mediterranean Cities

6.3.1 Scale and Purpose of the Analysis

The analysis of urban climate and heat island intensity of an urban area can be carried out with different experimental and modelling techniques, depending on the purpose and the scale of interest. For the purpose of building energy and comfort studies, we are interested in knowing and modelling the "canopy-level urban heat island" (CUHI) and the "surface urban heat island (SUHI)".[4]

Regarding the CUHI, we ideally would need to know the vertical profile of air temperature and wind speed in urban canyons. In fact, the temperature in urban canyons is not uniform and generally warmer near the ground and lower at the top, due to thermal gradient, solar radiation trapping and wind obstruction. However, either measuring or modelling the temperature and wind profiles in urban canyons is not an easy task. Furthermore, building performance simulations (BPS) are performed using weather files with hourly values of the climate variables for the typical meteorological year for the location (normally based on measurements at the city's airport weather station). In order to improve the accuracy of the energy simulations for urban buildings, the "rural" (airport) air temperature can be modified using average canyon air temperatures. For this reason, the experimental studies aimed at measuring urban air temperature for energy performance analysis use sensors installed at approximately mid-height within urban canyons (Kolokotroni et al. 2006; Santamouris et al. 2001; Zinzi et al. 2018), assuming that this is the most representative point to capture the average temperature of urban canyons. Similarly, tools such as UWG,[5] developed for including UHI effects into weather files for building energy modelling, assume uniform air temperature in urban canyons, considering the canyon mid-height point to calculate the values.

Air temperature is less variable on the horizontal plane around a building, since it is the result of the average characteristics of the neighbourhood or, better, the

[4] The different types of urban heat island corresponding measurement methods are described in the chapter "The Energetic Basis of the Urban Heat Island" by Mills, Futcher and Stewart (Chap. 3).

[5] The model is described in the chapter "The Urban Weather Generator Model: Physics-Based Microclimate Simulation for Performance-Oriented Urban Planning" by J. Mao, L. Nordford (Chap. 12).

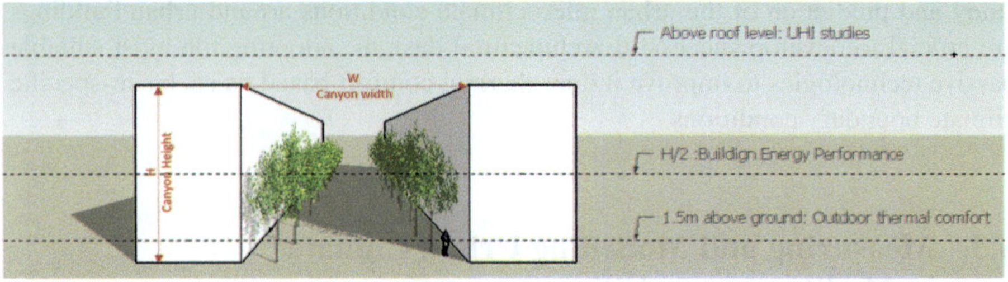

Fig. 6.2 Location of monitoring equipment within urban canyons to measure urban air temperature for different purposes such as the analysis of outdoor thermal comfort, building energy performance or local UHI intensity. (Image elaborated by the authors)

"local climate zone[6]" (Stewart and Oke 2012). As opposite to air temperature, other climate variables such as solar irradiation, surface temperature and wind speed are likely to vary significantly in very short distances, depending on the location within the urban canyon and the detailed three-dimensional geometry of buildings and other urban elements. These variables are crucial for analysing outdoor thermal comfort[7] and their temporal and spatial distribution needs to be carefully assessed using high-resolution microclimate models (i.e. ENVI-met, SOLENE-Microclimat[8] or RayMan-SkyHelios[9]) or using appropriate methods to determine the mean radiant temperature from multiple measurements (Johansson et al. 2014). Therefore, the experimental studies oriented to outdoor thermal comfort in urban canyons are carried out at the street level (1.5–2 m above ground level) and consider both the CUHI and the SUHI, which concur to determine the outdoor thermal comfort sensation.

The studies aimed at assessing the dynamics of the urban energy fluxes and the magnitude of the atmospheric urban heat island at the local and urban scales use equipment located above the roof level, in the roughness sublayer (Fig. 6.2).

In line with the topic of this book, the next sections will focus on representative experimental and numerical studies carried out in Mediterranean cities that provide useful knowledge and tools for improving building energy modelling and thermal comfort analysis in this region.

---

[6]The concept of the local climate zones is described in "The Energetic Basis of the Urban Heat Island" by Mills, Futcher and Stewart (Chap. 3).

[7]The complexity of outdoor thermal comfort assessments and the variables involved are described in the chapter "Thermal Comfort in Urban Spaces" by M. Nikolopoulou (Chap. 4).

[8]The model is described in the chapter "The SOLENE-Microclimate Model: Potentiality for Comfort and Energy Studies" by Musy et al. (Chap. 13).

[9]The model is described in the chapter "RayMan and SkyHelios Model" by Matzarakis et al. (Chap. 16).

Fig. 6.3 Cities where the UHI has been measured in the Mediterranean basin (map data: Google Earth, Landsat/Copernicus)

6.3.2 *Urban Heat Island and Microclimate Studies in Representative Mediterranean Cities*

In the last 20 years many experimental and numerical studies of the urban heat island intensity have been carried out in the Mediterranean basin (Fig. 6.3 and Table 6.1), reporting strong UHI intensity both at daytime and night-time and throughout the year. All the cities analysed are classified as Mediterranean climate in the Köppen-Geiger system (Kottek et al. 2006) and are located at latitudes between 43.7° N (Florence, Italy) and 31.2° N (Alexandria, Egypt).

6.3.2.1 Maximum Canopy-Layer UHI Intensity

Concerning energy and comfort studies, both the daytime and night-time UHI intensity are relevant, the former to assess the peak cooling loads of buildings occupied during the daytime (i.e. office buildings), and the latter to assess night cooling ventilation potential and indoor thermal comfort in residential buildings.

The studies carried out in Athens reported that the maximum absolute UHI intensity—namely the maximum observed air temperature difference between the urban fixed stations and the rural one—occurred during daytime, reaching more than 10 °C (Santamouris et al. 2001; Santamouris 2016); at night-time, the maximum UHI intensity is lower than 5 °C at all stations.

The UHI intensity was found to be higher during daytime also in Chania, Crete, reaching a maximum absolute value of 8 °C; in this coastal city, the measurements highlighted that the daytime UHI intensity is quite variable depending on wind speed, while the night-time UHI intensity is more stable, varying between 1.5 °C and 2 °C (Kolokotsa et al. 2009).

Table 6.1 Review of the methods used in UHI and microclimate studies carried out in representative Mediterranean cities

Country/city	Study	Method				
		Air temperature observations				
		Street level	Canyon mid-height	Above roof	Land surface temperature	Numerical study
Greece						
Athens	Santamouris et al. (2001)	×	×			
	Livada et al. (2002)		×			
	Mihalakakou et al. (2001)		×	×		× (data-driven model)
	Mihalakakou et al. (2004)		×			× (data-driven model)
	Giannopoulou et al. (2011)	×[a]	×[a]			
Thessaloniki	Giannaros and Melas (2012)		×			
Chanía (Crete)	Kolokotsa et al. (2009)		×			
Volos	Papanastasiou and Kittas (2012)		×			
Cyprus						
Nicosia	Theophilou and Serghides (2015)	×				
Italy						
Rome	Colacino and Lavagnini (1982)			×		
	Bonacquisti et al. (2006)	×				× (UHSM)
	Cantelli et al. (2011)			×		× (RAMS-LEAF3)
	Pelliccioni et al. (2012)	×		×		
	Cantelli et al. (2014)			×		× (RAMS-STEB)
	Zinzi et al. (2018)		×			
	Salvati et al. (2016)	×		×		× (UWG)
	Salvati et al. (2019)	×				× (UWG)
Bari	Martinelli et al. (2020)	×				

(continued)

Table 6.1 (continued)

Country/city	Study	Method				
		Air temperature observations				
		Street level	Canyon mid-height	Above roof	Land surface temperature	Numerical study
Florence	Petralli et al. (2011)	×				
Spain						
Barcelona	Moreno-Garcia (1994)	×		×		
	Salvati et al. (2017b)	×		×		
	Salvati et al. (2019)			×		× (UWG)
	Serra et al. (2020)				× (MODIS)	
	Martin-Vide and Moreno-Garcia (2020)			×		
Sevilla	Romero Rodríguez et al. (2020)	×		×[a]		
Israel						
Tel Aviv	Saaroni et al. (2000)	×		×		
Egypt						
Alexandria	Hassaan (2008)				× (LANDSAT)	
Algeria						
Constantine	Bourbia and Boucheriba (2010)	×				

[a]The sensor height is not specified in the reference paper

In Bari, the absolute maximum observed UHI intensity reached 6.6 °C in June and was higher than 4 °C for more than 30% of the summer days (Martinelli et al. 2020).

In Barcelona, the maximum UHI intensity measured at the roof level occurred at night-time, with an average monthly maximum of 2.8 °C in winter and 1.7 °C in summer; however at the street level, in summer, the UHI intensity occurred during both daytime and night-time, reaching a maximum intensity of 4.3 °C (Salvati et al. 2017b).

In Rome, a maximum UHI intensity close to 7 °C was detected during daytime in summer in two urban canyons in high-density neighbourhoods (Zinzi et al. 2018). In Sevilla a maximum UHI intensity of more than 7 °C was measured at 06:00 h in the city centre (Romero Rodríguez et al. 2020).

6.3.2.2 Seasonal Variability

Many studies highlighted the seasonal variability of the UHI intensity in Mediterranean cities.

Mihalakakou et al. (2001, 2004) reported the following variation of the maximum average daytime UHI in Athens: +7.5 °C in summer, +5.1 °C in autumn, + 3.7 °C in winter and +4.6 °C in spring. Similar results were found in Thessaloniki, where the maximum observed UHI varied between 2 °C and 4 °C in the warmer months and between 1 °C and 3 °C in the cold part of the year (Giannaros and Melas 2012).

The UHI intensity was found to be higher in summer than in winter also in Rome, where Colacino and Lavagnini (1982) showed that the mean temperature difference between the city centre and the rural area was +4.3 °C in summer and +2.5 °C in winter and a similar seasonal trend was reported more recently by Zinzi et al. (2018). The prevailing anticyclonic conditions for the Mediterranean region in the summer months can explain why the UHI intensity is generally stronger in summer than in winter in most of the cities.

As opposite to the previous cases, the UHI intensity was found to be stronger in winter than in summer in Barcelona. Studies carried out by Martin-Vide and Moreno-Garcia (2020), Moreno-Garcia (1994) and Salvati et al. (2017b) showed a higher UHI intensity in the winter months (+3–4 °C on average) compared to the summer months (+2–2.5 °C on average) in Barcelona. This is most probably due to the thermoregulatory effect of the sea, determining a very small daily temperature range (T_{max}–T_{min}) and reduced peak temperatures in summer (Fig. 6.4). Furthermore, these values are based on roof-level temperatures. A measurement campaign in street canyons showed high UHI intensity in summer at the street level also in Barcelona (Fig. 6.5), during both the daytime and the night-time.

6.3.2.3 Intra-urban Air Temperature Differences Due to Building Density and Land Use

Intra-urban temperature differences due to change in land cover, land use and building density were observed in all the cities.

In Athens, Livada and Giannopoulou (Giannopoulou et al. 2011; Livada et al. 2002) showed that the urban air temperature is lower in areas with high percentage of green areas and it is maximum in the dense city centre and in industrial areas with very low vegetation cover. The mean temperature difference between the hottest and the coolest spots across the city ranged between +3 °C and +5.3 °C during daytime and between +1.3 °C and +2.3 °C during night-time.

In Florence, cold and hot spots were observed across the city, with mean temperature difference of 2 °C and higher differences in summer and autumn (Petralli et al. 2011).

At the street level, in Barcelona, the canyon geometry determines significant air temperature differences from the sunset to the early hours of the morning, with differences up to 2.2 °C in favour of the narrower canyons (Fig. 6.5). The maximum daytime UHI intensity is reached in the low aspect ratio canyons (H/W ~ 0.3), while

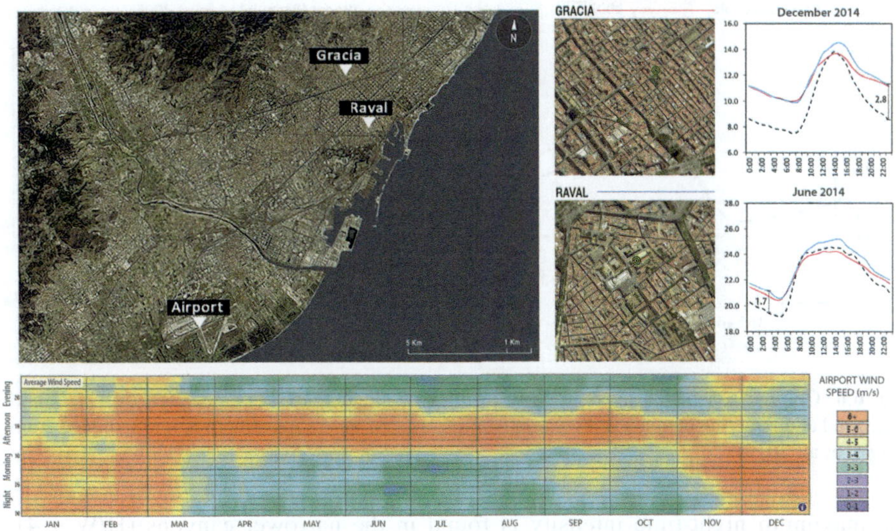

Fig. 6.4 Top: Daily cycle of air temperature in winter and summer months in two dense neighbourhoods of Barcelona; sensors located above roof level at the urban stations (adapted from Salvati et al. 2017b). Bottom: Wind speed at the airport site, showing the consistent effect of the sea breeze from April to October between 12 pm and 6 pm

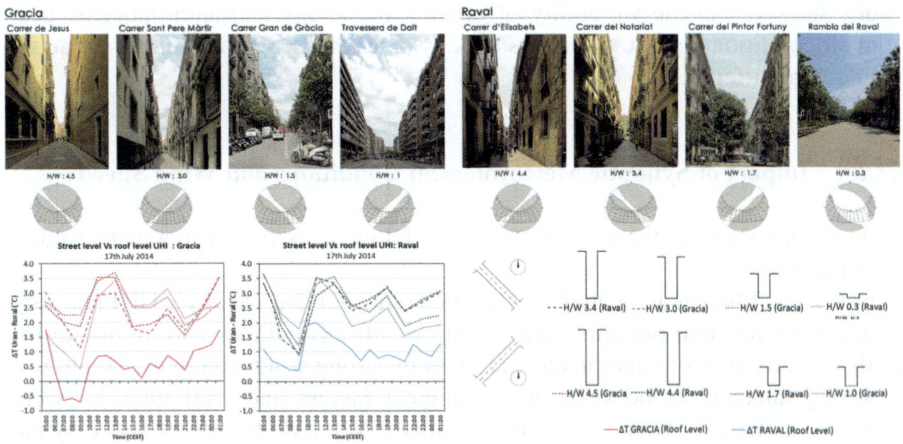

Fig. 6.5 Temperature difference in street canyons compared to the airport measurements at the same time during one hot summer day in Barcelona. (The field measurements have been published by Salvati et al. 2017b)

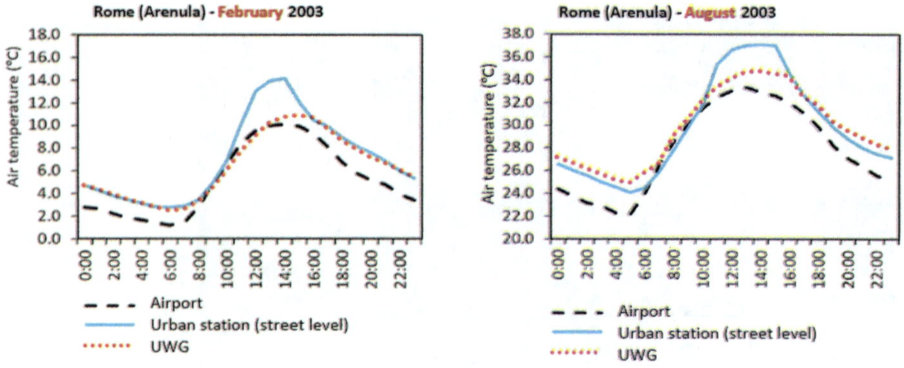

Fig. 6.6 Comparison of the average daily profiles of air temperature calculated by UWG and measured at a street level at a fixed weather station located in Via Arenula in Rome (adapted from Salvati et al. 2019)

the maximum night-time intensity is found in the narrower canyons (H/W ~ 4). Similarly, in Bari, higher daily maximum temperature is observed in the low-density urban areas while higher night-time air temperature in the denser urban areas (Martinelli et al. 2020).

Salvati et al. (2019) carried out a numerical study for Rome and Barcelona using the "Urban Weather Generator" model (Bueno et al. 2013; Mao et al. 2017), validated with air temperature observations from urban fixed stations (Fig. 6.6). The results showed that the variability of urban morphology and anthropogenic heat from air-conditioning systems across the city determines significant intra-urban air temperature differences.

6.3.2.4 Impact of Synoptic Meteorological Conditions and Wind Speed

Some studies analysed the variability of the UHI intensity for varying meteorological conditions.

In Athens, Mihalakakou et al. (2001, 2004) showed that the city's UHI intensity is maximum for high-pressure anticyclonic weather conditions and minimum for south-westerly flow circulation characterised by strong winds (more than 6 m/s).

Strong interactions between meteorological factors and UHI intensity were found also in Chania, Crete, due to the peculiar position of the city, bounded by the Aegean Sea to the North and the White Mountains (about 2000 m altitude) to the south. Here, Kolokotsa et al. (2009) reported that the UHI intensity varies significantly with wind speed and direction; the northern winds expand the UHI front, while the western winds contribute to the UHI mitigation in the city.

In Volos, a medium-size coastal town, the daily maximum UHI intensity was observed to be positively correlated with solar radiation and relative humidity in summer and negatively correlated with wind speed and relative humidity in winter (Papanastasiou and Kittas 2012). In Thessaloniki, the UHI intensity decreases significantly for wind speed higher than 4 m/s (Giannaros and Melas 2012).

In large coastal cities, the moderating effect of the sea is clear in the neighbourhoods close to the shoreline, where smaller diurnal air temperature variation are measured in comparison to more interior urban and rural locations; this was found in Tel Aviv (Saaroni et al. 2000), Bari (Martinelli et al. 2020) and Alexandria (Hassaan 2008). In Alexandria, the moderating effect of the sea on surface temperatures was clear up to about 300 m from the shoreline. Similarly, the moderating effect of the sea was found to vanish at few blocks' distance from the coast in Bari (Martinelli et al. 2020).

In Tel Aviv, the Mediterranean sea breeze (ranging between 3 and 6 m/s on the measurement days) was found to reduce the night-time summer temperatures at the roof level, but not at the street level. This result is similar to what was found in Barcelona, where the daytime summer street-level air temperature in central districts was up to 2 °C higher than the air temperature at the roof level in the same location (Salvati et al. 2017b). The influence of the sea and land breeze circulation on the vertical profile of the boundary layer above the metropolitan area of Rome has also been detected (Cantelli et al. 2014; Leuzzi and Monti 1997) as well as a significant temperature difference between the maximum daytime air temperature observed at roof level and street level in summer (Salvati et al. 2016).

6.3.3 Wind Speed in Urban Canyons

Much less experimental studies on the air circulation in street canyons have been carried out.

An important experimental campaign was carried out in Athens in the summer of 2001 by Georgakis and Santamouris (2005, 2006; Santamouris et al. 2008) for the URBVENT European research project (Ghiaus et al. 2005). The field study was aimed at developing simplified wind canyon models to assess the ventilation potential of buildings in urban context. The wind speed and direction at different points and at different heights within and above urban canyons with aspect ratio between 1.7 and 3.25 were performed. Based on the measurements, semi-empirical models were created to calculate the wind speed at different heights in street canyons, considering the direction and speed of the undisturbed wind and the geometry and orientation of the canyon. These models have been recently applied to assess the impact of urban context on the ventilation potential of buildings located in different areas across Rome (Salvati et al. 2020).[10]

[10] It has to be noted that the URBEVENT empirical models were applied to Rome on the assumption of similarities in climate and urban geometry with Athens, where the measurements were carried out. However, the application of simplified models to calculate the wind flow in cities is very limited, due to the many site-dependant variables involved and the complexity of urban fabric geometry. A complete overview of the topic is provided in the chapter "Air Circulation in Urban Areas" by Di Bernardino et al. (Chap. 10).

6.4 Urban Climate and Building Energy Performance

Many studies highlighted the negative impact of the UHI intensity on the energy demand of urban buildings in comparison to locations out of the city. The negative impact is due to the rise in cooling energy demand due to urban warmth, which is normally higher than the reduction of the heating demand in winter.

Santamouris et al. (2001) used Athens' urban air measurements to perform energy simulations using TRNSYS for an office building with cooling set point of 26 °C and compared the performance with respect to non-urban conditions. The simulations showed a rise of about 120% of the building monthly cooling load and a decrease of the heating load by 38% in the city centre with respect to the suburban location, resulting in a net negative impact of urban climate on building energy demand.

Based on 3-year measurements in four urban sites in Rome, Zinzi et al. (2018) showed that the heating degree days decrease by 18% while the cooling degree days increase by 157% in the city centre due to the urban heat island effect. This entailed a reduction of the heating loads by 18% for office buildings and by 21% for residential buildings, while the cooling loads increased by 53% and 74% for office and residential buildings, respectively. For non-cooled residential buildings, the UHI effect determined a significant increase of the overheating hours as compared to the same building in the countryside, due to the reduced cooling potential of night ventilation in urban context (Zinzi and Carnielo 2017). The study also showed that the cooling energy performance of non-insulated buildings is more influenced by urban climate as compared to insulated ones.

Although many studies analysed the impact of higher urban temperatures on the building energy performance, very few analysed the net energy impact considering other urban climate modifications. In fact, the air temperature increase in urban areas is just one of the urban climate modifications that affect building thermal performance. The complex geometry of the urban fabric determines different kinds of climate modifications, including variabilities in UHI intensity, obstruction of solar radiation and modifications of surface temperatures and wind speed in urban canyons (Fig. 6.7). All these climate modifications have an impact on the building energy performance, by modifying the indoor-outdoor heat transfer through the

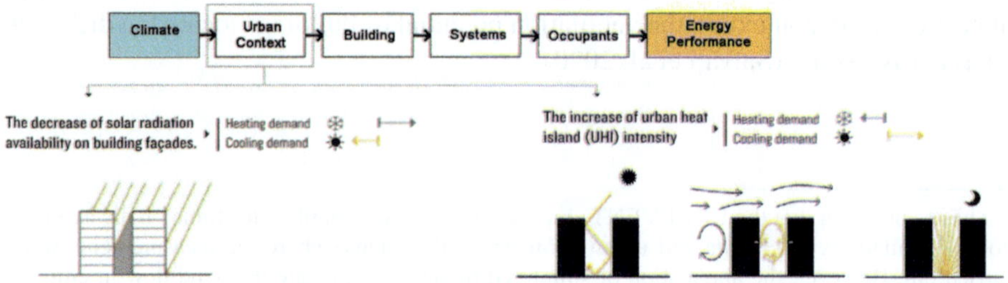

Fig. 6.7 Climate modifications determined by urban context and corresponding positive or negative impact on the cooling and heating energy performance of buildings. (Source: elaborated by the authors based on Ratti et al. 2005)

envelope, the solar gains, the ventilation and infiltration energy gains/losses and the infrared exchange in comparison to an open-rural environment.

Furthermore, these climate modifications have opposite energy impacts. The reduction of solar radiation on buildings' facades in a dense area has a positive impact on the sensible cooling loads and a negative one on the heating loads. At the same time, the air temperature and surface temperature increase and the wind speed reduction in urban canyons have the opposite effect, namely a reduction of the heating and an increase of the cooling demand due to lower outdoor-indoor temperature differences, reduced ventilation rates and reduced infrared radiation exchange with the environment. An example of the net energy balance of these opposite effects is presented in the next paragraphs for residential buildings in Barcelona and Rome.

6.4.1 Urban Morphology and Solar Radiation Availability on Facades and Roofs in Barcelona

Considering the crucial role played by solar radiation in the Mediterranean context, the impact of urban geometry on the solar radiation availability on the building envelope is the first necessary step for a correct assessment of the building energy performance in an urban context.

The solar radiation availability on facades and roofs on the iconic urban blocks of the "Eixample" in Barcelona was analysed by Curreli et al. (2016). The study showed the variability of the *sun factor* (i.e. incident direct solar energy per unit of surface in kWh/m^2) of the different components of the building envelope (roofs, external facades, courtyard walls and residual elements) in that urban texture. The sun factor was calculated with the software Helidon (Beckers and Masset 2011), a tool for solar energy analysis at the urban scale. The results showed a huge difference in the sun factor of vertical surfaces and horizontal ones and also significant differences among the external walls facing the street or the courtyard (Fig. 6.8).

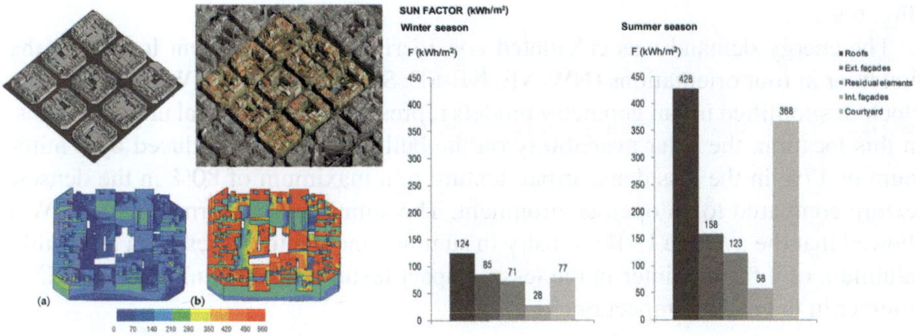

Fig. 6.8 Seasonal sun factor FR (kWh/m^2) for the different components of the building envelope in the "Eixample" district in Barcelona. (Analysis and figures by Curreli et al. (2016), available at https://doi.org/10.3390/en9070544)

The analysis of the *solar gain* (kWh) distribution on the envelope showed the dominant role of roofs compared to vertical surfaces in harvesting solar radiation in such a compact urban structure: roofs collect 33% and 43% of the total radiation incident on the envelope in winter and summer, while the street-facing walls receive 24% in winter and 17% in summer and the courtyard facades 14% in winter and 9% in summer. This also highlights a different interaction between urban morphology and solar radiation in summer and winter, due to the seasonal variability of solar elevations.

6.4.2 Solar Access and UHI Intensity: Assessing the Net Energy Impact on Urban Buildings

The high impact of urban geometry on solar access suggests that the impact of the UHI intensity on the energy demand of urban building may vary depending on the building's solar gains. This was shown by Salvati et al. (2017b), by investigating the relative impact of the UHI intensity on the cooling demand of a residential building in Barcelona for varying solar access conditions. Energy simulations performed using urban air temperatures and airport air temperatures showed that the UHI effect determines a higher relative increase in the sensible cooling demand in the apartments with lower solar gains and higher absolute increase in the apartments with higher solar gains.

The net energy impact of urban compactness considering solar access and UHI intensity was explored for buildings located in different urban textures of Rome and Barcelona in another study (Salvati et al. 2017a). The study was carried out using the Urban Weather Generator (UWG) for the estimation of the UHI intensity and EnergyPlus for the solar radiation availability on building facades and the energy demand calculation of representative apartments. The overall energy impact of the two urban climate modifications on the building heating, cooling and annual energy demand was calculated and correlated to the compactness of the urban textures (Fig. 6.9).

The energy demand was calculated considering a test apartment located at the first floor in four orientations (NW-NE, NE-SE, SE-SW and SW-NW) in the middle block of simplified urban geometry models representative of the real urban textures. In this location, the solar availability on the building facade is reduced by a minimum of 17% in the less dense urban texture to a maximum of 80% in the densest texture compared to an open environment. The simulation performed using UWG showed that the average UHI intensity in summer and winter varies from a monthly minimum of 1 °C in winter in the less compact texture to a maximum of 3.4 °C in summer in the most compact one (Fig. 6.9).

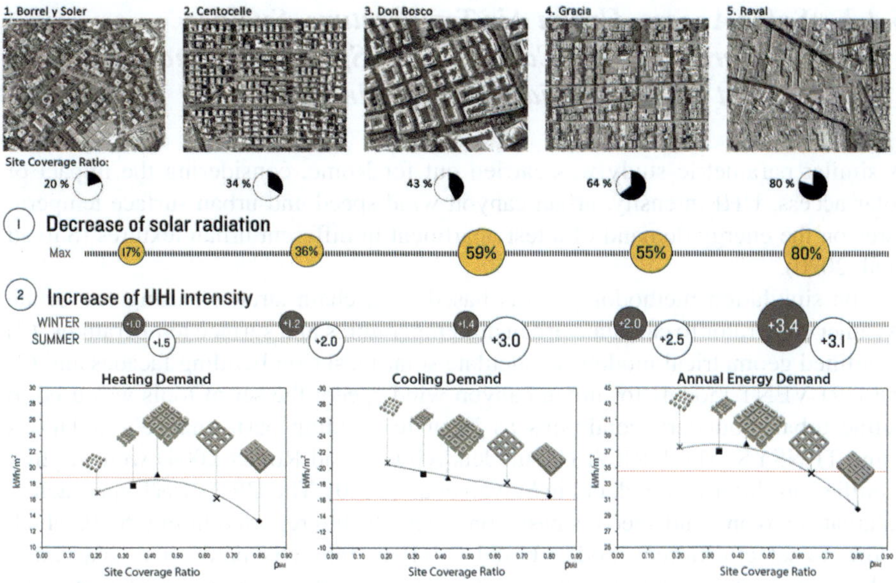

Fig. 6.9 Building heating, cooling and annual energy demand as a function of the urban texture compactness (site coverage ratio—ρ_{bld}), considering the decrease of solar radiation due to shadows and the UHI intensity in winter and summer in the textures analysed. The red line in the graphs is the energy demand of the same apartment in an open-rural context. (Images adapted from Salvati et al. (2016). Originals available at https://doi.org/10.1016/j.egypro.2017.07.303)

The net impact of these two climate modifications on the energy demand of a sample apartment was found to vary significantly with the texture compactness. The results showed a robust ($R^2 = 0.95$) negative linear correlation between the cooling energy demand and the urban compactness; this means that the building cooling energy demand decreases with the increase of urban compactness, because the obstruction of the solar radiation has a greater positive impact than the increase of the UHI intensity. A non-linear relationship was found between the heating demand and the urban compactness showing that, in winter, the best energy behaviour is found for the texture having medium compactness level. This is explained by the fact that a large reduction of the solar gains in winter has a negative impact on the heating demand but, at the same time, lower level of compactness also entails lower UHI intensity, which is beneficial in winter. The overall annual energy demand resulted in favour of more compact urban textures in this climate context.

6.4.3 Solar Access, Urban Air Temperature, Surface Temperatures and Canyon Wind Speed: Net Energy Impact on Residential Buildings in Rome

A similar parametric study was carried out for Rome, considering the impact of solar access, UHI intensity, urban canyon wind speed and urban surface temperatures on the energy demand of a test apartment in different urban textures (Salvati et al. 2020).

The simulation methodology was based on a chain strategy, using (1) Urban Weather Generator to calculate urban air temperatures and surface temperatures, (2) simplified geometrical models to calculate solar masks on building facades and (3) the URBVENT models for urban canyon wind speed; the set of tools was used to define urban boundary conditions as input to building performance simulations using TRNSYS. The URBVENT models (Ghiaus and Roulet 2004) were slightly modified, as described in detail in Salvati et al. (2020). The URBVENT approach to calculate canyon wind speed is based on the algorithm reported in Fig. 6.10, which allows to select different models based on urban geometry parameters and undisturbed wind speed and direction with respect to the canyon orientation. Five

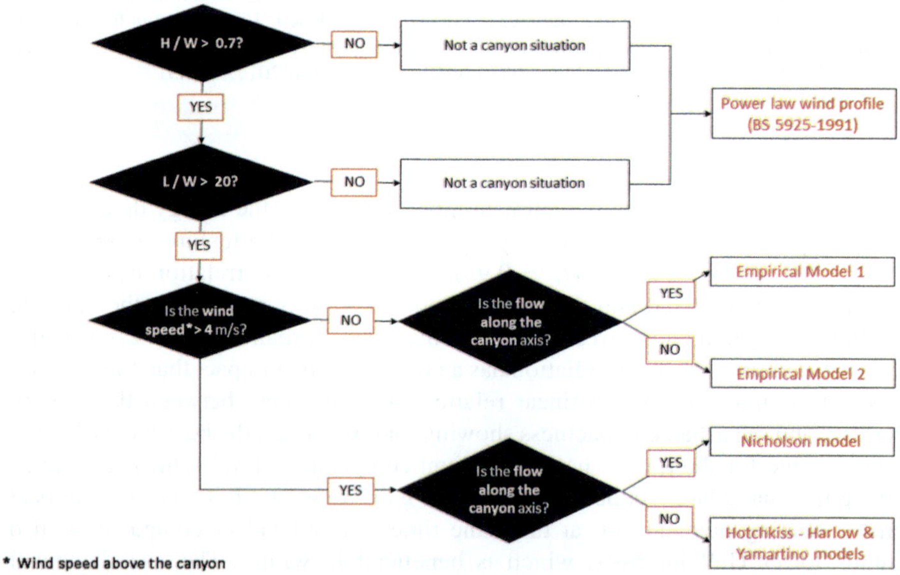

Fig. 6.10 Urban canyon empirical models and algorithm for their application developed for the URBVENT project (Christian Ghiaus et al. 2004). (The figure is from A. Salvati, M. Palme, G. Chiesa and M. Kolokotroni (2020) Built form, urban climate and building energy modelling: case-studies in Rome and Antofagasta, Journal of Building Performance Simulation, 13:2, 209–225, DOI: https://doi.org/10.1080/19401493.2019.1707876, where the models and their applicability are described in detail. The calculation spreadsheet is publicly available at the link https://doi.org/10.17633/rd.brunel.11371272.v1)

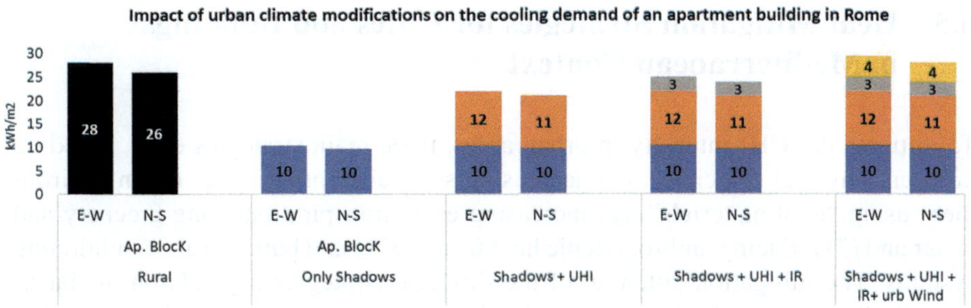

Fig. 6.11 Single impact of different climate modifications determined by an urban context on the cooling energy demand of an apartment building in Rome. (Adapted from A. Salvati, M. Palme, G. Chiesa and M. Kolokotroni (2020) Built form, urban climate and building energy modelling: case-studies in Rome and Antofagasta, Journal of Building Performance Simulation, 13:2, 209–225, DOI: https://doi.org/10.1080/19401493.2019.1707876)

calculation methods correspond to different situations: (1) low-density urban areas (i.e. no-canyon situation), (2) canyon situation with low wind speed parallel to the canyon axis or (3) perpendicular to the canyon axis and (4) canyon situation with undisturbed wind speed >4 m/s parallel to the axis or (5) perpendicular to the canyon axis. This procedure allows to generate hourly urban wind speed data to substitute the undisturbed wind speed values in the weather files for energy simulation.

The microclimate results for the urban context in comparison to an open-rural environment showed that the night-time air temperature is higher, the surface temperature is lower during daytime and higher during night-time and the wind speed is significantly reduced, as well as the solar radiation availability on walls. The energy impact of these climate modifications on the cooling demand of an apartment building located in the city centre is depicted in Fig. 6.11.

The obstruction of solar radiation in dense urban textures determines a significant reduction of the cooling demand (shading systems were modelled for the building in both urban and rural contexts). However, the UHI effect has also a large impact, slightly larger than the reduction due to shadows. A further increase of the cooling demand in urban context is also determined by the reduced infrared exchange with the environment due to higher urban surface temperatures and lower sky view factors and by the reduced ventilation rates determined by lower wind speed in urban canyons. Therefore, the overall energy impact of urban context results to be negative in the summer season, determining an increase of the cooling energy demand. The energy impact of these climate modifications is similar in trend but opposite in sign in the winter season. Therefore, the resulting annual energy demand in urban context is very similar to the one in the rural context, but the share of the heating and cooling demand is quite different, with an increase in cooling and decrease in heating needs in urban context.

Different outcomes can be found for different urban morphologies and locations within the city as well as in different climate regions and depending on the building function (occupancy patterns) and internal gains (Futcher et al. 2013, 2018).

6.5 Heat Mitigation Strategies for Cities and Buildings in Mediterranean Context

To mitigate the UHI intensity in urban areas, three main strategies exist, based on different physical principles: (1) decreasing solar absorption in the urban environment using "cool materials", (2) increasing evapotranspiration using greenery and water and (3) reducing anthropogenic heat from traffic and building air-conditioning systems. The mitigation potential of the different strategies depends on the background climate, namely the amount of solar radiation and precipitation in the hot season and the characteristics of the urban fabric and the buildings' envelope performance (thermal mass and insulation) and HVAC systems. In this last section of the chapter, we briefly discuss how the characteristics of the Mediterranean context, in terms of climate and urban morphology, may influence the potential of some of the UHI mitigation techniques and what strategies should be implemented at the building scale to improve the cooling performance and reduce the need of air-conditioning systems.

6.5.1 *Increasing Surface Albedo and Urban Albedo*

The first strategy is aimed at decreasing solar absorption in the urban environment by increasing the albedo of surfaces and the global albedo of urban areas. Morini et al. (2018) used the mesoscale Weather Research and Forecasting Model (WRF[11]) coupled with a multilayer Urban Canopy Model to investigate the impact of increasing surface albedo in Rome. The simulations showed that the impact of surface albedo (roofs, walls and paving) on air temperature depends on urban morphology and that, in some cases, increasing surface albedo may have some undesired negative impact on urban microclimate. In fact, urban morphology may reduce the cooling potential of reflective materials due to solar obstruction and radiation trapping phenomena: first, urban morphology decreases the solar radiation upon urban surfaces, roads and facades in particular; as a consequence, the beneficial effect of high-albedo materials is reduced in dense urban textures. Furthermore, urban canyons tend to trap the multiple reflections of solar radiation, vanishing the positive effect of reduced surface temperatures. Studies in the Mediterranean region and other high-radiation regions have found that increasing the reflectivity of building facades and paving may actually increase the mean radiant temperature at the street level, with a negative impact on the outdoor thermal comfort (Alchapar and Correa 2016; Erell et al. 2014; Salata et al. 2015). For these reasons, cool roofs have the highest mitigation potential in compact Mediterranean cities, due to the combined

[11] An overview of the modelling capabilities of WRF coupled with urban canopy models is provided in the chapter "The Coupling of the Weather Research and Forecasting Model with the Urban Canopy Models for Climate Simulations" by Jandaghian and Berardi (Chap. 11).

effect of urban morphology (i.e. elevated roof coverage) and surface albedo in increasing the global urban albedo (Santamouris et al. 2018; Synnefa et al. 2008; Yang and Li 2015). Conversely, the impact of reflective paving and reflective walls should be assessed case by case, as it may vary substantially based on the site urban geometry.

6.5.2 Increasing Vegetation and Comparison Among Different Mitigation Strategies

The second strategy is based on the increase of vegetation and water in urban areas and it also depends on urban morphology and climate background. Due to the dense and compact urban form of Mediterranean cities, the possibility to increase vegetation and trees in a significant amount is scarce, particularly in the historical neighbourhoods. Furthermore, Apreda et al. (2020) showed that increasing vegetation cover does not always determine an improvement of microclimate in Mediterranean climates, depending on the urban morphology. Conversely, in North American cities with similar climate such as Sacramento or Los Angeles, this strategy along with the increase of urban albedo proved to be feasible and successful due to a very different—less compact—urban morphology (Akbari et al. 2001).

In the Mediterranean climate, the mitigation potential of green infrastructure is also limited by the small amount of precipitation and water availability during the summer season, especially in southern dryer locations. In Israel, for instance, it was found that the beneficial effect of daytime temperature reduction is limited to the irrigated green areas (Spronken-Smith and Oke 1998), while large grassy areas can even cause higher daytime temperatures (Potchter et al. 2006) and additional energy expenditures due to the requirement of large quantities of water (Saaroni et al. 2018).

Battista et al. (2019) simulated the impact of urban vegetation, cool pavements, shading systems and water stretch on the outdoor thermal comfort in a square in Rome. The results showed a higher potential of the canopy-shading system to mitigate air temperature, followed by the water stretch, increase of trees and use of cool paving. Interestingly, when the UTCI[12] index was analysed, only the shading system turned out to provide a significant improvement of the outdoor thermal comfort, confirming the dominant role played by solar radiation with respect to air temperature in relation to thermal sensation in this climate zone.

---

[12] The different thermal indices developed to assess outdoor thermal comfort are described in the chapter "RayMan and SkyHelios Model" by Matzarakis et al. (Chap. 16).

6.5.3 Reduce Anthropogenic Heat Generation and Increase Passive Cooling Design

A third, less investigated mitigation and adaptation strategy for urban warmth is the reduction of anthropogenic heat from the buildings' HVAC systems and transport systems. A parametric analysis carried out using UWG showed that the air-conditioning systems may be responsible of up to 1.3 °C rise in urban temperature in summer in the Mediterranean context (Salvati et al. 2017c). Similar findings were found for Singapore (Mohegh et al. 2018), Paris (Tremeac et al. 2012), Phoenix (Salamanca et al. 2014) and Berlin (Jin et al. 2020).

Another parametric study comparing the beneficial effect of different mitigation measures in Mediterranean climates carried out by Palme et al. (2019) confirmed the important role played by anthropogenic heat generation in worsening outdoor thermal comfort.

6.5.4 Passive Architectural Solutions for Mediterranean Climates

Looking at the architecture of the past in the southern regions of Europe, many lessons can be learned on how to cope with heat in buildings. The strategies that have been developed over the years in the Mediterranean architecture can take the shape of very different architectural solutions according to local circumstances as regards climate, sociology, available construction technology, cultural tradition and so on. We do not intend to make an exhaustive list of them, but just report a selection of those which best illustrate the wisdom of popular architecture and that should be further developed and applied also in contemporary buildings to improve summer thermal comfort (Coch 1998; Coch and Serra 1996; Coch et al. 1998; Serra 1999):

- Protection from the sun by means of external barriers (vegetation, overhangs, external shutters, louvres and lattices)
- Protection from the sun by means of blinds (*Mediterranean blind*)
- Protection from the sun by means of white finishes on roofs (cold selective surfaces)
- Protection from the sun by means of ventilated air chambers (double roofs)
- Cross-ventilation with continuous openings
- Generated ventilation with wind chimneys, solar chambers or wind conduits
- Treated ventilation with wet surfaces
- Ventilation with underground conduits
- Partial sinking of the building underground
- Opaque, mobile surfaces to block openings
- Insulation in the external side of walls

Another crucial characteristic of any architectural solution to control the indoor environmental conditions is its *flexibility*, which is instead very lacking in contemporary buildings.

As we have already mentioned, temperate climates often have very variable conditions. This complexity can be handled using flexible architectural systems, with elements or combinations of elements that can easily change their environmental action according to the external weather conditions. The most recommended of these flexible systems are:

- Mobile shading systems, such as the typical louvre blind that allows the entry of radiation and ventilation to be controlled simply and conveniently.
- Mobile insulation in the openings, shutters, curtains, etc., which enables the flow of heat and light to be regulated, above all in winter.
- Window types that can be completely opened, permitting maximum control of ventilation and allowing the free passage of air and sunlight when appropriate.
- Intermediate spaces between indoor and outdoor areas, which can generate favourable microclimates: These can also be occupied at different times of the day and the year, thus adding to the building's functional possibilities.

6.6 Conclusion

The many topics discussed in this chapter intended to highlight the complexity of the environmental performance of urban areas, with a specific focus on the peculiar characteristics of compact Mediterranean cities. The aim was to describe the many aspects that need to be considered when approaching the urban scale, even if the objective of the analysis is restricted to "just" the thermal performance of buildings and urban spaces.

In the Mediterranean climate, thermal comfort is a function of solar radiation, humidity and air temperature. These variables are profoundly affected by the dense and compact urban forms typical of this region. Many urban climate studies have been carried out in the last years, bringing more insights on the physics of urban climate phenomena and the correlation with key urban and geographical features. The review of the studies carried out in the Mediterranean climates showed some similarities among the cities in this region. The temperature in urban areas can be significantly higher than surrounding rural areas both at daytime and night-time; the maximum daytime air temperature is often (not always) found in less dense urban areas that have higher solar access, but the maximum night-time air temperature is always found in the densest urban areas in the city centre. Significant intra-urban air temperature differences were observed in all the cities, depending on the location (i.e. city centre or suburban areas or distance from the seashore), the presence of vegetation and, ultimately, the density of the urban area, that plays the major role in most of the cases. Another common finding for large coastal cities is that the moderating effect of the sea and sea breeze on the summer air temperatures is noticeable at the roof level, but it is vanished at the street level at few blocks' distance from the shoreline. Nevertheless, a strong dependence of the UHI intensity on the dominant wind speed and direction was observed in many Mediterranean cities and should be carefully analysed in any future urban climate study.

The UHI intensity, wind reduction and radiation trapping in urban canyons have negative impact on outdoor thermal comfort and building energy performance in this context. The studies presented in the chapter also highlighted that a holistic and multiscale approach is needed to assess the net impact of urban climate on the performance of urban buildings. To this aim, many climate modelling tools already exist, as mentioned throughout the chapter. Due to the complexity of the phenomena involved, there is no complete or perfect tool yet for an effective integration of urban climate modelling into building energy performance analysis and architectural and urban design. Coupling or chaining techniques of multiple simulation tools need to be implemented to investigate the interrelationships between buildings and microclimate in urban areas, with many limitations on the real applicability in the design practice. However, huge advances have been done in the very last years in this field and the knowledge gap between urban climatology and urban and architectural design is shrinking. As architects and building designers, we will keep contributing with our research for the development of more design-oriented tools that could effectively support the work of practitioners in the development of climate and environmentally conscious buildings and cities.

References

Akbari, H., Pomerantz, M., & Taha, H. (2001). Cool surfaces and shade trees to reduce energy use and improve air quality in urban areas. *Solar Energy, 70*(3), 295–310. Retrieved from http://www.sciencedirect.com/science/article/pii/S0038092X0000089X.

Alchapar, N. L., & Correa, E. N. (2016). The use of reflective materials as a strategy for urban cooling in an arid "OASIS" city. *Sustainable Cities and Society, 27*, 1–14. https://doi.org/10.1016/j.scs.2016.08.015.

Apreda, C., Reder, A., & Mercogliano, P. (2020). Urban morphology parameterization for assessing the effects of housing blocks layouts on air temperature in the Euro-Mediterranean context. *Energy and Buildings, 223*, 110171. https://doi.org/10.1016/j.enbuild.2020.110171.

Bastin, J. F., Clark, E., Elliott, T., Hart, S., Van Den Hoogen, J., Hordijk, I., et al. (2019). Understanding climate change from a global analysis of city analogues. *PLoS One, 14*(10), 1–13. https://doi.org/10.1371/journal.pone.0224120.

Battista, G., de Lieto Vollaro, R., & Zinzi, M. (2019). Assessment of urban overheating mitigation strategies in a square in Rome, Italy. *Solar Energy, 180*, 608–621. https://doi.org/10.1016/j.solener.2019.01.074.

Beckers, B., & Masset, L. (2011). *Heliodon 2 documentation*. Retrieved from www.heliodon.net.

Bonacquisti, V., Casale, G. R., Palmieri, S., & Siani, A. M. (2006). A canopy layer model and its application to Rome. *The Science of the Total Environment, 364*, 1–13. https://doi.org/10.1016/j.scitotenv.2005.09.097.

Bourbia, F., & Boucheriba, F. (2010). Impact of street design on urban microclimate for semi-arid climate (Constantine). *Renewable Energy, 35*(2), 343–347. https://doi.org/10.1016/j.renene.2009.07.017.

Bueno, B., Norford, L., Hidalgo, J., & Pigeon, G. (2013). The Urban Weather Generator. *Journal of Building Performance Simulation, 6*(4), 269–281. https://doi.org/10.1080/19401493.2012.718797.

Cantelli, A., Monti, P., & Leuzzi, G. (2011). An investigation of the urban heat island of Rome through a canyon based subgrid scheme. *International Journal of Environment and Pollution, 1/2/3/4*(47), 239–247.

Cantelli, A., Monti, P., & Leuzzi, G. (2014). Development and integration of a subgrid urban surface scheme in a limited area model. *International Journal of Environment and Pollution, 55*(1–4), 230–237. https://doi.org/10.1504/IJEP.2014.065918.

Coch, H. (1998). Bioclimatism in vernacular architecture. *Renewable and Sustainable Energy Reviews, 2*, 67–87. https://doi.org/10.1016/S1364-0321(98)00012-4.

Coch, H., & Serra, R. (1996). Summer comfort solution in Mediterranean areas. *Renewable Energy, 8*, 128–132. https://doi.org/10.1016/0960-1481(96)88831-8.

Coch, H., Serra, R., & Isalgué, A. (1998). The Mediterranean blind: less light, better vision. *Renewable Energy, 15*, 431–436.

Colacino, M., & Lavagnini, A. (1982). Evidence of the urban heat island in Rome by climatological analyses. *Archives for Meteorology, Geophysics, and Bioclimatology Series B, 31*(1–2), 87–97. https://doi.org/10.1007/BF02311344.

Curreli, A., Serra-Coch, G., Isalgue, A., Crespo, I., & Coch, H. (2016). Solar energy as a form giver for future cities. *Energies, 9*(7), 544. https://doi.org/10.3390/en9070544.

Erell, E., Pearlmutter, D., Boneh, D., & Kutiel, P. B. (2014). Effect of high-albedo materials on pedestrian heat stress in urban street canyons. *Urban Climate, 10*, 367–386. https://doi.org/10.1016/j.uclim.2013.10.005.

Futcher, J., Kershaw, T., & Mills, G. (2013). Urban form and function as building performance parameters. *Building and Environment, 62*, 112–123. https://doi.org/10.1016/j.buildenv.2013.01.021.

Futcher, J., Mills, G., & Emmanuel, R. (2018). Interdependent energy relationships between buildings at the street scale. *Building Research and Information, 46*, 829–844. https://doi.org/10.1080/09613218.2018.1499995.

Gasparrini, A., Guo, Y., Hashizume, M., Lavigne, E., Zanobetti, A., Schwartz, J., et al. (2015). Mortality risk attributable to high and low ambient temperature: a multicountry observational study. *The Lancet, 386*(9991), 369–375. https://doi.org/10.1016/S0140-6736(14)62114-0.

Georgakis, C, & Santamouris, M. (2005). *Canyon effects: Calculation of wind speed in an urban street canyon with the aid of a semi-empirical model based on experimental data, (May)*. pp. 117–124.

Georgakis, C., & Santamouris, M. (2006). Experimental investigation of air flow and temperature distribution in deep urban canyons for natural ventilation purposes. *Energy and Buildings, 38*(4), 367–376. https://doi.org/10.1016/j.enbuild.2005.07.009.

Ghiaus, C., & Roulet, C.-A. (2004). *URBVENT Natural ventilation in urban areas—Potential assessment and optimal façade design. Workpackage 1: Soft computing of natural ventilation potential. Final Report*. https://doi.org/10.4324/9781849772068

Ghiaus, C., Allard, F., Santamouris, M., Georgakis, C., Roulet, C.-A., Germano, M., … Roche, L. (2004). *URBVENT WP1 final report: soft computing of natural ventilation potential.*

Ghiaus, C, Allard, F., Santamouris, M., Georgakis, C., Roulet, C., Germano, M., … Roche, L. (2005). Natural ventilation of urban buildings - summary of URBVENT project. *International Conference "Passive and low energy cooling for the built environment,"* 1(May), 1109. Retrieved from http://repositorium.sdum.uminho.pt/bitstream/1822/5076/1/Ghiaus_CI_2005.pdf.

Giannaros, T. M., & Melas, D. (2012). Study of the urban heat island in a coastal Mediterranean city: The case study of Thessaloniki, Greece. *Atmospheric Research, 118*, 103–120. https://doi.org/10.1016/j.atmosres.2012.06.006.

Giannopoulou, K., Livada, I., Santamouris, M., Saliari, M., Assimakopoulos, M., & Caouris, Y. G. (2011). On the characteristics of the summer urban heat island in Athens, Greece. *Sustainable Cities and Society, 1*(1), 16–28. https://doi.org/10.1016/j.scs.2010.08.003.

Hassaan, M. A. (2008). Mapping urban heat islands using remotely sensed data Case study: Alexandria city, Egypt. *Bulletin of Faculty of Arts, University of Alexandria*, (October).

Hsie, T.-S. (2008). Achieving Urban Sustainability through Urban Morphology Analyses and Optimum Ventilation. *PLEA 2008—25th Conference on Passive and Low Energy Architecture, Dublin*, (October).

IPCC. (2014). Climate change 2014 synthesis report the core writing team core writing team technical support unit for the synthesis report. Climate change 2014: Synthesis report. *Contribution of Working Groups I, II and III to the Fifth Assessment Report of the Intergovernmental Panel on Climate Change*.

Jin, L., Schubert, S., & Mohamed Hefny Salim, C. S. (2020). *Impact of air conditioning systems on outdoor thermal environment during summer in Berlin, Germany (June)*. https://doi.org/10.3390/ijerph17134645

Johansson, E., Thorsson, S., Emmanuel, R., & Krüger, E. (2014). Instruments and methods in outdoor thermal comfort studies—The need for standardization. *Urban Climate, 10*(P2), 346–366. https://doi.org/10.1016/j.uclim.2013.12.002.

Kolokotroni, M., Giannitsaris, I., & Watkins, R. (2006). The effect of the London urban heat island on building summer cooling demand and night ventilation strategies. *Solar Energy, 80*(4), 383–392. https://doi.org/10.1016/j.solener.2005.03.010.

Kolokotsa, D., Psomas, a., & Karapidakis, E. (2009). Urban heat island in southern Europe: The case study of Chania, Crete. *Solar Energy, 83*(10), 1871–1883. https://doi.org/10.1016/j.solener.2009.06.018.

Kottek, M., Grieser, J., Beck, C., Rudolf, B., & Rubel, F. (2006). World map of the Köppen-Geiger climate classification updated. *Meteorologische Zeitschrift, 15*(3), 259–263. https://doi.org/10.1127/0941-2948/2006/0130.

Leuzzi, G., & Monti, P. (1997). Breeze analysis by Mast and Sodar measurements. *Il Nuovo Cimento, C, 20*, 343–359.

Linares, C., Díaz, J., Negev, M., Martínez, G. S., Debono, R., & Paz, S. (2020). Impacts of climate change on the public health of the Mediterranean Basin population—Current situation, projections, preparedness and adaptation. *Environmental Research, 182*, 109107. https://doi.org/10.1016/j.envres.2019.109107.

Livada, I., Santamouris, M., Niachou, K., Papanikolaou, N., & Mihalakakou, G. (2002). Determination of places in the great Athens area where the heat island effect is observed. *Theoretical and Applied Climatology, 71*(3–4), 219–230. https://doi.org/10.1007/s007040200006.

Mao, J., Yang, J. H., Afshari, A., & Norford, L. K. (2017). Global sensitivity analysis of an urban microclimate system under uncertainty: Design and case study. *Building and Environment, 124*, 153–170. https://doi.org/10.1016/j.buildenv.2017.08.011.

Martinelli, A., Kolokotsa, D.-D., & Fiorito, F. (2020). Urban heat island in Mediterranean coastal cities: The case of Bari (Italy). *Climate, 8*, 79.

Martin-Vide, J., & Moreno-Garcia, M. C. (2020). Probability values for the intensity of Barcelona's urban heat island (Spain). *Atmospheric Research, 240*, 104877. https://doi.org/10.1016/j.atmosres.2020.104877.

Mihalakakou, G., Flocas, H. a., Santamouris, M., & Helmis, C. G. (2001). Application of neural networks to the simulation of the heat island over Athens, Greece, using synoptic types as a predictor. *Journal of Applied Meteorology, 41*(5), 519–527. https://doi.org/10.1175/1520-0450(2002)041<0519:AONNTT>2.0.CO;2.

Mihalakakou, G., Santamouris, M., Papanikolaou, N., Cartalis, C., & Tsangrassoulis, A. (2004). Simulation of the urban heat island phenomenon in Mediterranean climates. *Pure and Applied Geophysics, 161*, 429–451. https://doi.org/10.1007/s00024-003-2447-4.

Mohegh, A., Levinson, R., Taha, H., Gilbert, H., Zhang, J., Li, Y., et al. (2018). Observational evidence of neighborhood scale reductions in air temperature associated with increases in roof albedo. *Climate, 6*(4). https://doi.org/10.3390/cli6040098.

Moreno-Garcia, M. C. (1994). Intensity and form of the urban heat island in Barcelona. *International Journal of Climatology, 14*, 705–710. https://doi.org/10.1002/joc.3370140609.

Morganti, M. (2018). *Ambiente costruito mediterraneo. Forma densità ed energia*. Monfalcone (Gorizia): Edicom Edizioni.

Morini, E., Touchaei, A. G., Rossi, F., Cotana, F., & Akbari, H. (2018). Evaluation of albedo enhancement to mitigate impacts of urban heat island in Rome (Italy) using WRF meteorological model. *Urban Climate, 24*, 551–566. https://doi.org/10.1016/j.uclim.2017.08.001.

Natanian, J., Aleksandrowicz, O., & Auer, T. (2019). A parametric approach to optimizing urban form, energy balance and environmental quality: The case of Mediterranean districts. *Applied Energy, 254*, 113637. https://doi.org/10.1016/j.apenergy.2019.113637.

Oke, T. R. (1987). *Boundary layer climates* (2nd ed.). London: Taylor & Francis.

Oke, T. R., Mills, G., Christen, A., & Voogt, J. A. (2017). *Urban Climates*. Cambridge: Cambridge University Press. https://doi.org/10.1017/9781139016476.

Palme, M., Clemente, C., Cellurale, M., Carrasco, C., & Salvati, A. (2019). Mitigation strategies of the urban heat island intensity in Mediterranean climates: Simulation studies in Rome (Italy) and Valparaiso (Chile). *IOP Conference Series: Earth and Environmental Science, 323*(1). https://doi.org/10.1088/1755-1315/323/1/012025.

Papanastasiou, D. K., & Kittas, C. (2012). Maximum urban heat island intensity in a medium-sized coastal Mediterranean city. *Theoretical and Applied Climatology, 107*(3–4), 407–416. https://doi.org/10.1007/s00704-011-0491-z.

Pelliccioni, A., Monti, P., Gariazzo, C., & Leuzzi, G. (2012). Some characteristics of the urban boundary layer above Rome, Italy, and applicability of Monin–Obukhov similarity. *Environmental Fluid Mechanics, 12*(5), 405–428. https://doi.org/10.1007/s10652-012-9246-3.

Petralli, M., Massetti, L., & Orlandini, S. (2011). Five years of thermal intra-urban monitoring in Florence (Italy) and application of climatological indices. *Theoretical and Applied Climatology, 104*(3–4), 349–356. https://doi.org/10.1007/s00704-010-0349-9.

Potchter, O., Cohen, P., & Bitan, A. (2006). Climatic behaviour of various urban parks during hot and humid summer in the Mediterranean city of Tel Aviv, Israel. *International Journal of Climatology, 26*, 1695–1711. https://doi.org/10.1002/joc.

Ratti, C., Baker, N., & Steemers, K. (2005). Energy consumption and urban texture. *Energy and Buildings, 37*(7), 762–776. https://doi.org/10.1016/j.enbuild.2004.10.010.

Rojas-Fernández, J., Galán-Marín, C., Roa-Fernández, J., & Rivera-Gómez, C. (2017). Correlations between GIS-based urban building densification analysis and climate guidelines for Mediterranean courtyards. *Sustainability (Switzerland), 9*, 12. https://doi.org/10.3390/su9122255.

Romero Rodríguez, L., Sánchez Ramos, J., Sánchez de la Flor, F. J., & Álvarez Domínguez, S. (2020). Analyzing the urban heat Island: Comprehensive methodology for data gathering and optimal design of mobile transects. *Sustainable Cities and Society, 55*, 102027. https://doi.org/10.1016/j.scs.2020.102027.

Saaroni, H., Ben-Dor, E., Bitan, A., & Potchter, O. (2000). Spatial distribution and microscale characteristics of the urban heat island in Tel Aviv, Israel. *Landscape and Urban Planning, 48*(1–2), 1–18. https://doi.org/10.1016/S0169-2046(99)00075-4.

Saaroni, H., Amorim, J. H., Hiemstra, J. A., & Pearlmutter, D. (2018). Urban Green Infrastructure as a tool for urban heat mitigation: Survey of research methodologies and findings across different climatic regions. *Urban Climate, 24*, 94–110. https://doi.org/10.1016/j.uclim.2018.02.001.

Salamanca, F., Georgescu, M., Mahalov, A., Moustaoui, M., Wang, M., Georgescu, M., et al. (2014). Anthropogenic heating of the urban environment due to air conditioning. *Journal of Geophysical Research: Atmospheres RESEARCH, 119*, 5949–5965. https://doi.org/10.1002/2013JD021225.

Salata, F., Golasi, I., Vollaro, A. D. L., & Vollaro, R. D. L. (2015). How high albedo and traditional buildings' materials and vegetation affect the quality of urban microclimate. A case study. *Energy and Buildings, 99*, 32–49. https://doi.org/10.1016/j.enbuild.2015.04.010.

Salvati, A., Coch, H., & Cecere, C. (2015). Urban morphology and energy performance: The direct and indirect contribution in Mediterranean climate. In M. Cucinella, G. Pentella, A. Fagnani, & L. D'Ambrosio (Eds.), *PLEA2015 architecture in (R) evolution—31st International PLEA conference*. Bologna, Italy: PLEA.

Salvati, A., Coch Roura, H., & Cecere, C. (2016). Urban heat island prediction in the Mediterranean context: An evaluation of the urban weather generator model|Predicción urbana de la isla de calor en el contexto mediterráneo: Una evaluación del modelo generador de tiempo urbano. *Architecture, City and Environment, 11*, 32. https://doi.org/10.5821/ace.11.32.4836.

Salvati, A., Coch, H., & Morganti, M. (2017a). Effects of urban compactness on the building energy performance in Mediterranean climate. *Energy Procedia, 122*, 499–504. https://doi.org/10.1016/j.egypro.2017.07.303.

Salvati, A., Coch Roura, H., & Cecere, C. (2017b). Assessing the urban heat island and its energy impact on residential buildings in Mediterranean climate: Barcelona case study. *Energy and Buildings, 146*, 38–54. https://doi.org/10.1016/j.enbuild.2017.04.025.

Salvati, A., Palme, M., & Inostroza, L. (2017c). Key parameters for urban heat island assessment in a Mediterranean context: A sensitivity analysis using the urban weather generator model. *IOP Conference Series: Materials Science and Engineering, 245*, 8. https://doi.org/10.1088/1757-899X/245/8/082055.

Salvati, A., Monti, P., Coch Roura, H., & Cecere, C. (2019). Climatic performance of urban textures: Analysis tools for a Mediterranean urban context. *Energy and Buildings, 185*, 162–179. https://doi.org/10.1016/j.enbuild.2018.12.024.

Salvati, A., Palme, M., Chiesa, G., & Kolokotroni, M. (2020). Built form, urban climate and building energy modelling: case-studies in Rome and Antofagasta. *Journal of Building Performance Simulation, 13*(2), 209–225. https://doi.org/10.1080/19401493.2019.1707876.

Santamouris, M. (2016). Innovating to zero the building sector in Europe: Minimising the energy consumption, eradication of the energy poverty and mitigating the local climate change. *Solar Energy, 128*, 61–94. https://doi.org/10.1016/j.solener.2016.01.021.

Santamouris, M., Papanikolaou, N., Livada, I., Koronakis, I., Georgakis, C., Argiriou, A., & Assimakopoulos, D. (2001). On the impact of urban climate on the energy consumption of buildings. *Solar Energy, 70*(3), 201–216. https://doi.org/10.1016/S0038-092X(00)00095-5.

Santamouris, M., Georgakis, C., & Niachou, A. (2008). On the estimation of wind speed in urban canyons for ventilation purposes - Part 2: Using of data driven techniques to calculate the more probable wind speed in urban canyons for low ambient wind speeds. *Building and Environment, 43*(8), 1411–1418. https://doi.org/10.1016/j.buildenv.2007.01.042.

Santamouris, M., Haddad, S., Saliari, M., Vasilakopoulou, K., Synnefa, A., Paolini, R., et al. (2018). On the energy impact of urban heat island in Sydney: Climate and energy potential of mitigation technologies. *Energy and Buildings, 166*, 154–164. https://doi.org/10.1016/j.enbuild.2018.02.007.

Serra, R. (1999). *Arquitectura y climas (I)*. Barcelona: Gustavo Gili, SA.

Serra, C., Lana, X., Martínez, M. D., Roca, J., Arellano, B., Biere, R., et al. (2020). Air temperature in Barcelona metropolitan region from MODIS satellite and GIS data. *Theoretical and Applied Climatology, 139*(1–2), 473–492. https://doi.org/10.1007/s00704-019-02973-y.

Spronken-Smith, R. a., & Oke, T. R. (1998). The thermal regime of urban parks in two cities with different summer climates. *International Journal of Remote Sensing, 19*, 2085–2104. https://doi.org/10.1080/014311698214884.

Stewart, I. D., & Oke, T. R. (2012). Local climate zones for urban temperature studies. *Bulletin of the American Meteorological Society, 93*(12), 1879–1900. https://doi.org/10.1175/BAMS-D-11-00019.1.

Synnefa, A., Dandou, A., Santamouris, M., Tombrou, M., & Soulakellis, N. (2008). On the use of cool materials as a heat island mitigation strategy. *Journal of Applied Meteorology and Climatology, 47*(11), 2846–2856. https://doi.org/10.1175/2008JAMC1830.1.

Theophilou, M. K., & Serghides, D. (2015). Estimating the characteristics of the urban heat island effect in Nicosia, Cyprus, using multiyear urban and rural climatic data and analysis. *Energy and Buildings, 108*, 137–144. https://doi.org/10.1016/j.enbuild.2015.08.034.

Tremeac, B., Bousquet, P., de Munck, C., Pigeon, G., Masson, V., Marchadier, C., et al. (2012). Influence of air conditioning management on heat island in Paris air street temperatures. *Applied Energy, 95*, 102–110. https://doi.org/10.1016/j.apenergy.2012.02.015.

Yang, X., & Li, Y. (2015). The impact of building density and building height heterogeneity on average urban albedo and street surface temperature. *Building and Environment, 90*, 146–156. https://doi.org/10.1016/j.buildenv.2015.03.037.

Zinzi, M., & Carnielo, E. (2017). Impact of urban temperatures on energy performance and thermal comfort in residential buildings. The case of Rome, Italy. *Energy and Buildings, 157*, 20–29. https://doi.org/10.1016/j.enbuild.2017.05.021.

Zinzi, M., Carnielo, E., & Mattoni, B. (2018). On the relation between urban climate and energy performance of buildings. A three-years experience in Rome, Italy. *Applied Energy, 221*, 148–160. https://doi.org/10.1016/j.apenergy.2018.03.192.

Chapter 7
Enhancing Energy Performance and Comfort of Built Environment in Tropical Climates

Mahendra Gooroochurn and Renganathan Giridharan

7.1 Introduction

Rapid urbanisation is taking place in tropics in response to growing population. Tropics also witnesses large-scale flooding and drought as a consequence of climate change, with urban areas facing greater risks. These urban areas patronise high-density developments as a solution for urban growth. On most occasions, lack of compatibility of these development typologies results in increased cooling load (Giridharan and Emmanuel 2018).

On the other hand, most of these urban areas lack appropriate climate data for necessary modelling and analysis, barring few places like Singapore and Hong Kong (Giridharan 2016). This is also one of the reasons for incomplete understanding of the energy needs for space conditioning associated with high-density developments. This lack of understanding has major implications on energy budgeting and human health in tropics (Giridharan and Emmanuel 2018; Thompson et al. 2018). Further, there are designers and planners who are passionate about urban climate and energy usage but unable to get necessary guidance and tools to support their goals. However, on most occasions, these designers and planners tend to use subjective measures to evaluate their development proposals under complex climate premises (Schiller and Evans 1990–1991), leading to unforeseen risks.

Environment-conscious designers and planners, who take on-board the urban environmental issues, work at a higher order level as conceptualised in approaches such as the integrated and regenerated (Golany 1996; Givoni 1998; Giridharan

M. Gooroochurn (✉)
Faculty of Engineering, Mechanical and Production Engineering Department, University of Mauritius, Moka, Mauritius
e-mail: M.Gooroochurn@uom.ac.mu

R. Giridharan
Kent School of Architecture, Marlowe Building, University of Kent, Canterbury, UK
e-mail: G.Renganathan@kent.ac.uk

M. Palme, A. Salvati (eds.), *Urban Microclimate Modelling for Comfort and Energy Studies*, https://doi.org/10.1007/978-3-030-65421-4_7

2016). These approaches have major implications on planning and design optimisation, especially in developing passive cooling strategies. Further, there is growing evidence that architectural designs based on passive strategies may not be sufficient to bring thermal comfort in tropics, especially in the context of high-density urban areas even if they are scientifically worked out (Villadiego and Velay-Dabat 2014). Therefore, current urban development and construction practices cannot be improved in a bespoke and scientific manner to achieve sustainable environments in tropical countries without an in-depth and focused understanding of urban climate and its associated impact on comfort and energy (Giridharan and Emmanuel 2018; Rajagopalan et al. 2014; Ng et al. 2015; Doan and Kusaka 2016).

To understand the changes in urban climate and energy usage, a better quantification of the impact of density, land use, thermo-physical characteristics, geometry, local energy consumption, anthropogenic heat release, heat source and heat sink is essential (Oke 1988; Givoni 1998; Santamouris 2001; Emmanuel 2005; Giridharan et al. 2007; Yang et al. 2010; Ng et al. 2012; Dimoudi et al. 2013). This should be well linked to corresponding indoor parameters. For example, at urban design level, decisions are taken along the lines of variation in thermal mass, skyline and density. In urban design, thermal mass is not conceptualised in terms of the contents of a wall, but rather a relationship between height and total floor area (Knowles 1977; Giridharan et al. 2007). Similarly, the skyline is assessed as a function of massing, not based on turbulence and drag effect. In an individual space, thermal comfort is controlled using thermal transmittance of the wall. The equivalent initiative at urban design could come in the form of aspect ratio, which is a function of the ratio of building height to street width.

At this higher level, it is difficult to quantify them in an order and a level similar to atmospheric science parameters, especially in the complex urban setting. On the other hand, if we keep the parameters too general, meaningful optimisation is even more difficult (Nichol and Wong 2005; Giridharan et al. 2007), especially for quantifying and assessing anthropogenic heat and its impact. It is essential to bring the link between indoor and outdoor comfort, and associated energy consumption. Although the relationship between building and its energy consumption is well researched, urban fabric and its impact on energy are not well understood due to the complexity of the interactions of multiple urban variables, monitoring difficulties and complex urban climate. Moreover, the available information is not well incorporated into the design and planning process. Further, especially in tropical island nations, due to weak economic conditions, there is limited conversion of research findings into planning practice and policy (Giridharan and Emmanuel 2018). This conversion is essential considering the level of integration that is needed to successfully maintain the passive and the low-carbon technology measures for cooling both indoor and outdoor environments in tropics, especially considering the high level of humidity.

In the current form of urbanisation, the building techniques employed are becoming increasingly generic and systematic using standardised materials and layout. With the advent of active energy systems like mechanical ventilation, artificial lighting and air conditioning, less consideration is placed on the indoor space conditions achievable passively leading to buildings with increased carbon footprint when air-conditioned or otherwise poor thermal comfort. Access to daylight and views has also been sacrificed in this new design paradigm, although these have been found to contribute to occupant well-being and health.

This chapter presents the main research findings available in literature related to the factors influencing building energy performance and thermal comfort and results obtained from practical realisation of energy efficiency and passive measures, based on which research gaps are identified and a road map for improving energy efficiency and comfort in tropical climates is proposed.

7.2 Tropical Climate and Characteristics of Urban Fabric

The tropical climate is generally considered to be hot and humid (Oke 1987). However, for a meaningful assessment, the sub-classifications advocated by Köppen should be used (Roth 2007). These sub-classifications are wet climate (Singapore), wet/dry climate (Mumbai), monsoon climate (Jakarta) and highland climate (Mexico City). This is an indication that water availability (moisture) will be one of the main determinants of tropical climate (Giridharan and Emmanuel 2018). In tropical region, precipitation is increasingly becoming aligned to urban growth over sea breeze and coastline curvature (Gonzalez 2005). Generally, urban areas are drier than surrounding rural areas. However, in tropics, on some occasions, the absolute humidity in cities could be higher than surroundings due to stagnation of air, poor drainage systems and still water bodies (Sharmin et al. 2015; Oke 1987).

The tropical climate is not subject to major seasonal and diurnal variability like the temperate climate except in subtropical areas. On the other hand, characteristics of vegetation and solar altitude dictate substantial spatial comfort variations on most occasions (Adebayo 1990–1991; Oke 1987; Giridharan et al. 2008; Chow and Roth 2006; Yang et al. 2010). Unlike in temperate climates, solar altitude does not dramatically change in a tropical climate from season to season. Furthermore, for the critical part of the day, say 11 a.m. to 3 p.m., the sun is positioned almost directly overhead. This intense exposure drives designers to create a variety of shaded spaces. At times, these spaces are dynamic in nature. If vegetation and solar altitude are not exploited effectively, there will be major implications on the outdoor as well as indoor comfort as evidenced in the planned Hanoi capital city (Qaid et al. 2016) and Putrajaya in Malaysia (Moser 2010; Qaid and Ossen 2015; Ahmed et al. 2015). By and large, tropical urban climate problems could be linked to mismatch in street pattern, building geometry, density, inter-building spaces, roughness length, urban permeability, urban albedo, ground coverage, vegetation and anthropogenic heat (Giridharan et al. 2008; Emmanuel 2005; Ng 2012). Therefore, like temperate climate, the urban morphology is key to both indoor and outdoor comfort in tropics. However, the character of tropical urban morphology should respond to solar radiation and high relative humidity.

The canyon geometry and buildings form the urban morphology. Urban morphology is primarily based on features such as building form, skyline, plot size, land use and street pattern. These factors interact with meteorological variables and modify the climate (Oke 1987; Givoni 1998; Santamouris 2001). The degree of interaction of meteorological parameters on most occasions depends on canyon geometry. For example, Rajagopalan et al. (2014) found that the chaotic nature of urban canyon in fast-developing Muar (Malaysia) has reduced the ventilation. The urban canyon geometry (form) can be classified as compact, dispersed or clustered (Golany 1996). The compact and clustered forms are suitable for hot-dry and cold-dry climates while the dispersed form is suitable for hot and humid climates (the tropics). However, compact and clustered forms could be found in tropics, especially in developed places like Singapore and Hong Kong. These developed places follow multiple intensive land use strategy to vitalise the compact form (Lau et al. 2005). However, developing countries in the tropics also follow compact forms. In developing countries, compactness is largely articulated in terms of footprint. Therefore, Huang et al. (2007) are of the opinion that urban form in developing world is more compact than developed world. However, these developments do not generally achieve or meet other aspects of compactness such as distribution of amenities and other infrastructure as articulated by Lau et al. (2005).

The lack of standardisation of tropical urban compactness poses problems in decoding its overall effect on urban heat islands (UHI). In general, compact building geometry such as stand-alone towers linked by sky bridges, towers on podium and blocks with multiple courtyards has great potential to shade the urban environment (Yang et al. 2010; Giridharan et al. 2008; Lau et al. 2010; Chen et al. 2012). Nevertheless, the cooling effects of building shade become increasingly significant as inter-building distance decreases due to the resultant low level of exposure to strong direct radiation. Further, Sharmin et al. (2015) have shown that compact urban canyon geometry with aspect ratio between 2.4 and 3.5 could provide good thermal comfort.

Traditionally, tropical urban areas follow a linear development (Fig. 7.1) to enhance natural ventilation for cooling. Urban layouts are based on low density and large inter-building spaces. The buildings have high-pitched roofs and high internal height to drain rainwater and enhance air circulation, respectively. Paths and streets are shaded predominantly by vegetation.

Putrajaya (Fig. 7.2) is a modern planned tropical city (Siong 2006), where the low-rise environments experience high temperatures compared to high-rise environments due to alignment of the streets with the sun path (Qaid et al. 2016). Furthermore, in the low-rise environment, the buildings are arranged linearly, forming a walling effect to the incoming wind, while in high-rise areas, buildings cast shadow and streets are perpendicular to enhance wind movement. Even before the completion of this city plan, maximum temperature has reached 37 °C. Similar situation could be noticed in recently developed high-rise high-density environments in Colombo (Rathnayake et al. 2019). Qaid et al. (2016) are of the opinion that the Putrajaya city will face even more alarming conditions upon completion in 2025 due to increase in built-up area and population, evidencing lack of consideration for

Fig. 7.1 Historical image of City of Victoria (source: Writes 1908)

thermal comfort and associated energy implications. In most tropical cities, urban morphology is a reflection of weak planning controls and building regulations (Giridharan et al. 2008; Ng 2012; Giridharan and Emmanuel 2018).

The unregulated rapid urbanisation is changing outdoor conditions; most places which historically had high temperatures around 35 °C are now experiencing 40 °C during summer. Some cities in India have even registered temperatures around 50 °C. Furthermore, in the fast-developing tropics, low-lying lands, flood plains and green areas are being built to accommodate the ever-increasing population. The situation could be improved by mapping the urban climate with a regular feed-in mechanism and developing an urban morphology that respects the local thermal comfort requirements. Urban morphology features should be manipulated to create distinct physical areas within a neighbourhood that generate positive and negative UHI influences such that those areas with positive UHI during the day will create a

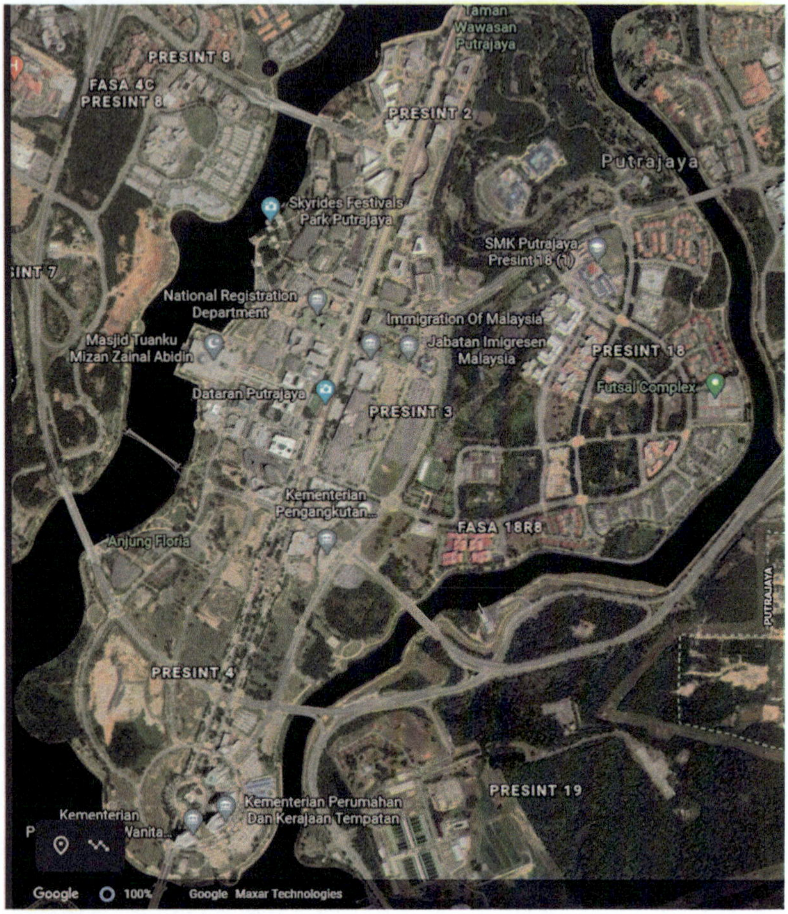

Fig. 7.2 Putrajaya—a tropical planned city (source: Google Earth 2020)

negative effect at night and vice versa by creating high-sky-view-factor and low-sky-view-factor areas, and similarly high-albedo and low-albedo areas within the same neighbourhood. Such an environment will enable better cross-ventilation for both day and night. However, these positive and negative environments should be carefully linked so that they are not subjected to the urban walling effect and consequent increases in UHI (Giridharan et al. 2007). Moreover, mixing of density and ground coverage in terms of green and hard surface could create these positive and negative effects in urban environment, and enhance ventilation rate (Givoni 1998; Giridharan et al. 2008).

7.3 Space, Massing Typology and Landscape

In tropics, high degree of integration is needed between indoor and outdoor spaces without major obstructions to enhance ventilation. Therefore, the transitional spaces are as important as indoor and outdoor spaces (Ahmed 2003). To formulate any passive or low-carbon strategy, one needs to understand the limitations at least at site and neighbourhood levels which changes from context to context. In Hong Kong, on-site and neighbourhood-level design variables within 15 and 300 m radius, respectively, will have an impact on outdoor thermal comfort at a point (Giridharan et al. 2007). This point will impact any indoor space it is coming in contact with, especially the fenestration on the façade. Sometimes these transitional spaces are explicit. For example, ventilation in high-rise high-density environment largely depends on site coverage, inter-building distance and height of the buildings (Yuan and Ng 2012).

Similarly, small design details, such as a topographical level change as small as 0.5 m along with adequate shading (say 20–25% tree cover within 1000 m^2), could enhance the cooling potential of ventilation and produce tangible reduction in outdoor temperature in high-rise high-density environments (Giridharan et al. 2008). In Malaysia, it is found that clustered trees are more effective in reducing the surface temperature compared to scattered isolated trees (Ahmed et al. 2015). In tropics, vegetation has the capacity to reduce the thermal impact both day and night (Giridharan and Emmanuel 2018). However, one must look at this in conjunction with air pollution, mainly the dust impact. Linking indoor to highly polluted outdoor spaces through passive strategies could lead to uncomfortable indoor environment. However, more context-specific research is needed since there is uncertainty in the radius of impact. For example, Niu et al. (2015) reported that thermal variations take place within 200 m without explicitly articulating the boundary conditions. On the other hand, Ignatius et al. (2015) have stated that on-site variable should be assessed within 50 m radius of the point under consideration. More importantly, information in this regard is very weak for low-rise and medium-density, and low-rise and low-density, environments.

The on-site and neighbourhood constraints and opportunities in terms of air speed, air temperature and moisture content will determine the overall massing typology. For example, Fig. 7.3 presents the massing typologies in Hong Kong's residential environment. Giridharan et al. (2008) found that on most occasions these massing typologies are not compatible to on-site and neighbourhood design variables. However, open linear type of Repulse Bay development is well integrated to the context taking into account both on-site and neighbourhood-level design variables. Such integration has great potential to incorporate passive and low-carbon strategies for indoor space conditioning. Further, in the tropics, both shading and ventilation are equally important. Especially in high-density settings, courtyard built forms should be 'connected' to the street (in terms of air passage) at ground level, while windows and walls should be shaded with 'permanent' ventilation openings (Tablada et al. 2009). At the same time, evidence from subtropics suggests

MASSING TYPE	TYPICAL PLAN	EXAMPLE
Internal Courtyard	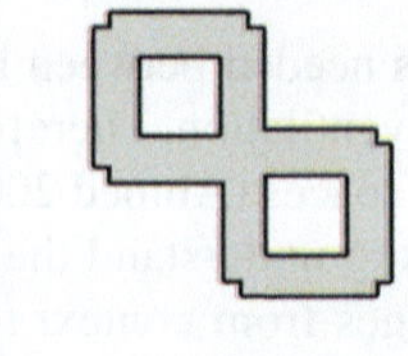	
Closed Linear	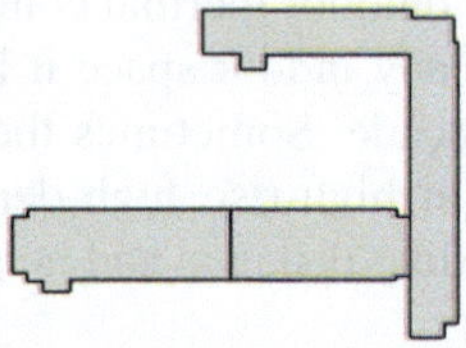	
Open Linear	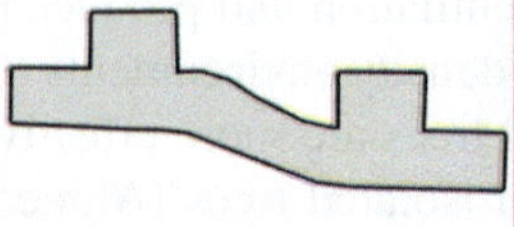	
Open Re-entrant	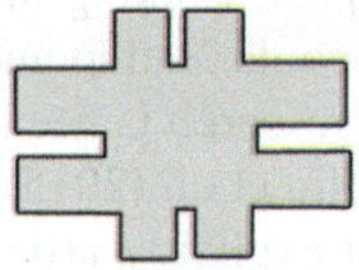	
Open Cross	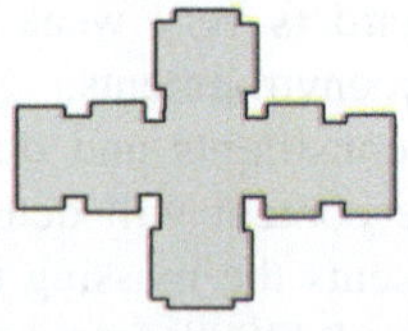	
Open Y	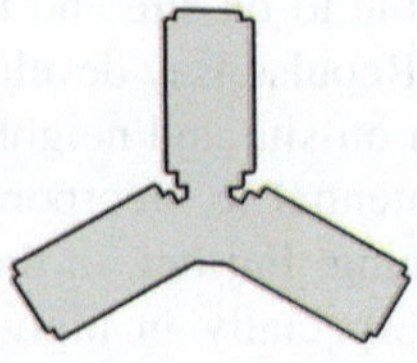	

Fig. 7.3 Massing typology (source: Giridharan et al. 2008)

that at higher densities, the distribution of wind velocity around the buildings became polarised, and weak wind regimes began to dominate (Giridharan and Emmanuel 2018). On most occasions, this happens due to inappropriate massing typology.

Indoor and outdoor thermal conditions will be affected by landscape characteristics such as density of plants, types of plants (trees, shrubs, grass), size, shape of trees and shrubs, and location of plants, trees, etc. Geographical and sky conditions also play a significant role in the impact of vegetation on outdoor thermal conditions. However, in high-rise high-density environments, the effect of vegetation is very intimately linked to the total environmental setting through mutual shading, multiple reflection, etc. (Oke 1987). For example, approximately 40% vegetation within a 1000 m^2 area in a low-rise low-density environment (approximately a net density of 250 persons/ha) could reduce UHI by more than 1.0 °C. But in a high-rise high-density environment (net density of 2400 persons/ha) the potential will be limited to below 1.0 °C in the absence of modification to on-site variables such as sky view factor, altitude and height-to-floor area ratio (Giridharan et al. 2008). Level changes (altitude difference) within the landscape are an important factor in lowering UHI at locations with vegetation. Even a level change as small as 0.5 m along with 20–25% tree cover within a 1000 m^2 area could produce tangible results for both daytime and nocturnal UHI reduction. Further, there should be frequent level changes rather than one or two drops.

In high-rise high-density environments, on-site variables such as sky view factor and altitude could negate the natural behaviour of tree and shrub cover. However, off-site variables such as anthropogenic heat created by traffic and density in the immediate surroundings have the potential to negate the impact of sky view factor and altitude, especially at night. There is also a great relationship between cloud cover (change in solar radiation intensity) and on-site design variables. During partially cloudy days, in certain locations in Hong Kong with more than 40% vegetation within 1000 m^2 area, a 30–40% (high) tree cover will result in at least 1 °C reduction in temperature compared to 0–20% (low) tree cover in a similar setting (Giridharan et al. 2008). However, in similar setting, but at night, a high tree cover resulted only in a 0.5 °C reduction compared to a low tree cover setting. This suggests that pocket parks with permeability in the massing are more effective than large centralised open spaces for high-rise high-density environments. However, when pocket parks in a permeable urban setting are integrated to a large centralised open space, the cooling potential is much higher.

In tropics, it is a common practice to use water body as part of on-site design feature, which can bring thermal relief mainly during early morning hours, but generally creates undesirable effects on microclimate during day and night in the absence of adequate wind velocity (Morris et al. 2016). The people in tropics are generally sensitive to convective and radiative changes at high relative humidity, which is around 90%. Therefore, to make the water very effective, it has to be combined with the right amount of shading. However, shade becomes the most important factor in driving thermal comfort during the hottest periods of the day. On the other hand, even the non-shaded spaces will be acceptable for public, if that space has high wind speed, say anything above 2 m/s. This could be witnessed in some seafront hard-surfaced open spaces in Hong Kong. In general, in tropics, most people could adapt to outdoor temperature as high as 35 °C and participate in outdoor activities (Makaremi et al. 2012).

The stressful outdoor conditions could have major impact on the indoor spaces, especially in naturally ventilated buildings. More specifically, outdoor space comfort has indirect effect on the perception of indoor space comfort (Ahmed 2003). The issues are manifold in social housing in developing countries, especially in post-disaster housing. For example, on most occasions, in Sri Lanka, indoor temperature was found to be above 30 °C in the post-disaster social housing (Coorey et al. 2017). This is largely due to lack of context-specific passive and low-carbon strategies. This could have a major effect on the well-being and productivity of the people, and lead to increased use of active space cooling systems.

In general, in tropics, uncomfortable outdoor spaces have increased the demand for space-conditioning energy which in turn has a direct effect on UHI. A typical office building cluster modelled by Liu et al. (2011) showed that the largest heat island intensity contributed by space-conditioning systems can reach 0.7 °C at midday with a daily average rise of 0.5 °C, and a low temperature set point can further increase anthropogenic heat and outdoor air temperature. Hsieh et al. (2007) have pointed out that in the subtropical Taipei the heat discharged from space-conditioning units raised the outside temperature between 0.5 °C and 2 °C during evenings (7 p.m. to 2 a.m.).

The effect is further modified by materials and geometry of buildings and location of the heat rejection points with respect to the outdoor environment. In the tropics, these changes in outdoor temperature not only have a year-round negative effect on the neighbourhood but also create a positive feedback loop with space-conditioning energy demand. In general, cooling of buildings using space-conditioning systems will lead to dramatic changes in the neighbourhood energy demand, which in turn acts as a feed-in mechanism for further local and regional climate change.

7.4 Practical Realisation of Passive Measures and Its Energy Implications in Buildings

Tinker et al. (2004) performed a study in Malaysia on a modern low-income house and a traditional Malay house by recording the thermal conditions. They concluded that the traditional Malay house has a better design and achieved better thermal comfort. Similarly traditional houses in Mauritius were known to be well shaded by the use of terraces (Fig. 7.4), which regulate heat transfer to the interior spaces, but this has given way to more resistant concrete constructions, which are not necessarily optimised for thermal comfort and energy efficiency.

Fig. 7.4 Traditional creole house with shading and terrace (top). Terrace providing outdoor living space and shading of interior spaces (bottom)

A blend of contemporary and past passive design methods is possible to satisfy the structural, comfort and energy functions. Various approaches for passive design have been listed by Geetha and Velraj (2012) in a taxonomy with three main categories, namely (1) solar and heat protection, (2) heat modulation and (3) heat dissipation, all aimed to improve the energy efficiency of the building.

Due to the aesthetic appeal of glazing, its increased use in modern architecture, especially in high-rise buildings, has been witnessed, but the associated thermal and visual discomfiture and high energy use during operation have come up as indications of poor design. A lot of research effort has been expended to understand the impact of glazed openings on building physics, given their added contribution to daylight penetration and access to external views for which high visible light transmittance (VLT) and low solar heat gain coefficient (SHGC) glass is desirable to minimise energy use in buildings.

The window-to-wall ratio (WWR) has received particularly high interest as a design parameter for optimising thermal comfort or energy use, where setting of suitable ranges for the WWR has been attempted in different contexts. Alwetaishi (2019) assessed thermal comfort by conducting experiments in schools in three cities of Saudi Arabia and concluded that WWR can be maximised up to 0.1 in hot humid climates. Su and Zhang (2010) studied the impact of WWR in Shanghai in hot summer and cold winter climate zones and showed that lower U-value glazing is more effective than WWR for reducing the life cycle environmental impact of a building project, showing that the impact of glazed openings needs to be carefully assessed in terms of its size, placement and material properties.

Large seasonal differences in temperature may require flexible passive design measures, which can take the form of automated systems, but preferably simple adjustable measures which can be operated by the user (Fig. 7.5). The added complexity, cost and maintenance requirement are drawbacks of automated systems, but they may prove their worth in projects where user intervention may not be possible or desirable.

Direct sunlight must exclusively be kept out of the building in tropical climates in summer times using suitable external or internal shading devices. Since the diffused and reflected components of daylight carry little heating effect (around 1 W per 100 lumens), it represents a fraction of the heat gains from artificial light fixtures, and therefore is preferable to artificial lighting, with the added advantage of better uniformity of illumination levels and suitability for human beings. Daylight design typically breaks a window into vision and daylight glazing (Fig. 7.6). Vision glazing provides connection of the occupant to the outside, and is typically situated between 75 cm and 225 cm above the floor level and daylight glazing is the portion of the window found above eye level (225 cm) which is exclusively used to provide interior illumination. In this configuration, light shelves can be used to shade the

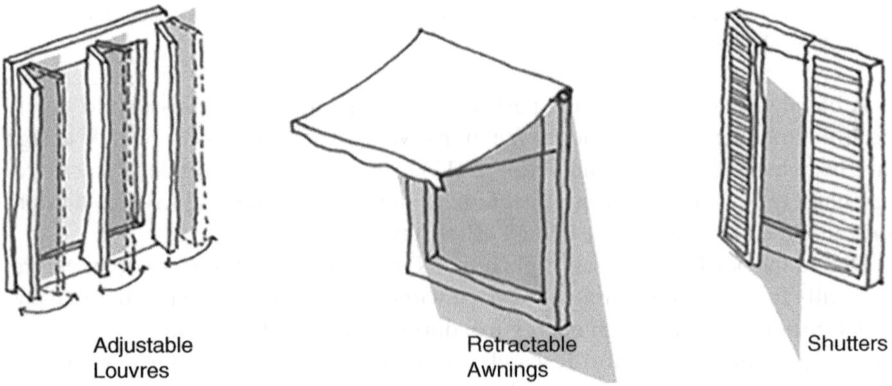

Fig. 7.5 External shading devices

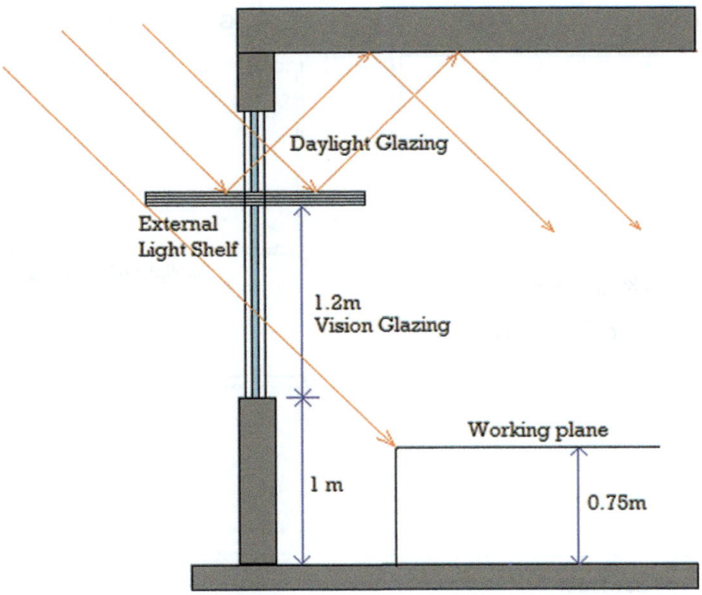

Fig. 7.6 Daylight and vision glazing

vision glazing from direct solar radiation while increasing the penetration of daylight into the interior spaces via daylight glazing and reflecting techniques.

An alternative to using windows to decrease the lighting load and foster daylight penetration is the use of skylights. Being on the horizontal or inclined plane of the roof allows them to admit sunlight directly from the sky dome, thus casting higher illuminance level, up to three times more than windows, over a larger extent of interior area, with the added benefit of providing a more uniform light distribution. Al-Obaidi et al. (2014a) reported that skylight installations are better suited for cold regions where both the lighting and heating brought about by direct sunlight are desirable. Therefore, their use in the predominantly warm tropical context will require skylight designs which control the penetration of direct sunlight (as opposed to the skylight and atrium types, Fig. 7.7) and through careful positioning block or admit direct sunlight penetration as required (for example, the saw tooth, roof monitor and clerestory designs).

The other main components of a building through which heat gains occur are the roof and wall. Heat energy is firstly absorbed by these two components through radiation from the sun and then transferred through the material by conduction. Depending on their thermal mass, these surfaces can be made to act as an energy storage during the daytime, which can be released at night. The profile of building occupancy also becomes an important consideration. For example, offices operating between 9 a.m. and 5 p.m. can benefit from the delayed heat transfer brought about by thermal mass, but this can be detrimental in residential spaces as the heat amassed during the day leads to discomfort in the evening. Thus choice of the appropriate

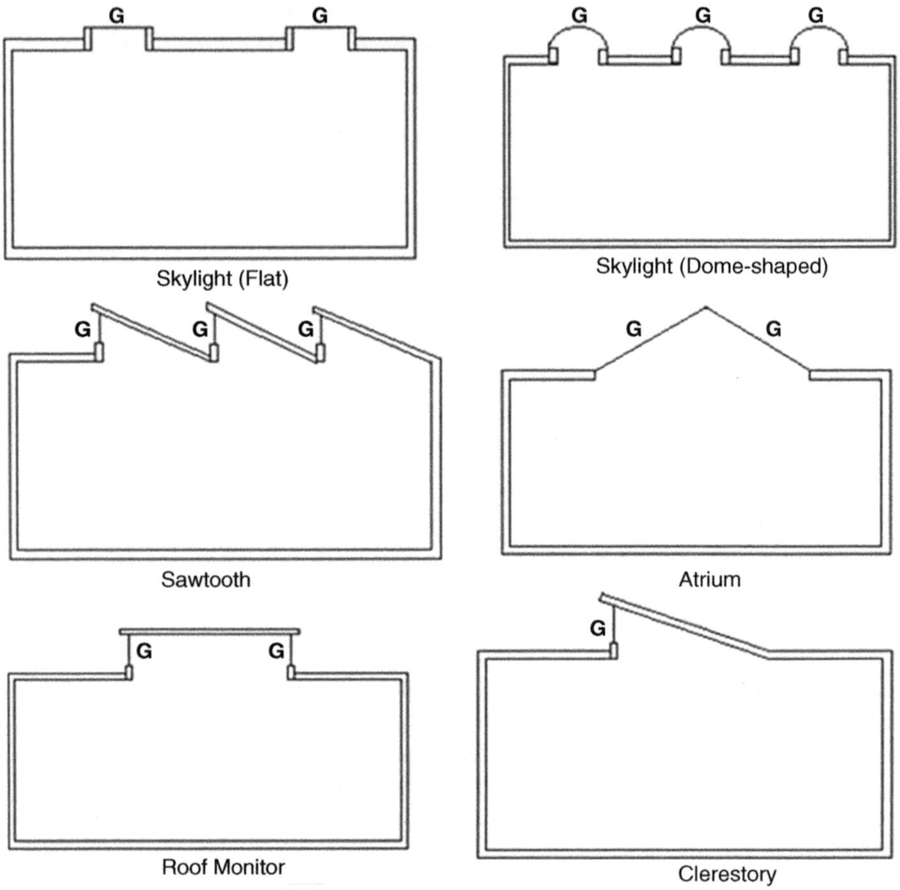

Fig. 7.7 Skylight configurations

material which has thermal properties to store and emit heat energy is another area which holds great potential to cope with the milder diurnal temperature experienced by tropical countries.

For example, a case study of two types of houses in the locality of Sydney by Albayyaa et al. (2019) showed that construction with higher thermal mass (R-value) has a dominant effect on reducing cooling and heating energy requirements, with a decrease of 35% reported for the use of concrete slabs, and up to 58% reduction by combining high thermal mass and passive measures.

On the other hand, Ouedraogo et al. (2012) found that using glazing with lower U-value and material with higher thermal mass is not economically viable for the region of Burkina Faso, a hot arid country in West Africa, as the decrease in cooling load is minimal. The most profitable measure was found to be shading devices. The results showed that the simple use of curtains can result in as much as 40% decrease in cooling load, again lending support for simple and effective shading measures to control heat gains to the interior spaces.

Being the most exposed part of a building, the roof, especially when flat, admits significant amount of heat gains to the interior. Among the various methods available for roofs, the use of reflective and radiative (cool) roof strategies is a simple one (Al-Obaidi et al. 2014b). When selecting a cool roof, the solar reflectivity and thermal emissivity need to be carefully specified. A comparison between the two properties revealed that solar reflectivity has a more pronounced effect on decrease in absorbed sunlight. Cool roof can also be combined with a moving air cavity to not only reflect back solar radiation but also move heat away through convection and natural ventilation (Saw et al. 2018). These studies show the absolute need to regulate heat transfer through the roof, especially for uninsulated and flat roofs, which are usually found in tropical climates. Shanshan et al. (2016) calculated the payback period for a cool roof to be only 1.4 months.

Another measure of increasing interest to control heat gains through the roof and walls is phase-change materials (PCM). Piselli et al. (2018) showed the promising effectiveness of combining a cool roof polyurethane-based membrane and organic paraffin as PCM when compared to a traditional bitumen-based membrane, where a 9 °C reduction in temperature was recorded. However, PCM still remains a costly technology for the milder tropical climates, difficult to economically justify, but research into bespoke materials, especially those that can be sourced locally, can make it a viable solution in the future.

To further improve the indoor conditions for occupants, an increase in air change per hour (ACH) is beneficial in the warm tropical context, both in promoting good indoor quality and in providing cooling through the 'chill' effect of air movement, typically at less than 1 m/s. For the hot and humid climate of Singapore, a study by Hien et al. (2019) showed the effective use of stack and cross-ventilation with high roof height for one- or two-storey buildings. Another method to increase natural ventilation is the use of a wind tower to promote cross-ventilation. A wind tower causes the cooler outdoor air to rush into the building also causing draught, hence improving comfort level (Dehghani-sanij et al. 2015). The same buoyancy effect can be created in a double-skin façade, achieved by having two walls separated by a thin ventilated layer of air. This configuration is particularly interesting for tropical climates as the heat accumulated in the thermal mass of the building fabric can be effectively vented out.

Gooroochurn et al. (2019) proposed a space-saving ventilated façade using hollow-cavity concrete blocks and reported a maximum drop of 6 °C in internal wall surface temperature for the west façade compared to an unventilated façade. Huang et al. (2017) found that the heat flux carried away by the airflow in the cavity ranges from a minimum of 10 W/m^2 during low light intensity to a maximum of 269 W/m^2 during peak sun hours, confirming the ability of the ventilated cavity to extract heat energy from the building fabric.

Another on-site natural resource which can be used to promote good indoor thermal conditions is the earth's (almost infinite) thermal capacity. Several researches have been carried out to investigate the possibility of an earth-air heat exchanger

(EAHE) buried under the ground (Singh et al. 2018; Vidhi 2018). Liu et al. (2019) performed on-site experiment with a 15.5 m long U-tube heat exchanger using air as moving fluid and air temperature at the outlet was found to range from 22.4 °C to 24.4 °C in summer and from 16.0 °C to 18.0 °C in winter. Serageldin et al. (2016) carried out a similar study under Egyptian weather conditions, and obtained 27 °C and 32.9 °C for the outlet and ambient temperatures, respectively, during summer conditions. They also favoured the more cost-effective PVC pipes despite slightly better performance with metal pipes. The integration of this technology in the tropical context will require a shift in the construction industry to bring in the needed expertise for the design and installation of such systems with due consideration of condensation due to humid conditions.

7.5 Microenvironment: Passive Measures for Building External Environment and Neighbourhood

Landscaping can be used to shield both vertical surfaces, namely walls and glazing, and horizontal and inclined surfaces such as roofs and terraces. Gómez-Muñoz et al. (2010) investigated the effect of shading of trees on walls in the city of La Paz in Mexico. The results obtained show that the tree can stop most of the heat energy on the façades. Berry et al. (2013) conducted a similar study in the University of Melbourne Burnley Campus using potted plants placed at equal distance in the north and west of two small single-room green weatherboard buildings surrounded by grass, positioned so that they do not shade each other. A third, unshaded building was used as control. The reduction in ambient and surface temperature was unequivocal and they noticed that the distance between the trees and building is critical in determining the shade coverage, especially for tree canopies which do not exceed the roofline of the building.

Similarly Morakiny et al. (2013) performed an experimental investigation in Nigeria on two buildings, one of them shaded by tall trees. The climate at the project location was classified as tropical with the temperature ranging between 21.4 °C and 31.1 °C annually. They found that under any sky condition, the outdoor temperature of the unshaded building was higher than the shaded one by 3.1–5.4 °C and that the positioning of the tree is fundamental for thermal comfort and energy conservation.

Green roofs are also landscaping solutions which can be applied to roof surfaces, and typically consist of the layers illustrated in Fig. 7.8. The installation of green roofs needs to be performed in compliance to relevant standards, so as to preserve the integrity of the roof surface and any associated waterproofing membrane, to ensure safety for users and avoid water ingress.

Feitosa and Wilkinson (2018) investigated the effect of green roofs in two similar houses in Rio de Janeiro, Brazil, and Sydney, Australia, and found a 6 °C and 9 °C drop in maximum temperature, respectively. Jim and Tsang (2011) evaluated the

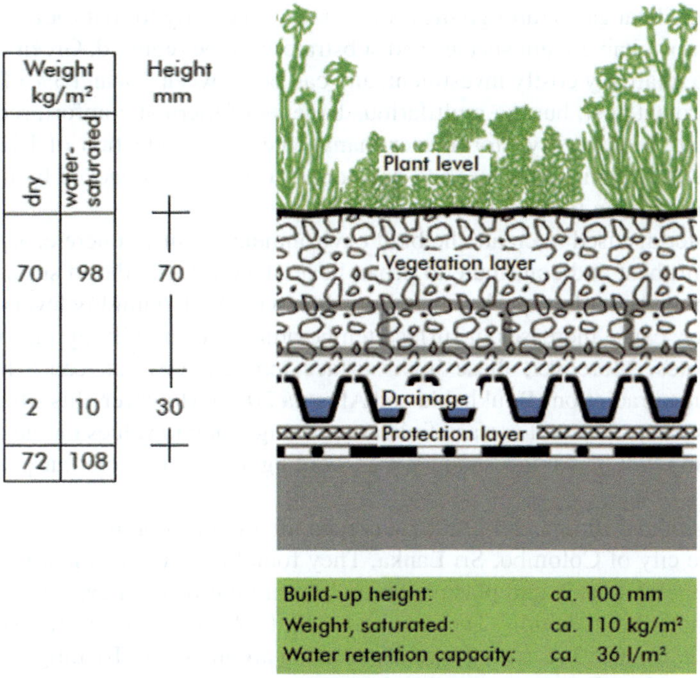

Fig. 7.8 Green roof layers (source: ZinCo GmbH, Germany)

ecological energetics of green roofs in the tropical and humid climate of Hong Kong. They measured the seasonal energy flux above the canopy in the range of 400–550 W/m² while 170–270 W/m² was recorded below the canopy; this further confirms the insulation provided by green roofs to control heat gains and promote thermal comfort. Morau et al. (2012) obtained similar results in Reunion Island, with a maximum average drop of 6.8 °C recorded between the top of the green roof substrate layer and the roof surface over the summer months from October to February, and Mungur et al. (2020) recorded a maximum drop of 9 °C in internal ceiling surface temperature with a green roof as compared to a bare roof over the winter and mild summer conditions from July to October in Mauritius. Both Reunion and Mauritius are Small Island Developing States (SIDS) with tropical climates.

Teemusk and Mander (2009) analysed the effect of green roofs with different substrates in the city of Tartu in Estonia, where the climate is warm, temperate and humid. They concluded that a green roof of 100 mm entirely covered with plants will efficiently mitigate the seasonal surface temperature fluctuations on the roof. They observed that the surface temperature under the plants was a maximum of 30 °C lower than the hard surface and this varied with the leaf area index (LAI) of the plant.

Green roofs represent an interesting passive measure to ward off the intense high-elevation summer sun of the tropical climate for improving thermal comfort and reducing heat gains through the roof surface, especially for flat roofs. However, research into suitable plant species and substrates will be required. Green roofs still represent a relatively costly investment, and can be between 3% and 5% of the total building project cost, but the multifarious benefits of thermal comfort, reduced air conditioning, biodiversity, stormwater management and reduction of UHI can be readily used to convince owners and developers, especially if backed by incentives in the form of green loans and rebates.

The materials used to create the built environment, mainly concrete, asphalt and granite, emit heat and contribute to urban heat island effects. Paved surfaces combined with the lack of evaporative cooling, due to the high humidity levels of tropical climates, cause increases in surface temperature, thus modifying the prevailing microclimates. Moreover, these high surface temperatures can cause high heat transfer due to radiation (Boukhabla and Alkama 2012). However, this situation can be addressed by choice of surface finishes with high albedo values on the one hand and planting tall trees for shading and cooling through evapotranspiration on the other.

Johansson and Emmanuel (2006) surveyed the thermal comfort in various locations in the city of Colombo, Sri Lanka. They found that surface materials such as asphalt, concrete, brick and plaster, which have a high thermal capacity, absorbed most of the solar radiation. The maximum PET observed during the study was between 33 °C and 50 °C with the peak occurring around noon. To mitigate this heat island effect, they found that shading from tree canopies in the street canyons and movement of air in between the building should be improved.

At neighbourhood level Coutts et al. (2015) conducted a study in three different streets with different tree canopy coverage in Melbourne, Australia. They found that trees played an important role in mitigating urban air temperature and the cooling effect of trees was more prevailing in open and shallow street canyons. Studies by Nasir et al. (2015) and Shahidan et al. (2012) in Malaysia confirmed the benefit of tall trees, albeit an increase in humidity in the former research. Yang et al. (2018) found that the combination of grass and trees proved to be more beneficial as the maximum surface temperature dropped by 10 °C, the air temperature by 0.75 °C and the mean radiant temperature by 16 °C in a study in Singapore.

In Malaysia, Ghaffarianhoseini et al. (2015) showed that covering 75% of the courtyard with trees yielded the highest temperature decrease, between 2.5 °C and 3.3 °C, but with a drastic decrease in wind speed and flow. They also deduced that the use of grass in the courtyard provided very limited effect in the temperature change, namely 0.13 °C decrease, compared to courtyards with bare ground.

Overall, the findings from literature show great potential in reimagining the spaces in the tropical context, to build on past good practices, which seem to have been dropped in favour of more structurally sound constructions, but which can be revived using modern construction and engineering techniques. The passive design approach should be central to this re-imagination as it holds the key to high performance of buildings in the true sense, as opposed to high-tech solutions which are resource intensive. As illustrated in Fig. 7.9, the ideal case would be to have a building layout with the longer axis oriented along the east-west direction, so that the

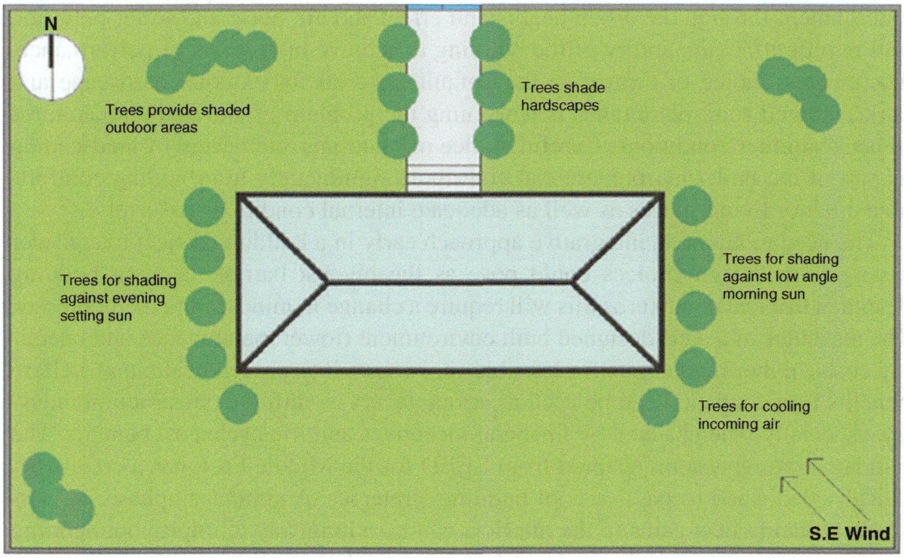

Fig. 7.9 Representative case of an ideal plan of building layout and external passive features for southern hemisphere

high-elevation sun can be readily regulated, and the low-angle sun can be effectively blocked by tall trees. In the southern hemisphere, the south façade is not visited by the sun, and hence can have large glazed areas. However, on most occasions, the geometry of land does not allow this ideal case, and any specific building design therefore requires careful planning of spaces and application of passive design principles to deal with individual interior spaces. This approach will yield buildings with better thermal comfort and energy performance in the tropical context.

7.6 Challenges and Barriers for Energy-Efficient and Comfortable Spaces in Tropical Climates

The engineering of indoor spaces with acceptable environmental conditions for the occupants needs to have a strong focus on passive design principles as opposed to relying on active energy systems. However, the uniqueness of urban and building projects and the complexity of modern urban and architecture design present an optimisation problem with multifarious factors to be considered. For this reason, urban and building design principles cannot be overly prescriptive, especially when it comes to high-rise and high-density setting; a prescriptive approach can work well for the residential sector for individual small housing units.

The design of the external environment needs to be coordinated with the building itself, as the building lives in its environment, and any impoverishment of that environment through the urban heat island effect and air, noise and water pollution will in turn affect the ability of the building to achieve high levels of performance. The current choice of tarmac as the default material for external hardscape and dark-coloured roof membranes is worsening the problem of elevated temperatures in peak summer conditions. Careful choice of materials and design of landscaping as part of the built environment can contribute significantly to providing comfortable outdoor living spaces as well as adequate internal conditions passively.

The need to adopt an integrative approach early in a building project to consider synergies in design choices could pose as the biggest barrier to green building design in tropical contexts as this will require a change in mindset and methodology. The attributes of a well-designed built environment (lower maintenance and operating costs, higher market rate for leasable space as well as proven health and welfare benefits for occupants) can be used as strong levers to shift the construction industry, especially when backed by financial incentives and life cycle cost benefits. The Estidama rating system, adapted from LEED for the Middle East, has a credit specifically dedicated to passive cool building strategies. A similar emphasis on passively reducing heat gains to the interiors can go a long way to improving building energy performance and thermal comfort in tropical contexts.

Due to the complexity of the interaction of design variables on target outputs like energy consumption, air quality, thermal comfort and daylight factors, it has become necessary to use software packages to carry parametric analyses. Training in the application of building physics and passive design methods for the local context needs to equip learners with knowledge and skills to use urban and building energy modelling and dynamic simulation software packages. The lack of trained manpower can be addressed through specialised training programmes, for example short courses or master programmes.

Fundamental to the correct and efficient use of software packages for energy-efficient building design is the availability of representative weather parameters for the different climate zones of the region. However, the need for weather data is not restricted to its requirement by software packages only; it is a fundamental prerequisite for any sustainable design effort, and the unavailability of or inaccessibility to weather files for tropical climates remains a serious barrier to permit any possible effort for setting passive design guidelines or for carrying out building optimisation for a specific project.

The absence of standard weather files for a given tropical country cascades down to failure to set suitable benchmarks for energy performance and definition of climate zones. In the absence of climate zones, it becomes challenging to set a reference or notional building for a given location. The definition of reference or notional buildings has been fundamental to designing more energy-efficient buildings, for example the ASHRAE 90.1 performance rating method or the UK Part L regulations.

Finally, to lead by example, and to increase confidence in the adoption of passive design principles in the general population and the construction industry, demonstration projects should be constructed to showcase green practices, as has been effectively done in several countries. Governments can lead by example by greening public building projects to showcase the implementation of bespoke passive design elements to the research, student and industry communities; simply promulgating passive design principles based on research findings tends to be perceived as academic and impractical, and hence does not have the desired market penetration and commercial value.

References

Adebayo, Y. R. (1990–1991). Consideration for climate-sensitive design in tropical Africa. *Energy and Buildings, 15*(1–2), 15–21.

Ahmed, K. S. (2003). Comfort in urban spaces: Defining the boundaries of outdoor thermal comfort for the tropical urban environments. *Energy and Buildings, 35*(1), 103–110.

Ahmed, A., Ossen, D., Jamei, E., Manaf, N., Said, I., & Ahmed, M. (2015). Urban surface temperature behaviour and heat island effects in a tropical planned city. *Theoretical and Applied Climatology, 119*, 493–514.

Albayyaa, H., Hagare, D., Saha, S., & S. (2019). Energy conservation in residential buildings by incorporating passive solar and energy efficiency design strategies and higher thermal mass. *Energy and Buildings, 182*, 205–213. ISSN 0378-7788.

Al-Obaidi, K. M., Ismail, M., & Rahman, A. M. A. (2014a). A study of the impact of environmental loads that penetrate a passive skylight roofing system in Malaysian buildings. *Frontiers of Architectural Research, 3*(2), 178–191. ISSN 2095-2635.

Al-Obaidi, K. M., Ismail, M., & Rahman, A. M. A. (2014b). Passive cooling techniques through reflective and radiative roofs in tropical houses in Southeast Asia: A literature review. *Frontiers of Architectural Research, 3*(3), 283–297. ISSN 2095-2635.

Alwetaishi, M. (2019). Impact of glazing to wall ratio in various climatic regions: A case study. *Journal of King Saud University-Engineering Sciences, 31*(1), 6–18. ISSN 1018-3639.

Berry, R., Livesley, S. J., & Aye, L. (2013). Tree canopy shade impacts on solar irradiance received by building walls and their surface temperature. *Building and Environment, 69*, 91–100.

Boukhabla, M., & Alkama, D. (2012). Impact of vegetation on thermal conditions outside, thermal modeling of urban microclimate, Case study: The street of the republic, Biskra. *Energy Procedia, 18*, 73–84.

Chen, L., Ng, E., An, X., Ren, C., Lee, M., Wang, U., & He, Z. (2012). Sky view factor analysis of street canyons and its implications for daytime intra-urban air temperature differentials in high-rise, high-density urban areas of Hong Kong: A GIS-based simulation approach. *International Journal of Climatology, 32*, 121–136.

Chow, W. T. I., & Roth, M. (2006). Temporal dynamics of the urban heat island of Singapore. *International Journal of Climatology, 26*, 2243–22460.

Coorey, S., Perera, N., Giridharan, R., & Emmanuel, R. (2017). *Thermal Comfort study in post disaster housing in the southern coast of Sri Lanka*. Moratuwa: University of Moratuwa.

Coutts, A. M., White, E. C., Tapper, N. J., Beringer, J., & Livesley, S. J. (2015). Temperature and human thermal comfort effects of street trees across three contrasting street canyon environments. *Theoretical and Applied Climatology, 124*(1–2), 55–68.

Dehghani-sanij, A., Soltani, M., & Raahemifar, K. (2015). A new design of wind tower for passive ventilation in buildings to reduce energy consumption in windy regions. *Renewable and Sustainable Energy Reviews, 42*, 182–195. ISSN 1364-0321.

Dimoudi, A., Kantzioura, A., Zoras, S., Pallas, C., & Kosmopoulos, P. (2013). Investigation of urban microclimate parameters in an urban center. *Energy and Buildings, 64*, 1–9.

Doan, Q. V., & Kusaka, H. (2016). Numerical study on regional climate change due to the rapid urbanization of greater Ho Chi Minh City's metropolitan area, over the past 20 years. *International Journal of Climatology, 36*, 3633–3650.

Emmanuel, R. (2005). Thermal comfort implication of urbanisation in a warm humid city: The Colombo Metropolitan Region (CMR), Sri Lanka. *Building and Environment, 40*, 1591–1601.
Feitosa, R. C., & Wilkinson, S. J. (2018). Attenuating heat stress through green roof and green wall retrofit. *Building and Environment, 140*, 11–22.
Geetha, N. B., & Velraj, J. (2012). Passive cooling methods for energy efficient buildings with and without thermal energy storage—A review. *Energy Education Science and Technology Part A: Energy Science and Research, 29*, 913–946.
Ghaffarianhoseini, A., Berardi, U., & Ghaffarianhoseini, A. (2015). Thermal performance characteristics of unshaded courtyards in hot and humid climates. *Building and Environment, 87*, 154–168.
Giridharan, R. (2016). Urban climate modelling - challenges in tropics. In R. Emmanuel (Ed.), *Urban climate challenges in the tropics: Rethinking the design basis and opportunities* (pp. 255–299). London: Imperial College Press.
Giridharan, R., & Emmanuel, R. (2018). The impact of urban compactness, comfort strategies and energy consumption on tropical urban heat island intensity: A review. *Sustainable Cities and Society, 40*, 677–687.
Giridharan, R., Lau, S. S. Y., Ganesan, S., et al. (2007). Urban design factors influencing urban heat island intensity in high-rise high-density environments of Hong Kong. *Building and Environment, 42*, 3669–3684.
Giridharan, R., Lau, S. S. Y., Ganesan, S., & Givoni, B. (2008). Lowering the outdoor temperature in high-rise high density residential developments of coastal Hong Kong: The vegetation influence. *Building and Environment, 43*, 1583–1595.
Givoni, B. (1998). *Climate considerations in building and urban design*. New York, NY: John Wiley & Sons.
Golany, G. S. (1996). Urban design morphology and thermal performance. *Atmospheric Environment, 30*, 455–465.
Gómez-Muñoz, V., Porta-Gándara, M., & Fernández, J. (2010). Effect of tree shades in urban planning in hot-arid climatic regions. *Landscape and Urban Planning, 94*(3–4), 149–157.
Gonzalez, J. (2005). Urban heat islands developing in coastal tropical cities, Eos, Transactions. *American Geographical Union, 86*(42), 397–412.
Google Earth Image of Putraya City. (2020). *Google earth, Google Maxar Technologies, USA*. Retrieved 28 August, 2020, from https://earth.google.com/web/search/Putrajaya,+Malaysia/.
Gooroochurn, M., Shamachurn, H., Surnam, Y., & Cadersa, A. (2019). Investigation and in situ measurement of dynamic thermal performance of concrete block in Mauritian context, *Technical Report, Mauritius Research and Innovation Council.*
Hien, W. N., Erna, T., & Sukma, A. A. (2019). *Utilization of natural ventilation for hot and humid Singapore* (pp. 165–184). Singapore: Springer Singapore. ISBN 978-981-13-7519-4.
Hsieh, C. M., Aramaki, T., & Hanaki, K. (2007). The feedback of heat rejection to air conditioning load during the night-time in subtropical climate. *Energy and Buildings, 39*, 1175–1182.
Huang, J., Lu, X. X., & Sellers, J. M. (2007). A global comparative analysis of urban form: Applying spatial metrics and remote sensing. *Landscape and Urban Planning, 82*(4), 184–197.
Huang, J., Yu, J., Xu, X., Ye, H., Xiong, C., Wang, J., & Tian, L. (2017). A semi-dynamic heat transfer model of hollow block ventilated wall for thermal performance prediction. *Energy and Buildings, 134*, 285–294. ISSN 0378-7788.
Ignatius, M., Wong, N. H., & Jusuf, S. K. (2015). Urban microclimate analysis with consideration of local ambient temperature, external heat gain, urban ventilations and outdoor thermal comfort in the tropics. *Sustainable Cities and Society, 19*, 121–135.
Jim, C. Y., & Tsang, S. (2011). Ecological energetics of tropical intensive green roof. *Energy and Buildings, 43*(10), 2696–2704.
Johansson, E., & Emmanuel, R. (2006). The influence of urban design on outdoor thermal comfort in the hot, humid city of Colombo, Sri Lanka. *International Journal of Biometeorology, 51*(2), 119–133.

Knowles, R. L. (1977). *Energy and form: An ecological approach to urban growth*. New York, USA: The MIT press.
Lau, S. S. Y., Giridharan, R., & Ganesan, S. (2005). Multiple intensive land use: Case studies in Hong Kong. *Journal of Habitat International, 29*(3), 527–546.
Lau, S. S. Y., Yang, F., Tai, J., Wu, X. L., & Wang, J. (2010). The study of summer-time heat island, built form and fabric in a densely built urban environment in compact Chinese cities: Hong Kong. *Guangzhou, International Journal of Sustainable Development, 14*, 30–48.
Liu, J., Ma, F., & Li, Y. (2011). The effect of anthropogenic heat on local heat island intensity and the performance of air conditioning systems. *Advanced Materials Research, 250–253*, 2975–2978.
Liu, Z., Yu, Z. J., Yang, T., Li, S., Mankibi, M. E., Roccamena, L., Qin, D., & Zhang, G. (2019). Experimental investigation of a vertical earth-to-air heat exchanger system. *Energy Conversion and Management, 183*, 241–251. ISSN 0196-8904.
Makaremi, N., Salleh, E., Jaafar, M. Z., & Ghaffaria Hoseini, A. (2012). Thermal comfort conditions of shaded outdoor spaces in hot and humid climate of Malaysia. *Building and Environment, 48*, 7–14.
Morakiny, T. E., Balogun, A. A., & Adegun, O. B. (2013). Comparing the effect of trees on thermal conditions of two typical urban buildings. *Urban Climate, 3*, 76–93.
Morau, D., Libelle, T., & Garde, F. (2012). Performance evaluation of green roof for thermal protection of buildings in reunion Island. *Energy Procedia, 14*, 1008–1016.
Morris, K. I., Chan, A., Ooi, M. C., Oozeer, M. Y., Akbar, Y. A., & Morris, K. J. K. (2016). Effect of vegetation and waterbody on the garden city concept: An evaluation study using a newly developed city, Putrajaya, Malaysia. *Computers, Environment and Urban Systems, 58*, 39–51.
Moser, S. (2010). Putrajaya, Malaysia's new federal administrative capital. *Cities, 27*, 285–297.
Mungur, M., Poorun, Y., Juggurnath, D., Ruhomally, Y. B., Rughooputh, R., Dauhoo, M. Z., Khoodaruth, A., Shamachurn, H., Gooroochurn, M., Boodia, N., Chooneea, M., & Facknath, S. (2020). A numerical and experimental investigation of the effectiveness of green roofs in tropical environments: The case study of Mauritius in mid and late winter. *Energy, 202*, 117608.
Nasir, A. R., Ahmad, S. S., Zain-Ahmed, A., & Ibrahim, N. (2015). Adapting human comfort in an urban area: The role of tree shades towards urban regeneration. *Procedia-Social and Behavioral Sciences, 170*, 369–380.
Ng, E. (2012). Towards planning and practical understanding of the need for meteorological and climate information in design of high-density cities: A case study of Hong Kong. *International Journal of Climatology, 32*, 582–598.
Ng, E., Chen, L., Wang, Y., & Yuan, C. (2012). A study on the cooling effects of greening in high-density: An experience from Hong Kong. *Building and Environment, 47*, 256–271.
Ng, W. Y., Chau, C. K., Powell, G., & Leung, T. M. (2015). Preferences for street configuration and street tree planting in urban Hong Kong. *Urban Forestry & Urban Greening, 14*, 30–38.
Nichol, J., & Wong, M. S. (2005). Modelling urban environment quality in a tropical city. *Landscape and Urban Planning, 73*, 49–58.
Niu, J., Liu, J., Lee, T. C., Lin, Z., Mak, C., Tse, K. T., Tang, B. S., & Kwok, K. C. S. (2015). A new method to assess spatial variations of outdoor thermal comfort: Onsite monitoring results and implications for precinct planning. *Building and Environment, 91*, 263–270.
Oke, T. R. (1987). *Boundary layer climate*. USA: Methuen.
Oke, T. R. (1988). The urban energy balance. *Progress in Physical Geography, 12*, 471–508.
Ouedraogo, B., Levermore, G., & Parkinson, J. (2012). Future energy demand for public buildings in the context of climate change for Burkina Faso. *Building and Environment, 49*, 270–282. ISSN 0360-1323.
Piselli, C., Castaldo, V. L., & Pisello, A. L. (2018). How to enhance thermal energy storage effect of PCM in roofs with varying solar reflectance: Experimental and numerical assessment of a new roof system for passive cooling in different climate conditions. *Solar Energy, 192*, 106–119. ISSN 0038-092X.

Qaid, A., & Ossen, D. (2015). Effects of asymmetrical street aspect ratios on microclimates in hot, humid regions. *Internal Journal of Biometeorology, 59*, 657–677.
Qaid, A., Lamit, H. B., Ossen, D. R., & Shahminan, R. N. R. (2016). Urban heat island and thermal comfort conditions at micro-climate scale in a tropical planned city. *Energy and Buildings, 133*, 577–595.
Rajagopalan, P., Lim, K. C., & Jamei, E. (2014). Urban heat island and wind flow characteristics of a tropical city. *Solar Energy, 107*, 159–170.
Rathnayake, N. U., Perera, N. G. R., & Emmanuel, R. (2019). *Anthropogenic heat implications of Colombo core area development plan*. Leuven, Belgium: International Conference on Environment Science and Engineering (ICESE).
Roth, M. (2007). Review of urban climate research in (sub) tropical regions. *International Journal of Climatology, 27*, 1859–1873.
Santamouris, M. (Ed.). (2001). *Energy and climate in the urban built environment*. London, UK: James & James.
Saw, L. H., Yew, M. C., Yew, M. K., Ng, T. C., Chen, K. P., Rajkumar, D., & Beh, J. H. (2018). Experimental analysis on the active and passive cool roof systems for industrial buildings in Malaysia. *Journal of Building Engineering, 19*, 134–141. ISSN 2352-7102.
Schiller, S. D., & Evans, J. M. (1990–1991). Bridging the gap between climate and design at the urban and building scale: Research and application. *Energy and Buildings, 15*(1–2), 51–55.
Serageldin, A. A., Abdelrahman, A. K., & Ookawara, S. (2016). Earth-air heat exchanger thermal performance in Egyptian conditions: Experimental results, mathematical model, and computational fluid dynamics simulation. *Energy Conversion and Management, 122*, 25–38. ISSN 0196-8904.
Shahidan, M. F., Jones, P. J., Julie, G., & Salleh, E. (2012). An evaluation of outdoor and building environment cooling achieved through combination modification of trees with ground materials. *Building and Environment, 58*, 245–257.
Shanshan, T., Zhang, Z., & Yu, H. (2016). Life cycle analysis of cool roof in tropical areas. *Procedia Engineering, 169*, 392–399. ISSN 1877-7058.
Sharmin, T., Steemers, K., & Matzarakis, A. (2015). Analysis of microclimate diversity and outdoor thermal comfort perceptions in the tropical megacity, Dhaka, Bangladesh. *Building and Environment, 94*, 734–750.
Singh, R., Sawhney, R., Lazarus, I., & Kishore, V. (2018). Recent advancements in earth air tunnel heat exchanger (EATHE) system for indoor thermal comfort application: A review. *Renewable and Sustainable Energy Reviews, 82*, 2162–2185. ISSN 1364-0321.
Siong, H. C. (2006). *Putrajaya - administrative centre of Malaysia - Planning concept and implementation, Sustainable urban development and governance conference*. Seoul, South Korea: SungKyanKwan University.
Su, X., & Zhang, X. (2010). Environmental performance optimization of window-to-wall ratio for different window type in hot summer and cold winter zone in china based on lifecycle assessment. *Energy and Buildings, 42*(2), 198–202. ISSN 0378-7788.
Tablada, A., De Troyer, F., Blocken, B., Carmeliet, J., & Verschure, H. (2009). On natural ventilation and thermal comfort in compact urban environments—The Old Havana case. *Building and Environment, 44*, 1943–1958.
Teemusk, A., & Mander, Ü. (2009). Green roof potential to reduce temperature fluctuations of a roof membrane: A case study from Estonia. *Building and Environment, 44*(3), 643–650.
Thompson, R., Horngold, R., Page, L., & Waite, T. (2018). Association between high ambient temperature and heat waves with mental health outcomes: A systematic review. *Public Health, 161*, 171–191.
Tinker, J.A., Ibrahim, S.H., & Ghisi, E. (2004). An evaluation of thermal comfort in typical modern low income housing in Malaysia. In *Thermal Performance of Exterior Envelopes of Whole Buildings IX*.
Vidhi, R. (2018). A review of underground soil and night sky as passive heat sink: Design configurations and models. *Energies, 11*(11), 2941.

Villadiego, K., & Velay-Dabat, M. A. (2014). Outdoor thermal comfort in a hot and humid climate of Colombia: A field study in Barranquilla. *Building and Environment, 75*, 142–152.

Writes, A. (1908). *Twentieth century impressions of Hong-Kong, Shanghai, and other Treaty Ports of China*. London, Lloyd's Greater Britain Pub. Co. Retrieved from https://archive.org/details/twentiethcentury00wriguoft/page/146/mode/2up.

Yang, F., Lau, S. S. Y., & Qian, F. (2010). Summertime heat island intensities in three high-rise housing quarters in inner-city shanghai China: Building layout, density and gareenery. *Building and Environment, 45*, 115–134.

Yang, W., Lin, Y., & Li, C. Q. (2018). Effects of landscape design on urban microclimate and thermal comfort in tropical climate. *Advances in Meteorology, 2018*, 1–13.

Yuan, C., & Ng, E. (2012). Building porosity for better urban ventilation in high-density cities—A computational parametric study. *Building and Environment, 50*, 176–189.

Chapter 8
Urban Microclimatic Conditions in Arid Climates

Irene Marincic and Jose Manuel Ochoa

8.1 Introduction

Urban heat island (UHI) effects are considered one of the most important problems of the twenty-first century (Shalaby 2011). By combining the civilization growth phenomenon, the increase of land use (Ali et al. 2017), and the effects of climate change, urban warming is a consequence that is currently difficult to control in many cities, and a real challenge for urbanists, architects, landscape architects, and authorities.

The increase of urban temperature directly impacts people's thermal comfort outdoors. Furthermore, the thermal conditions of outdoor spaces around buildings can affect indoor climate, thermal comfort of the occupants, and energy consumption needed in these buildings (Önder and Akay 2014). This last consequence leads to an increased use of air-conditioning, more greenhouse gas emissions, and increase of UHI. This problem is even greater in cities built in the desert, where the local climate presents extremely hot conditions, especially during the summer period.

We will address our study in the Sonoran Desert in North America, which includes mainly Sonora in Mexico and Arizona in the United States, as well as part of California (USA) and part of Baja California (Mexico). This desert has different regions and sub-climates, but in most cases climatic conditions are very harsh. For example, the climate of Hermosillo city (Mexico), where this study is focused as a representative city of the Sonoran Desert, is characterized by high solar radiation levels, clear skies the whole year, and high temperature oscillations daily and during the different seasons. Summers are very warm, with daily temperatures between 25–30 °C and 40–45 °C, and relative humidity between 15% and 50%. Summer wind is usually warm, so it is not useful for passive cooling or for a better outdoor thermal comfort. Winters are mild, with minimum temperatures from 0 °C to 7 °C

I. Marincic (✉) · J. M. Ochoa
Department of Architecture and Design, University of Sonora, Hermosillo, Mexico
e-mail: irene.marincic@unison.mx; josemanuel.ochoa@unison.mx

M. Palme, A. Salvati (eds.), *Urban Microclimate Modelling for Comfort and Energy Studies*, https://doi.org/10.1007/978-3-030-65421-4_8

and maximum temperatures between 25 °C and 30 °C. During 5 or 6 months per year, the use of air-conditioning inside buildings is almost constant (day and night).

Under these harsh circumstances, local people adapt their way of life by taking a "siesta" (nap) during the afternoon hours, a very common practice. In general, the necessary physical activities and movements, such as walking, are also done very slowly. Thermal comfort surveys of local people show a wide comfort range and very high indoor neutral temperatures (T_n) (Marincic et al. 2013; Ochoa and Marincic 2016), compared with other climate comfort studies (Givoni 1998; Gómez-Azpeitia et al. 2014). Behavioral, social, and cultural factors, including people's expectations and acclimatization, can help overcome adverse thermal conditions indoors and outdoors, but in many cases, they may also decrease work productivity and, most importantly, risky health conditions make it very difficult to live in the urban desert.

Thermal sensation outdoors varies from indoors because of people's lower thermal comfort expectations. But anyway, thermal conditions, especially in summer, are extreme. Thermal comfort conditions in outdoor spaces are also important for social reasons: public spaces are naturally appropriate to get together, talk, and hang out with people between activities. If outdoor conditions are not barely comfortable, people can only walk fast to reach their car and leave.

Conscious and well-designed outdoor spaces can not only contribute to habitable outdoor spaces, but also control UHI temperatures and provide more energy-efficient buildings.

8.2 Desert Climate and Thermal Comfort

The adaptive comfort model (Nicol and Humphreys 2002) considers that people's thermal sensation depends on climatic parameters, i.e., temperature, humidity, radiant temperature, wind, and solar radiation (outdoors), among others, and individual characteristics and situations, such as age, gender, clothing, activity, and subjective issues, such as behavior, expectations, and acclimatization (Nikolopoulou et al. 2004). This approach takes into account not only the physical interaction between the subject and the environment, but also their long-term psychological and physiological interactions, incorporating the effects of acclimatization and the decisions that people can and indeed make to improve thermal sensation.

Although all aspects are important, subjective issues have a vast relevance, mainly in outdoor spaces, and especially for people adapted to local climates. The thermal sensation of this population group can be surprisingly different as expected because of their adaptation to local conditions (Marincic et al. 2005). In hot climates, people's neutral temperature is higher than expected, and if the climate has a wide range of temperature variations, such as the case of desert climates, the comfort range will also be wider (Marincic et al. 2013; Ochoa and Marincic 2016) than that reported for temperate climates, because neutral temperatures are highly related to the outdoor average annual temperature and the comfort range depends on temperature oscillations.

Thermal comfort sensation of people can be very different depending on comfort expectations, which are usually very high in permanent acclimated buildings, with a narrow range of temperature and humidity variations.

In contrast, in naturally ventilated buildings, and even more in outdoor spaces, people have fewer expectations and have more tolerance to extreme temperatures. They tend to expect less comfort as in indoor spaces, compared with people habituated to more temperate climates or constant acclimated environments. Indoor comfort has been much more studied than outdoor comfort, maybe because outdoor comfort does not usually impact energy consumption in acclimatization outdoors (Ochoa 2009). Although it is not the subject that we are going to study, considering indoor comfort in a specific climate, it can be expected that people will probably have less expectations outdoors and therefore they will tolerate more extreme conditions. In a comfort survey in indoor spaces carried out in the mentioned city (Marincic et al. 2012, 2013; Ochoa and Marincic 2016; Gómez-Azpeitia et al. 2014), the thermal comfort in low-cost dwellings was analyzed. The comfort range as a result of the adaptive comfort survey was between 29.7 °C and 34.5 °C in the summer (neutral temperature T_n = 32.2 °C) and between 23.5 °C and 31.3 °C in the winter (T_n = 26.9 °C). In terms of acclimatization, most houses included evaporative coolers and a few of them, window unit air conditioners in only one room of the dwelling. Regarding the thermal comfort outdoors, preliminary results in a field survey in the city of Hermosillo (Ochoa and Marincic 2005) indicate an outdoor neutral temperature during the summer of 36.2 °C. This could be a reasonable result, considering the obtained indoor neutral temperature.

Behavioral factors can also have a high impact on thermal sensation. Some decades ago, when the use of air-conditioning was not generalized, people had to deal with and survive in the desert urban climate using their common sense. During summer nights at their homes, people slept on moistened cots in patios or rooftops facing the clear sky, taking advantage of passive evaporative cooling and radiative cooling strategies. Nowadays, many people use evaporative coolers or, whenever possible, air-conditioning devices. However, basic outdoor space behavioral strategies must be put into action for protection from high levels of solar radiation by using hats, umbrellas, or even a folder (if there is nothing else) (Figs. 8.1 and 8.3). In addition, walking fast among tree shadows, pergolas, and outdoor corridors is also a common practice. Regardless of the high temperatures, people do not usually wear very-low-Clo clothing. For example, they tend to use long-sleeve shirts for protection from solar radiation, and they also frequently use dark clothing for the same reason (Figs. 8.1, 8.2, 8.3, and 8.4).

Aside from the high impact of solar radiation, surface temperature of outdoor space limits related to radiant temperature (and black globe temperature) have an impact on peoples' thermal sensation outdoors. So keeping away from hot surfaces, such as metal roofs and sunny walls, is also a good strategy to mitigate high thermal sensation.

Fig. 8.1 Person protected from solar radiation with an umbrella

Fig. 8.2 People under a tree shadow

Undeniably, the use of water sources to improve passive evaporative cooling could be a successful strategy for certain temperature and humidity conditions. However, this strategy is not widely used in the region, and especially in public spaces, mainly because of low water availability and to avoid maintenance costs.

On the other hand, when formulating cooling strategies for indoor or outdoor spaces in desert climates, we must take into account that some climatic parameters can have a different effect than expected. For example, hot summer winds do not refresh or improve the thermal sensation (Ochoa and Marincic 2005; Ochoa 2017), so it is important to identify which climatic variables and design parameters have more impact (and viability to apply) on microclimatic conditions and on people's thermal comfort.

Fig. 8.3 Cyclist resting under a palm tree shadow

Fig. 8.4 People gathered in a shadowed space

8.3 Impact of Vegetation on Microclimatic Spaces

Most cities in the Sonoran Desert region follow a low-density urban model and their constructions are scattered in the territory. The majority of these cities are made up of low-rise buildings and the main streets are wide due to the excessive use of individual cars. Examples of this model include cities such as Phoenix, Tucson, Mexicali, and Hermosillo, among others.

In general, urban morphology, and mostly flat topography (in the urban areas) as well as the relationship of building height to street width, allows solar radiation to hit almost during the entire day on urban surfaces and to create urban spaces that are not very friendly for pedestrians, considering the extreme climate. At the human scale, the results are few shadows, long walking distances, and few suitable microclimatic spaces to remain outdoors, regardless of the fact that in the mentioned cities, the application of microclimatic design strategies is quite different, and in some of them, efforts to create friendly outdoor spaces can be appreciated.

The climate, city morphology, and urban materials have a relevant impact on UHI effects (Palme et al. 2016), as well as the anthropogenic heat generated by cars, air conditioners, industrial activities, and other heat sources (Önder and Akay 2014). At a smaller scale, the adequate design of outdoor spaces, including the thermal design of buildings (especially the envelope), has an impact on urban climate. In addition, the results can be seen in the short and medium terms on the people's quality of life. This implies improving issues like health, social interactions, and energy consumptions in buildings (Shima et al. 2015).

As mentioned earlier, the design of outdoor spaces has a great impact on microclimatic temperatures and outdoor thermal comfort. Although the comfort sensation is a combination of factors such as radiant temperature (long-wave radiation emitted by the surrounding surfaces), air humidity, and wind speed, air temperature in the shade and solar radiation are the factors that most influence when evaluating thermal comfort in outdoor spaces, specifically in predominantly dry climates. The above has been demonstrated in field studies carried out in our case study in the city of Hermosillo (Ochoa and Marincic 2005). Different design factors, such as space proportions and dimensions, distance between buildings, shading possibilities, surface materials, presence of water sources, and use of vegetation (Önder and Akay 2014), among other aspects, have an impact on peoples' comfort sensation and on outdoor habitability. Desert urban climates have an important cooling potential by landscaping design and urban design strategies, according to studies in different Phoenix neighborhoods (Sonoran Desert), which show that urban form and landscaping design can lower the midafternoon temperatures during summer (Middel et al. 2014). Other studies in the city of Phoenix explore the impact of vegetation shading and different pavement materials to lower the air temperature, to provide better outdoor comfort for pedestrians, and to contribute to the UHI mitigation (Rosheidat and Harvey Bryan 2010).

As an example, different microclimatic spaces will be shown and analyzed in order to compare how different applied strategies work in a desert climate, and to evince how these strategies can influence thermal conditions. The spaces shown in Figs. 8.5, 8.6, and 8.7 are located in the urban area of Hermosillo city, Mexico, as mentioned in the Sonoran Desert. Microclimatic variable measurements are compared with urban and rural meteorological station data in order to visualize differences in climate conditions.

The most challenging season in this climate is definitely summer. Outdoor climatic conditions during the warm period (almost 6 months of the year) are extremely hot and living outside is exceedingly difficult. Comparative data in a summer day

Fig. 8.5 Case 1: Entrance to a building. Aerial view (left) and street view (right)

Fig. 8.6 Case 2: Little garden square. Aerial view (left) and street view (right)

Fig. 8.7 Case 3: Parking lot. Aerial view (left) and street view (right)

are presented, to visualize the differences between climatic variables in the aforementioned microclimatic spaces and other locations within the same city.

Comparing the microclimatic conditions among the three spaces previously described and the city climate (urban) and rural climate (desert), differences in air temperature, relative humidity, and black globe temperature can be observed, where the last variable was compared only in the three microclimatic spaces. The wind speed in the microclimatic spaces varies extensively and it has gusts that are difficult to analyze, so wind measurements are not presented here but the variable was qualitatively evaluated in the site and later commented. Previous outdoor comfort studies in the city (Ochoa and Marincic 2005) also reveal that the wind does not have a clear relation with peoples' thermal sensation; the reason could be that occasionally wind gusts are very warm and do not help lower the thermal sensation (Ochoa and Marincic 2005).

In case 1 (Fig. 8.5), the space partially limited by three vertical surfaces and the floor is shown. It does not have any solar radiation protection. Two vertical surfaces are made out of a white-painted plastered brick wall; one of them is partially shaded by trees. The other vertical surface (entrance) is mainly composed of metal terracotta plates and terracotta cement boards. In this space, air temperature, solar radiation, and radiant temperature are the main variables to consider. The orientation of the façade is west, and the only shades present are the ones coming from the same building and the adjacent buildings at certain times of the day. The ventilation is low because there is a low possibility of cross ventilation due to the geometry of the space. Regarding the surface materials that limit the space, in addition to the described walls, there are concrete and gravel floor materials. Vegetation largely covers the north wall (adjacent building) and a small part of the ground. Vegetation partially protects from infrared radiation emitted by the different surfaces; thereby, it limits the radiant temperature into the space.

Case 2 (Fig. 8.6) is a small garden square with vegetation consisting of trees, shrubs, and ground cover vegetation, but also with concrete and gravel surfaces on floors. Measurements are taken in the proximity of a wooden and metal bench, which is shaded during part of the day by a tree. There are shrubs, grass, and other vegetation that partially limit the emission of the infrared radiation from the floor. It is important to point out that trees are important to shadow the resting spaces and floors and that it is an open space, where the wind can run freely.

Case 3 (Fig. 8.7) is a parking lot with asphalt floor. It is a large space, without sun protection devices, so solar radiation strikes the floor the entire day, without nearby walls and with free wind circulation.

The day measurements were taken, the meteorological air temperature in the city reached a maximum of about 45.7 °C, while the maximum horizontal solar radiation was 949 W/m^2, and it was a completely sunny day, like many others in the region (Fig. 8.8). Effect of UHI in the city can be appreciated in the air temperature differences of both meteorological stations.

It can be noticed that the air temperature (Fig. 8.8) in the microclimatic spaces is slightly lower than the temperature of the meteorological stations (depending on the daytime), and the temperature in the small square is the lowest. In general, at the maximum air temperature time, that was at about 4 p.m., the difference between square temperature and meteorological station temperature was about 1–1.5 °C lower. Analyzing the measured black globe temperature (Fig. 8.9), which is highly related to the radiant temperature, it can be clearly seen that the black globe temperature is lower in the square, which contributes to a better comfort sensation. At

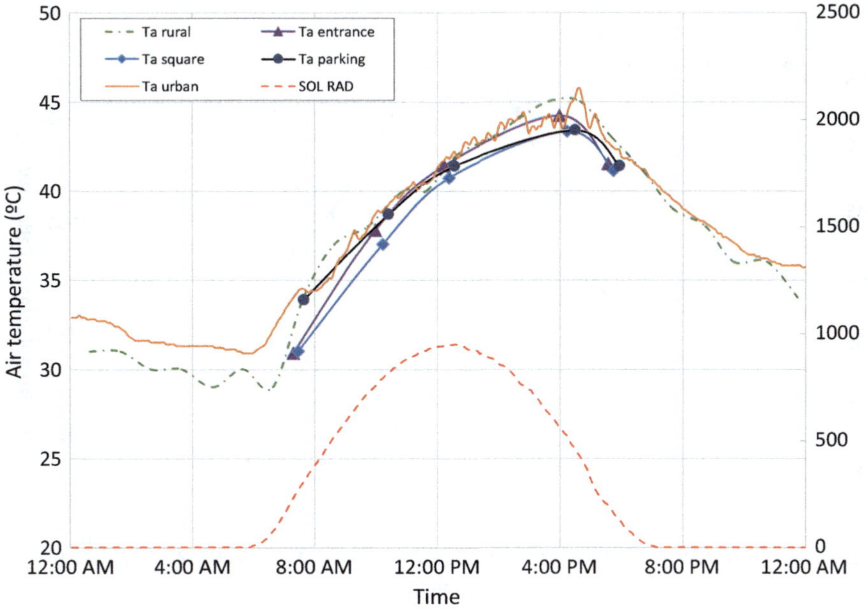

Fig. 8.8 Air temperature in three microclimatic spaces, urban meteorological station, and rural meteorological station near the city (desert) as a function of daytime. Solar radiation is also included as reference of the daylight length

noon, the square black globe temperature was more than 11 °C lower than the parking lot black globe temperature.

Regarding humidity, in the meteorological stations only relative humidity measurements are available. However, comparing relative humidity in rural and urban meteorological stations, it can be assumed that with slightly different air temperatures between both stations, there are appreciable differences in relative humidity (Fig. 8.10). This indicates that absolute humidity (water content in air) would be lower in rural than in urban areas. It is also inferred that these differences in humidity between both cases are due to water sources such as well-irrigated urban vegetation, condensation of air-conditioning devices (in the city of Hermosillo it is usual to drain them on green areas or directly to the street), as well as other human activities. In the microclimatic cases, the presence of vegetation is probably the most important cause of humidity differences.

Apart from the mentioned sources, there are no water sources such as fountains, cascades, or ponds or any other microclimatic contribution to evaporative cooling effect, which in this climate would benefit the thermal comfort. The comparison of relative humidity measurements is shown in Fig. 8.10.

The microclimatic humidity values are close to the urban weather station humidity (Fig. 8.10). However, the parking relative humidity is the lowest value probably due to the lack of vegetation (there are only a few small trees).

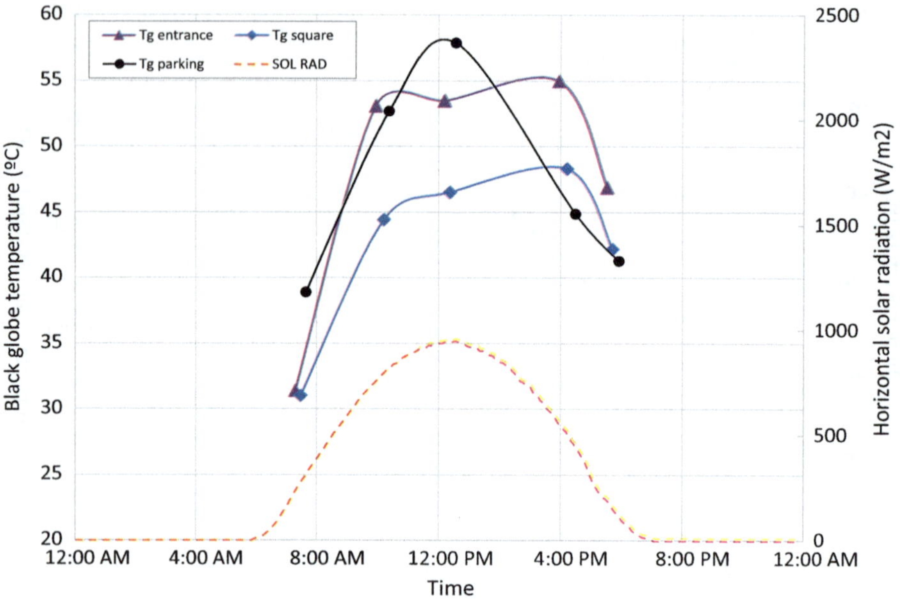

Fig. 8.9 Black globe temperature in three microclimatic spaces as a function of daytime. Solar radiation is also included as reference of the daylight length

The surface materials that limit the spaces and the presence of vegetation that protects with shades the surfaces and prevents the reflection of long-wave radiation played an important role in the results of the microclimatic measurements. It is well known that superficial vegetation temperatures, even in the sun, are similar to air temperature. In the case of very hot climates, with many hours of high radiation levels during the day, shading the space and controlling surface temperature with vegetation can be the most relevant strategies to cope with the climate and have an acceptable thermal sensation in microclimatic spaces.

The surface temperatures of the three cases at about 4:00 p.m. on a summer day can be seen in Figs. 8.11, 8.12, and 8.13, as infrared photographs (right), together with visible photographs (left). Although perspective infrared pictures do not have the same precision as those taken perpendicular to the surface (García Nevado 2018), qualitative and comparative analyses can be useful to help dimension the range of temperatures that can be reached in these spaces.

In the case of the entrance to the building space, which is west oriented (Fig. 8.11), the horizontal concrete surface exposed to the sun had a temperature of about 58 °C, and in the shadow areas, about 52 °C. The perforated terracotta painted plate, part of the sunny front façade, reached more than 46 °C; the solid terracotta plate more than 65 °C; and the terracotta cement board, about 62 °C; the superficial temperature of the white brick wall, at the shadow, was about 46 °C. In contrast, the vegetal ground cover with sun exposure had superficial temperatures between 37 °C and 38 °C, depending on the color, and at the shadow, about 35 °C. Shadowed bushes

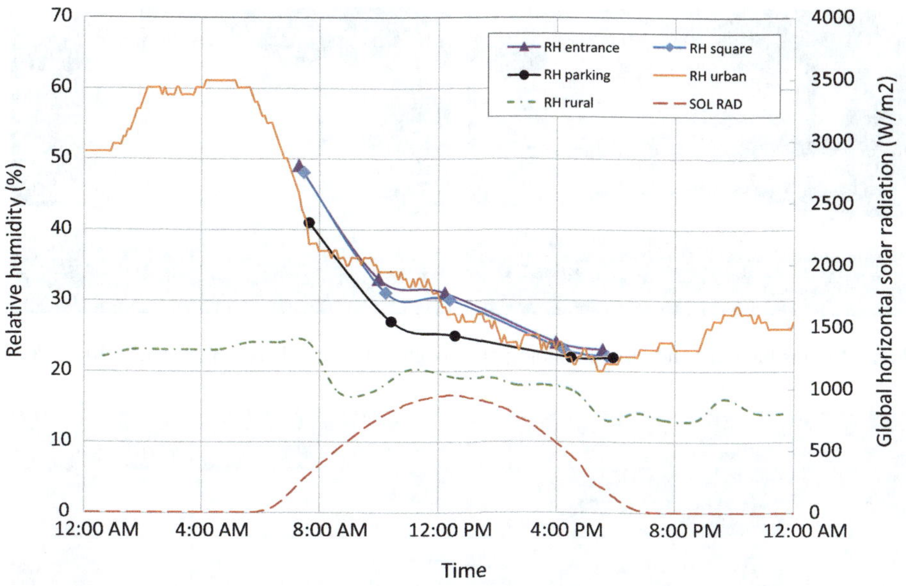

Fig. 8.10 Relative humidity in three microclimatic spaces, urban meteorological station, and rural meteorological station near the city (desert) as a function of daytime. Solar radiation is also included as reference of the daylight length

Fig. 8.11 Visible (left) and infrared photography (right) of case 1. False color scale is 30–70 °C

had a temperature of about 34 °C. The microclimate air temperature at this time was about 44.2 °C. Vegetation at the sun, therefore, maintains a superficial temperature around air temperature at the microclimatic space, or even lower, and the shaded vegetation, about 10 °C lower than air temperature.

Analyzing the infrared photography around the bench at the little square (Fig. 8.12), also at 4:00 p.m., it was possible to measure the temperature of concrete and ground cover gravel at 48–59 °C. It is remarkable that gravel can reach higher temperatures than concrete, because of the higher area exposed to solar radiation. Shaded gravel, in this case, registered a superficial temperature of 45.5 °C. Gravel

Fig. 8.12 Visible (left) and infrared photography (right) of case 2. False color scale is 30–70 °C

Fig. 8.13 Visible (left) and infrared photography (right) of case 3. False color scale is 30–70 °C

is used frequently as surface coverage in outdoor spaces due to the need for preventing rising dust from winds, and due to its low cost and almost zero maintenance.

Vegetation surfaces registered temperatures between 32 °C (vegetal ground cover) and 43 °C (grass) under solar radiation. Shaded grass was at about 37.5 °C. At this time, local air temperature was about 43.3 °C. Again, vegetation at the sun maintained a superficial temperature around air temperature and the shaded vegetation, around 10 °C lower. A difference of 10 °C (or more) of superficial temperature lower than air temperature implies that the use of vegetation really has a significant impact on the possibility of making more livable outdoor spaces in desert climates.

In these last two examples (cases 1 and 2), the ground surfaces shown are composed of several materials, which have different thermal properties, such as concrete, gravel, and vegetation. Depending on these properties and the superficial temperatures shown, a landscape design considering microclimatic effects can be planned.

The microclimate at the parking lot (Fig. 8.13), as can be expected, was not favorable. At 4:00 p.m., while air temperature was 43.4 °C, the ground surface temperature was near 66 °C in the case of asphalt and about 50 °C in the shaded concrete floor near the building, both under the sun. There is no vegetation and no shading devices in this zone. There are only a few small trees in the rest of the parking lot, which are too small to shade.

At first instance, because of the limited water availability in the region, vegetation would not seem to be a viable strategy to control the microclimate conditions in desert environments; however, there are other options, such as the use of endemic vegetation and vegetation adapted to the local environment. Urban greening is an important mitigation strategy for the UHI, but it must also be considered that the efficiency of this strategy depends on the appropriate selection of the type of vegetation, the species, and its distribution in the space. Choosing the plant species carefully and providing adequate controlled irrigation, vegetation can be very useful for landscape design and microclimatic design of outdoor spaces, even in urban deserts. These species need to withstand high levels of solar radiation, be resistant to sporadic strong winds (the region is a hurricane zone), and preferably survive with low amounts of water.

A relevant microclimatic strategy is to shade outdoor spaces with structures or with vegetation. Another design consideration is the geometry of space and its surface limits to avoid successive reflections of long-wave radiation among vertical and horizontal surfaces.

Regarding surface materials, it is desirable that traditional materials, such as concrete and gravel, be shaded. In the case of walls, these can be shaded by trees, shrubs, or vines. In the case of floors, interspersed concrete, tile or gravel floors, and vegetation covers can be used to lower the radiant temperature. Vegetation is also useful to decrease long-wave radiation reflections from floors or other buildings, through bushes, for example. There are also important urban furniture materials like benches made out of materials with low heat absorption and low thermal capacity.

During the winter, the climatic conditions in the city are very benign as described and the outdoor spaces are close to being comfortable most of the day, considering the lower expectations of people in outdoor spaces. During the hours of the day when the outdoor spaces can be inhabited, the minimum temperatures are no less than 12 °C and the maximum temperatures rarely exceed 30 °C in this period. Measurements carried out during the winter will not be presented here.

8.4 Microclimatic Spaces Around Buildings

The spaces around buildings, for example, those between constructions, front yards and backyards, garages, and parking spaces, are outdoor living spaces that must be carefully designed in order to have or look for most comfortable conditions. Further on, the microclimate generated in these spaces has a thermal impact on the heat transfer between the outdoors and indoor building space.

Particularly in desert environments, the better the design of the spaces around buildings, the lower the heat transfer to the interior, and thus there are more possibilities of having a comfortable indoor environment with lower air-conditioning energy consumption. Improving outdoor design strategies for an existing space can be simple, such as shading or properly choosing the floor material.

In low-cost housing developments, mostly tract housings (which are the largest number of buildings in some cities), in addition to the austere design and construction, there is an absence of outdoor space design, such as home access, garage, backyard, and outdoor laundry. In most cases, the appearance of outdoor spaces is that of residual space between houses. Over time, some owners try to arrange these spaces, but in most cases, probably due to scarce resources or interest, these spaces remain neglected. This implies in many cases dirt floors, lack of shadows, and lack of vegetation. As mentioned, poor design and particularly no intention of thermal design of outdoor spaces affect not only comfort but also housing thermal behavior and energy consumption for acclimatization. As known, an adequate microclimate design in spaces around buildings is a relevant thermal design strategy that impacts indoor thermal behavior.

Next, the evaluation of several strategies applied to low-cost housings (less than 40 m^2 construction area) in Hermosillo city will be shown, where the redesign possibilities and modifications of outdoor spaces are restricted because of the limited space and the scarcity of resources. In spite of this, it is possible to visualize the differences from the impact of simple strategies applied to outdoor spaces on the indoor electricity consumption.

The types of acclimatization used in these homes mainly include evaporative coolers (evaporative cooling effect) and secondly window unit air conditioners, both usually used in only one room of the house. The cooler is the most used device due to the relatively low investment and operation costs. The average electricity consumption of all analyzed houses is about 3100 kWh/year. From the total electricity consumption in houses that use evaporative coolers, it is estimated that 24% corresponds to acclimatization. Those houses that use window unit air conditioners spend 40% from total electricity consumption in acclimatization (Marincic et al. 2009). The remaining consumption in each case is used for lighting and other electrical devices and appliances.

Dwellings are generally delivered with outdoor areas without shading and almost all the exterior floors with natural soil. In certain cases, concrete footprints for the car are also added at the entrance of the house (Fig. 8.14). Over time, owners may plant vegetation and, in other cases, increase the concrete surface. Also, several homeowners may add shading to the garage area using mesh shade or tarp-awnings or a roof made of any available construction material (metal sheet, board, concrete), considering the availability of resources. Modifications in outdoor design, such as the ones mentioned previously, have an impact on the microclimate and on the thermal loads transferred to the interior of the buildings. Therefore, the analysis of how these effects influence the energy saving possibilities by air-conditioning was performed. From the multiple combinations of floor materials observed, including concrete, natural soil, gravel, tiles, grass, etc., we analyzed the most frequent combinations and their relationship with the electrical consumption. The relationship between the type of exterior floor and the average annual electrical consumption of each type of house in a specific situation is shown in Fig. 8.15.

Fig. 8.14 Typical low-cost dwelling constructed in the region (Marincic et al. 2010). View of the façade (left) and floor plan (right)

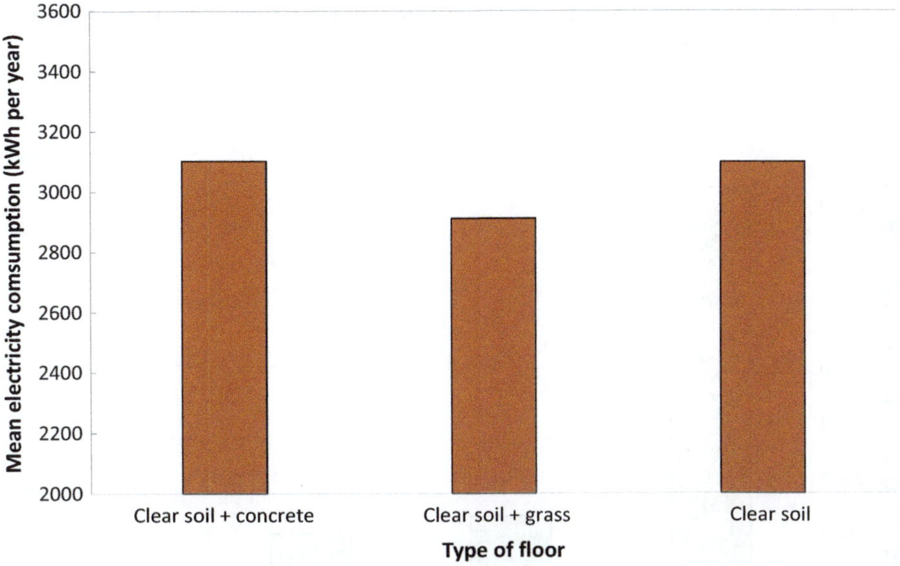

Fig. 8.15 Average annual electricity consumption in relation to the floor material in outdoor areas (own work based on Marincic et al. 2010)

As shown in Fig. 8.15, the intervention in the small available outdoor area by modifying the type of floor has an effect on the decrease in electricity consumption, particularly with the use of grass in some areas. Although this type of vegetation is not ideal for the desert climate, due to its high water consumption, it is estimated that with other type of vegetation coverings adapted to the region, similar results could be achieved. In this case, the function of the vegetation would be, on the one hand, to reflect less infrared radiation from the soil to the building during the hours of solar incidence, and on the other hand, vegetation does not store heat that could be emitted later, and its superficial temperature decreases when the air temperature drops. It is necessary to highlight the high heat capacity and emissivity of dry and clear soil (without vegetation) that affects the microclimate and the thermal loads towards the building.

Regarding shading improvements outdoors, in some cases, small trees are planted and sometimes awnings and tarps are placed as garage roofs. In few cases a garage with more permanent materials is built. In Fig. 8.16, these three observed solutions and the original situation, without any shading or only the possible shadow of the neighbor's house, can be observed. The major impact of the shading on the average annual electricity consumption of the houses is clearly shown.

Shading of the exterior areas has a great impact on the decrease of electricity consumption within the building, and the major decrease corresponds to the shade of vegetation elements, since they have the advantage of not storing or emitting heat.

Although in desert climates the maintenance and conservation of vegetation elements are laborious and constant, the adequate selection of suitable plant species adapted to the region makes maintenance more viable and allows a thermal benefit advantage from their use.

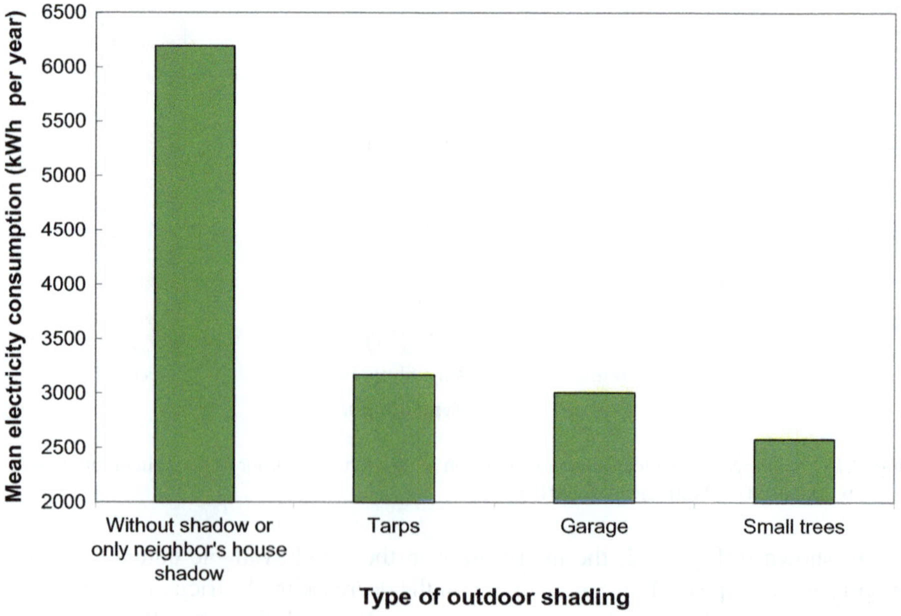

Fig. 8.16 Average annual electricity consumption in relation to the type of shading in outdoor areas (own work based on Marincic et al. 2010)

8.5 Final Comments

UHI is a complex phenomenon that involves factors such as the shape, density, and materials of the urban tissue as well as anthropogenic activities among others. The result is a modification of the urban microclimate that affects the thermo-hygrometric conditions of outdoor space and the energy behavior of buildings, and therefore the quality of life of the city's inhabitants (Román et al. 2017).

Although there is not necessarily a relationship between UHI and outdoor thermal comfort (OTC), some of the factors that cause UHI and strategies to mitigate it are similar to strategies to improve OTC (Irmak et al. 2017; Evola et al. 2017), especially in hot dry climates.

In this chapter, outdoor microclimatic conditions and particularities in a desert climate have been described and commented, highlighting the harsh climate conditions to consider during the summer period. Also, their impact on outdoor comfort and indoor energy consumption in buildings has been commented. Some examples of thermal conditions in small microclimatic outdoor spaces have been shown, demonstrating how certain thermal design strategies, especially the use of vegetation, have an impact on climatic variables that can modify thermal conditions and thus peoples' thermal sensation in these spaces.

Finally, outdoor spaces around low-cost dwellings in a tract housing development have been analyzed and design improvements made by the owners, such as the implementation of shading and the change of floor materials, have been described. These modifications have been related with the electricity consumption of the corresponding house, considering that 24–40% of this consumption is related to air-conditioning, used during almost half of the year. Also, in these cases, the use of vegetation as a design microclimatic element shows its benefits and advantages in outdoor spaces.

In order to incorporate vegetation in outdoor space design, considering operational purposes, it is necessary to select suitable vegetation species for each region, endemic or locally adapted, and also to plan an efficient irrigation system. Even in desert climates with limited water availability, the space design with vegetation and their maintenance is a great investment to improve microclimatic climates and energy consumption inside buildings and thus these are considered as actions that contribute to mitigating a city's urban heat island.

Acknowledgments Some measuring instruments were acquired with the support of Fondo Sectorial SENER-CONACyT project. The field surveys are part of the project "Comfort and thermal energy efficiency in low-cost dwellings in Mexico: regions of warm dry and warm humid climates" and have been supported by the Mexican federal funds of the National Housing Council and the National Council of Science and Technology (CONAVI and CONACyT). It is also important to thank the students who collaborated in the measurement process and information research.

References

Ali, S. B., Patnaik, S., & Madguni, O. (2017). Microclimate land surface temperatures across urban land use/land cover forms. *Global Journal of Environmental Science and Management, 3*(3), 231–340.

Evola, G., Gagliano, A., Fichera, A., Marletta, L., Martinico, F., Nocera, F., & Pagano, A. (2017). UHI effects and strategies to improve outdoor thermal comfort in dense and old neighborhoods. *Energy Procedia, 134*, 692–701.

García Nevado Elena. (2018). *Uso de la perspectiva para una evaluación térmica global de la calle*. PhD Thesis, Universitat Politècnica de Catalunya, Spain.

Givoni, B. (1998). *Climate considerations in building and urban design*. New York: John Wiley & Sons.

Gómez-Azpeitia, L. G., Bojórquez-Morales, G., Ruiz, R. P., Marincic, I., González, E., & Tejeda, A. (2014). Extreme adaptation to extreme environments in hot dry, hot sub-humid and hot humid climates in Mexico. *Journal of Civil Engineering and Architecture, 8*(8s), 929–942.

Irmak, M., Yilmaz, S., & Dursun, D. (2017). Effect of different pavements on human thermal comfort conditions. *Atmosfera, 30*(4), 355–366.

Marincic, I., Ochoa, J. M., & Isalgué, A. (2005). Thermal comfort educational software for hot climates. In M. Santamouris (Ed.), *1st International Conference on Passive and low energy cooling for the built environment Palenc 2005* (Vol. I, pp. 309–314). Santorini, Greece: Heliotopos Conferences.

Marincic, I., Ochoa, J. M., Alpuche, M. G., & Vázquez, E. (2009). Perfil del usuario de la vivienda económica en Hermosillo y patrones de consumo de energía eléctrica. In: *XXXIII Semana Nacional de Energía Solar*, Guadalajara, Mexico, October 2009, pp. 53–57.

Marincic, I., Ochoa, J. M., & Alpuche, M. G. (2010). La vivienda económica en Hermosillo y el consumo de energía eléctrica. In: *XXXIV Semana Nacionasl de Energía Solar*, Guanajuato, Mexico, October 4–9 2010.

Marincic, I., Ochoa, J. M., Alpuche, M. G., & Gómez-Azpeitia, G. (2012). Adaptive thermal comfort for occupants of low-cost dwellings in hot dry climate. *Journal of Civil Engineering and Architecture, 6*(3), 356–363.

Marincic, I., Ochoa, J. M., & Alpuche, M. G. (2013). La vivienda económica en Hermosillo: Diagnóstico para mejorar las condiciones de confort térmico y eficientar el uso de la energía. In D. C. Á. Ramírez, S. A. Orozco, & F. C. Canela (Eds.), *Procesos de certificación ambiental de las edificaciones sustentables* (pp. 39–63). Jalisco, Mexico: Centro Universitario de Arte, Arquitectura y Diseño de la Universidad de Guadalajara y Secretaria de Medio Ambiente y Desarrollo Sustentable del Edo. de Jalisco.

Middel, A., Häb, K., Brazel, A. J., Martin, C. A., & Guhathakurta, S. (2014). Impact of urban form and design on mid-afternoon microclimate in Phoenix Local Climate Zones. *Landscape and Urban Planning, 122*, 16–28.

Nicol, J. F., & Humphreys, M. A. (2002). Adaptive thermal comfort and sustainable thermal standards for buildings. *Energy and Buildings, 34*(6), 563–572.

Nikolopoulou, M., Lykoudis, S., & Kikira, M. (2004). Thermal comfort models for urban spaces. In M. Nikolopoulou (Ed.), *Designing open spaces in the urban environment: A bioclimatic approach* (pp. 2–6). Greece: Centre for Renewable Energy Sources.

Ochoa, J. M. (2009). *Ciudad vegetación e impacto climático*. Spain: Erasmus ediciones.

Ochoa, J. M. (2017). La vegetación como elemento de control bioclimáticos en espacios exteriores. In P. Elias & V. Fuentes (Eds.), *Estudios de Arquitectura Bioclimática Vol XIII*. Mexico: Universidad Autónoma Metropolitana.

Ochoa, J. M., & Marincic, I. (2005). Thermal comfort in urban spaces: The case of very warm and dry climate. In M. Santamouris (Ed.), *1st International Conference on Passive and low energy cooling for the built environment Palenc 2005* (Vol. II, pp. 785–789). Santorini, Greece: Heliotopos Conferences.

Ochoa, J. M., & Marincic, I. (2016). La habitabilidad de la vivienda económica en México: Análisis para el clima cálido seco. In C. Rueda (Ed.), *Apuntes de la vivienda mínima en México* (pp. 149–163). Mexico: Universidad de Guadalajara.

Önder, S., & Akay, A. (2014). The roles of plants on mitigating the urban heat islands' negative effects. *International Journal of Agriculture and Economic Development, 2*(2), 18–32.

Palme, M., Carrasco, C., & Lobato, A. (2016). Quantitative analysis of factors contributing to urban heat island effect in cities of Latin-American Pacific Coast. In *4th International Conference on Countermeasures to Urban Heat Island.* Singapore: National University of Singapore.

Román, E., Gómez, G., & Luxán, M. D. (2017). La isla de calor en Madrid y su influencia en el confort urbano. In *Proceedings of the 3rd International congress on sustainable construction and eco-efficient solutions* (pp. 467–508). Seville: Universidad de Sevilla, Escuela Técnica Superior de Arquitectura. Retrieved from http://hdl.handle.net/11441/59050.

Rosheidat, A., & Harvey Bryan, H. (2010). *Optimizing the effect of vegetation for pedestrian thermal. Comfort and urban heat island mitigation in a hot arid urban environment.* New York, NY: Fourth National Conference of IBPSA-USA.

Shalaby, A. S. (2011). Urban heat island and cities design: A conceptual framework of mitigation tools in hot-arid regions. *Journal of Urban Research JUR, 8*, 42–63.

Shima, T., Danial, M. P., & Arezou, S. (2015). Urban design guidelines to mitigate urban heat island (UHI) effects in hot-dry cities. *Journal Teknologi (Sciences & Engineering), 74*(4), 119–124.

Chapter 9
Integrating Urban Climate Knowledge: The Need for a New Knowledge Infrastructure to Support Climate-Responsive Urbanism

Gerald Mills and Julie Futcher

9.1 Introduction

Cities affect climate at a hierarchy of scales and decisions that are made to manage these climates do not always recognise the interdependencies between these scales. This is partly because the relationships between the climate drivers at city scales are complex but it is also a product of the history of urban climate research and application that has resulted in a fragmented knowledge base that is difficult to integrate. As a result, solutions to one climate problem at one scale can present a problem elsewhere; as an example, air-conditioning systems to cool an indoor space contribute to heating the outdoors. In this short contribution we outline some of the climate issues that arise at urban scales from narrowly focussed solutions to urban energy management. In particular, we focus on the relationship between the indoor and the outdoor climates in cities and argue for the development of a body of knowledge that integrates urban climate science knowledge across multiple fields of study. This is needed to support comprehensive policies that seek to create more sustainable cities that create liveable and healthy indoor/outdoor city climates while contributing to global sustainability. Here, our focus is on the fundamental link between natural and anthropogenic energy exchanges and climate impacts in cities and we draw upon examples from the City of London to support our points.

G. Mills (✉)
School of Geography, UCD, Dublin, Ireland
e-mail: gerald.mills@ucd.ie

J. Futcher
Urban Generation, London, UK

M. Palme, A. Salvati (eds.), *Urban Microclimate Modelling for Comfort and Energy Studies*, https://doi.org/10.1007/978-3-030-65421-4_9

9.2 Cities, Energy and Climate Changes

Rees and Wackernagel (2008) argued that although cities cannot be sustainable, they are a key to achieving sustainability. This proposition captures the conundrum posed by cities; on the one hand, as individual entities, the land-cover change and resource demands of cities means that they rely on the productivity of landscapes elsewhere and, on the other hand, their relatively compact form and high population density offer opportunities for efficiency that would not be possible otherwise. Currently, the urbanised landscape of the Earth's land is less than 3%, yet cities are collectively responsible for over 70% of CO_2 emissions (Seto et al. 2014). This suggests that much of the global mitigation policies should be focussed on cities as major drivers of global climate change where there are management systems that can implement change at relevant scales. However, until very recently, the focus of climate change policies has been at national levels and on the major sources of energy demand (e.g. transport, industry and buildings), even though much of this demand is spatially 'bundled' in cities. The importance of urban areas in the global context has changed considerably with the emergence of international consortiums of cities that are focussed on mitigation and adaptation strategies and encourage knowledge exchange.[1] The most recent IPCC assessment report recognises the potential role of human settlements by linking aspects of urban form and functions and states that key 'drivers of energy and GHG emissions are density, land use mix, connectivity, and accessibility. These factors are interrelated and interdependent. Pursuing one of them in isolation is insufficient for lower emissions' (Seto et al. 2014). Compact city policies that encourage higher population and built densities are advocated to make cities more energy efficient. In practice this is often used to support policies for the more efficient use of serviced land by building closely and vertically.

Separately from global climate change concerns, the climates in cities have been a subject of study for a very long time. We might usefully divide these into studies of indoor and outdoor climates and the (mis)management of natural and anthropogenic energy fluxes.

9.2.1 *Indoor Climates*

Much of the energy consumed by a city is used in buildings for managing the internal climate and supporting the occupants. In many buildings the large proportion of this energy is used for space heating and cooling to balance internal and external energy loads. The amount of energy required depends on the ambient climate and

[1] For example, C40 is a network of the world's megacities committed to addressing climate change. C40 supports cities to collaborate effectively, share knowledge and drive meaningful, measurable and sustainable action on climate change (https://www.c40.org/).

the degree of control on the indoor climate to ensure thermal comfort (De Dear 2004). Innovations in architectural and engineering practices have revolutionised building and building practice such that building form and function can be considered to have been liberated from its climate-driven form (e.g. Lehmann 2010). Historically, settlements accounted for natural resources in their layout, which regulated building height and spacing to permit (or limit) access to daylight and sunshine. The result of these historical design decisions is often embedded in the urban landscape including road width, plot size and building dimensions. Modern buildings have minimised their dependency on passive resources through technological innovations that rely increasingly on mechanical controls. As a result, in large buildings (such as modern office blocks and apartment buildings) great emphasis is placed on the envelope as the interface with the outdoors and its ability to regulate energy gains and losses to the building. The heating, ventilation and air-conditioning (HVAC) systems are designed to manage the internal climate within acceptable limits. In current thinking, a building can achieve high efficiency through this fabric-first approach that achieves energy efficiency through the design of envelope and the efficiency of HVAC systems. This approach can be complemented by availing of renewable energy (such as photovoltaic cells) on-site.

Buildings are significant consumers of energy and as a result are a focus of climate change mitigation;[2] in the EU buildings accounted for 40% of the final energy consumption and 60% of electricity consumption in 2016, two-thirds of which was expended in residential buildings.[3]

9.2.2 Outdoor Urban Climates

The urban effect on the outdoor climate in cities is an outcome of characteristics of form and function. Form describes the land cover (e.g. the fractions of vegetative and impervious cover), the fabric (the characteristics of manufactured materials used to construct the paving and buildings) and the geometry (the corrugated nature of the urban surface). Urban functions describe the throughput of materials, energy and water that are needed to sustain the urban system. Many of these resources pass through the city in degraded form and much is expended as waste gases and particulates into the overlying atmosphere. Together, urban form and functions modify the exchange processes at the surface-air interface and generate extreme spatial and temporal variations in microclimates across the urban landscape. These effects extend through a deep (>1 km) layer of the overlying air but are most profound near the ground, in the spaces between buildings and below roof level (known as the urban canopy layer). The best known of these urban climate effects is the urban heat

[2] See Building Regulation Standard (Conservation of fuel and power: Approved Document L) at https://www.gov.uk/government/publications/conservation-of-fuel-and-power-approved-document-l.

[3] https://www.odyssee-mure.eu/publications/policy-brief/buildings-energy-efficiency-trends.html.

island (UHI), which results in warmer surface and air temperatures within the canopy by day (surface) and night (air). The UHI impacts the public health, especially during warm weather events, and impacts the heating/cooling energy needs of buildings. The urban impact on climate varies considerably across the urbanised landscape but the magnitude is usually greatest in city centres characterised by high building densities, little green cover and high daytime occupancy. Although there is a substantial body of literature on how urban planning and design decisions affect the local climate (e.g. Grimmond et al. 2010), there is little evidence that this knowledge is incorporated into practice (e.g. Mills et al. 2010); this has been referred to as a post-war knowledge circulation failure (Hebbert and Mackillop 2013).

The links between global and local climate changes are synergistic and many of the projections of global climate change, such as increased frequency of heatwaves, will be enhanced by the UHI (Li and Bou-Zeid 2013). Climate change policies on the outdoor urban climate largely focus on increasing the adaptive capacity to offset projected changes through various greening strategies.

9.2.3 Redefining the Urban Canopy Layer

The partitioning of the near-surface urban environment into indoor and outdoor climates with distinct management policies can result in undesireable climatic outcomes. In the urban realm, the emphasis on buildings as independent entities with responsibility for managing its internal energy use can have unintended impacts on the nearby outdoor spaces and other buildings. Similarly changes to the urban context (by modifying building dimensions and/or greening outdoor spaces) surrounding an individual building will impact its ability to meet its energy needs efficiently. Balancing the needs of both the indoors and outdoors in an urban setting requires an integrated perspective that can bring together expertise that is fragmented into many different fields of study (e.g. building engineers, architects, meteorologists, designers and planners), each of which has its own language and methods that inhibits communication. A potential starting point is to simply consider the outdoor and indoor climates in cities as parts of a single layer of the atmosphere linked by natural (and enhanced) energy/mass exchanges and extend the definition of the meteorological concept of 'urban canopy layer' (UCL) to include the indoor space.

The etymology of the UCL in boundary-layer meteorology is based on its equivalence with the architecture of forests, which separates the climates found above and below the leafy forest canopy. From the vantage above the forest canopy, the relevant plane where energy and momentum exchanges are concentrated is located above the ground at the top of the canopy. Below the canopy top, exchanges at the ground surface are greatly altered by the overlying leafy 'roof' which limits short-wave radiation receipt and long-wave radiation loss. Within the forest canopy layer it is not possible to understand the climate impact of trees in a forest simply as the sum of the individual trees, as their interaction through mutual exchanges (e.g.

shading and sheltering) produces a unique climate that affects the responses of each tree.

The equivalent canopy layer in urban meteorology is defined as the outdoor space between the buildings below roof level. Apart from the different 'architecture' of the canopy itself, buildings are hollow and are a ubiquitous energy source as heat is added/removed as needed to maintain an indoor temperature suited to its needs. Ironically, to ensure a near-uniform microclimate, the building waste energy must be disposed into the outdoors contributing to the urban climate effect, which in turn impacts the energy needs of the building itself. Moreover, the mutual shadowing and sheltering of buildings in an urbanised landscape affect the energy performance of each building. Finally, unlike the forest, much of the ground between the buildings is paved and is also an enhanced energy source as a result of waste energy emitted by traffic.

Fundamentally, the layer between the ground and the rooftop level is a zone of human occupation. Redefining the urban canopy layer to include both the indoor and outdoor space below roof level would overcome many of the difficulties that arise because of the partitioning of this space into separate areas of study. The boundary at the top level of this UCL includes the solid roofs of buildings and the open interface between them. The wall facets are the shared boundary separating the indoors and outdoors and, like the top of the UCL, can be open to air exchanges (infiltration, natural ventilation and HVAC intake and exhaust), radiative transfer and heat conduction. While the indoor environment is highly regulated, the outdoor space may be partially managed (through landscaping, choice of fabric, traffic control) or unregulated. This perspective has the advantage of describing the entire urban landscape using the same context; the challenge is to assemble the underpinning scientific knowledge. Here we outline a perspective and methodology that is used to explore the relationship between urban form and energy (mis)managment in indoor and outdoor environments.

9.3 Urban Climate Management

Urban form management on the wider environment can be evaluated and usefully applied to the study of indoor and outdoor climates has been demonstrated Urban form, that is the dimensions and layout of buildings, has been shown to have a dramatic impact on the outdoor climate between builidngs and to impact the ambient environment of individual buildings (e.g. Ratti et al. 2005; Salat 2009; Kolokotroni et al. 2012; Futcher et al. 2018; Salvati et al. 2020). Nonetheless, current methodologies for assessing urban energy sustainability focusses on individual buildings, which can achieve impressive energy credentials without considering their impacts on neighbouring buildings or adjacent outdoor spaces (e.g. loss of sunshine, wind effects, etc.). In most jurisdictions there is no legal framework or guidance for conducting a microclimatic assessment of these impacts and no basis for examining the aggregate impact of buildings on the atmospheric environment. In the UK the exception is the right to daylight, which is enshrined in law as an easement (that is, a right to cross over someone else's land for a specific purpose), and must be taken into account when a new construction affects the daylight resources of neighbouring buildings. However, there are no mandatory criteria to assess potential impacts on the surrounding urban landscape more generally and nor are assessments part of planning and/or policy guidance. As a result, city landscapes are developing in such a way that the emerging morphology, which will have significant long-term impacts for on the outdoor and indoor climate are not taken into account.

The current approach to the environmental management of our cities focusses on the aspects of indoor and outdoor spaces (e.g. zero-carbon buildings, vehicle emission standards, urban greening) often in isolation. An integrated approach would account for the interdependencies between urban built form and impacts. It would also consider the urban commons and the use, preservation and access to our collective shared resources (e.g. daylight, ventilation, air quality) to create healthy environments and encourage more sustainable urban practices. A shared understanding of the urban climate at all scales (macro to micro in both the vertical and horizontal) requires clear definitions of building energy interdependencies and a common set of methods and teminology to support knowledge exchange. Urban building energy models (Reinhart and Davila 2016), which permit the analysis of neighbourhoods rather than buildings and account for the mutual interactions between buildings and outdoor spaces as a shared environment, offer a pathway toward an integrated science. Ideally these urban building energy models would be coupled with climate models that could simulate the environmental impacts of design decisions.

9.3.1 Case Study: The City of London

Our case on the need for an integrated science of the urban landscape for climate management has been formed through studies of the outdoor impacts of the emerging urban landscape in the City of London, which occupies a space of just over 3 km^2 or less than 0.2% of the Greater London area (Table 9.1).

It is distinguished within London by its occupation patterns that are strongly linked to commercial functions; during the daytime its population swells to nearly 500,000 but there are just over 8000 residents. The reliance on commercial functions is seen in the intensity of energy consumption, much of it for space cooling in buildings that have large internal energy gains. The desire to maximise floor space and the availability of modern construction techniques have seen a radical change in the historic urban form including:

- Tall and very tall buildings that occupy small plots and are inserted into a relatively low-lying urban setting
- Deep-plan buildings that fill large floor plates and replace courtyard forms

Many of these building types use curtain wall systems that permit large uninterrupted areas of glazed facades (Fig. 9.1). While all of these buildings are designed

Table 9.1 A comparison of the energy and occupancy profiles for Greater London (32 boroughs and City of London) and the City of London. Data are from 2014 and sourced from https://data.london.gov.uk/

Property	Greater London		City of London	
Area (km^2)	1594.69		3.15	
Daytime population	8,676,835		360,075	
Residential population	8,538,689		8072	
Population density (day) Persons per km^2	5441		114,330	
Population density (residential) Persons per km^2	5354		2563	
Energy density GWh per km^2	82.78		1033.07	
Energy use (kWh) per capita (day)	1521.45		903.58	
Energy use (kWh) per capita (residential)	1546.06		40,306.77	
Total energy use (GWh)	132,013.5		3253.6	
Commercial energy use (GWh)	48,279.3	36.6%	2986.0	91.7%
Domestic energy use (GWh)	53,249.2	40.3%	57.4	1.8%

Fig. 9.1 High-rise buildings of the City of London by Tristan Surtel (25 April 2018) at https://commons.wikimedia.org/wiki/File:City_of_London_seen_from_Tower_Bridge.jpg#file under a Creative Commons Attribution 4.0

Fig. 9.2 Examples in the City of London that illustrate the relationship between buildings, energy management and outdoor spaces: (**a**) shows a canopy extension which limits the impact of fast winds that have been displaced downwards toward the ground; (**b**) shows a covering placed over the south-facing, parabolic shaped glazed façade of a building which redirected and focussed solar radiation onto adjacent streets (the façade was subsequently refurbished); (**c**) shows a set of tall buildings that channel air through a pedestrianised street and require vegetation to make the space more comfortable; (**d**) shows a configuration that has a Venturi effect on airflow as it is accelerated through a narrow gap between two very tall buildings; (**e**) shows a residential apartment block with embedded (stationary) wind turbines that rely on a common resource; and (**f**) shows a tall structure with embedded photovoltaic cells along its south-facing façade, which has subsequently become overshadowed by a neighbouring building. Each of the buildings shown here is an exemplar of energy-efficient buildings that meet carbon goals (credit: Futcher and Mills)

to meet stringent energy regulations, their combined environmental impact on their surroundings is often negative. Figure 9.2 shows images of several of these buildings that illustrate the challenges to developing on-site renewable energy, which is not a private resource, in dense urban settings; the failure to consider the impacts of redirected winds and solar radiation on climates at the ground; and the *ad hoc* architectural and landscaping responses to mitigate deleterious outdoor outcomes, some of which are predictable (such as the Venturi effect).

Over the last 5 years, the authors have run Urban Climate Walks through the City of London as an opportunity to engage with a range of urban academics, practitioners and students with interests in architecture, energy management, outdoor comfort and air quality, design and planning. The walk treats the participants as mobile 'weather stations' and links their sensory faculties to climate processes and observations (Mills et al. 2018). For this reason, the route is designed through a heterogenous landscape characterised by variations in:

1. Street widths, building heights and orientation
2. Traffic flows including vehicles and pedestrians
3. Building dimensions, fabrics and uses
4. Green surface cover and street plantings

The discussions that take place during the walk have convinced us that while all the experts ostensibly study the same urban environment, they do so from distinct perspectives that inhibits effective communication. For example, the well-known UHI phenomenon is frequently misunderstood in terms of type (surface or air), timing (daytime or night-time) and cause (natural energy exchanges or anthropogenic heating).

9.4 Conclusions

The urban environment is a spatially complex system with mutual dependencies such that altering a part has ramifications for other components of the system. Currently, our methods for addressing climate changes, which are often focussed on energy management, are narrowly focussed and do not address the wider environmental consequences of small-scale interventions, such as new building developments within an existing neighbourhood. We need a more comprehensive set of tools that integrate existing urban knowledge to address the challenges of creating more sustainable cities.

References

De Dear, R. (2004). Thermal comfort in practice. *Indoor Air, 14*, 32–39.

Futcher, J., Mills, G., & Emmanuel, R. (2018). Interdependent energy relationships between buildings at the street scale. *Building Research and Information, 46*(8), 1. Special issue urban form, density and microclimate.

Grimmond, C. S. B., Roth, M., Oke, T. R., Au, Y. C., Best, M., Betts, R., Carmichael, G., Cleugh, H., Dabberdt, W., Emmanuel, R., & Freitas, E. (2010). Climate and more sustainable cities: Climate information for improved planning and management of cities (producers/capabilities perspective). *Procedia Environmental Sciences, 1*, 247–274.

Hebbert, M., & Mackillop, F. (2013). Urban climatology applied to urban planning: A postwar knowledge circulation failure. *International Journal of Urban and Regional Research, 37*(5), 1542–1558.

Kolokotroni, M., Rena, X., Davies, M., & Mavrogianni, A. (2012). London's urban heat island: Impact on current and future energy consumption in office buildings. *Energy and Buildings, 47*, 302–311.

Lehmann, S. (2010). *The principles of green urbanism: Transforming the city for sustainability*. London: Earthscan.

Li, D., & Bou-Zeid, E. (2013). Synergistic interactions between urban heat islands and heat waves: The impact in cities is larger than the sum of its parts. *Journal of Applied Meteorology and Climatology, 52*(9), 2051–2064.

Mills, G., Cleugh, H., Emmanuel, R., Endlicher, W., Erell, E., McGranahan, G., Ng, E., Nickson, A., Rosenthal, J., & Steemer, K. (2010). Climate information for improved planning and management of mega cities (needs perspective). *Procedia Environmental Sciences, 1*, 228–246.

Mills, G., Futcher, J., & Luo, Z. (2018). *Teaching urban climatology through field observations. PLEA 2018: Smart and healthy within the two-degree limit*. 2018 Dec 10: 1226.

Ratti, C., Baker, N., & Steemers, K. (2005). Energy consumption and urban texture. *Energy and Buildings, 37*, 762–776.

Rees, W., & Wackernagel, M. (2008). Urban ecological footprints: Why cities cannot be sustainable—and why they are a key to sustainability. In *Urban ecology* (pp. 537–555). Boston, MA: Springer.

Reinhart, C. F., & Davila, C. C. (2016). Urban building energy modeling–A review of a nascent field. *Building and Environment, 97*, 196–202.

Salat, S. (2009). Energy loads, CO_2 emissions and building stocks: Morphologies, typologies, energy systems and behaviour. *Building Research and Information, 37*(5–6), 598–609.

Salvati, A., Palme, M., Chiesa, G., & Kolokotroni, M. (2020). Built form, urban climate and building energy modelling: Case-studies in Rome and Antofagasta. *Journal of Building Performance Simulation, 13*(2), 209–225.

Seto, K. C., Dhakal, S., Bigio, A., Blanco, H., Delgado, G. C., Dewar, D., Huang, L., Inaba, A., Kansal, A., Lwasa, S., McMahon, J. E., Müller, D. B., Murakami, J., Nagendra, H., & Ramaswami, A. (2014). Human settlements, infrastructure and spatial planning. In O. Edenhofer, R. Pichs-Madruga, Y. Sokona, E. Farahani, S. Kadner, K. Seyboth, A. Adler, I. Baum, S. Brunner, P. Eickemeier, B. Kriemann, J. Savolainen, S. Schlömer, C. von Stechow, T. Zwickel, & J. C. Minx (Eds.), *Climate change 2014: Mitigation of climate change. Contribution of Working Group III to the Fifth assessment report of the intergovernmental panel on climate change*. Cambridge/New York, NY: Cambridge University Press.

Part II
Urban Climate Modelling and Simulation: Physics and Tools

Chapter 10
Air Circulation in Urban Areas

Annalisa Di Bernardino, Olga Palusci, Agnese Pini, Giovanni Leuzzi, Marco Cacciani, Armando Pelliccioni, and Paolo Monti

10.1 Introduction

From a fluid dynamic point of view, the airflow above the Earth's surface is comparable to that developing above an unevenly warm rough surface. On a large scale and away from the Earth's surface, winds are mainly governed by Coriolis and pressure forces. These forces affect the intensity and direction of the synoptic (or geostrophic) wind, which defines the boundary condition for the underlying boundary layer.

The wind speed at the surface must necessarily be zero due to the viscosity of the air. Exhaustive studies, observations and theories concerning atmospheric flows can be found in several textbooks including those of Stull (1988), Garrat (1992), Holton (1992) and Jacobson (2005).

Atmospheric flows can be classified according to the spatial scales of the phenomena involved and range from the planetary scale (>10,000 km) to the microscale (1 mm–1 km). These include the synoptic scale (500–10,000 km), the mesoscale (2–2000 km) and subsequent subdivisions such as the urban scale (1 m–100 km) encompassing towns and neighbouring areas, the district scale (10 m–1 km), the urban canyon scale (~10 m) and the human scale (1 cm–1 m). The physical

A. Di Bernardino · M. Cacciani
Dipartimento di Fisica, Università di Roma "La Sapienza", Rome, Italy
e-mail: annalisa.dibernardino@uniroma1.it; marco.cacciani@uniroma1.it

O. Palusci · A. Pini · G. Leuzzi · P. Monti (✉)
Dipartimento di Ingegneria Civile Edile e Ambientale, Università di Roma "La Sapienza", Rome, Italy
e-mail: olga.palusci@uniroma1.it; agnese.pini@uniroma1.it; giovanni.leuzzi@uniroma1.it; paolo.monti@uniroma1.it

A. Pelliccioni
INAIL-DIMEILA, Monte Porzio Catone, Italy
e-mail: a.pelliccioni@inail.it

M. Palme, A. Salvati (eds.), *Urban Microclimate Modelling for Comfort and Energy Studies*, https://doi.org/10.1007/978-3-030-65421-4_10

processes associated with these scales are complex, especially in the presence of reservoirs and mountain ranges and when chemical, biological, geological and physical transformations are considerable (Fernando et al. 2001).

The atmospheric boundary layer (ABL) structure is influenced by soil properties, e.g. aerodynamic roughness, emissivity, albedo, heat capacity, thermal conductivity and water availability. The set of these parameters determines the surface temperature, which, in turn, regulates long-wave radiation towards the space, molecular heat flux into the ground and turbulent exchanges of heat and water vapour with the overlying air. The ABL depth varies considerably over time and space (usually from 100 to 2000 m) and evolves with the diurnal cycle of the solar radiation (Fig. 10.1a). Over land surfaces in high-pressure regions, the ABL is generally subdivided into several layers. At the bottom, there is the surface layer (10–100 m deep), where the turbulent fluxes are approximately constant. Above, the convective (or well-mixed) boundary layer and the stable (or nocturnal) boundary layer characterize daytime and night-time hours, respectively (Fig. 10.1b).

Although terrain effects are present even in nominally flat areas, the ABL is broadly classified in literature into that existing over flat terrain and that existing over complex terrain. When the ground is neither flat nor uniform, the ABL can be modified in terms of heating, moisture and momentum. Typical examples are geographically generated local winds, e.g. slope flows and sea breezes (Simpson 1994) as well as urban heat island (UHI) circulations (e.g. Fan et al. 2019).

UHIs are defined as the warmth produced by cities (Oke 1982). They may form above an urban area during both daytime and night-time, showing peculiar characteristics with respect to the ABL generally existing in rural areas. The difference in terrain coverage between urban and rural areas is mainly responsible for night-time and daytime urban-rural temperature anomalies that can even exceed 10 °C in the case of large cities (e.g. Roth et al. 1989; Santamouris 2001; Ravanelli et al. 2018). Possible causes for UHI formation are (a) reduced vegetation and increased paved

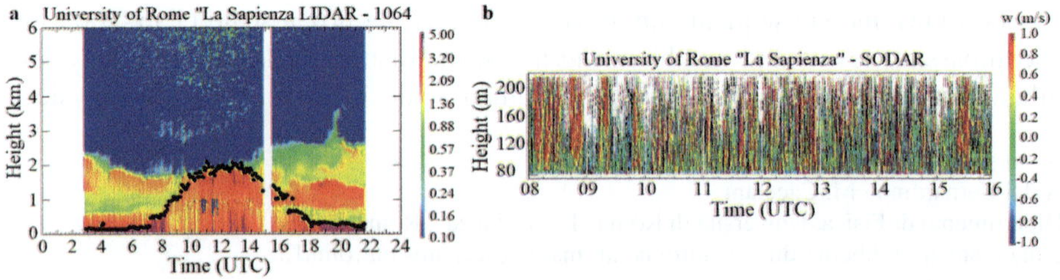

Fig. 10.1 (**a**) Example of LIDAR (light detection and ranging) scans of the backscattered light from the atmosphere on the 22nd of June 2005. The LIDAR is installed on the rooftop of the Department of Physics of the University of Rome "La Sapienza", Italy (Pichelli et al. 2014). The intensity of the backscatter (colour) is proportional to the aerosol concentration. The black dots denote the ABL height inferred from the backscatter vertical gradient. (**b**) Example of time history of the vertical velocity (w) profile measured by a SODAR (sonic detection and ranging) apparatus, placed next to the LIDAR. The features in red are signatures of thermals of warmer air rising from the ground during periods of strong convection, i.e. coherent vortical structures having diameter and depths of the order of the ABL depth (Stull 1988)

terrain that lower the potential for evapotranspiration which, in turn, implies that energy is converted into sensible rather than latent heat; (b) properties of urban materials, such as albedo, thermal emissivity, thermal conductivity and heat capacity; (c) urban geometry, which enhances the trapping of short- and long-wave radiations; and (d) anthropogenic heat (Oke 1982).

Knowledge of the wind flow structure within and above an urban area is of considerable interest in many respects, e.g. urban energy conservation (Blocken et al. 2011), air pollution (Ottosen et al. 2019), indoor air quality (Lai et al. 2015) and human comfort (Salvati et al. 2017b). Unfortunately, owing to the complexity of the geometric configurations involved, the wind flow is largely site dependent, thus limiting the application of simple models and formulations to predict wind flow in cities (Zajic et al. 2011).

This chapter illustrates the influence of city morphology on the flow field within urban areas. After a brief description of the gross features of ABLs above cities, the main characteristics of the wind flow around an isolated building and in correspondence with street canyons and groups of buildings are described and discussed.

10.2 Wind Flows Above Urban Canopies

Buildings and other urban infrastructure substantially change the dynamic and thermodynamic fields. Figure 10.2 shows a schematic of an urban environment and the modifications induced by buildings on the undisturbed wind.

Wind flow over an urban canopy, which is defined as the layer that extends from the ground up to the mean building height, is governed by the geometrical, thermal and radiative characteristics of the site. The urban boundary layer (UBL) is defined as the portion of the ABL between the surface and the height at which the underlying city no longer affects the airflow.

The similarity theory (Monin and Obukhov 1954) has given rise in the past to the profusion of considerable efforts in the search for general laws for average wind and air temperature profiles suitable for ABL analysis in the various stability conditions. Most studies reported in the literature show the considerable progress made hitherto, but also highlight the absolute shortage of laws valid in general for the urban environment.

The particular nature of city centres does not allow the development of robust parameterizations and theories for the velocity and temperature, as is the case of flat terrain, where the vertical profile of the average wind speed U reads

$$U = \frac{u_*}{k}\left[\ln\left(\frac{z}{z_0}\right) + \Psi\left(\frac{z}{L}\right)\right], \tag{10.1}$$

where z is the height, u_* the friction velocity (i.e. the scaling velocity, which is related to the drag force at the surface) and $k = 0.41$ the von Karman constant. L and

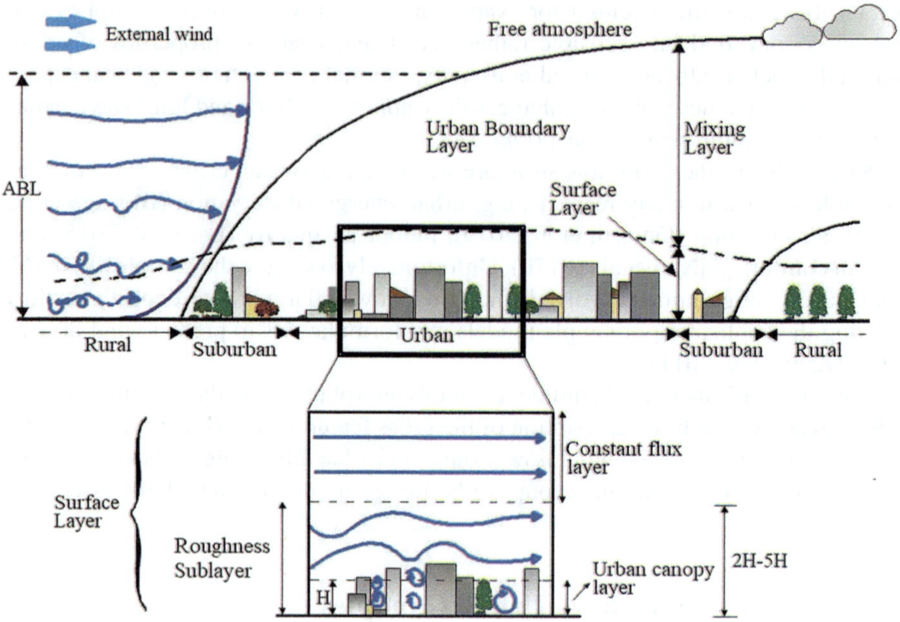

Fig. 10.2 Sketch of the UBL structure (modified from Fernando 2010)

z_0 are the Obukhov length and the aerodynamic roughness, respectively, and assume the role of scaling lengths. The latter is usually parameterized by means of descriptive land-use types (e.g. Britter and Hanna 2003), while L is proportional to the height above the surface at which buoyant factors first dominate over mechanical (shear) production of turbulence (e.g. Stull 1988). $\Psi(z/L)$ is a universal dimensionless function that equals zero in neutral conditions (Businger et al. 1971).

Conversely, an open question is the determination of the wind speed profile above cities. For example, within the roughness sublayer, i.e. the portion of the UBL immediately above the urban canopy (Fig. 10.3), Eq. (10.1) is not applicable. In such sublayer, the flow is three-dimensional, non-homogeneous and strongly influenced by vegetation, buildings and other roughness elements that constitute the urban canopy (Rotach 1999). Above urban or vegetation canopies, a region of constant flux generally exists (the constant flux layer), and Eq. (10.1) now assumes the form

$$U = \frac{u_*}{k}\left[\ln\left(\frac{z-d_0}{z_0}\right) + \Psi\left(\frac{z}{L}\right)\right], \tag{10.2}$$

where d_0 is the displacement height, i.e. the effective height of the ground due to the vertical flow displacement through the canopy. Both z_0 and d_0 are linked to the surface roughness elements. While u_* to be used in Eq. (10.2) is commonly measured within the constant flux layer, z_0 and d_0 are generally estimated by using the

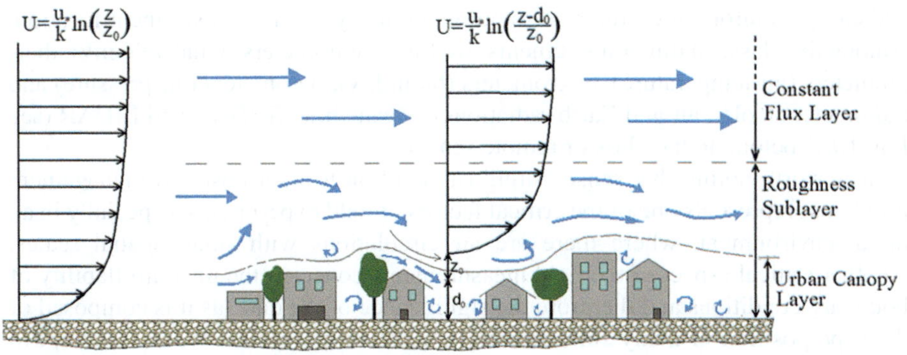

Fig. 10.3 Sketch of the wind profile above an urban area in neutral conditions (modified from Britter and Hanna 2003)

morphometric or anemometric method (Kent et al. 2017). In UBL studies, Eq. (10.2) is generally applied also within the roughness sublayer, where the condition of momentum flux constancy is never satisfied (Cheng and Castro 2002). An alternative formulation of (10.2) has been proposed recently by Pelliccioni et al. (2015), in which z_0 varies with height.

The picture described above is made even more complicated by the presence of complex terrain, in which airflows are mainly driven by local pressure gradients and thermal forcing (Whiteman 2000; Fernando 2010). This is the case of cities located, for example, in valleys surrounded by mountains (e.g. Giovannini et al. 2011, 2013), in which the airflow is determined mainly by terrain-induced perturbations to geostrophic flow (e.g. Valerio et al. 2017), or in coastal areas (e.g. Zhou et al. 2019).

10.3 Field Campaigns, Laboratory Experiments and Numerical Modelling

In order to solve practical problems in urban fluid mechanics area, we have to comply with the turbulent nature of wind flows. Randomness and non-linearity are just two of the intrinsic characteristics of turbulence that make the solution of the governing equations of fluid flows impossible to achieve analytically. Nevertheless, in order to solve urban fluid mechanics problems, three methods have been developed: field experiments, laboratory measurements and numerical simulations.

Field experiments could be considered in principle the ideal method since they provide values of the meteorological variable collected in the site of interest. Moreover, data coming from measurement campaigns may also be used to validate numerical simulations and laboratory experiments. Measurement instruments can be classified as either direct sensors or remote sensors. The former are placed on instrument platforms (e.g. masts and towers) and measure the parameters of interest at the location of the sensor. The latter measure waves (electromagnetic or sound) and can give information referred to points far away from the instrument location. Among the classical direct instruments, we find anemometers (wind velocity), thermometers (air temperature), hygrometers (humidity), barometers (air pressure) and radiometers (solar, air and Earth radiations). Meanwhile, SODAR and LIDAR (see Fig. 10.1) belong to the class of remote sensors.

It is worth noting that sensor siting (in addition to their cost and management problems) represents one of the critical factors of field experiments, especially in an urban environment, where there are air circulations with small spatial scales. Another critical aspect of the field measurements concerns the uncontrollability of boundary conditions and therefore the significance of the data, as it is composed of the superposition of many simultaneous effects and processes. We can attempt to focus on one specific process by the careful selection of a weather pattern (e.g. fair weather). In this regard, it is necessary to consider that a measurement station within the urban canopy layer is dominated by microscale processes. It would be advisable to avoid siting in parks, if the aim is to collect data representative of the urban area. The choice of a location that is not affected by microclimatic effects too site-specific is essential.

Another important problem is the repeatability, which strictly does not occur in reality. In this respect, field measurements are fundamentally different from laboratory experiments and computational fluid dynamics (CFD), in which the boundary conditions can be easily controlled and repeatability should be straightforward. The variability and uncontrollability of field measurement conditions imply that validation of laboratory results and CFD simulations with field measurements only makes sense when the latter have been obtained based on quasi-steady boundary conditions (e.g. in the presence of persistent winds).

For years, laboratory simulations have been the only alternative to direct field measurements. They create an artificial turbulent domain in the laboratory (wind tunnel or water channel, see, e.g., Fig. 10.4), where only a limited number of processes or boundary conditions act on the flow. One of the advantages of using laboratory models lies in the fact that it is generally easy to control the boundary conditions as well as to investigate the role played by the non-dimensional variables governing the flow field. However, it should be noted that sometimes it is complicated (or even impossible) to attain the matching of non-dimensional parameters of models with their full-scale flow counterparts (e.g. Di Bernardino et al. 2015a). For example, the major restriction in laboratory simulations of the whole UHI is caused by the difficulty in reproducing the small UBL height–UHI diameter ratio observed in real UHIs (Lu et al. 1997; Cenedese et al. 2000). Further problems and limitations are due to the low Reynolds number (the ratio between inertial and viscous forces) attainable in the laboratory compared with that of real UHIs.

CFD is a well-established tool for solving urban fluid mechanics problems. Despite some drawbacks still present in its fundamental formulation (e.g. Rodi 1995), it is routinely employed to study wind flow and pollutant dispersion in urban areas (Blocken 2014). CFD solves the equations of fluid mechanics that describe the dynamics and thermodynamics of the gases in the atmosphere. Unlike analytical models, which provide a solution in each point of the domain, numerical models

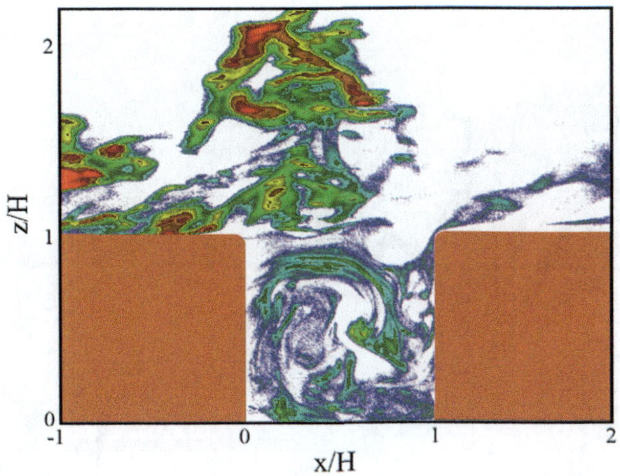

Fig. 10.4 Sketch of the instantaneous pollutant concentration field simulated in a water channel experiment simulating flow and dispersion in a street canyon. The flume is directed to the right. Each feature in brown represents a building $H = 2$ cm tall. The pollutant is emitted from a point source mimicking a chimney stack placed on the rooftop of one of the buildings located upwind of the building on the left (not included in the picture). The concentration is in arbitrary units (red refers to the maximum value, while white denotes zero). The pollutant enters the cavity mostly due to flow instability occurring at the interface between the cavity and the outer region (see Sect. 10.4.2) (the data set used refers to the experiment described in Di Bernardino et al. 2015b)

discretize the flow field to determine a solution only in discrete points of a numerical grid. The continuous differential equations that govern the flow field are approximated by a large number of algebraic equations describing local processes around the discrete points.

A number of different types of simulation approaches exist. Direct numerical simulation (DNS) solves all the scales of the turbulent fluctuating motions and the flow can be obtained in every detail of its very complex behaviour. Since only motions of scales larger than the mesh size can be resolved, the number of grid points required for solving the turbulent flow in the whole domain is too large and DNS cannot be used for practical applications in urban climate processes. Large eddy simulations (LES) and Reynolds-averaged Navier-Stokes (RANS) are the most common alternatives to DNS.

With RANS, all stochastic turbulent fluctuations are filtered out and only averaged equations are solved. In essence, only the mean flow field is explicitly solved and all scales of turbulence are modelled by solving additional balance equations for the closure (e.g. for the turbulence kinetic energy and its rate of dissipation). Large eddy simulations solve the large, energy-containing eddies in the flow field, which are usually anisotropic and contain most of the turbulence kinetic energy, while the small-scale eddies are modelled with a subgrid-scale model. LES intrinsically performs better than RANS and requires finer grids compared to RANS and therefore the computational cost is considerably higher.

Fig. 10.5 CFD model of part of the Sapienza University of Rome. The planar area in the figure is about 1 km^2, while the maximum building height is 32 m. The grid was generated using the surface-grid extrusion technique (van Hooff and Blocken 2010) and consists of about 26.5×10^6 hexahedral cells. The red star denotes the building roof where the SODAR and LIDAR apparatus described in Fig. 10.1 are located

For more information on research methods currently used for urban fluid mechanics see the recent review by Moonen et al. (2012) (Fig. 10.5).

10.4 Wind Flow Inside Building Canopies

As mentioned above, airflow through real building canopies is extremely complex and depends on several geometrical parameters, e.g. arrangement and shape of the buildings, and thermal and radiative properties of the materials. Furthermore, it is also strongly affected by direction and turbulence characteristics of the approaching flow (Salizzoni et al. 2009, 2011).

Knowledge of how an isolated building or a group of buildings affect the local wind is important for several reasons; for example, it allows us (a) to estimate the way in which people could be buffeted by the wind at ground level or on aerial walkways; (b) to model concentration of pollutants emitted by urban sources (e.g. vehicles and chimneys); and (c) to analyse the influence of wind on heating and ventilating systems and thus to estimate their performance properly.

Although a few pioneering efforts to modelling the spatially averaged wind speed profile within the canopy have been proposed in the past (see, e.g., Cionco 1965; Macdonald 2000; Bentham and Britter 2002; Di Sabatino et al. 2008), the complexity of the problem usually requires site-specific measurements performed by means of expensive field campaigns (e.g. Allwine et al. 2002), numerical simulations (e.g. Blocken 2015) or scaled physical models (e.g. Kastner-Klein and Rotach 2004). Alternatively, investigation of simple, basic configurations provides a useful contribution to the interpretation and physical comprehension of the basic mechanisms involved and can also give useful guidance for interpreting and sometimes estimating flows through real building canopies (Zajic et al. 2011).

10.4.1 Wind Past an Isolated Cubic Obstacle

The cube is the simplest idealization of a real building and it has been extensively investigated in the past both numerically and experimentally (e.g. Li and Meroney 1983; Martinuzzi and Tropea 1993; Murakami 1993; Snyder 1994; Santos et al. 2009; Tominaga and Stathopoulos 2010; Gousseau et al. 2011).

Despite the simple geometry, the flow pattern past a cube is very complex even when the approaching wind is perpendicular to one of the façades. The shear in the approaching wind causes a downward flow over the lower portion of the upwind façade of the building and a "horse-shoe vortex" at the upwind base, which wraps around the building near the ground extending also further downstream (Fig. 10.6). The flow separates at the upwind edges, producing separation zones on the rooftop and on the lateral sides of the building. At the downwind edges, the flow separates again, producing a cavity region and the associated bow vortex. The flow there interacts with the current merging from the roof and the sidewalls as well as with the horseshoe vortex. These vortex structures play a significant role in the physical processes governing pedestrian comfort and pollutant dispersion (e.g. Tominaga and Stathopoulos 2009; Bazdidi-Tehrani and Jadidi 2014). Their complex shape and structure should be considered in designing field campaigns in urban areas, where the siting and exposure of meteorological sensors are often, or to put it better, always, an incredibly complicated problem.

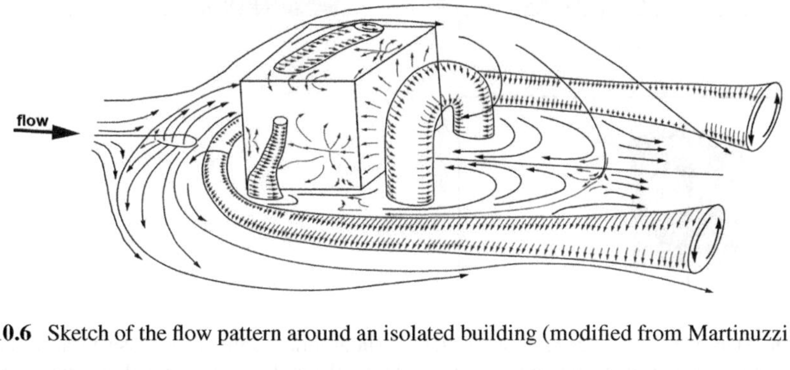

Fig. 10.6 Sketch of the flow pattern around an isolated building (modified from Martinuzzi 1992)

Figure 10.7 shows an example of the vector field of the mean velocity along a vertical plane passing through the centreline of a cubic obstacle. The eddy structure *A* refers to the bow vortex, while *B* is the signature of the horseshoe vortex located upstream of the cube. The recirculation region is about 1.3*H* long, where *H* is the building height and *R* the reattachment point. A third vortical structure (*C*) can also be seen and is located on the first half of the rooftop, while *S* is the stagnation point forming on the upwind façade.

The three panels in Fig. 10.8 show the mean concentration fields of a passive tracer emitted from a point source located upwind and downwind of the building

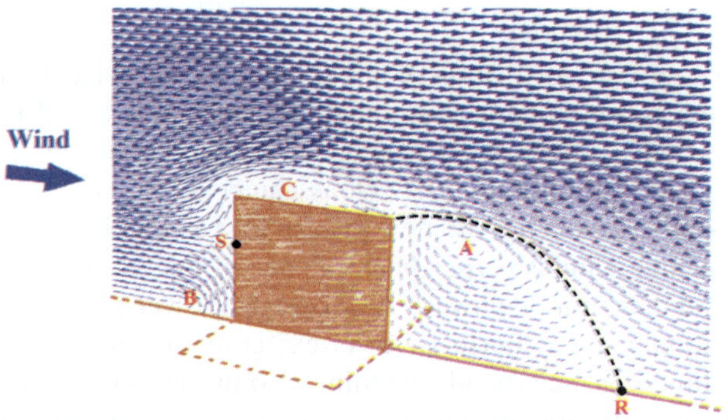

Fig. 10.7 Flow field around an isolated cubic obstacle. The approaching flow is directed rightward. Vectors indicate the velocity magnitude (in arbitrary units) along the vertical section parallel to the streamwise velocity passing through the obstacle centreline. The black dashed line refers to the upper limit of the recirculating region (the data set used refers to the water channel experiment described by Di Bernardino et al. 2017)

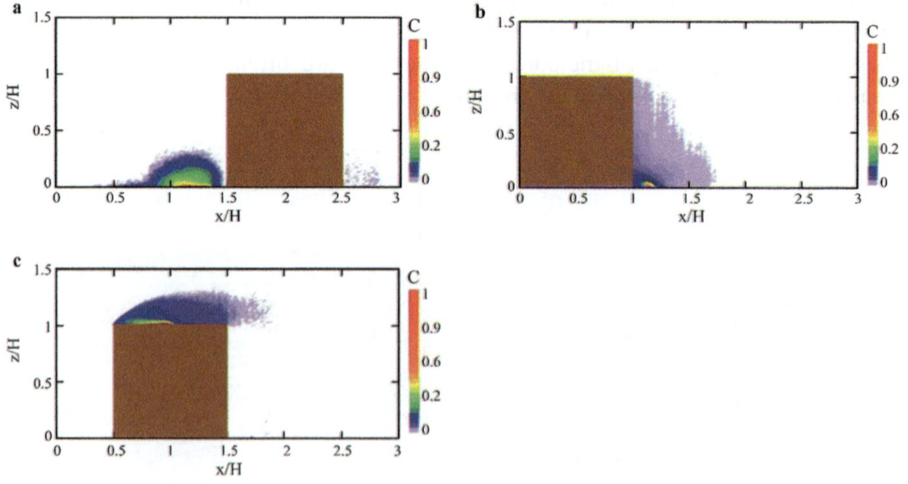

Fig. 10.8 Map of the mean concentration field C (in arbitrary units) of a passive tracer emitted from a point source located (**a**) upwind of the building, (**b**) in the lee of the building and (**c**) at the rooftop. The three panels refer to the same vertical section and data set of Fig. 10.7

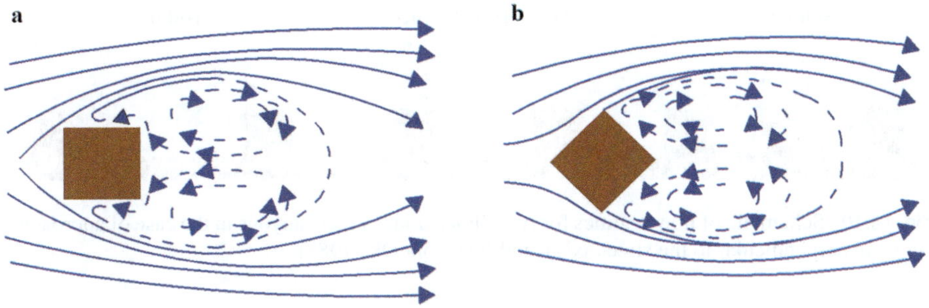

Fig. 10.9 Sketch of the flow patterns around an isolated cubic building along a horizontal plane in the case of approaching wind (**a**) perpendicular to a façade and (**b**) forming an angle of 45° with the façade (modified from Oke 1987)

and on the rooftop. Passive tracers have been extensively used in the past in identifying coherent structure in turbulent flows and can help in the flow analysis both in field campaigns (e.g. Leo et al. 2016) and in the laboratory (e.g. Pournazeri et al. 2012). When the source is upwind of the building (Fig. 10.8a), the pollutant is partially advected upstream and remains mainly trapped within the horseshoe vortex. Then, the tracer moves laterally and is carried downwind. Only a small part of the tracer penetrates into the cavity region in the lee of the building.

Similarly, when the tracer is emitted from the rooftop source (Fig. 10.8c), it remains on average confined in the vortex located in the upwind half of the cube top. For the case of the source located within the cavity region (Fig. 10.8b), the concentration field is quite different: the mean concentration shows much lower values because of the higher pollutant dilution caused by the bow vortex. Furthermore, the polluted plume moves upwind rising towards the upper levels of the building.

It is worthwhile remembering that the sketches depicted above can be considered representative only of cases of wind perpendicular to one of the façades. For all other wind directions, the airflow pattern changes dramatically and depends on the angle between the approaching flow and the façade (Fig. 10.9).

10.4.2 Wind in Street Canyons

The street canyon is defined as a geometric entity developed mainly along the street axis (in theory, of infinite length) and having the characteristic U-shape along the cross section. The street canyon is considered as an archetype for more complex and realistic urban geometries. One of the geometrical parameters to be considered is the aspect ratio $AR = H/W$, i.e. the ratio of the height of the buildings, H, to the spacing between buildings, W. Based on past studies conducted in wind tunnels and water channels, Oke (1987) summarized the nature of the flow in urban canopies in terms of AR (Fig. 10.10), viz. the skimming flow ($AR > 0.7$), in which a single vortex develops within the street canyon; the wake interference flow $0.3 < AR < 0.7$,

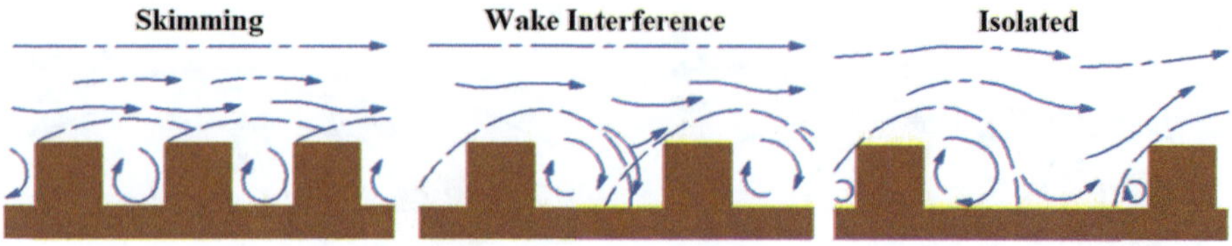

Fig. 10.10 Schematic of flow regimes for two-dimensional street canyons in the case of approaching wind perpendicular to the street axis (modified from Oke 1988)

which allows the development of two counterrotating vortexes; and the isolated obstacle regime ($AR < 0.3$), where the flow strictly resembles that observed for the isolated building case.

It is worth mentioning that the three flow conditions hold strictly in a neutrally stable atmosphere and when wind blows perpendicularly to the street axis. In all other cases, the flow pattern may change substantially due to wind canalization (e.g. Soulhac et al. 2008) and buoyancy effects. The latter can come into play when the wind speed is small (Kim and Baik 1999, 2001; Fernando 2010). Both these factors play an essential role in UHI dynamics (Martilli et al. 2002) and cannot be neglected in UBL modelling (e.g. Cantelli et al. 2015; Salvati et al. 2019).

A sketch of the wind circulation for the skimming flow with $AR = 1$ in the case of uniform building height is depicted in Fig. 10.11. The former is one of the archetypal flow configurations adopted in street flow studies (e.g. Baik et al. 2000; Kovar-Panskus et al. 2002) and represents the basic flow field for any further analysis concerning more realistic canyon geometry. The external wind (from left to right) flows almost parallel to the building rooftop above the canopy and separates from the flow inside the cavity, within which a main vortex forms. A counterrotating recirculating region located in the upper part of the façade of the leeward building and two other smaller vortices, located at the bottom corners of the canyon, complete the flow picture (see also Fig. 10.12 for a sketch of the average velocity streamlines). The instantaneous concentration field depicted in Fig. 10.11a highlights the Kelvin–Helmholtz billowing at the cavity top, triggered by dynamic instability that, in turn, is generated by the high shear levels at the top of the canyon. These billows govern mass and momentum transfer between the cavity and the external flow (see, e.g., Jaroslawski et al. 2019 and references cited therein).

The interaction between canyon and outer flow is greater for the wake interference regime, where the size of the secondary vortex located at the bottom of the leeward building grows with AR (e.g. Ferrari et al. 2017). Flow separation at the cavity top is however still present. Case $AR = 0.5$ shows two well-defined adjacent vortices: the downstream one is by far the larger and rotates clockwise, while the upstream one is smaller, occupying nearly one-fourth of the canyon and rotating counterclockwise (Fig. 10.12). For this geometry, the outer flow has difficulty in penetrating the inter-element space. As a result, it skims and remains nearly parallel to the roofs, thus preventing the formation of the recirculating region at the building top seen previously for the isolated cubic building (Figs. 10.6 and 10.7). This

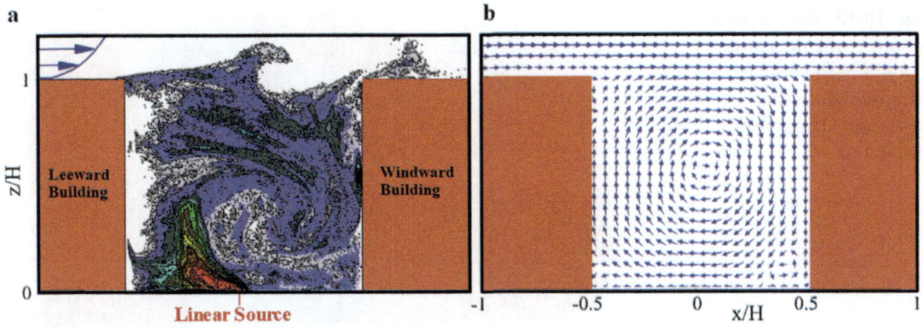

Fig. 10.11 (**a**) Colour map of the instantaneous concentration field (in arbitrary units; red is the maximum concentration value, white indicates zero) of a pollutant emitted from a linear source at ground level within an $AR = 1$ street canyon. (**b**) As in (**a**), but for the mean velocity vectors (in arbitrary unit). The external flow is rightward (the data set used refers to the water channel experiments described by Di Bernardino et al. 2018)

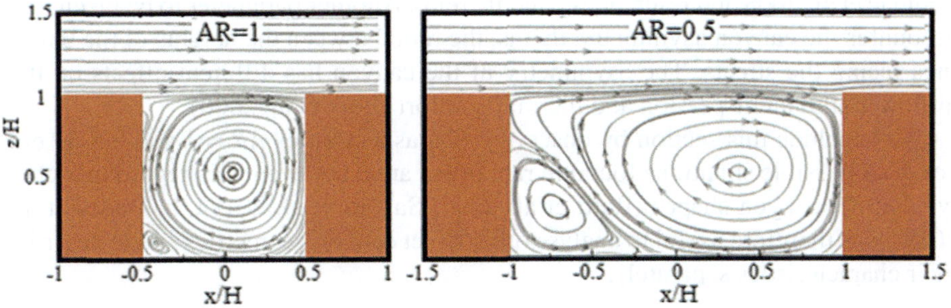

Fig. 10.12 Examples of streamlines of the mean velocity. The external flow comes from the left (adapted from Nardecchia et al. 2018)

feature causes poor canyon ventilation, even for strong external winds, and it is one of the main causes of air quality deterioration and air stagnation in cities. Canyon ventilation is essentially governed by the dynamics of the shear layer at the canyon top, which forms because of the large velocity gradients occurring in this area (see also Fig. 10.4). The shear layer plays a key role in the exchange of air and scalars between the canopy and the overlying region (Louka et al. 2000; Takimoto et al. 2011; Di Bernardino et al. 2018).

Similarly to what happens in the case of isolated building flows, wind direction plays a significant role in the flow field inside the cavity. In principle, if the external wind has a component along the street axis, the flow topology changes dramatically within the canyon. It is no longer two-dimensional and is characterized by an irregular helix (Fig. 10.13).

The main difference with the case of the approaching wind perpendicular to the street axis is the presence of a wind component parallel to the street axis itself. This channelling effect increases close to the ground and causes the progressive

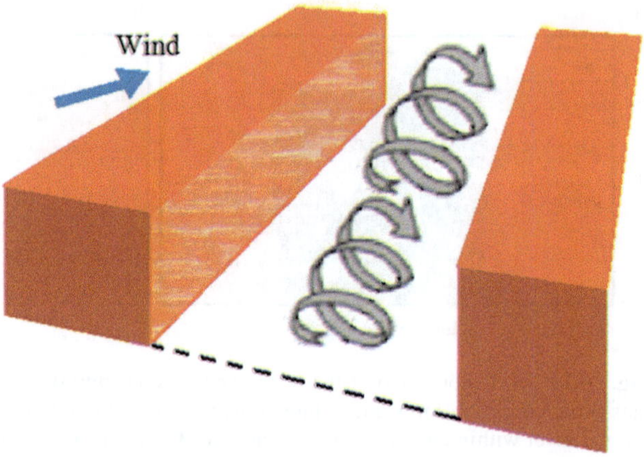

Fig. 10.13 Sketch of flow topology inside a street canyon in the case of external wind not perpendicular to the street axis

reduction of the angle between the wind direction and the street axis (see, e.g., Soulhac et al. 2008 and references cited therein). Another parameter to be taken into account is the canyon asymmetry, that is, the ratio between the heights of the buildings facing the streets. The asymmetry of the canyon has different effects on the airflow, pollutant dispersion and thermal comfort (Baik et al. 2000).

We conclude this section by observing that aspect ratios $AR = 1$ and 0.5 do not cover all possible situations found in real urban area, not even with regard to archetypal street canyon shapes (Ratti et al. 2006; Salvati et al. 2017a, b; Badas et al. 2019). It is therefore useful to analyse other street canyon geometries and to describe their characteristics separately.

10.4.2.1 Effects of Aspect Ratio and Building Rooftop Shape

In principle, a decrease in AR is accompanied by an increase in canyon ventilation, and vice versa. Figure 10.14 illustrates the streamlines of the average velocity for three skimming flow configurations. While for $AR = 1.33$ no substantial differences with $AR = 1$ appear, further increases of AR give rise to multi-vortex configurations (see Zajic et al. 2011 and references cited therein). For $AR = 2$, the two vortices at the bottom seen for $AR = 1.33$ merge to form a larger counterclockwise structure. Two large separated regions therefore appear inside the canyon. Further increases in AR lead to the formation of additional vortices.

For example, for $AR = 4$ three vortices form, with the lowest of them of smaller size. The configuration of narrow canyons is particularly interesting for the investigation of dispersion phenomena and air ventilation (e.g. Cheng and Liu 2011) in that the bottom region is practically disconnected from the outer flow. In essence, the narrow the street canyon the lower the canyon ventilation. For this reason, it is fundamental to consider a correct urban planning to minimize the unwanted effects of pollutant accumulation or air stagnation.

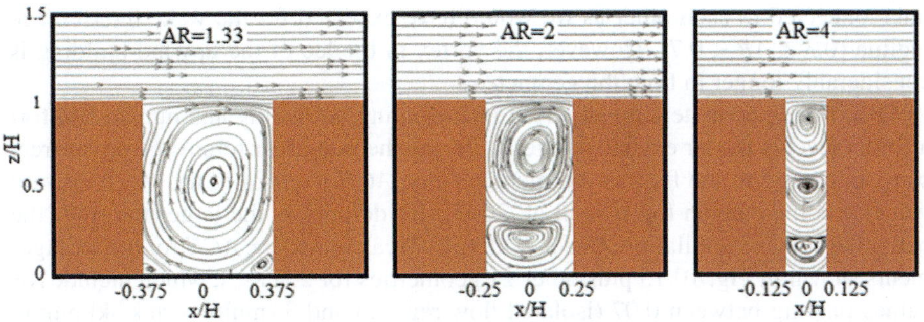

Fig. 10.14 Streamlines of the mean velocity magnitude for the skimming flow regime. The external flow comes from the left (adapted from Nardecchia et al. 2018)

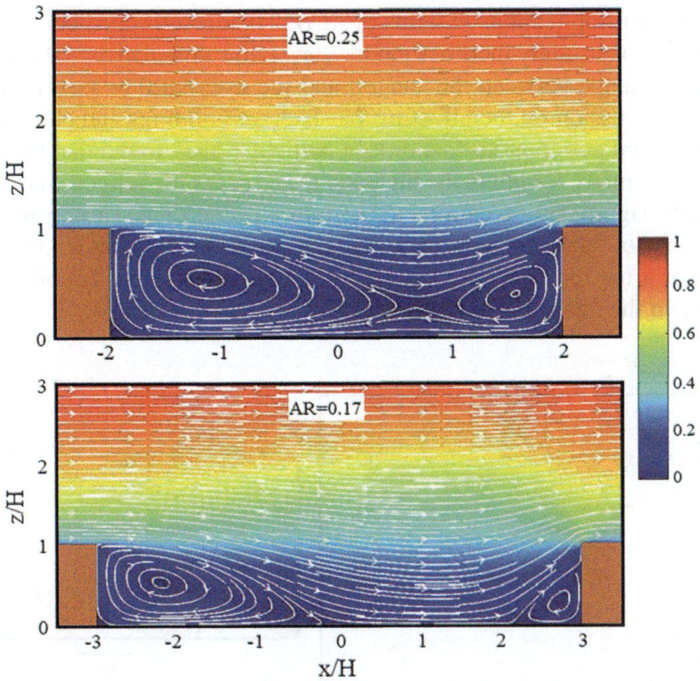

Fig. 10.15 Examples of the mean velocity magnitude normalized with the free stream velocity for the isolated flow regime (colours). The lines denote streamlines (adapted from Badas et al. 2017)

In contrast, reductions in the value of AR are accompanied by a progressive increase in canyon ventilation. Figure 10.15 shows velocity maps obtained numerically by Badas et al. (2017) for $AR = 0.25$ and 0.17 (colours represent mean velocity magnitude made non-dimensional by the free stream velocity). Although both of them are considered as belonging to the isolated flow regime ($AR < 0.3$), the two maps show that the two vortical structures within the canyon are in the first case still interconnected to each other as we would have expected for the wake interference regime ($0.3 < AR < 0.7$). However, the vortex in the lee of the upwind building is considerably longer in both the geometries.

One of the parameters mostly used in evaluating air quality and human comfort at street level is the air exchange rate, *ACH*, i.e. the rate of air removal from a street canyon. Owing to the flow two-dimensionality, *ACH* is equal to the mean upward velocity at the canyon top (Liu et al. 2005). By definition, the larger the *ACH* the better the canyon ventilation. Badas et al. (2017) calculated the *ACH* for the arrangements shown in Fig. 10.15 plus other 21 geometries for 23 cases, which include *AR* values ranging between 0.07 (isolated flow regime) and 3 (multi-vortex skimming flow regime). The ACH shows relatively large values for low aspect ratios, a region of low values for large *AR*, where it is about ¼ compared to the maxima, and an intermediate trend in between (Fig. 10.16).

Badas et al. (2017) also focused on assessing the effect of gable roofs on the flow regimes and their implications in terms of *ACH*. Their results corroborated the idea that a pitched roof strongly modifies the low-level flow and increases the turbulence and the air exchange between the canyon and the external flow (see Ferrari et al. (2019) for a discussion on the effect of the shape of buildings and chimney stacks on ventilation and pollutant dispersion). In particular, they found that the pitch has a positive effect on *ACH* both in the high and low aspect ratio ranges. Such a growth is particularly significant (even a factor of three) in the range of narrow canyons, mainly because of the larger contribution of the mean flow compared to that present in the case of a flat roof (Fig. 10.17). Note that the curves of *ACH* collapse onto a single curve in the intermediate, descending range of aspect ratios.

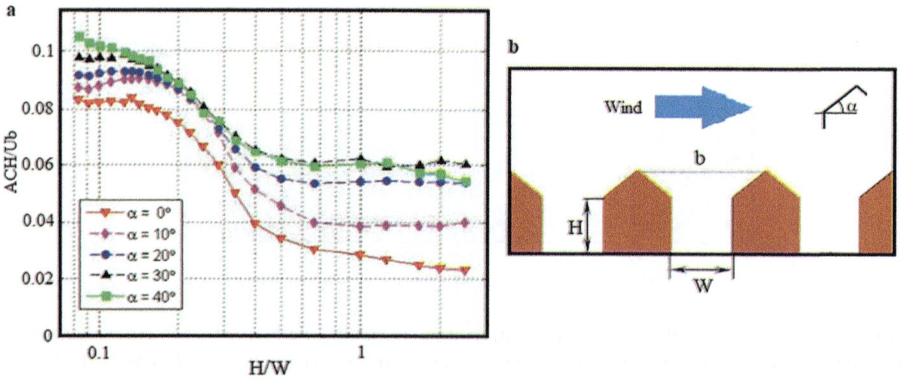

Fig. 10.16 (**a**) Non-dimensional *ACH* versus *H*/*W* for two-dimensional street canyons with planar rooftop (roof slope $\alpha = 0^{\circ}$, red line). The other curves refer to *ACH* calculated for different α. *U* indicates the free stream velocity. (**b**) Sketch of the computational domain (adapted from Badas et al. 2017)

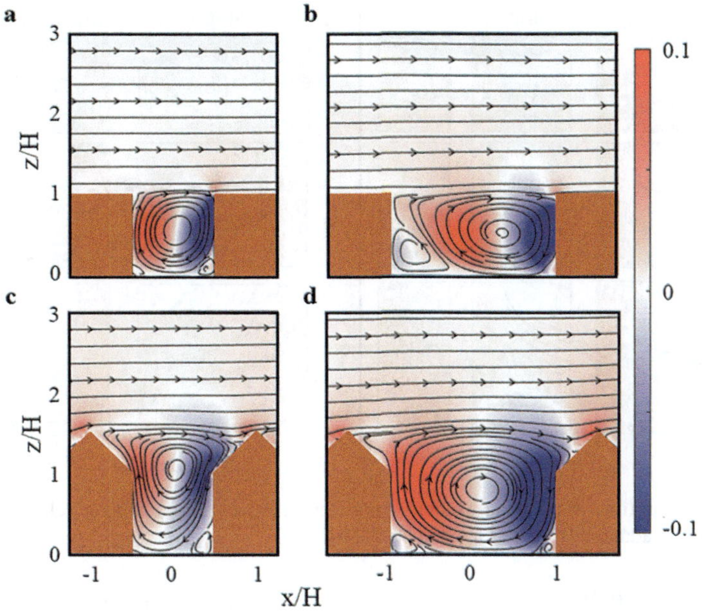

Fig. 10.17 Streamlines of the mean velocity field for different aspect ratios and roof shapes. Colours indicate the vertical wind velocity component made non-dimensional by the free stream velocity for (**a**) $AR = 1$ and $\alpha = 0$, (**b**) $AR = 0.5$ and $\alpha = 0$, (**c**) $AR = 1$ and $\alpha = 45°$ and (**d**) $AR = 0.5$ and $\alpha = 45°$ (adapted from Garau et al. 2018)

10.4.2.2 Effect of Building Height Variation

The role played in the flow topology by the ratio between the height of the windward and the leeward building, $R = H_W/H_L$, has been extensively investigated in the past (e.g. Zajic et al. 2011 and references cited therein). With regard to the case of flat rooftops, Nardecchia et al. (2018) focused on the combined effect of the variability of both R and AR by considering six values of R and five ARs. In particular, they analysed the step-up configurations, where the leeward building ($H_L < H_W$) is shorter than the windward building, and the step-down configurations ($H_L > H_W$). Figure 10.18 shows examples of streamlines for step-up (R = 1.5, 2 and 2.5) and step-down (R = 0.67, 0.5 and 0.4) geometries and AR = 0.5, 1.5 and 4.

For the step-up configurations ($R > 1$) for $AR = 0.5$ and 1.5 the flow fields do not change much from that seen for $R = 1$ (Fig. 10.11). The main vortex increases in size and moves upward as R grows. Regarding the skimming flow cases for $AR = 1.5$, their dependence on R is definitely greater. There is a progressive ejection of the upper vortex from the canyon into the overlying layer going from $R = 1.5$ to 2.5. In terms of air ventilation, this is reflected in the fact that ACH does not depend significantly on R when $AR = 0.5$ and 1 (Fig. 10.19), while a clear decrease in ACH for increasing R occurs for the other aspect ratios. Taller windward buildings allow lower vertical mass transfer between the canyon and the overlying region. On the

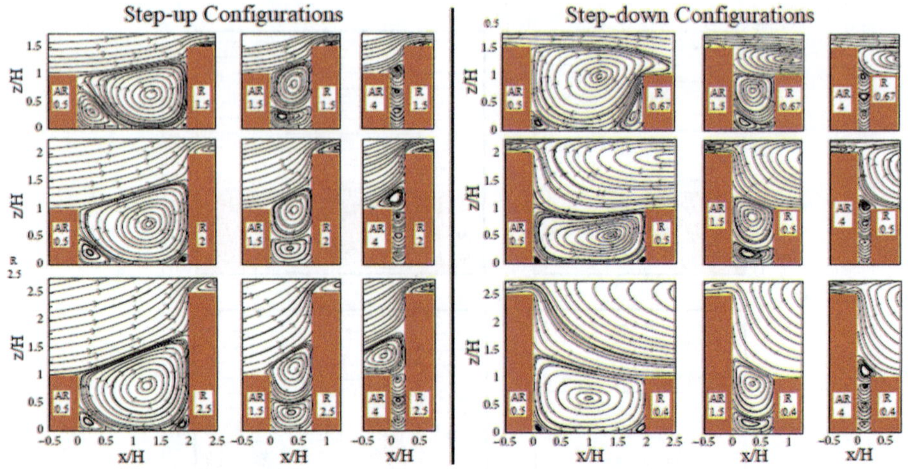

Fig. 10.18 Streamlines of the mean velocity for different aspect ratios (*AR*) and relative height (*R*) of the buildings (adapted from Nardecchia et al. 2018)

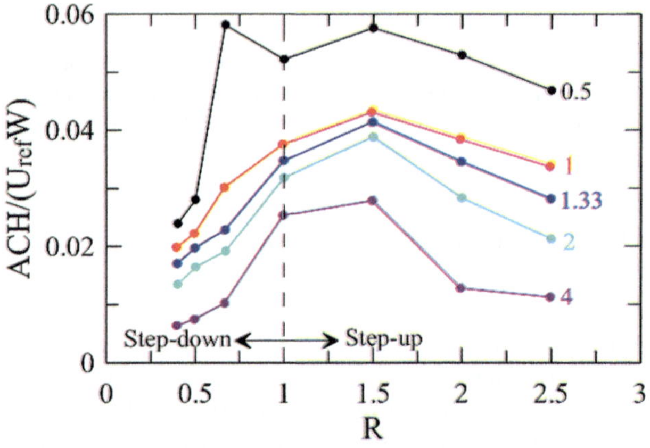

Fig. 10.19 Non-dimensional *ACH* versus *R*. Numbers near the curves denote *AR* (adapted from Nardecchia et al. 2018)

other hand, the step-down configurations ($R < 1$) show a wide clockwise vortex placed over the canyon and the top of the windward building. Overall, the lower the R the smaller the *ACH*, with the exception of the case ($AR = 0.5 - R = 0.67$) for which there is only a large vortical structure that occupies both the canyon and the overlying region up to $z = H_L$. The latter configuration corresponds with the largest *ACH* calculated for all the cases analysed. In contrast, for all the other step-down configurations, the main vortex (or the two or more vortices) remains confined within the canyon. The latter represents the main difference between step-up and step-down configurations and it certainly has great influence on pollutant concentration, particularly at street level.

From the point of view of air quality analysis, the development of secondary vortices in the lower corners of the canyon for $AR = 0.5$ and 1 should determine an accumulation of pollutants near the sidewalk, regardless of R. For $AR = 1.5$ and 2, the presence of the two counterrotating vortices further limits the canyon ventilation, especially at ground level. For $AR = 4$, the vertically aligned multiple vortex configuration strongly inhibits the exchange of air with the higher layers and paves the way to stagnation of pollutants at pedestrian level.

We conclude this subsection by showing the effects on the flow pattern of street canyons having flat roof on one street side and gable roofs on the opposite side. The results found numerically for several rooftop geometries by Xie et al. (2005a, b) show a substantial resemblance of A and B cases to the step-up and step-down configurations seen above for $AR = 1$ and $R = 1.5$ and 0.67, respectively (Fig. 10.20). On the other hand, cases C and D show quite clearly the role played by the approaching wind direction even in cases of winds perpendicular to the street axis. This fact is further evidence that the knowledge of prevailing winds can be extremely useful during earlier design stage.

10.4.3 Groups of Buildings

As mentioned earlier, the turbulent nature of the urban flow makes it impossible to analytically solve the wind field in urban environments. Similarly to the street canyon case, most CFD and laboratory studies on groups of buildings consider idealized situations like the regular arrangement of obstacles depicted in Fig. 10.21a, i.e. an approaching wind perpendicular to the façades of a group of regular cubes mimicking a real urban complex. Usually, the geometrical parameters used to classify

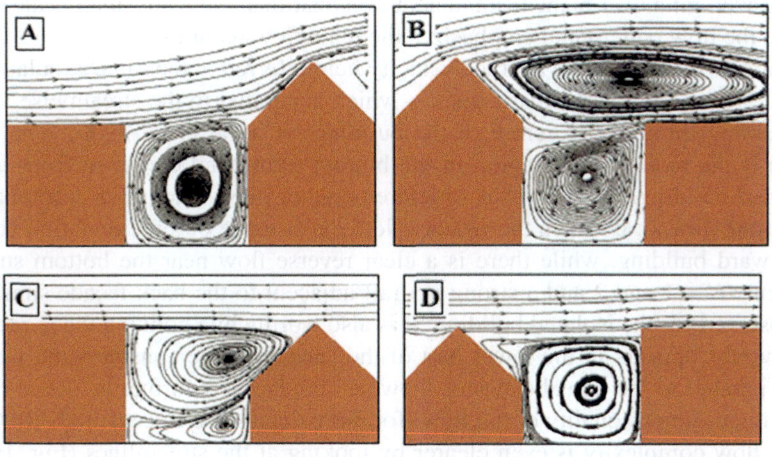

Fig. 10.20 Influence of roof shape on streamlines in street canyon (adapted from Xie et al. 2005a, b)

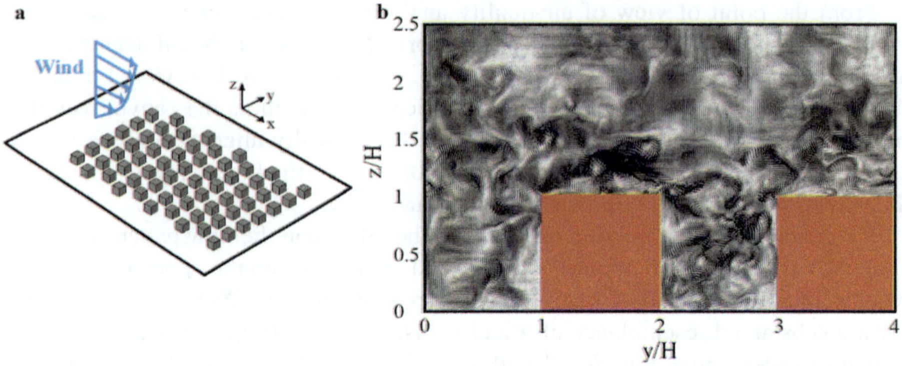

Fig. 10.21 (**a**) Sketch of archetypal array of cubes usually adopted in CFD simulations and laboratory experiments and (**b**) instantaneous snapshot of the velocity field in a vertical *y*–*z* plane perpendicular to the approaching mean flow (which is out of the page) obtained for a staggered building array with $\lambda_P = 0.25$ by means of a DNS (adapted from Coceal et al. 2006)

urban canopies are the plan area density, λ_P, and the frontal area density, λ_F, defined, respectively, as the plan area and frontal area occupied by buildings divided by the total area of the land on which they are located. We remind the reader that λ_P and λ_F are just two of the several parameters involved in urban canopy classification. For additional information see Grimmond and Oke (1999), Ratti et al. (2006) and Stewart and Oke (2012).

The instantaneous snapshot of the velocity field depicted in Fig. 10.20b helps to elucidate the complexity of the turbulent flow within a group of staggered buildings. The wind field is strongly inhomogeneous and vortical structures (with circulation in both clockwise and counterclockwise sense) are present throughout the flow. In terms of average quantities, the velocity pattern is undoubtedly simpler in so much that the averaging over many instantaneous realizations filters out small-scale structures in the flow pattern and emphasizes the main flow features.

Figure 10.22 depicts the average velocity field referred to the same simulation of Fig. 10.21, but for a vertical *x*–*z* plane—which is parallel to the streamwise velocity—passing through the centre of the buildings (Coceal et al. 2006). The main feature is the recirculation region in the bottom right-hand corner in front of the windward building. Such vortical structure is rather persistent and is accompanied by a strong downdraft down the façade. Note the absence of a recirculation behind the leeward building, while there is a clear reverse flow near the bottom surface between $x/H = 1$ and 2 and a strong updraft adjacent to the back façade similar to that observed for the isolated building. It is also worthwhile noticing that a recirculation vortex appears in the lower part of the canopy if one considers the vertical planes parallel to the one above (not shown). This fact is a further evidence of the strong three-dimensionality of the flow (for more details see Coceal et al. 2006).

The flow complexity is even clearer by looking at the streamlines (Fig. 10.23) corresponding to the horizontal average velocity referred to an aligned cube array for two external wind directions (0° and 45°) and a staggered array for 45° (Coceal

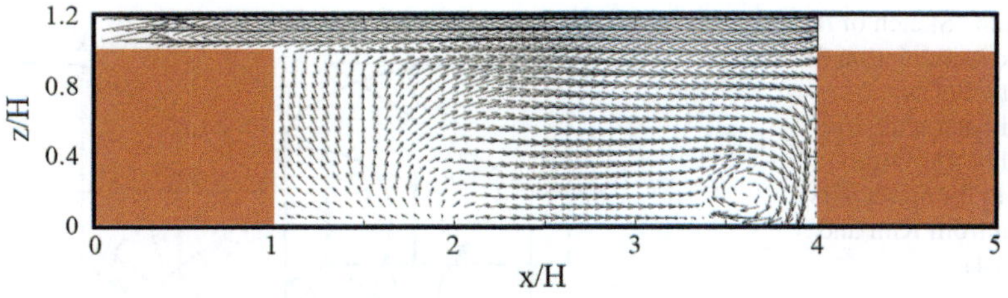

Fig. 10.22 Average velocity field in a vertical *x*–*z* plane through the middle of the cubes of the array in Fig. 10.21. The streamwise velocity is now directed rightward (adapted from Coceal et al. 2006)

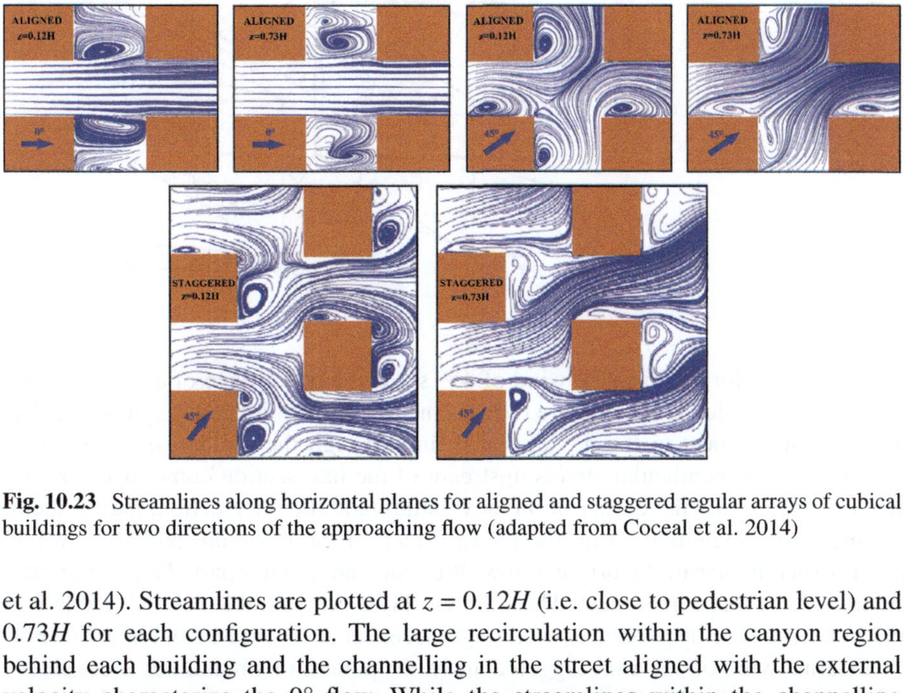

Fig. 10.23 Streamlines along horizontal planes for aligned and staggered regular arrays of cubical buildings for two directions of the approaching flow (adapted from Coceal et al. 2014)

et al. 2014). Streamlines are plotted at $z = 0.12H$ (i.e. close to pedestrian level) and $0.73H$ for each configuration. The large recirculation within the canyon region behind each building and the channelling in the street aligned with the external velocity characterize the 0° flow. While the streamlines within the channelling region are similar at the two heights, the vortical structures behind the buildings are essentially three-dimensional and change considerably with height.

As shown by Kim and Baik (2004), the flow coming into the street canyon curls up around the so-called portal vortex, whose ends are located near the lower edges of the windward building (Fig. 10.24a). The portal vortex is symmetric with respect to the centre of the street canyon and its horizontal axis is perpendicular to the external wind direction. The same authors showed that for external winds not perpendicular to the façade, the horizontal size of the portal vortex changes and a horseshoe vortex forms around the façades of the leeward building. When the incident wind angle is 45°, the flow is diagonally symmetric behind the windward building (Fig. 10.24b).

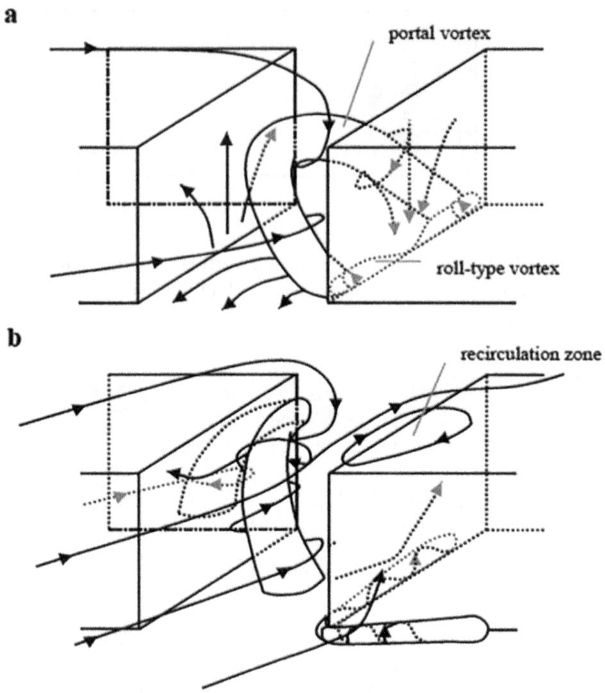

Fig. 10.24 Sketch of the mean flow circulation for external wind (**a**) perpendicular and (**b**) inclined of 45° with respect to the street axis (adapted from Kim and Baik 2004)

Streamlines for the external wind of 45° show the occurrence of a recirculation behind each façade. Such a feature reduces in size with height and is symmetrically situated with respect to the 45° diagonal line through the buildings. Streamlines from the two perpendicular streets upstream of the intersection come together and then diverge around the corner of the next building at the intersection, part of them feeding into a recirculation on the leeward face of the upstream buildings in the street perpendicular to the original flow direction and another part channelling into the far side of that same street. These flows play a considerable role in ventilation and in near-source dispersion as well (e.g. Coceal et al. 2014; Mei et al. 2017). For the external wind of 45° and staggered array, the pattern of streamlines is considerably more complicated. There is no longer symmetry with respect to the 45° diagonal line throughout the array and the flow topology is modified substantially.

10.5 Conclusions

The wind flows described in the preceding sections are only a limited part of cases that can be seen in real cities. Vegetation (e.g. trees and green roofs), canyon asymmetries, squares, viaducts as well as thermal inhomogeneity of the urban canopy layer are just some of the features normally found in cities that make the picture even more complicated and particularly difficult to generalize. Despite the continuous effort of many researchers, the progressive lowering of the meteorological instrumentation costs, and the increase in computing power, the knowledge of wind circulation in cities is still far from being exhaustive. The lack of analytical laws that provide wind speed even in simple cases such as an isolated building is just one example of the complexity of the phenomenon. This must spur on the scientific community to enrich its knowledge of an important and fascinating topic such as urban fluid mechanics.

References

Allwine, K. J., Shinn, J. H., Streit, G. E., Clawson, K. L., & Brown, M. (2002). Overview of URBAN 2000: A multiscale field study of dispersion through an urban environment. *Bullettin of the American Meteorological Society, 83*, 521–536.

Badas, M. G., Ferrari, S., Garau, M., & Querzoli, G. (2017). On the effect of gable roof on natural ventilation in two-dimensional urban canyons. *Journal of Wind Engineering and Industrial Aerodynamics, 162*, 24–34.

Badas, M. G., Salvadori, L., Garau, M., Querzoli, G., & Ferrari, S. (2019). Urban areas parameterization for CFD simulations and cities air quality. *International Journal of Environment and Pollution, 66*, 5–18.

Baik, J.-J., Park, R.-S., Chun, H.-Y., & Kim, J.-J. (2000). A laboratory model of urban street canyon flows. *Journal of Applied Meteorology, 39*, 1592–1600.

Bazdidi-Tehrani, F., & Jadidi, M. (2014). Large eddy simulation of dispersion around an isolated cubic building: Evaluation of localized dynamic k_{SGS}-equation sub-grid scale model. *Environmental Fluid Mechanics, 14*, 565–589.

Bentham, T., & Britter, R. E. (2002). Spatially averaged flow velocity within large groups of obstacles. *Atmospheric Environment, 37*, 2037–2043.

Blocken, B. (2014). 50 years of computational wind engineering: Past, present and future. *Journal of Wind Engineering and Industrial Aerodynamics, 129*, 69–102.

Blocken, B. (2015). Computational fluid dynamics for urban physics: Importance, scales, possibilities, limitations and ten tips and tricks towards accurate and reliable simulations. *Building and Environment, 91*, 219–245.

Blocken, B., Stathopoulos, T., Carmeliet, J., & Hensen, L. M. (2011). Application of computational fluid dynamics in building performance simulation for the outdoor environment: An overview. *Journal of Building Performance Simulation, 4*, 157–184.

Britter, R. E., & Hanna, S. (2003). Flow and dispersion in urban areas. *Annual Review of Fluid Mechanics, 35*, 469–496.

Buccolieri, R., Salizzoni, P., Soulhac, L., Garbero, V., & Di Sabatino, S. (2015). The breathability of compact cities. *Urban Climate, 13*, 73–93.

Businger, J. A., Wyngaard, J. C., Izumi, Y., & Bradley, E. F. (1971). Flux-profile relationships in the atmospheric surface layer. *Journal of the Atmospheric Sciences, 28*, 181–189.

Cantelli, A., Monti, P., & Leuzzi, G. (2015). Numerical study of the urban geometrical representation impact in a surface energy budget model. *Environmental Fluid Mechanics, 15*, 251–273.

Cenedese, A., Miozzi, M., & Monti, P. (2000). A laboratory investigation of land and sea breeze regimes. *Experiments in Fluids, 29*(Suppl. 1), S291–S299.

Cheng, H., & Castro, I. P. (2002). Near wall flow over urban-like roughness. *Boundary-Layer Meteorology, 104*, 229–259.

Cheng, W. C., & Liu, C.-H. (2011). Large-eddy simulation of flow and pollutant transports in and above two-dimensional idealised street canyons. *Boundary-Layer Meteorology, 139*, 411–437.

Cionco, R. M. (1965). Mathematical model for air flow in a vegetative canopy. *Journal of Applied Meteorology, 4*, 517–522.

Coceal, O., Thomas, T. G., Castro, I. P., & Belcher, S. E. (2006). Mean flow and turbulence statistics over groups of urban-like cubical obstacles. *Boundary-Layer Meteorology, 121*, 491–519.

Coceal, O., Goulart, E. V., Branford, S., Thomas, T. G., & Belcher, S. E. (2014). Flow structure and near-field dispersion in arrays of building-like obstacles. *Journal of Wind Engineering and Industrial Aerodynamics, 125*, 52–68.

Di Bernardino, A., Monti, P., Leuzzi, G., & Querzoli, G. (2015a). On the effect of the aspect ratio on flow and turbulence over a two-dimensional street canyon. *International Journal of Environment and Pollution, 58*, 27–38.

Di Bernardino, A., Monti, P., Leuzzi, G., & Querzoli, G. (2015b). Water-channel study of flow and turbulence past a 2D array of obstacles. *Boundary-Layer Meteorology, 155*, 73–85.

Di Bernardino, A., Monti, P., Leuzzi, G., Sammartino, F., & Ferrari, S. (2017). Experimental investigation of turbulence and dispersion around an isolated cubic building. Proc. 18th Int. Conf. Harmonisation within Atmospheric Dispersion Modelling for Regulatory Purposes, 9–12 October, Bologna, Italy.

Di Bernardino, A., Monti, P., Leuzzi, G., & Querzoli, G. (2018). Pollutant fluxes in two-dimensional street canyons. *Urban Climate, 24*, 80–93.

Di Sabatino, S., Solazzo, E., Paradisi, P., & Britter, R. (2008). A simple model for spatially-averaged wind profiles within and above an urban canopy. *Boundary-Layer Meteorology, 127*, 131–151.

Di Sabatino, S., Leo, L. S., Cataldo, R., Ratti, C., & Britter, R. E. (2010). Construction of digital elevation models for a southern European city and a comparative morphological analysis with respect to northern European and North American cities. *Journal of Applied Meteorology and Climatology, 49*, 1377–1396.

Fan, Y., Hunt, J., Wang, Q., Yin, S., & Li, Y. (2019). Water tank modelling of variations in inversion breakup over a circular city. *Building and Environment, 164*, 106342.

Fernando, H. J. S. (2010). Fluid dynamics of urban atmospheres in complex terrain. *Annual Review of Fluid Mechanics, 42*, 365–389.

Fernando, H. J. S., Lee, S. M., Anderson, J., Princevac, M., Pardyjak, E., & Grossman-Clarke, S. (2001). Urban fluid mechanics: Air circulation and contaminant dispersion in cities. *Environmental Fluid Mechanics, 1*, 107–164.

Ferrari, S., Badas, M. G., Garau, M., Seoni, A., & Querzoli, G. (2017). The air quality in narrow two-dimensional urban canyons with pitched and flat roof buildings. *International Journal of Environment and Pollution, 62*, 347–368.

Ferrari, S., Badas, M. G., Garau, M., Salvadori, L., Seoni, A., & Querzoli, G. (2019). On the effect of the shape of the buildings and chimneystacks on ventilation and pollutant dispersion. *EPJ Web Conferences, 213*, 02017.

Garau, M., Badas, M. G., Ferrari, S., Seoni, A., & Querzoli, G. (2018). Turbulence and air exchange in a two-dimensional urban street canyon between gable roof buildings. *Boundary-Layer Meteorology, 167*, 123–143.

Garrat, J. R. (1992). *The atmospheric boundary layer*. Cambridge, UK: Cambridge University Press.

Giovannini, L., Zardi, D., & de Franceschi, M. (2011). Analysis of the urban thermal fingerprint of the city of Trento in the Alps. *Journal of Applied Meteorology and Climatology, 50*, 1145–1162.

Giovannini, L., Zardi, D., & de Franceschi, M. (2013). Characterization of the thermal structure inside an urban canyon: Field measurements and validation of a simple model. *Journal of Applied Meteorology and Climatology, 52*, 64–81.

Gousseau, P., Blocken, B., & van Heijst, G. J. F. (2011). CFD simulation of pollutant dispersion around isolated buildings: On the role of convective and turbulent mass fluxes in the prediction accuracy. *Journal of Hazardous Materials, 194*, 22–434.

Grimmond, C. S. B., & Oke, T. R. (1999). Aerodynamic properties of urban areas derived from analysis of urban surface form. *Journal of Applied Meteorology, 38*, 1261–1292.

Holton, J. R. (1992). *An introduction of dynamic meteorology*. New York: Academic Press.

Jacobson, M. Z. (2005). *Fundamentals of atmospheric modelling*. Cambridge: Cambridge University Press.

Jaroslawski, T., Perret, L., Blackman, K., & Savory, E. (2019). The spanwise variation of roof-level turbulence in a street-canyon flow. *Boundary-Layer Meteorology, 170*, 373–394.

Kastner-Klein, P., & Rotach, M. W. (2004). Mean flow and turbulence characteristics in an urban roughness sublayer. *Boundary-Layer Meteorology, 111*, 55–84.

Kent, C. W., Grimmond, S., Barlow, J., Gatey, D., Kotthaus, S., Lindberg, F., & Halios, C. H. (2017). Evaluation of urban local-scale aerodynamic parameters: Implications for the vertical profile of wind speed and for source areas. *Boundary-Layer Meteorology, 164*, 183–213.

Kim, J.-J., & Baik, J.-J. (1999). A numerical study of thermal effects on flow and pollutant dispersion in urban street canyons. *Journal of Applied Meteorology, 38*, 1249–1261.

Kim, J.-J., & Baik, J.-J. (2001). Urban street canyon flows with bottom heating. *Atmospheric Environment, 35*, 3395–3404.

Kim, J.-J., & Baik, J.-J. (2004). A numerical study of the effects of ambient wind direction on flow and dispersion in urban street canyons using the RNG k-E turbulence model. *Atmospheric Environment, 38*, 3039–3048.

Kovar-Panskus, A., Louka, P., Sini, J. F., Savory, E., Czech, M., Abdelqari, A., et al. (2002). Influence of geometry on the mean flow within urban street canyons—A comparison of wind tunnel experiments and numerical simulations. *Water, Air, & Soil Pollution: Focus, 2*, 365–380.

Lai, D., Karava, P., & Chen, Q. (2015). Study of outdoor ozone penetration into buildings through ventilation and infiltration. *Building and Environment, 93*, 112–118.

Leo, L. S., Thompson, M. J., Di Sabatino, S., & Fernando, H. J. S. (2016). Stratified Flow Past a Hill: Dividing Streamline Concept Revisited. *Boundary-Layer Meteorology, 159*, 611–634.

Li, W.-W., & Meroney, R. N. (1983). Gas dispersion near a cubical model building. Part I. Mean concentration measurements. *Journal of Wind Engineering and Industrial Aerodynamics, 12*, 15–33.

Liu, C.-H., Leung, D. Y. C., & Barth, M. C. (2005). On the prediction of air and pollutant exchange rates in street canyons of different aspect ratios using large-eddy simulation. *Atmospheric Environment, 39*, 1567–1574.

Louka, P., Belcher, S. E., & Harrison, R. G. (2000). Coupling between air flow in streets and the well-developed boundary layer aloft. *Atmospheric Environment, 34*, 2613–2621.

Lu, J., Arya, S. P., Snyder, W. H., & Lawson, R. E., Jr. (1997). A laboratory study of the urban heat island in a calm and stably stratified environment. Part I: Temperature field. *Journal of Applied Meteorology, 36*, 1377–1391.

Macdonald, R. W. (2000). Modelling the mean velocity profile in the urban canopy layer. *Boundary-Layer Meteorology, 97*, 25–45.

Martilli, A., Clappier, A., & Rotach, M. W. (2002). An urban surface exchange parameterisation for mesoscale models. *Boundary-Layer Meteorology, 104*, 261–304.

Martinuzzi, R, 1992. *Experimentelle Untersuchung der umströmung wandgebundener, rechteckiger, prismaticher hindernisse*. Ph.D. Dissertation, Universitat Erlangen-Nurnberg.

Martinuzzi, R., & Tropea, C. (1993). The flow around surface-mounted, prismatic obstacles placed in a fully developed channel flow: (data bank contribution). *Journal of Fluids Engineering, 115*, 85–92.

Mei, S. J., Hu, J. T., Liu, D., Zhao, F. Y., Li, Y., Wang, Y., & Wang, H. Q. (2017). Wind driven natural ventilation in the idealized building block arrays with multiple urban morphologies and unique package building density. *Energy and Buildings, 155*, 324–338.

Monin, A. S., & Obukhov, A. M. (1954). Basic laws of turbulent mixing in the atmosphere near the ground. *Trudy Geofiz. Inst. AN SSSR, 24*(151), 163–187.

Moonen, P., Defraeye, T., Dorer, V., Blocken, B., & Carmeliet, J. (2012). Urban Physics: Effect of the micro-climate on comfort, health and energy demand. *Frontiers of Architectural Research, 1*, 197–228.

Murakami, S. (1993). Comparison of various turbulence models applied to a bluff body. *Journal of Wind Engineering and Industrial Aerodynamics, 46–47*, 21–36.

Nardecchia, F., Di Bernardino, A., Pagliaro, F., Monti, P., Leuzzi, G., & Gugliermetti, L. (2018). CFD analysis of urban canopy flows employing the V2F model: Impact of different aspect ratios and relative heights. *Advances in Meteorology, 2018*, 2189234.

Oke, T. R. (1982). The energetic basis of the urban heat island. *Quarterly Journal of the Royal Meteorological Society, 108*, 1–24.

Oke, T. R. (1987). *Boundary layer climates* (2nd ed.). London: Routledge.

Oke, T. R. (1988). Street design and urban canopy layer climate. *Energy and Buildings, 11*, 103–113.

Ottosen, T.-B., Ketzel, M., Skov, H., Hertel, O., Brandt, J., & Kakosimos, K. E. (2019). Micro-scale modelling of the urban wind speed for air pollution applications. *Scientific Reports, 9*, 14279.

Pelliccioni, A., Monti, P., & Leuzzi, G. (2015). An alternative wind profile formulation for urban areas in neutral conditions. *Environmental Fluid Mechanics, 15*, 135–146.

Pichelli, E., Ferretti, R., Cacciani, M., Siani, A. M., Ciardini, V., & Di Iorio, T. (2014). The role of urban boundary layer investigated with high-resolution models and ground-based observations in Rome area: A step towards understanding parameterization potentialities. *Atmospheric Measurements Technology, 7*, 315–332.

Pournazeri, S., Princevac, M., & Venkatram, A. (2012). Rise of Buoyant Emissions from Low-Level Sources in the Presence of Upstream and Downstream Obstacles. *Boundary-Layer Meteorology, 144*, 287–308.

Ratti, C., Di Sabatino, S., & Britter, R. (2006). Urban texture analysis with image processing techniques: Winds and dispersion. *Theoretical and Applied Climatology, 84*, 77–90.

Ravanelli, R., Nascetti, A., Cirigliano, R. V., Di Rico, C., Leuzzi, G., Monti, P., & Crespi, M. (2018). Monitoring the impact of land cover change on surface urban heat island through Google Earth Engine: Proposal of a global methodology, first applications and problems. *Remote Sensing, 10*, 1488.

Rodi, W. (1995). Introduction to the numerical simulation approaches in wind engineering. In J. E. Cermak, A. G. Davenport, E. J. Plate, & D. X. Viegas (Eds.), *Wind climate in cities* (NATO ASI Series (Series E: Applied Sciences)) (Vol. 277). Dordrecht: Springer.

Rotach, M. W. (1999). On the influence of the urban roughness sublayer on turbulence and dispersion. *Atmospheric Environment, 33*, 4001–4008.

Roth, M., Oke, T. R., & Emery, W. J. (1989). Satellite-derived urban heat islands from three coastal cities and the utilization of such data in urban climatology. *International Journal of Remote Sensing, 10*, 1699–1720.

Salizzoni, P., Soulhac, L., & Mejean, P. (2009). Street canyon ventilation and atmospheric turbulence. *Atmospheric Environment, 43*, 5056–5067.

Salizzoni, P., Marro, M., Soulhac, L., Grosjean, N., & Perkins, R. J. (2011). Turbulent transfer between street canyons and the overlying atmospheric boundary layer. *Boundary-Layer Meteorology, 141*, 393–414.

Salvati, A., Coch Roura, H., & Cecere, C. (2017a). Assessing the urban heat island and its energy impact on residential buildings in Mediterranean climate: Barcelona case study. *Energy and Buildings, 146*, 38–54.

Salvati, A., Palme, M., & Inostroza, L. (2017b). Key parameters for urban heat island assessment in a Mediterranean context: A sensitivity analysis using the Urban Weather Generator Model. *IOP Conferences Series: Material Sciences and Engineering, 245*(8), 082055.

Salvati, A., Monti, P., Coch Roura, H., & Cecere, C. (2019). Assessing the urban heat island and its energy impact on residential buildings in Mediterranean climate: Barcelona case study. *Energy and Buildings, 185*, 162–179.

Santamouris. (2001). *Energy and climate in the urban built environment*. London, UK: James & James Ltd.

Santos, J. M., Reis, N. C., Jr., Goulart, E. V., & Mavroidis, I. (2009). Numerical simulation of flow and dispersion around an isolated cubical building: The effect of the atmospheric stratification. *Atmospheric Environment, 43*, 5484–5492.

Simpson, J. E. (1994). *Sea breeze and local winds*. Great Britain: Cambridge University Press.

Snyder, W. H. (1994). Downwash of plumes in the vicinity of buildings: A wind-tunnel study. In P. A. Davies & M. J. V. Neves (Eds.), *Recent research advances in the fluid mechanics of turbulent jets and plumes* (NATO ASI Series (Series E: Applied Sciences)) (Vol. 255). Dordrecht: Springer.

Soulhac, L., Perkins, R. J., & Salizzoni, P. (2008). Flow in a street canyon for any external wind direction. *Boundary-Layer Meteorology, 126*, 365–388.

Stewart, I. D., & Oke, T. R. (2012). Local climate zones for urban temperature studies. *Bullettin of the American Meteorological Society, 92*, 1879–1900.

Stull, R. B. (1988). *An introduction to boundary layer meteorology*. The Netherlands: Kluwer Academic Publisher.

Takimoto, H., Sato, A., Barlow, J. F., Moriwaki, R., Inagaki, A., Onomura, S., & Kanda, M. (2011). Particle image velocimetry measurements of turbulent flow within outdoor and indoor urban scale models and flushing motions in urban canopy layers. *Boundary-Layer Meteorology, 140*, 295–314.

Tominaga, Y., & Stathopoulos, T. (2009). Numerical simulation of dispersion around an isolated cubic building: Comparison of various types of k-ε models. *Atmospheric Environment, 43*, 3200–3210.

Tominaga, Y., & Stathopoulos, T. (2010). Numerical simulation of dispersion around an isolated cubic building: Model evaluation of RANS and LES. *Building and Environment, 45*, 2231–2239.

Valerio, G., Cantelli, A., Monti, P., & Leuzzi, G. (2017). A modeling approach to identify the effective forcing exerted by wind on a prealpine lake surrounded by a complex topography. *Water Resources Research, 53*, 4036–4052.

van Hooff, T., & Blocken, B. (2010). Coupled urban wind flow and indoor natural ventilation modelling on a high-resolution grid: A case study for the Amsterdam ArenA stadium. *Environmental Modelling & Software, 25*, 51–65.

Whiteman, C. D. (2000). *Mountain meteorology: Fundamentals and applications*. UK: Oxford University Press.

Xie, X., Huang, Z., & Wang, J. S. (2005a). Impact of building configuration on air quality in street canyon. *Atmospheric Environment, 39*, 4519–4530.

Xie, X., Huang, Z., Wang, J. S., & Xie, Z. (2005b). The impact of solar radiation and street layout on pollutant dispersion in street canyon. *Building and Environment, 40*, 201–212.

Zajic, D., Fernando, H. J. S., Calhoun, R., Princevac, M., Brown, M. J., & Pardyjak, E. R. (2011). Flow and turbulence in an urban canyon. *Journal of Applied Meteorology and Climatology, 50*, 203–223.

Zhou, Y., Guan, H., Huang, C., Fan, L., Gharib, S., Batelaan, O., & Simmons, C. (2019). Sea breeze cooling capacity and its influencing factors in a coastal city. *Building and Environment, 166*, 106408.

Chapter 11
The Coupling of the Weather Research and Forecasting Model with the Urban Canopy Models for Climate Simulations

Zahra Jandaghian and Umberto Berardi

11.1 Introduction

The urbanization leads to alteration in the land surface characteristics and contributes to the alteration of climate (Akbari and Kolokotsa 2016). Built structures affect the transfer of momentum, turbulence, and thermal among the lower boundary layer (surfaces) and upper boundary layer (atmosphere) of a city (Huang et al. 2019; Li and Zhou 2019). The regional meteorological conditions are influenced by the interactions occurring among surface characteristics, planetary boundary layer (PBL), and physical processes in the atmosphere (Berardi 2016; Roberge and Sushama 2018), which is already showing to impact the energy demand, and will do so even more in the future (Berardi and Jafarpur 2020).

It is well known that the surface and ambient temperatures are higher in urban areas compared to their neighbors, while the UHI has been reported to be responsible for human thermal discomfort and increasing building energy demand (Vahmani and Ban-Weiss 2016; Kim et al. 2018; Berardi and Taleghami 2018). The UHII is calculated by land surface and air temperature differences between urban and rural areas.

Worldwide an increasing interest exists around the importance of predicting the UHI phenomenon to investigate UHI mitigation strategies (Jato-Espino 2019). Mesoscale models are applied to illustrate the impacts of physical parameterizations and urban characteristics on the meteorological processes (Jandaghian and Akbari

Z. Jandaghian
Department of Architectural Science, Ryerson University, Toronto, ON, Canada
e-mail: zjandaghian@ryerson.ca

U. Berardi (✉)
BeTOP Lab, Faculty of Engineering and Architecture, Ryerson University, Toronto, ON, Canada
e-mail: uberardi@ryerson.ca

M. Palme, A. Salvati (eds.), *Urban Microclimate Modelling for Comfort and Energy Studies*, https://doi.org/10.1007/978-3-030-65421-4_11

2018a; Doan et al. 2019; Jandaghian and Berardi 2020b). For this scope, the Weather Research and Forecasting (WRF) model is employed to predict the surface and atmospheric phenomenon on regional climate. However, to precisely estimate the UHI intensity, the differences among surface or near-surface ambient temperatures must be investigated in the urban region and its neighborhoods. Thus, it has been proposed to couple the WRF with the urban canopy models (UCMs) to evaluate the extent of spatial and temporal variations in atmospheric variables (WRF-UCMs).

The WRF incorporates three types of urban canopy models (UCMs) to estimate the heat emissions and moisture fluxes from the urban to the atmosphere: slab (SB-UCM), single layer (SL-UCM), and multilayer (ML-UCM) models. Previous research demonstrated the effects of urbanizations on regional climate using the WRF-UCMs (Lin et al. 2008; Flagg and Taylor 2011; Vahmani and Ban-Weiss 2016; Jandaghian and Akbari 2018b). Results illustrated the potential of the WRF-UCMs to simulate the exchanges among the urban surfaces and the meteorological conditions that affect the regional climate. The SB-UCM and ML-UCM indicated similar results for the air temperature and wind speed during summer in Houston (Salamanca et al. 2011). Kusaka et al. showed that the SL-UCM results are more reliable compared to the SB-UCM (2012). It is worth mentioning that the proper choice of the UCMs is conditioned to the compatibility of the models, computational performance, and data required for the regional climate analyses. Therefore, to assess the impacts of urbanizations, a proper urban canopy model needs to be selected.

This chapter aims to describe the WRF and the UCM simulation preparations and processes and to shed light on the capabilities of the WRF-UCMs to simulate the UHI intensity in the GTA during the 2011 and 2018 heat wave periods. This chapter intends to provide comprehensive knowledge regarding mesoscale simulation processes as well as the urban canopy modeling for future urban climate studies.

11.2 Preparations for Mesoscale Simulations

Meteorological models use surface characteristics and initial and boundary conditions to solve a set of conservation equations, simulate the advection and diffusion of pollutants, and predict meteorological conditions, such as the air temperature, moisture, or wind speed. The online Weather Research and Forecasting (WRF) model considers a variety of meteorological and physical characteristics (cloud convection, water cycle, planetary boundary layer, and atmospheric and surface radiation) to predict weather conditions. The WRF is a mesoscale numerical weather prediction (NWP) system which is fully compressible and non-hydrostatic. The WRF can be coupled with the urban canopy models (UCMs) to simulate the thermal and humidity cycles from canopies to the atmosphere. The UCMs represent urban areas to precisely predict ambient temperature, wind speed, relative humidity, precipitation, and short-wave and long-wave radiations.

Figure 11.1 shows the flowchart of WRF with the urban canopy model for urban climate simulations. The terrestrial data and weather-gridded data are obtained from North America Regional Reanalysis (NARR). The preprocessing of WRF (called WPS) is used to define the domain and period of simulation and interpolates the NARR data into the domain of interest.

Figure 11.2 shows the simulation approach to accomplish the goals of WRF simulations consisting of preparation and processes phases. The preparation includes compiling and coupling of models and simulation steps, such as defining domain and period of simulation and collecting input data and measurement data. The processes refer to WRF and UCM simulations. Besides, data analysis is the main part of the process, including comparing the simulation results with measurements.

11.2.1 Compiling and Coupling of the Models

WRF simulations require significant preparation and computer resources. At the onset, one should make sure that the computer has sufficient memory capacity and a fast processing system to compile, couple, and carry out various simulations promptly. The first step to start simulations is to collect and couple the preprocessing system (WPS), data assimilation (DA), and advanced research WRF solver (ARW-WRF).

- **WRF Preprocessing System (WPS)**—WPS is applied primarily to predict real data. In WPS, the domain is defined, the terrestrial data is mapped, and the meteorological parameters are interpolated to the domain.
- **WRF Data Assimilation (WRF-DA)**—WRF-DA is used to inject observed data into the domain produced by the WPS. It accounts to engross the initial condi-

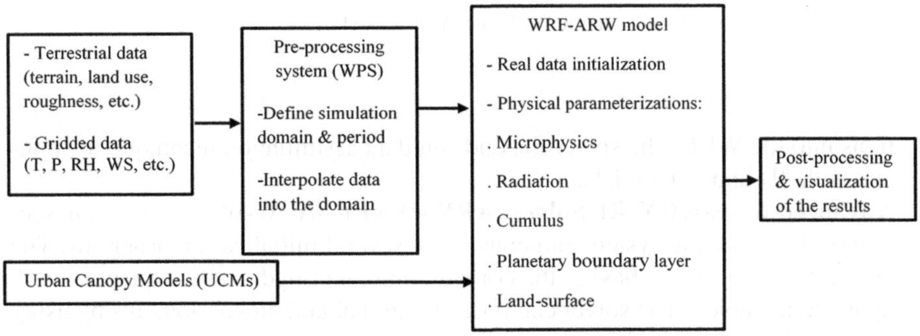

Fig. 11.1 Flowchart of WRF coupled with the urban canopy models (UCMs) (*T* temperature, *P* pressure, *RH* relative humidity, *WS* wind speed, *WPS* weather preprocessing system, *UCM* urban canopy model, *WRF* Weather Research and Forecasting model, *ARW* advanced research of the WRF)

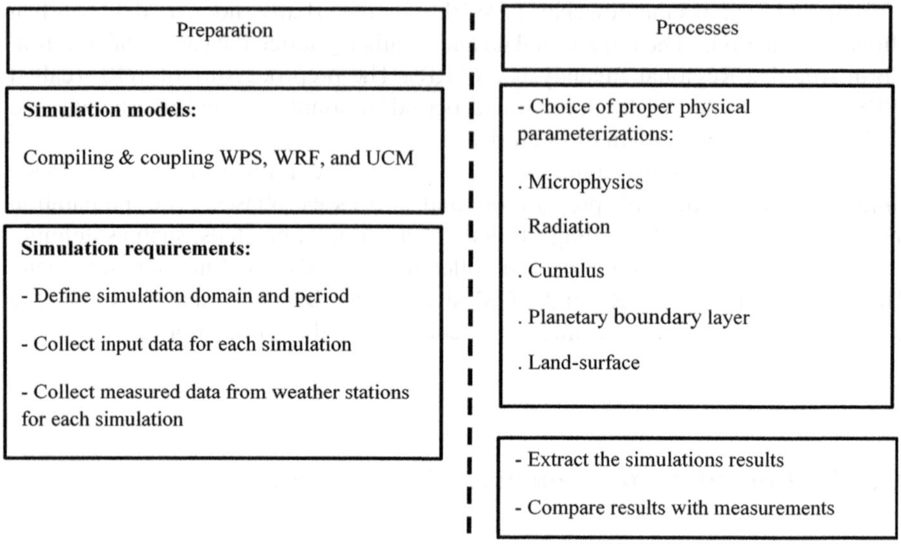

Fig. 11.2 Simulation preparation and processes (*WPS* weather preprocessing system, *WRF* Weather Research and Forecasting model, *UCM* urban canopy model)

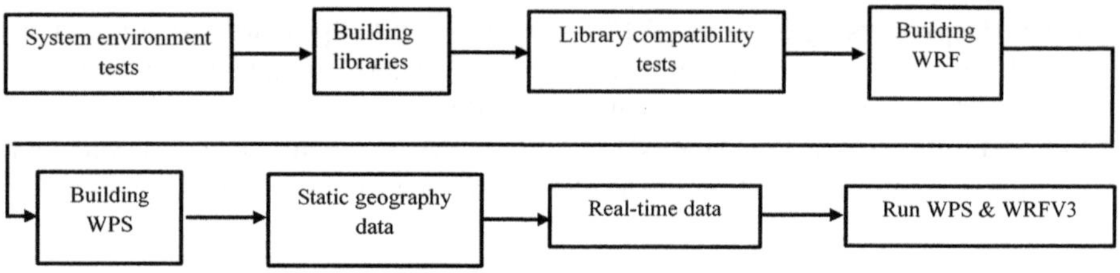

Fig. 11.3 Steps to compile and run the WPS and WRF models

tions into the WRF. The solver depends on data assimilation technique to simulate both 3D and 4D models.

- **Advanced Research WRF Solver (ARW-WRF)**—ARW-WRF is the main section of the modeling system and consists of several initialization programs. The key features consist of having the compressible, non-hydrostatic equations with hydrostatic choice. The solver can predict regional and global domains by using the mass-based terrain-following coordinates and vertical grid spacing.

Other features are the Runge-Kutta 2nd and 3rd order and 2nd to 6th order advection options, monotonic transport, and positive-definite advection option for moisture, scalar, tracer, and turbulent kinetic energy. Figure 11.3 shows the steps to compile and run WPS and WRF. Table 11.1 summarizes the description of these steps.

- **Building WRFV4.** After ensuring that all libraries are compatible with the compilers, one can now prepare to build WRFV4. First, the tar file should be downloaded from a verified source (NCAR) and unpacked in the preferred directory. Then, a configuration file should be created to compile. The compiler is selected to be serial or in parallel. For parallel, which is for real case simulations, there are three options: "smpar," "dmpar," and "dm+sm." The "dmpar" is the most proper

Table 11.1 Description of the steps to compile and run the WPS and WRF models

Steps to compile and run WPS and WRF models	Description
System environment tests	It is important to have the required compiler as gfortran. The WRF build system has scripts as the top level for the user interface as well
Building libraries	Various libraries should be installed, for example netcdf and Jasper. *These libraries must be installed with the same compilers as they will be used to install WRF and WPS*
Library compatibility tests	These tests are essential to verify that the libraries can work with the compilers that are to be used for the WPS and WRF builds
Building WRFV3	After ensuring that all libraries are compatible with the compilers, one can now prepare to build WRFV3. First, the tar file should be downloaded from a verified source and unpacked in the preferred directory. Then, a configuration file should be created to compile. The compiler is selected to be serial or in parallel
Building WPS	After building the WRF model, WPS program needs to be built. A tar file having the WPS source code is downloaded and unpacked. Then the WPS is compiled to be compatible with WRF. If the compilation is successful, there should be three executables in the WPS top-level directory that are linked to their corresponding directories
Static geography data	To initiate a real data case, the domain's physical location on the globe and the static information for that location must be created. This requires a data set that includes such fields as topography and land use categories. These data need to be downloaded and uncompressed
Real-time data	For real data cases, the WRF model requires up-to-date meteorological information for both a first condition and lateral boundary conditions. This meteorological data is traditionally a file that is provided by a previously run external model or analysis. For a semi-operational setup, the meteorological data is usually sourced from a global model, which allows locating the WRF model's domains anywhere on the globe
Run WPS and WRFV3	First, the WPS is executed by modifying its name list to reflect information that is required for the particular simulation. The geogrid will match the geographical data and define the simulation domain. The ungird unpacks necessary data on the simulation period. The met gird interpolates the weather and terrestrial data on the domain of interest. To simulate the WRF, the name list needs to be changed. The data provided by WPS should be connected and linked to the run directory

possibility; it has fewer errors and is more compatible with other programming languages. To check whether it was successful, the executable files—namely "wrf.exe," "real.exe," "ndown.exe," and "tc.exe"—need to be checked.

- **Building WPS.** After building the WRF model, WPS program needs to be built. A tar file containing the WPS source code is downloaded and unpacked. Then the WPS is compiled to be compatible with WRF. If the compilation is successful, there should be three executables in the WPS top-level directory, which are linked to their corresponding directories—namely "geogrid," "ungrib," and "metgrid."
- **Static Geography Data.** To start a real data case, the domain physical location on the globe and the static information for that location must be created. This requires a data set that includes such fields as topography and land use categories. These data need to be downloaded and uncompressed.
- **Real-Time Data.** For real data cases, the WRF model requires up-to-date meteorological information for both an initial condition and lateral boundary conditions. This meteorological data is traditionally a Grib file that is provided by a previously run external model or analysis. For a semi-operational setup, the meteorological data is usually sourced from a global model, which permits locating the WRF model domains anywhere on the globe. The National Centers for Environmental Prediction (NCEP) run the Global Forecast System (GFS) model four times daily (initializations valid for 0000, 0600, 1200, and 1800 UTC). This is a global, isobaric, 0.5-degree latitude/longitude, forecast data set that is freely available and is usually accessible +4 h after the initialization time. A single data file needs to be acquired for each requested period.
- **Run WPS and WRFV3.** First, the WPS is executed by changing its "namelist.wps" to reflect the information that is needed for the simulation. The "geogrid.exe" will match the geographical data and define the simulation domain. The "ungird.exe" unpacks necessary data regarding the simulation period. The "metgird.exe" interpolates the weather and terrestrial data in the domain of interest. To simulate the WRF, the "namelist.input" needs to be modified. The data provided by WPS should be connected and linked to the run directory. First, the "real.exe" is executed and then the "WRF.exe." The "error.rsl" file needs to be checked for any errors. If the execution was successful, then the required data should be extracted and analyzed. The physical parameterizations used in WRF include planetary boundary layer, short-wave and long-wave radiation, microphysics, cumulus, and land surface schemes.

11.2.2 Parametric Simulations of Physical Options

The physical parameterizations need to be carefully selected to predict weather conditions. A proper simulation platform provides a reliable opportunity to investigate the effects of UHI on meteorological parameters for environmental policymakers. A land surface model (LSM) predicts momentum and thermal and humidity cycles on land points through atmospheric feedbacks of other schemes. LSM updates surface parameters in every time step.

- **Microphysics Scheme.** The microphysics scheme predicts the water cycle in the domain of interest. Water vapor initiates cloud water and cloud ice to produce snow, graupel, and rain. A set of conservation equations are applied to predict the water budget during the modeling period. Table 11.2 presents a brief description of the different schemes of microphysics in WRF.
- **Cumulus Scheme.** Cumulus scheme evaluates the impacts of convective air movement on up-drafting and down-drafting of clouds (Grell and Devenyi 2002). It influences the vertical thermal and humidity cycles in the simulation. The model identifies the convection, owes flux equations, and benefits from the closure assumptions. Table 11.3 presents a brief description of the different schemes of cumulus in WRF.

Table 11.2 Parameterization schemes of microphysics model in WRF

Scheme	Description
Lin	Applies the conservation equations and predicts source and sink of snow, hail, and rain
SBU-YLin	Predicts ice and snow
Eta	Considers six species of water
WRF Single-Moment 6-class	Has more accurate dependency of snow to temperature
WRF Double-Moment 6-class	Considers the mixing ratio and number concentration as independent variables
Thompson	Predicts the mixing ratios of five hydrometeors and the concentration of cloud ice
Morrison	Considers five species of water and predicts the mixing ratio
Severe Storms Laboratory	Considers lightning in microphysical models
Goddard	Calculates the condensation and deposition of cloud water and cloud ice
Milbrandt-Yau	Uses gamma size distribution

Table 11.3 Parameterization schemes of cumulus model in WRF

Scheme	Description
Simplified Arakawa-Schubert	Considers the mass and energy balance in clouds. Surface rainfall is parameterized in the moisture balance equation
Betts-Miller-Janjic	Considers convective mixing and predicts the shallow clouds
Grell 3D	Divides parameterizations into dynamic control and feedback
Grell-Freitas	Predicts the cloud convection in high-resolution grid size simulations
Kain-Fritsch	Simulates the water cycle instabilities
New Simplified Arakawa-Schubert	Considers the convection-induced pressure gradient forcing in the momentum equation
Tiedtke	Considers the eddy transport of energy in prognostic equations
Zhang-McFarlane	Considers the exchange of unstable air change with adjacent layers

Table 11.4 Parameterization schemes of planetary boundary layer models in WRF

Scheme	Description
Mellor-Yamada-Janjic (MYJ)	Considers the viscous sublayer above water bodies
Bougeault-Lacarrere (BouLac)	Calculates the TKE in a prognostic equation

- **Planetary Boundary Layer (PBL) Schemes.** PBL scheme accounts for vertical flux exchange in a grid cell. PBL predicts the impacts of momentum, heat, and moisture fluxes. PBL is divided into three sublayers: viscous layer, surface layer, and transition layer. Table 11.4 presents a brief description of the different schemes of PBL in WRF.
- **Radiation Scheme.** The radiation scheme calculates the energy balance. The short-wave radiation from the sun reaches the urban surfaces. The urban surface receives short-wave radiation as well as long-wave radiation from the sky and other structures. Urban absorbs a part of the radiation and reflects the remaining. The energy that reaches the surface is a function of sky condition and solar zenith angle. The long-wave radiation that is emitted from the urban surfaces goes through clouds and atmospheric gas or pollution.

11.3 Urban Canopy Models

The three urban canopy models can be coupled with the solver of the WRFV4.0: slab (SB-UCM), single-layer (SL-UCM), and multilayer (ML-UCM). The SL-UCM and ML-UCM consist of the three urban categories: low-intensity residential (with more than 20–70% vegetation cover), high-intensity residential (with less than 20% vegetation cover), and industrial/commercial areas.

11.3.1 Slab of the UCM

The SB-UCM applies roughness variables to simulate momentum and turbulence. The SB-UCM is a one-dimensional model assuming the buildings as bulk parameterizations and thus indicating the zero-order effects of urban surfaces (Liu et al. 2006). Here, the albedo is constant and the vegetation fraction is not accounted for. The PBL calculates the mixing length and vertical wind distribution (Chen and Dudhia 2001). The SB-UCM disregards the variability of urban structures.

The WRF includes the SB-UCM in NOAH parameterization. In the SL-UCM, the roughness length for momentum, over the building wall and over the ground, is assumed to be 0.0001 m and 0.01 m, respectively. Here, the albedo is 0.15, and the heat capacity for roof, wall, and road is 2.6 $Jm^{-3} K^{-1}$. The thermal conductivity of urban surfaces is 3.14 $Wm^{-1} K^{-1}$.

11.3.2 Single Layer of the UCM

The SL-UCM was developed to calculate the street canyons and building structure (Kusaka et al. 2001; Kanda et al. 2005). The SL-UCM accounts for a single orientation of the two-dimensional streets and estimates the three urban classes with various thermal characteristics. The SL-UCM estimates the sensible heat fluxes by assuming the wind distribution and evaluates the radiation processes in urban geometry. The SL-UCM accounts for the anthropogenic heat as a fixed temporal profile (Chen et al. 2011). In addition, the model considers the radiation behavior as shadowing, reflections, and trapping in the domain. The surface energy budget is applied to calculate the surface skin temperature. The thermal conduction equation estimates the temporal pattern. The Monin–Obukhov similarity theory calculates the surface-sensible heat fluxes. The sensible heat and momentum fluxes are conducted in the WRF-NOAH model. The canyon drag coefficient and friction velocity are analyzed. The anthropogenic heat and its diurnal variation are considered. The SL-UCM also accounts for the horizontal and vertical wind distribution in the canopy. The heat absorbed by urban surfaces and the sensible fluxes are defined as follows. The definition for each symbol is presented at the beginning of the chapter:

$$H_{\mathrm{W}} = H_{\mathrm{W,S}} + R_{\mathrm{W}} \times H_{\mathrm{W,S}} \times F_{\mathrm{W}\rightarrow\mathrm{W}} + R_{\mathrm{G}} \times H_{\mathrm{G,S}} \times F_{\mathrm{G}\rightarrow\mathrm{W}} \tag{11.1}$$

$$H_{\mathrm{G}} = H_{\mathrm{G,S}} + R_{\mathrm{W}} \times H_{\mathrm{W,S}} \times F_{\mathrm{G}\rightarrow\mathrm{W}} \tag{11.2}$$

$$\mathrm{SHF}_{\mathrm{a}} = \frac{2\tilde{\mathrm{h}}\,\mathrm{SHF}_{\mathrm{W}} + \delta\mathrm{SHF}_{\mathrm{G}}}{\delta} \tag{11.3}$$

$$\mathrm{SHF}_{\mathrm{W}} = C_{\mathrm{W}}\left(T_{\mathrm{W}} - T_{\mathrm{a}}\right) \tag{11.4}$$

$$\mathrm{SHF}_{\mathrm{G}} = C_{\mathrm{G}}\left(T_{\mathrm{G}} - T_{\mathrm{a}}\right) \tag{11.5}$$

11.3.3 Multilayer of the UCM

The ML-UCM was developed to estimate the interaction of the buildings with the PBL (Martilli et al. 2002; Kondo et al. 2005) and thus named as the building effect parameterization (BEP). The ML-UCM simulates the three-dimensional canopy and calculates the vertical exchange of heat, moisture, and momentum (Salamanca et al. 2011). The model computes the TKE and energy and estimates heat emissions from the canopy by analyzing the drag force, diffusion factor, and radiations.

The ML-UCM accounts for the heterogeneous structure of the city and calculates the anthropogenic heat (AH) emissions that affect the UHI intensity estimation (Miao et al. 2009). The building energy model (BEM) is used to estimate the AH by calculating the exchanges of energy between the interior of buildings and the outdoor atmosphere. The BEM accounts for the diffusion of heat from building surfaces, radiation exchanged through windows, long-wave radiation exchanged between indoor surfaces, and heat generations from occupants and equipment as well as air-conditioning, ventilation, and heating systems (Jandaghian and Berardi 2020a).

11.4 WRFV4.0 Configurations

The initial and boundary conditions were derived from the North American Regional Reanalysis (NARR) (Mesinger et al. 2006). The NOAH-LSM evaluates the boundary conditions for the meteorological model. The USGS 24-category is applied for the LULC data. The physical parameterization in the solver of the WRF is explained in the previous section. For these simulations, the Mellor-Yamada-Janjic scheme (Janjic 2002), the Goddard scheme, the rapid radiative transfer model—RRTMG (Iacono et al. 2008), and the Lin and Colle's (2011) and the Grell and Devenyi's (2002) schemes are, respectively, used to estimate the planetary boundary layer, short-wave and long-wave radiations, microphysics, and cumulus models. The moisture, scalars, and TKE are activated for model stability.

11.4.1 Simulation Domain and Episode

The Greater Toronto Area (GTA) is located at ~43.7° N and ~79.3° W and has more than six million inhabitants. The semi-continental climate of the GTA causes cold, long winters and hot, humid summers. High temperature and occurrences of heat waves in summer episode have negative influences on dwellers' health and comfort, especially on the vulnerable people as infants, elderly, and those with preexisting health conditions (Berardi 2016; Wang et al. 2016). Here, the simulations are composed of four two-way nested domains with the horizontal resolution of 9 km, 3 km, 1 km, and 0.333 km, respectively. The number of vertical layers is 51 eta level. Two heat wave periods are selected. The 2011 and 2018 heat wave periods started on the 17th and 1st of July, respectively, and lasted for 5 consecutive days. The 3 days of these periods are assessed as having the highest hourly maximum temperature (Jandaghian and Berardi 2020a).

11.4.2 Model Performance Evaluation

The WRF-UCM performance is evaluated by comparing the simulated results against measured data collected from urban and rural weather stations across the GTA in the heat wave period (21st–23rd of July) and 2018 heat wave period (3rd–5th of July). Four weather stations are selected: Toronto City Centre, Pearson International Airport, Buttonville, and Oshawa. To practice model evaluation, mean bias error (MBE), mean absolute error (MAE), and root mean square error (RMSE) are calculated.

11.5 Results and Analyses

11.5.1 Ambient Temperature

The model evaluation of 2-m air temperature in four weather stations is presented in Table 11.5 showing the SB-UCM, SL-UCM, and ML-UCM. These comparisons show that the WRF-UCM predicts the temporal changes of 2-m air temperature precisely. The average MBE shows that the SB-UCM underestimated the ambient temperature during both heat wave events by −1.4 °C in 2011 and −0.4 °C in 2018. The SL-UCM underestimated the air temperature in 2011 (~−1 °C), while slightly overestimated it in 2018 (~0.2 °C). The ML-UCM underestimated ambient temperature in 2011 (~−1 °C) and overestimated it in 2018 (~1 °C) during the simulation period. However, the SL-UCM predicted the air temperature better in rural areas compared to urban region because it disregarded the anthropogenic heat emissions in its calculations.

Figure 11.4 presents the hourly comparisons of the air temperature with the measured data. The temporal variation of ambient temperature illustrates the same trend in both the SB-UCM and the SL-UCM simulations, but adding the heat emission

Table 11.5 The MBE, MAE, and RMSE of air temperature (°C) in SB-UCM, SL-UCM, and ML-UCM at four weather stations over the GTA during the 2011 heat wave period (21st–23rd of July) and 2018 heat wave period (3rd–5th of July)

Station name	MBE (°C)			MAE (°C)			RMSE (°C)		
	SB	SL	ML	SB	SL	ML	SB	SL	ML
2011 Heat wave (21st–23rd of July)									
Toronto City Center	−1.3	−1.1	−1.1	1.1	1.0	1.2	1.4	1.3	1.1
Pearson Int. Airport	−2.1	−1.2	−1.1	2.6	2.6	2.1	3.2	3.2	2.0
Buttonville	−0.7	0.5	−0.5	1.0	0.9	1.0	1.3	1.1	1.1
Oshawa	−1.4	−1.9	−1.5	1.5	1.2	1.1	1.7	1.6	1.7
Average	**−1.4**	**−0.9**	**−1.0**	**1.6**	**1.4**	**1.1**	**1.9**	**1.8**	**1.4**
2018 Heat wave (3rd–5th of July)									
Toronto City Center	−0.2	−0.1	1.0	0.7	1.2	1.4	0.8	1.4	1.7
Pearson Int. Airport	0.2	0.4	1.6	1.4	1.1	2.1	1.8	1.5	2.5
Buttonville	−1.1	−0.6	1.0	1.2	0.7	1.0	1.2	0.8	1.0
Oshawa	−0.4	1.2	1.3	1.1	1.2	1.3	1.3	1.5	1.6
Average	**−0.4**	**0.2**	**1.2**	**1.1**	**1.1**	**1.5**	**1.3**	**1.3**	**1.7**

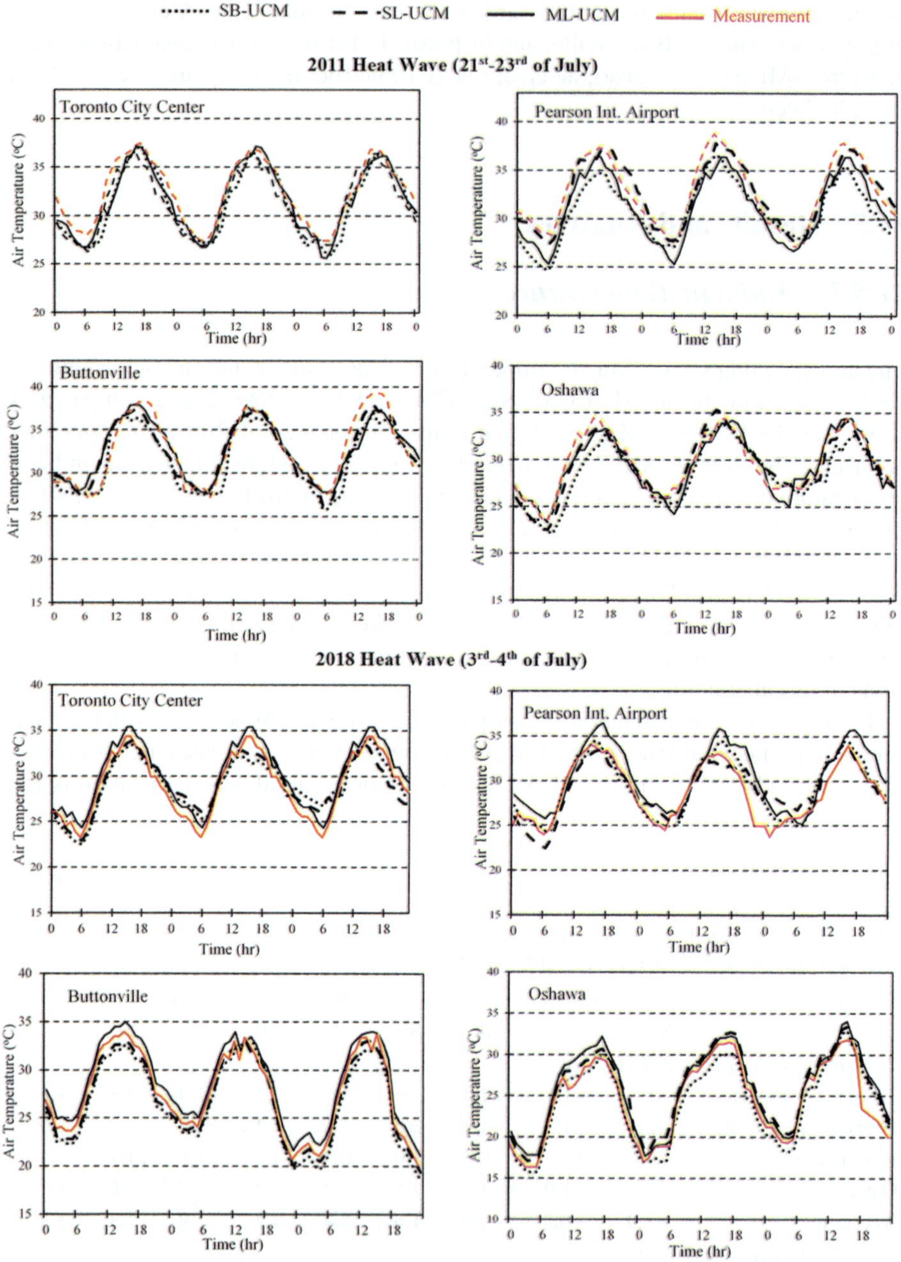

Fig. 11.4 Daily comparisons of 2-m air temperature (°C) in three UCMs vs. measured data from Toronto City Centre, Pearson Int. Airport, Buttonville, and Oshawa over the GTA during the 2011 heat wave period (21st–23rd of July) and 2018 heat wave period (3rd–5th of July)

from urban to the SL-UCM resulted in more accurate estimations. The correlation between the measurements and simulations is close to 0.9 and confirms the capability of the WRF-UCMs in forecasting air temperature efficiently. The results state that the SL-UCM and ML-UCM tend to slightly overpredict the temperature, but the hourly comparisons mark the more precise estimation of variables in the ML-UCM.

11.5.2 Wind Speed

Wind speed is underestimated in the SB-UCM (−2.4 m/s in 2011 vs. –0.4 m/s in 2018) and the SL-UCM (−1.6 m/s in 2011 vs. –0.8 m/s in 2018) simulations more compared to the ML-UCM (−1.5 m/s in 2011 vs. –0.1 m/s in 2018) (Table 11.6). The ML-UCM presents a better estimation of wind patterns. The reason is due to the capability of the ML-UCM to simulate the building height and street width. However, the daily comparisons of simulated results vs. measured data are not quite the same as the correlation close to 0.6. The reason is that the PBL cannot accurately predict the TKE and wind distribution for fine resolution grids. The largest underestimation of 10-m wind speed is seen in the SB-UCM because buildings and roughness impacts are dismissed. The average MAE of the UCMs in the domain is less or more the same for both heat wave events, but these criteria are slightly improved by applying the ML-UCM in urban canopy.

Table 11.6 The MBE, MAE, and RMSE of wind speed (m/s) in SB-UCM, SL-UCM, and ML-UCM at four weather stations over the GTA during the 2011 heat wave period (21st–23rd of July) and 2018 heat wave period (3rd–5th of July)

Station name	MBE (°C)			MAE (°C)			RMSE (°C)		
	SB	SL	ML	SB	SL	ML	SB	SL	ML
2011 Heat wave (21st–23rd of July)									
Toronto City Center	−1.5	−1.8	−2.2	2.4	3.8	4.6	3.4	4.3	5.1
Pearson Int. Airport	−4.5	−3.3	−3.6	5.9	5.2	5.2	6.8	6.0	6.0
Buttonville	−1.1	−0.7	−0.8	3.6	3.3	3.6	4.6	4.1	4.2
Oshawa	−2.6	−0.5	−0.5	4.1	2.8	2.9	4.8	3.3	3.0
Average	**−2.4**	**−1.6**	**−1.5**	**4.0**	**3.8**	**4.1**	**4.9**	**4.4**	**4.2**
2018 Heat wave (3rd–5th of July)									
Toronto City Center	0.4	−1.4	0.6	1.4	1.6	1.7	1.6	2.1	2.2
Pearson Int. Airport	−1.2	0.1	0.1	1.2	0.4	1.3	1.7	0.6	1.8
Buttonville	−1.3	−1.1	−0.6	1.3	1.1	1.1	1.8	1.7	1.7
Oshawa	0.7	−0.6	−0.4	1.0	0.8	0.6	1.6	1.3	0.9
Average	**−0.4**	**−0.8**	**−0.1**	**1.2**	**1.0**	**1.2**	**1.7**	**1.4**	**1.7**

11.5.3 Urban Heat Island Intensity

The urban heat island intensity (UHII) is considered as the difference in air temperature between urban and rural areas. In the mesoscale model, the ambient temperature depends on the surface temperature as urban fabric as well as sensible heat flux. The average daily, daytime, and nighttime urban-rural scale differences of ambient temperature are presented in Table 11.7. The results illustrate that the daily UHII is around 1.2–1.5 °C, while the daytime UHII is near 0.7–0.9 °C. The UHII is magnified during the night and reaches 2 °C and even 3 °C across the domain.

The averaged thermal pattern in the urban and rural areas of the GTA is presented in Fig. 11.5 during the simulations in 2011 and 2018 heat waves over the GTA. The daily pattern of air temperature illustrates that the nocturnal UHII is more evident, starting in the early evenings. The three UCMs simulate the changes in temperatures in urban and rural regions well. But the ML-UCM provides a more exact prediction of the UHII when compared against the measurements. It is worth mentioning that the location of the GTA close to the Lake Ontario has noticeable effects on regulating the air and skin temperatures even during the day and at around noon.

Table 11.7 The average daily, daytime, and nighttime urban heat island intensity of the SB-UCM, SL-UCM, ML-UCM, and measurement across the GTA during the 2011 heat wave period (21st–23rd of July) and 2018 heat wave period (3rd–5th of July)

Daily UHII				Daytime UHII				Nighttime UHII			
SB	SL	ML	Measured	SB	SL	ML	Measured	SB	SL	ML	Measured
2011 Heat wave (21st–23rd of July)											
1.35	1.35	1.46	1.34	0.85	0.82	0.91	0.89	3.11	3.16	2.94	2.11
2018 Heat wave (3rd–5th of July)											
1.47	1.35	1.39	1.26	0.77	0.73	0.71	0.58	2.28	1.92	2.09	1.88

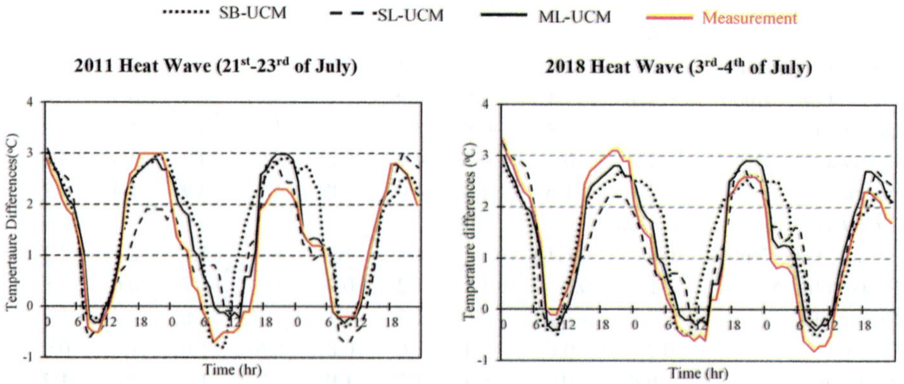

Fig. 11.5 Daily changes of 2-m air temperature (°C) in SB-UCM, SL-UCM, and ML-UCM results vs. measured data between urban and rural areas of the GTA during the 2011 heat wave period (21st–23rd of July) and 2018 heat wave period (3rd–5th of July)

11.6 Conclusions

To develop a platform to precisely analyze the urban climate and meteorological process in the GTA, a set of simulations are conducted. The WRF is coupled with the three urban canopy models separately. The simulations are run for two heat wave periods in 2011 and 2018. The results of these six base case simulations are compared with measured data from urban and rural weather stations. The aim to evaluate the integrated WRF-UCMs is to prepare the proper mesoscale model for urban climate simulations. The other goal is to accurately predict the impacts of urbanizations and climate change on urban structures, inhabitants, and environment. The growth of urbanization and climate change intensify the UHII impacts, increase the frequency and duration of the heat wave events, and deteriorate the regional air quality.

The WRF-UCMs forecast the temporal variation of air temperature reasonably well (the correlation between simulations and measurements is 0.9). In contrast, the prediction of wind speed requires more improvement in the mesoscale simulations (the correlation of the wind speed is 0.6). The SB-UCM underestimates the air temperature and wind speed because the model only estimates the zero-order impacts of structures and dismisses the vegetation cover.

The SL-UCM slightly overestimates air temperature and underestimates wind speed due to the miscalculations of the two-dimensional canopy, and estimation of fixed value for the anthropogenic heat emissions.

The ML-UCM overestimates air temperature and slightly underestimates wind speed. The ML-UCM simulation results are more reliable because it considers the three-dimensional urban surfaces and accounts for the vertical and horizontal exchange variables in urban canopy. The anthropogenic heat emissions are calculated and various building structures are considered in the urban canopy consisting of the influences of the radiation budget.

The UHI intensity is well characterized by the three UCMs. However, as it was expected, the ML-UCM simulates the diurnal variation of the ambient temperature more accurately. The analysis illustrated that the SL-UCM is reliable to simulate the urban climate, but for the precise evaluation of the UHII, the exact estimation in built-up structures is required that is satisfied via the ML-UCM model. It is worth mentioning that conducting the model with ML-UCM increases the computational time by 15%. These simulations illuminate the capability of the SL-UCM and ML-UCM in predicting the meteorological parameters and provide an urban climate platform to support assessing the impacts of UHI mitigation strategies in future studies.

Nomenclature

A	Area
C_p	Specific heat capacity
C_W	Convection heat transfer coefficient from walls to the canopy
E	Latent heat
$F_{W\rightarrow W}$	View factor from wall to wall
$F_{G\rightarrow W}$	View factors from ground to wall
g	Gravity
G	Convection heat transfer coefficient from the ground to the canopy
H_W	Total heat absorbed by walls
H_G	Total heat absorbed by ground
$H_{W,S}$	Heat absorbed by walls from sky
$H_{G,S}$	Heat absorbed by grounds from sky
$\tilde{h}$	Height of the wall
K	Von Karman constant
K_z	Turbulence coefficient
k	Unit vector in the z-direction
p	Pressure
q_i	Specific humidity
R_W	Reflectance of the wall
R_G	Reflectance of the ground
SHF_a	Sensible heat fluxes from the canopy to the atmosphere
SHF_W	Sensible heat fluxes from walls to the canopy
SHF_G	Sensible heat fluxes from the ground to the canopy
S_θ	Source/sink term for potential temperature
S_q	Source/sink term for specific humidity
T_W	Temperatures of wall
T_G	Temperatures of ground
T_a	Temperatures of atmosphere
u	Horizontal velocity
U	Advection
V	Wind speed
W	Vertical perturbation from turbulence
Z	Momentum roughness length
θ	Potential temperature
ρ	Density
δ	Width of the wall
Ω	Angular velocity of the earth

Acknowledgments Funding for this research was provided by the NSERC DG #2016-04904 and the Ontario Ministry of Research Innovation and Science (MRIS) through the ERA award.

References

Akbari, H., & Kolokotsa, D. (2016). Three decades of urban heat islands and mitigation technologies research. *Energy and Buildings, 133*, 834–842.

Berardi, U. (2016). The outdoor microclimate benefits and energy saving resulting from green roofs retrofits. *Energy and Buildings, 121*, 217–229.

Berardi, U., & Jafarpur, P. (2020). Assessing the impact of climate change on building heating and cooling energy demand in Canada. *Renewable Energy and Sustainable Reviews, 121*, 109681.

Berardi, U., & Taleghami, M. (2018). The effect of pavement characteristics on pedestrians' thermal comfort in Toronto. *Urban Climate, 24*, 449–459.

Chen, F., & Dudhia, J. (2001). Coupling an advanced land surface–hydrology model with the Penn State–NCAR MM5 modeling system. Part I: Model implementation and sensitivity. *Monthly Weather Review, 129*, 569–585.

Chen, F., Kusaka, H., Bornstein, R., Ching, J., Grimmond, C., Grossman-Clarke, S., Loridan, T., Manning, K., Martilli, A., Miao, S., Sailor, D., Salamanca, F., Taha, H., Tewari, M., Wang, X., Wyszogrodzki, A., & Zhang, C. (2011). The integrated Wrf/urban modelling system: Development, evaluation, and applications to urban environmental problems. *International Journal of Climatology, 31*, 273–288.

Doan, V. Q., Kusaka, H., & Nguyen, T. M. (2019). Roles of past, present, and future land use and anthropogenic heat release changes on urban heat island effects in Hanoi, Vietnam: Numerical experiments with a regional climate model. *Sustainable Cities and Society, 47*, 101479.

Flagg, D. D., & Taylor, P. A. (2011). Sensitivity of mesoscale model urban boundary layer meteorology to the scale of urban representation. *Atmospheric Chemistry and Physics, 11*, 2952–2972.

Grell, G. A., & Devenyi, D. (2002). A generalized approach to parameterizing convection combining ensemble and data assimilation techniques. *Geophysical Research Letters, 29*(14), 31–38.

Huang, Q., Huang, J., Yang, X., Fang, C., & Liang, Y. (2019). Quantifying the seasonal contribution of coupling urban land use types on urban heat island using land contribution index: A case study in Wuhan, China. *Sustainable Cities and Society, 44*, 666–675.

Iacono, M. J., Delamere, J. S., Mlawer, E. J., Shephard, M. W., Clough, S. A., & Collins, W. D. (2008). Radiative forcing by long-lived greenhouse gases: Calculations with the AER radiative transfer models. *Journal of Geophysical Research, 113*, 131–103.

Jandaghian, Z., & Akbari, H. (2018a). The effects of increasing surface albedo on urban climate and air quality: A detailed study for Sacramento, Houston, and Chicago. *Climate, 6*, 2–19.

Jandaghian, Z., & Akbari, H. (2018b). The effects of increasing surface reflectivity on heat-related mortality in the Greater Montreal Area, Canada. *Urban Climate, 25*, 135–151.

Jandaghian, Z., & Berardi, U. (2020a). Analysis of the cooling effects of higher albedo surfaces during heat waves coupling the Weather Research and Forecasting model with building energy models. *Energy and Buildings, 207*, 109–627.

Jandaghian, Z., & Berardi, U. (2020b). Comparing urban canopy models for microclimate simulations in Weather Research and Forecasting Models. *Sustainable Cities and Society, 55*, 102–025.

Janjic, Z. (2002). *Nonsingular implementation of the Mellor-Yamada level 2.5 scheme in the NCEP meso model*. Camp Springs, MD: National Centers for Environmental Prediction.

Jato-Espino, D. (2019). Spatiotemporal statistical analysis of the urban heat island effect in a Mediterranean region. *Sustainable Cities and Society, 46*, 101427.

Kanda, M., Kawai, T., Kanega, M., Moriwaki, R., Narita, K., & Hagishima, A. (2005). A simple energy balance model for regular building arrays. *Boundary-Layer Meteorology, 116*, 423–443.

Kim, H., Gu, D., & Kim, H. Y. (2018). Effect of f urban heat island mitigation in various climate zones in the United States. *Sustainable Cities and Society, 41*, 841–852.

Kondo, H., Genchi, Y., Kikegawa, Y., Ohashi, Y., Yoshikado, H., & Komiyama, H. (2005). Development of a multilayer urban canopy model for the analysis of energy consumption in a big city: Structure of the urban canopy model and its basic performance. *Boundary-Layer Meteorology, 116*, 395–421.

Kusaka, H., Kondo, H., Kikegawa, Y., & Kimura, F. (2001). A simple single-layer urban canopy model for atmospheric models: Comparison with multi-layer and slab models. *Boundary-Layer Meteorology, 101*, 329–358.

Kusaka, H., Fei, C., Tewari, M., Dudhia, J., Gill, D. O., & Duda, M. G. (2012). Numerical simulation of urban heat island effect by the WRF model with 4-km grid increment: An inter-comparison study between the urban canopy model and the slab model. *Journal of the Meteorological Society of Japan, 90*, 33–45.

Li, X., & Zhou, W. (2019). Spatial patterns and driving factors of surface urban heat island intensity: A comparative study for two agriculture-dominated regions in China and the USA. *Sustainable Cities and Society, 48*, 101518.

Lin, Y., & Colle, B. A. (2011). A new bulk microphysical scheme that includes riming intensity and temperature-dependent ice characteristics. *Monthly Weather Review, 139*, 1013–1035.

Lin, C., Chen, W., Liu, S. C., Liou, Y., Liu, G. R., & Lin, T. H. (2008). Numerical study of the impact of urbanization on precipitation over Taiwan. *Atmospheric Environment, 42*, 2934–2947.

Liu, Y., Chen, F., Warner, T., & Basara, J. (2006). Verification of a mesoscale data-assimilation and forecasting system for the Oklahoma City area during the joint urban 2003 field project. *Journal of Applied Meteorology and Climatology, 45*, 912–929.

Martilli, A., Clappier, A., & Rotach, M. (2002). An urban surface exchange parameterisation for mesoscale models. *Boundary-Layer Meteorology, 104*, 261–304.

Mesinger, F., Dimego, G., Kalnay, E., Mitchell, K., Shafran, P. C., Ebisuzaki, W., Jovic, D., Woollen, J., Rogers, E., Berbery, E. H., Ek, M. B., Fan, Y., Grumbine, R., Higgins, W., Li, H., Lin, Y., Manikin, G., Parrish, D., & Shi, W. (2006). North American regional reanalysis. *Bulletin of the American Meteorological Society, 87*(3), 343–360.

Miao, S., Chen, F., LeMone, M., Tewari, M., Li, Q., & Wang, Y. (2009). An observational and modeling study of characteristics of urban heat island and boundary layer structures in Beijing. *Journal of Applied Meteorology and Climatology, 48*(3), 484–501.

Roberge, F., & Sushama, L. (2018). Urban heat island in current and future climates for the island of Montreal. *Sustainable Cities and Society, 40*, 501–512.

Salamanca, F., Martilli, A., Tewari, M., & Chen, F. (2011). A study of the urban boundary layer using different urban parameterizations and high-resolution urban canopy parameters with WRF. *Journal of Applied Meteorology and Climatology, 50*, 1107–1128.

Vahmani, P., & Ban-Weiss, G. A. (2016). Impact of remotely sensed albedo and vegetation fraction on simulation of urban climate in WRF-urban canopy model: A case study of the urban heat island in Los Angeles. *Journal of Geophysical Research—Atmospheres, 121*, 1511–1531.

Wang, Y., Berardi, U., & Akbari, H. (2016). Comparing the effects of urban heat island effect mitigation strategies in the city of Toronto. *Energy and Buildings, 114*, 2–19.

Chapter 12
Urban Weather Generator: Physics-Based Microclimate Simulation for Performance-Oriented Urban Planning

Jiachen Mao and Leslie K. Norford

12.1 Introduction: A Dance Between Art and Science

Climate change is now pervasive in scientific, social and political communities. There is clearly a detectable global warming trend in the long-term climate records. Vast resources are being processed, transported and consumed in order to accommodate humans in our modern life. As a consequence, exhaustion of raw materials and transformation of chemical substances have generated an increasing level of hazardous residues that may pollute air, water and soil.

With over half of the global population residing in cities now and an arguably growing projection of urban settlements in the near future, the scale of this resource demand is profound and the subsequent problems are unprecedentedly intensified in every aspect. Motivated by these challenges, urban sustainability holds great attention to tackle energy and environmental issues. Buildings, which incorporate most human activities in the urban areas, are identified as having the highest economic mitigation potential of any other energy sector. Simulation models play an essential role in the decision-making process for performance-oriented building design and urban planning.

Mathematical models are central to high-level human endeavour to formally abstract, understand and change the world. On the other hand, simulation models are low-level computer-aided specializations to numerically implement such mathematical rigor, with various applications in physical observations, engineering systems, social sciences, etc. As we delve into the literature, we have found a rich

J. Mao (✉)
Laboratory for Information and Decision Systems, Massachusetts Institute of Technology, Cambridge, MA, USA
e-mail: j.mao@alum.mit.edu

L. K. Norford
Department of Architecture, Massachusetts Institute of Technology, Cambridge, MA, USA
e-mail: lnorford@mit.edu

M. Palme, A. Salvati (eds.), *Urban Microclimate Modelling for Comfort and Energy Studies*, https://doi.org/10.1007/978-3-030-65421-4_12

family of simulation paradigms for energy systems at the building and urban scale, ranging from physics-based computational approximation to data-driven black-box generalization. Thanks to a few ambitious initiatives, some software progress has been made to assist designers, planners and engineers in improving the sustainability of their districts and cities. Recent networking efforts have also produced a road map toward a broad introduction of simulation-based domain, or may have just opened a Pandora's box in the AEC industry.

These developments are still in their infancy, as buildings and cities are extremely complex systems that require a need for interdisciplinary research in simulation studies. One particular issue is the urban heat island (UHI) effect (Oke 1973), which states that the climate pattern in urban areas is different from that in rural areas due mainly to distinct built ecosystems. The UHI effect modifies the thermal response of buildings, especially naturally ventilated buildings, and may intensify the peak energy demand in summer. At the same time, the energy performance of buildings can also have an impact on the outdoor urban microclimate, mainly through waste heat emissions. Regardless of how successfully the UHI effect has been observed throughout the world, many people still fail to incorporate it into their simulation-based analyses of energy use or thermal comfort, since the weather data used in practice comes from the station measurements at an open site (e.g. airport) that do not accurately represent the local urban environment.

Ongoing developments in measurement and modelling of the UHI are adapting localized environmental data for simulation use. It is possible to adopt geospatially aware regional models to generate site-specific weather sequences by driving them with the observations from nearby stations. In particular, mesoscale models are considered state of the art in atmospheric weather prediction and are used for either research or practice (Grimmond et al. 2010). Popular simulation tools include computational fluid dynamics (CFD), Weather Research and Forecasting (WRF), ENVI-met, etc. However, due to the high computational cost, their feasibility is usually limited considering the time-constrained nature of real engineering applications. In addition, the accuracy of these models in conjunction with a spatially fine mesh depends strongly on the boundary conditions, for which detailed information is not easily available in common practice. Despite some promising evidence in ad hoc studies, a painfully high computational cost is usually paid for a gain in accuracy that cannot be guaranteed.

Perhaps the most interesting irony we should point out is that, during a design or planning stage, whenever a performance verification uses simulation to approximate the real behaviour of a system, one has to recognize that no system ever behaves as its model (which is de facto an idealization) predicts (Augenbroe 2012). Seeking the "best" simulation might actually be counterproductive since it would often lead to over-engineered models with time-consuming computation runs. Surprisingly, the need for sophisticated simulation seems to be less than what is usually assumed, owing to the presence of system uncertainties, lack of required information or preference of decision makers. A reasonably impressive principle of simulation in performance-oriented design for civil projects is, therefore, to promptly provide a variety of parametric calculations and rational feedback when an advanced simulation model cannot offer a more accurate answer.

With that in mind, Bueno (2012) proposed the Urban Weather Generator (UWG), a physics-based simulation paradigm, to quantify the energy interactions between buildings and urban climate. Since then, the UWG has been progressively developed and maintained at the Massachusetts Institute of Technology (MIT), along with a few intergovernmental and commercial collaborations. The UWG is now a stand-alone program to quickly generate localized microclimate sequences, which could be readily used as the urban weather data, based on rural meteorological information in public weather files and site-specific urban characteristics. At the same time, it has the potential to evaluate the building energy consumption at a neighbourhood or city scale, specifically accounting for the UHI effect. A typical run of annual calculation with an acceptable performance can be currently finished in *minutes*, given normal computational power. The UWG enjoys the marriage of building physics and urban climatology to enable a fresh way to express and explore a wide range of energy and environmental studies in an urban context—a bridge between abstraction and specialization, a dialogue between designer and engineer as well as a dance between art and science.

It is time to present what we have learned and achieved over the past decade. The purpose of this chapter is to explain some basic treatments that lead to the foundation of UWG, and specifically to cover how the underlying models behave with customized variations. A second purpose is to direct the practitioners to some of the recent open-source code and software tools that have come out of the UWG, as well as some of the historically relevant case studies that were used to validate the UWG mechanism. We consider this chapter as a formal statement to illustrate what the UWG is capable of at present and to suggest some possible improvements in the future.

This chapter is intended to be accessible to a general audience; that is, it is intended for researchers, engineers, designers and hopefully the people outside of our field. Therefore, while mathematical and physical arguments will be used, the goal is for the theoretical treatments to be intuitive and neat rather than technical and rigorous. Sometimes, specific details may be suppressed for general readability; interested reader may further look into the original documents and/or links that are referred. We expect that the reader in both academia and industry can benefit from this chapter.

12.2 Scientific Modelling: The UWG Innovation

We begin by reviewing the core of the UWG mechanism. In general, a complex physical system consists of multiple functional components, whose evolution can be characterized by differential and/or discrete equations in time and space domains. Some linear and non-linear treatments are then applied to different equations based on specific physical laws. In our case, modelling the urban microclimate is even more complicated due to the diverse time and space scales involved in meteorological study. The space scale, for example, can vary from a few meters (e.g. building) to several kilometres (e.g. neighbourhood). It is hardly possible to resolve all the scales online with one single model given the available computational resources in practice.

To make things efficient, one universal recipe is to introduce separate models with some synchronization between them to exchange boundary information. Accordingly, the system would no longer be differentiable as a whole, and a higher level of abstraction is needed to fully leverage such modularity. Inspired by this idea, the exchange of mass, momentum and energy in an urban context could then be implicitly specified by embedding an urban canopy layer and building energy model within a mesoscale atmospheric model. The boundary conditions of the mesoscale model may be further provided by some other models at a larger scale. Given great care for each model and their connectivity, such a synthesized simulation is able to generate sufficiently good energy and environmental data with reasonable computation time.

Following this rationale, the program architecture of UWG is mainly composed of four modules: the rural station model (RSM), the vertical diffusion model (VDM), the urban boundary layer model (UBLM) as well as the urban canopy and building energy model (UC-BEM). Loosely speaking, the UWG creates the UHI profile by processing necessary information from a rural station to an urban site via these four modules, as shown in Fig. 12.1a. The function interrelations and information

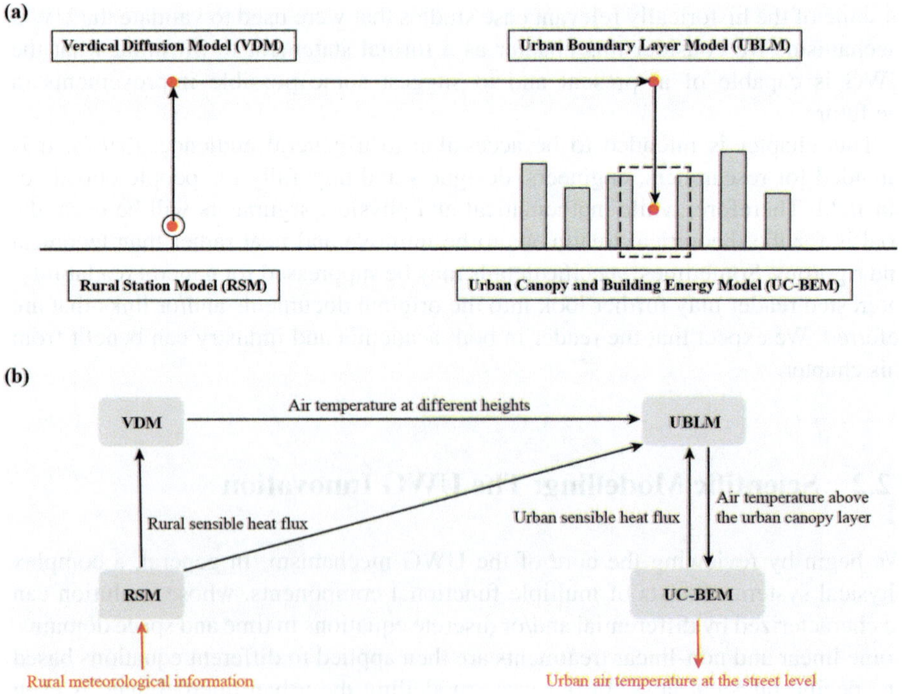

Fig. 12.1 Diagram of the basic UWG scheme. (**a**) Four coupled models in the UWG; (**b**) information exchanged among different models in the UWG

exchange between each module are illustrated in Fig. 12.1b. Related physical phenomena in one model can be implicitly forced using the information from the other model(s), so that all the models are coupled online in a coherent direction. We will now detail the essential treatments of each model within the UWG scheme.

12.2.1 Rural Station Model

The RSM (Bueno 2012; Bueno et al. 2013b) is a rural canopy model that reads hourly meteorological data measured at the rural site and calculates rural sensible heat flux, which will then be used by the VDM and UBLM.

The RSM is based on an energy balance at the soil surface, where a transient heat diffusion equation is formulated using a thermal network of constant resistances and capacitances to represent the heat storage and release from the ground. As shown in Fig. 12.2, by dividing the soil into discrete layers, the model solves a system of finite difference equations where:

For the first layer

$$d_1 (\rho c)_1 \frac{\partial T_1}{\partial t} = C_{1,2} (T_2 - T_1) + Q_{\text{surf}} \tag{12.1}$$

For each middle layer l

$$d_l (\rho c)_l \frac{\partial T_l}{\partial t} = C_{l,l+1} (T_{l+1} - T_l) + C_{l,l-1} (T_{l-1} - T_l) \tag{12.2}$$

And for the deepest layer n

$$d_{n-1} (\rho c)_{n-1} \frac{\partial T_{n-1}}{\partial t} = C_{n-1,n} (T_{\text{deep}} - T_{n-1}) \tag{12.3}$$

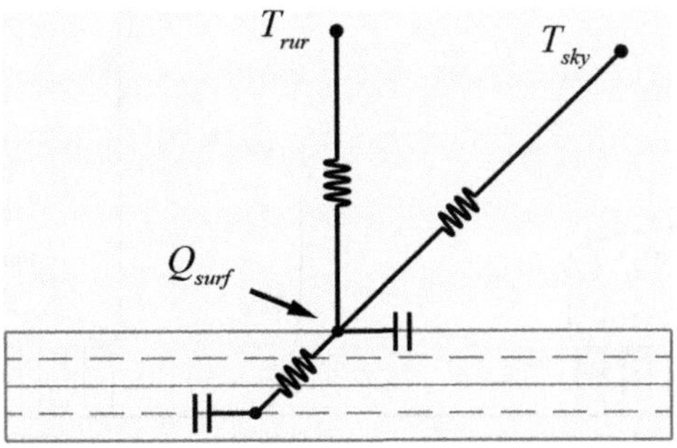

Fig. 12.2 The heat transfer process as a thermal network of constant resistances and capacitances in the RSM (mainly from Bueno et al. 2013b). A capacitance is associated with each temperature node. Nodes are connected via resistances. The surface heat flux is represented by an arrow

Surface sensible heat fluxes are calculated using a convective heat transfer coefficient (CHTC), which is a function of the air velocity above the canopy layer based on empirical correlation. Considering the accuracy requirement of UWG, latent heat fluxes due to the presence of vegetation (if any) are simply calculated as a prescribed fraction of the absorbed short-wave radiation.

12.2.2 Vertical Diffusion Model

The VDM (Bueno 2012; Bueno et al. 2013a) reads hourly meteorological data measured at the rural site and the rural sensible heat flux calculated by the RSM to derive vertical profiles of the air temperature above the rural area, which will then be used by the UBLM.

It is worth noting that the VDM and UBLM take the concept of potential temperature to evaluate the air temperatures at different heights without specifically including the pressure differences. The potential temperature (usually denoted as θ) of a parcel of fluid at a certain pressure is the temperature that the parcel would acquire if adiabatically brought to a standard reference pressure. As shown in Fig. 12.3, the VDM calculates the vertical profiles of quasi-steady potential temperature based on a heat diffusion equation:

$$\frac{\partial\theta(z)}{\partial t} = -\frac{1}{\rho(z)}\frac{\partial}{\partial z}\left[\rho(z)K_{\mathrm{d}}(z)\frac{\partial\theta(z)}{\partial z}\right] \tag{12.4}$$

where the lower bound is the temperature measured at the rural site $\theta(z_r)$, and the upper bound considers the fact that at a certain height (normally about 150–200 m) the profile of potential temperature is approximately uniform so that

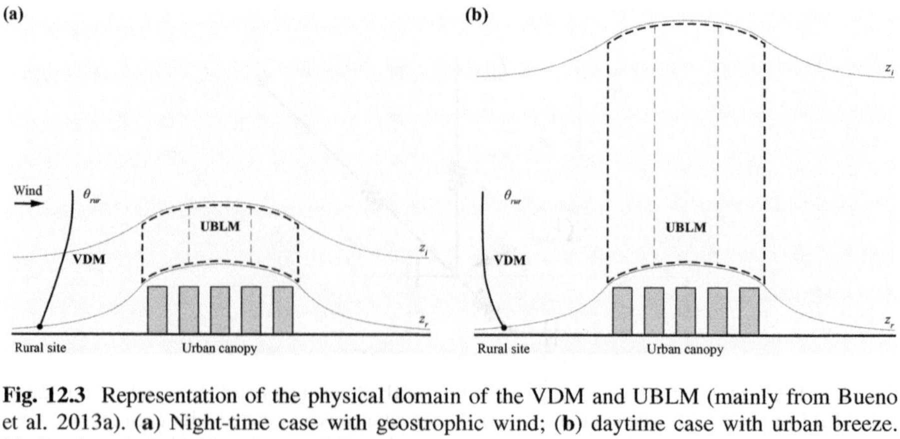

Fig. 12.3 Representation of the physical domain of the VDM and UBLM (mainly from Bueno et al. 2013a). (**a**) Night-time case with geostrophic wind; (**b**) daytime case with urban breeze. Idealized vertical profiles of potential temperature are shown at rural sites (not at scale)

$$\frac{\partial \theta}{\partial z}\bigg|_{z_{\mathrm{ref}}} = 0 \tag{12.5}$$

The difficulty of implementing this diffusion equation in the VDM lies in the estimation of the diffusion coefficient (K_{d}), due mainly to the difficulty of predicting stable boundary layers. This coefficient is currently established as a function of the turbulent kinetic energy at each vertical level:

$$K_{\mathrm{d}} = C_{\mathrm{k}} l_{\mathrm{k}} E^{1/2} \tag{12.6}$$

where l_{k} and E are obtained by iteratively solving a set of coupled functions based on specific atmospheric conditions. A comprehensive description of the VDM is presented in the appendix of Bueno (2012) and Bueno et al. (2013a), and a new formulation for certain aerodynamic treatments has been updated in the appendix of Bueno et al. (2014).

12.2.3 Urban Boundary Layer Model

The UBLM (Bueno 2012; Bueno et al. 2013a) calculates air temperatures above the urban canopy layer using the meteorological information from the rural site, the vertical profiles of air temperature from the VDM, the rural sensible heat fluxes from the RSM and the urban sensible heat fluxes from the UC-BEM.

The UBL is assumed as a region of well-mixed air and there is no significant heat exchange at the top of it. The model solves an energy balance for a control volume with imposed boundary conditions:

$$V_{\mathrm{ubl}} \rho c_{\mathrm{v}} \frac{d\theta_{\mathrm{ubl}}}{dt} = H_{\mathrm{urb}} + \int u_{\mathrm{ref}} \rho c_{\mathrm{p}} \left(\theta_{\mathrm{ref}} - \theta_{\mathrm{ubl}}\right) dA_{\mathrm{f}} \tag{12.7}$$

where θ_{ref} is the potential temperature outside the city derived from the VDM. The term on the left-hand side stands for the thermal inertia of the control volume, and the second term on the right-hand side represents the advection effect. The control volume is bounded by the blending height (z_r), where the impacts of individual obstacles on the vertical domain become horizontally blended (e.g. at the rural site), and the boundary layer height (z_i). These heights are the inputs to be determined before simulation.

As one might expect from common atmospheric knowledge, the UBL is affected by changes in the urban surface and there is a diurnal variation at the mesoscale level (Stull 2012). During night-time, the long-wave radiation exchange between the sky and the rural surface makes the boundary layer stratified and the night-time UBL is mixed at a low height (usually 50–100 m). During daytime, on the other hand, the solar radiation heats the surface and the resulting atmospheric layer is well mixed up to a higher altitude (usually 700–1500 m). In trying to capture the diurnal pattern with a pragmatic concern of the computational cost, the UBLM differentiates between night-time and daytime conditions,[1] as well as between the advection driven by geostrophic wind (forced effect) and by urban breeze circulation (buoyancy effect), as depicted in Fig. 12.3.

We sketch some basic developments of the advection effect, for which the UBL is divided along the windward direction into various control volume sections. In the presence of geostrophic wind, the reference velocity (u_{ref}) is set to the velocity outside the control volume, derived from a logarithmic vertical profile above the rural site via the VDM. The lateral area of heat exchange (A_f) considers the width of the urban area orthogonal to the wind direction. The temperature calculated for an upwind section in the UBL is used as the reference temperature for the subsequent downwind section. Under the circumstance of urban breeze, a characteristic circulation velocity is expressed as

$$u_{\mathrm{circ}} = \left[\frac{g\left(H_{\mathrm{urb}} - H_{\mathrm{rur}}\right) z_i}{\rho c_{\mathrm{p}} \theta_{\mathrm{ref}}} \right]^{1/3} \tag{12.8}$$

where H_{urb} and H_{rur} are obtained from calculations in the UC-BEM and RSM, respectively. The advection is assumed to be buoyancy driven if the circulation velocity is greater than the wind velocity outside the city. Of course, the reference velocity will then become the circulation velocity in the energy balance equation, and the lateral area of heat exchange includes its entire city perimeter. The temperature of the UBL section at the periphery of the city is used as the reference temperature for the central area of the city. A clearer and more formal description of the UBLM can be followed via Bueno (2012) and Bueno et al. (2013a, 2014).

Interestingly, there is a digression in the history of developing the UBLM that we should mention. After the original version (Bueno 2012; Bueno et al. 2013a) was built, Bueno et al. (2014) further included the radiation effects of water vapor and carbon dioxide in the energy balance of the UBL. However, this treatment was removed by Yang (2016) following an argument that the air is essentially transparent to the long-wave radiation emitted from its boundary surfaces with a body size of around 1 km in altitude. This argument has been numerically validated using simulations, and hence, the UBLM returned back to its original version again. Given the available evidence, it is not very clear that the above discussion settles one way or another which version is better in terms of the accuracy requirement associated with the UWG. It might be practically reasonable to rely on the current UBLM, while further research is needed to refine such debate as we proceed in the future.

[1] The eagle-eyed reader might note that, technically, solving only two different sets of equations for night-time and daytime conditions would somewhat lead to discontinuities during the transition periods in the morning and evening, respectively. Although some small jumps in the temperature profile have been observed in our numerical trials, we treat this as a minor issue from the perspective of practical implementation. The discontinuities can be attenuated by the thermal inertia of the air body in the UBL and can be further reduced by adaptively shifting the transition times between day and night.

12.2.4 Urban Canopy and Building Energy Model

The UC-BEM (Bueno 2012; Bueno et al. 2013b) calculates the urban weather information at the street level and estimates the building energy consumption at the neighbourhood scale, using the meteorological data measured at the rural site, the air temperature above the urban canopy layer calculated by the UBLM and the built characteristics of a specific urban site.

In order to be coupled with mesoscale atmospheric simulations online, the UC-BEM considers the well-established Town Energy Balance (TEB) scheme (Masson 2000) for modelling the urban canopy layer. The TEB maps the 3D urban geometry to a 2D canyon model consisting of a roof, a wall and a road, with average site characteristics. It calculates the climate conditions within a neighbourhood formed by identical urban configurations, where all the orientations are equally possible. Based on the energy balance, the urban canyon air is assumed as a well-mixed thermal zone that exchanges heat with the generic road, the generic wall and the atmosphere above the urban canopy. The underlying goal is to robustly capture the basic physical mechanisms that lead to the microclimate conditions in a specific urban site.

In decoding the physical machinery of the energy interactions within an urban environment, it would be remiss to not consider the anthropogenic heat fluxes, which generally include the heat released from building operation, traffic vehicles and human metabolism (Sailor 2011). The outdoor metabolic heat is disregarded in the current urban canopy model, as it is negligible compared to other heat sources. The traffic-generated heat, which normally remains at 8–15 W/m^2 in many cities, is incorporated as a prescribed profile in the canyon energy balance (Yang 2016). Finally, as the largest portion of the anthropogenic heat sources, the building operation is physically parameterized to allow a better representation by embedding a building energy model in the TEB scheme (Bueno et al. 2012a). This naturally justifies the foundation of the UC-BEM.

Abiding by this logic, it might be tempting to associate a detailed physics-based simulation (e.g. EnergyPlus) with the TEB scheme. However, this does not work well for our general purpose. The problem is that, for the many different building types in an urban site, it is computationally hard to include all the detailed simulations and to couple them with mesoscale atmospheric models online as one standalone program. Another reason lies in the fact that a sophisticated physics-based simulation requires more information than is usually available during a design or planning stage, and thus leads to an over-engineered method that exceeds the level of our expectation. In order to efficiently parameterize the essential energy nexus between buildings and urban climate, the UC-BEM considers a thermal network of constant resistances and capacitances (Bueno et al. 2012b), as shown in Fig. 12.4.

The RC model retains major design parameters that are typically considered in building energy and urban climate studies. To simplify the estimation of various building types within a neighbourhood, the commercial building reference data has been imported by Yang (2016) from the US DOE online database (Deru et al. 2011)

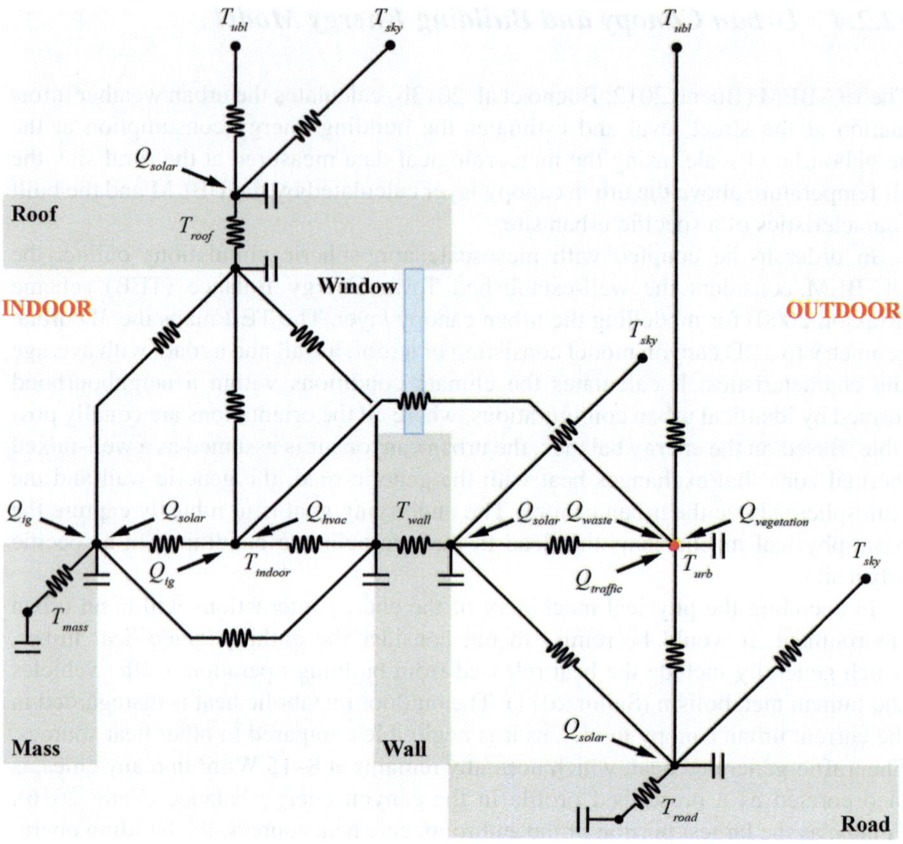

Fig. 12.4 Representation of the UC-BEM based on a thermal network of constant resistances and capacitances (mainly from Bueno et al. 2012b with revisions from Yang 2016). A capacitance is associated with each temperature node. Nodes are connected via resistances. The heat sources of each node are represented by arrows. The red node denotes the urban canyon air temperature. Subscript *ig* stands for internal heat gain

into the UC-BEM. This allows us to express a rich family of abstract building types that make up the urban area, instead of modelling each individual building. Accordingly, the BEM simulates hourly load behaviours for every building type using either the default values provided by the database or the customized values provided by the user (e.g. by updating the database) according to specific case. The goal is a better estimation of the building energy pattern at the urban scale and the temporal heat fluxes associated with the surrounding environment. The BEM considers a single building zone with a generic thermal mass and uses weighted average thermal parameters in the energy balance calculations to avoid additional computations for a trivial gain in accuracy. Many other assumptions and treatments have also been made to keep the model as simple as possible while maintaining the key features that represent the general system dynamics (mostly for the HVAC systems).

Based on the energy conservation principle, the energy flow of each node is related to all the heat fluxes that interact with it. This can be described, without loss of generality, as

$$C_j \frac{dT_j}{dt} = \sum_k \frac{1}{R_k}\left(T_k - T_j\right) + \Sigma Q_j \tag{12.9}$$

where C_j and T_j are the capacitance and temperature of node j, R_k and T_k are the resistance and temperature of node k associated with node j and Q_j is the heat flux acting on node j. A few iterations of the RC model are needed to calculate specific values.

To give a concrete idea of how the RC model actually works, we consider the urban canyon air temperature (i.e. the red node in Fig. 12.4) as an example. In the current version, the urban canyon energy balance accounts for the heat fluxes from the wall, window and road; the heat exchange between the canyon air and the above atmosphere simulated by the UBLM; the heat fluxes due to exfiltration from buildings; the anthropogenic heat sources (i.e. waste heat emissions and traffic-generated heat); as well as the heat flux from vegetation; so we have

$$\begin{aligned} V_{\text{can}} \rho c_{\text{v}} \frac{dT_{\text{urb}}}{dt} = & \sum_i A^i_{\text{wall}} h^i_{\text{wall}} \left(T^i_{\text{wall}} - T_{\text{urb}}\right) + \sum_i A^i_{\text{window}} U^i_{\text{window}} \left(T^i_{\text{indoor}} - T_{\text{urb}}\right) + \\ & A_{\text{road}} h_{\text{road}} \left(T_{\text{road}} - T_{\text{urb}}\right) + u_{\text{ex}} \rho c_{\text{p}} A_{\text{road}} \left(T_{\text{ubl}} - T_{\text{urb}}\right) + \\ & \sum_i \dot{V}^i_{\text{inf/vent}} \rho c_{\text{p}} \left(T^i_{\text{indoor}} - T_{\text{urb}}\right) + \sum_i Q^i_{\text{waste}} + Q_{\text{traffic}} + Q_{\text{vegetation}} \end{aligned} \tag{12.10}$$

where i denotes the building type i that is considered in the UC-BEM.

Likewise, an analogous energy balance is solved for each node in Fig. 12.4 based on the formulation of Eq. (12.9). The external surface temperatures of the wall, roof and road are also calculated by solving a set of heat diffusion equations as described in the RSM (Eqs. (12.1)–(12.3)). Similarly, CHTC is applied to evaluate the surface sensible heat fluxes. As required by the UBLM, the urban sensible heat fluxes are expressed as the sum of the heat exchange between the canyon air and the UBL and the convective heat fluxes from the roof, including the portion of waste heat emissions from the outdoor HVAC equipment located there. Once the quasi-steady state is obtained for both indoor and outdoor environments, the building energy consumption will be recalculated to incorporate the UHI effect.

Each component in the energy balance of the UC-BEM has been carefully characterized and updated in terms of its treatment. In retrospect, we have found that the UWG is indeed becoming more physically sound and more capable to handle increasingly detailed information. While it is hardly possible for us to detail all the formal arguments here to justify the underlying physical laws, we hope interested readers could look into their manifests following the work of Bueno et al. (2012a, b, 2013b, 2014), Bueno (2012), Yang (2016) and Mao (2018), especially some of the appendices.

In more recent independent work, Moradi et al. (2019) resolved a multilayer vertical profile of the urban environment and included an advanced tree model within the UC-BEM, which yields another simulation paradigm called Vertical City Weather Generator (VCWG). Given the boundary conditions from the rural area, an urban VDM calculates the quasi-steady thermal quantities within each control volume, which is obtained by splitting the urban computational domain into multiple layers. Essentially, the urban VDM could be approximated by a set of one-dimensional time-averaged momentum equations in the cross- and along-canyon components, which will be coupled with other models within the VCWG. As a result, the VCWG extends the UWG to initiate parametric investigation of design options for the urban environment at some spatiotemporal variations within the canyon.

12.3 Simulation Tool: The UWG Community

We now move from high-level semantics of the UWG to practical simulation programs and tools. Intuitively, boundary-coupled partial differential equations and discrete event equations are synchronized to form a thermal information network of constant resistances and capacitances within the UWG. Derivatives and integrals can be numerically solved using finite methods, while the RC model can be numerically addressed using a state-space formulation. Each component is specified in the computer program by (1) defining a class, (2) exposing its boundary conditions and (3) encapsulating the constraints between the boundary conditions, state variables and their derivatives. The ultimate synthesized program can be further integrated with other APIs to facilitate a more coherent workflow.

Currently, the UWG is built to process the hourly meteorological information in public EPW files mostly from rural stations. EPW files provide the default environmental boundary conditions that drive energy performance simulations at the building and urban scale. As one might encounter in practice, selecting a suitable weather data source when no station is close to the project site under study is often a form of art. Although there is no generally accepted protocol, the UWG initiates a better decision-making paradigm to some extent. As shown in Fig. 12.5, starting with the rural weather data provided in the EPW format and the specific urban characteristics, the UWG generates the simulated urban weather data (during either an entire calendar year or a subset thereof) that can be readily exported to the EPW format. In addition, the UWG outputs a fruitful set of energy information (e.g. surface temperature, heat flux, energy use, etc.) that can be further analysed for various purposes.

The UWG program has been under development since 2008, and some public releases of the open-source code were announced. Moreover, several attempts have been actively made to introduce the UWG as a software tool in many research and commercial communities (Nakano 2015; Awino 2019). While the UWG was nurtured at MIT, it is really the engagement of many people around the world that has made it a joy to use for urban design computing. It is also recognized as a

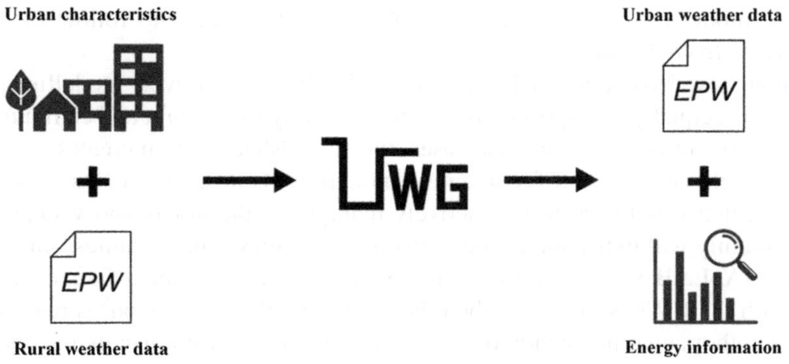

Fig. 12.5 General workflow of the UWG (as of 2019, the output of simulated energy information was only enabled in the Excel interface of the UWG MATLAB version and in the Dragonfly)

specific-purpose program, allowing it to be deployed not only as a prototype tool in many AEC companies, but also as a teaching tool in numerous educational institutions. A great example of this is the Dragonfly plug-in for Grasshopper, which makes it possible for designers to efficiently perform large-scale climate and UHI modelling within the Rhino interface. These use cases take advantage of the UWG's capabilities not only for numerical simulations, but also for building interactive APIs, database access and much more. We present some popular resources here.

12.3.1 *Open-Source Code: Fantastic Projects and Where to Find Them*

With the increasing demand in the day-to-day life of scientific modellers for both productivity and utility, the need for high-performance computing has become pressing. UWG is fast because of careful system design and right combination of suitably chosen models that work well with each other. For the practitioners and researchers to fully benefit from the UWG wisdom, we provide two projects written in MATLAB[2] and Python[3] that are available (as of 2019) under the MIT License for open-source program.

Each computational module is written following the theoretical developments summarized in Sect. 12.2. Users can depend on the existing numerical implementation or customize their own formulation within the computing ecosystem. Through adaptive code manipulations, the UWG can even be extended to couple with other dynamic models and algorithmic solvers to enable more advanced simulation-based

[2] UWG MATLAB version. https://github.com/Jiachen-Mao/UWG_Matlab.

[3] UWG Python version. https://github.com/ladybug-tools/uwg.

analysis, model predictive control, etc. We invite the reader to follow the basic instructions from the websites of these projects.

Although we have touched thus far on efficiently and intuitively modelling physical systems, equally as important is a convenient way to interpret and communicate the results obtained. For many use cases, the spreadsheet system creates a surprisingly versatile environment for scientific modelling. One reason for its impressive success is that it is trivial to interactively manipulate the inputs and visualize the results, mainly within the spreadsheet. In order to achieve such seamless interactivity, the MATLAB version[2] further incorporates a macro-enabled Excel interface along with the UWG scripts. On the other hand, the Python version[3] is ready to be coupled with many other standard computing libraries to customize interactive data analysis and visualization with the UWG engine.

Taking advantage of these interfaces, one can easily create modelling results that embed urban computing problems at scale. We believe that EPW formatting, spreadsheet operation and object-oriented programming provide a quite satisfying solution in many scenarios to the long-standing challenge of managing data for simulation use.

12.3.2 Dragonfly and So Much More: The Zootopia of Urban Design Computing

Capitalizing on the UWG code, Dragonfly[4] is a plug-in for Rhino and Grasshopper that allows users to model and estimate urban-scale climate and energy phenomena, such as the UHI effect and district energy use. This is accomplished with the help of several simulation engines, including the UWG Python version[3]. It also links to several public climate-related databases and satellite image data sets (e.g. the Landsat from NASA) for climate projection. Dragonfly intends to make large-scale climate variables easily accessible to the visual scripting interface of Grasshopper as well as the 3D visualization interface of Rhino. The latest executable version, including the program code (written in Python), is actively maintained within the Ladybug Tools,[5] which is available (as of 2019) under the GNU General Public License for open-source software.

Since the third millennium, the World Wide Web has increasingly propagated the mantra of distributed, concurrent, and collaborative development. A characteristic of many successful computational projects is that they enable modularization that consists of small discrete components with standardized interfaces. Each component is usually good at one particular task, thus providing a core of functions that could be further featured by a community. Structuring the code into small components allows each subsystem to be developed individually and then integrated into a

[4]Dragonfly. https://www.ladybug.tools/dragonfly.html.

[5]Ladybug Tools. https://www.ladybug.tools/.

larger system that yields something greater than their value in isolation (Wetter 2012). This modularization facilitates reusing code and distributing computation efforts. It also benefits from different experts on different tasks, which can often improve the quality of the whole computing platform.

In agreement with this philosophy, Dragonfly is further connected with other plugins having distinct functions (e.g. Ladybug, Honeybee, and Butterfly) to form the Ladybug Tools that supports energy and environmental design. Thanks to a joint effort of different validated simulation engines and a rich library of computing and visualizing functions, the Ladybug Tools makes it possible for practitioners to build sophisticated models within the Grasshopper ecosystem and to predict not only urban microclimate but also energy use, thermal comfort, and much more even without expert knowledge of particular simulation functionalities. Such extensive connectivity and interactivity further promote the use of the UWG and Dragonfly in urban design computing. As of July 2019, over 232,000 downloads of the Ladybug Tools plugins have been made, and they have now been utilized as teaching tools in dozens of universities and graduate schools around the world.

12.4 Case Study: A Tale of Five Cities

In what follows, we briefly describe some case studies that have been documented to validate the UWG mechanism during the course of its development. From a more pragmatic point of view, we intend to give an expectation of what the UWG is capable of compared with other atmospheric simulations. Please note that, given the stochastic and uncertain nature of real-world complex systems, some observations and conclusions might be case specific and thus cannot be generalized. Of course, there is no single model that can work for all the scenarios. We present a few specific interesting lessons that we have learned and some plausible reasons for the observed phenomena. A summary of related case studies can be seen in Table 12.1 and Fig. 12.6.

12.4.1 Europe

The case of Europe, in collaboration with the Centre National de Recherches Météorologiques (CNRM-GAME) in France, played an essential role during the initial development of the UWG cores and provided many perspectives on the urban climate situations in European cities (Bueno 2012). Major explorations were conducted for Toulouse (France), Athens (Greece) and Basel (Switzerland).

In summary, the measured data from Toulouse and Athens along with a calibrated EnergyPlus model were used to validate the UC-BEM scheme on the simulation of indoor air temperature and energy demand (Bueno et al. 2012a, b), while the field observations from Toulouse and Basel together with an advanced mesoscale atmospheric simulation were further compared to evaluate the performances of the

Table 12.1 Summary of the case studies used to validate the UWG (as of 2019)

Case study	Europe Toulouse, France	 Athens, Greece	 Basel, Switzerland	Singapore	Abu Dhabi, UAE
Location					
Latitude	43.48°	37.98°	47.33°	1.37°	24.49°
Longitude	1.30°	23.68°	7.35°	103.98°	54.37°
Climate zone	Humid subtropical	Mediterranean	Temperate oceanic	Tropical rainforest	Hot desert
Köppen climate classification	*Cfa/Cfb*	*Csa*	*Cfb*	*Af*	*BWh*
Measurement					
Time	Feb 2004–Mar 2005	May–Sept 2009	June–July 2002	Feb and July 2010	Oct 2016–Aug 2017
Urban canopy air temperature	×	×	×	×	×
Above-canopy air temperature	×		×		
Building energy use	×			×	×
Anthropogenic heat flux	×			×	×
Surface heat flux	×		×		
References	Bueno et al. (2012a, 2013a, b)	Bueno et al. (2012a)	Bueno et al. (2013a, b)	Bueno et al. (2014)	Mao et al. (2018); Santos et al. (2018)

Note: This table only presents the details of the measurements that have been used to validate the UWG scheme. There should be more energy and environmental information in each original case study than what we have included in the table

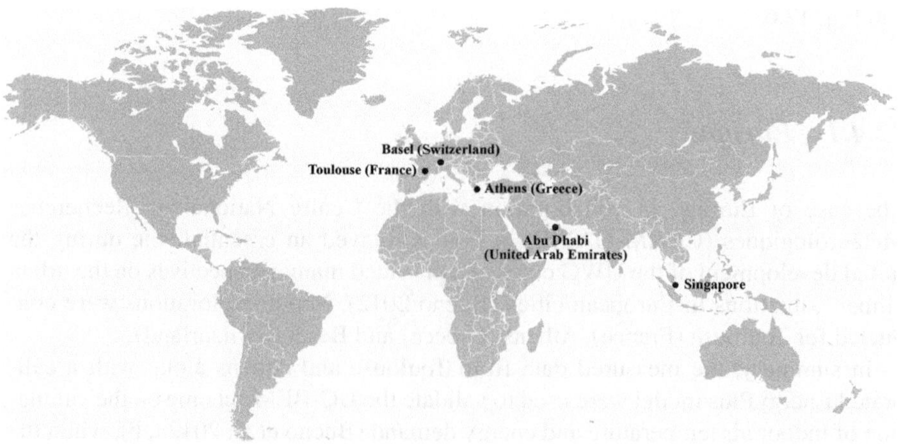

Fig. 12.6 Plot of the case studies used to validate the UWG (as of 2019) on a world map

VDM-UBLM scheme and UWG program in predicting the above-canopy and in-canopy outdoor air temperatures, respectively (Bueno et al. 2013a, b). These pilots have shown a reasonably good agreement, considering the many uncertainties associated with state-of-the-art urban climate and energy predictions. Therefore, the UWG makes it possible to carry out a long-term analysis of climate scenarios and design policies without having to run computationally expensive simulations.

Among the many achievements during the early developments and tests, the most important thing that we have learned is arguably how buildings and urban climate essentially interact with each other. A general, if slightly not rigorous, statement for such interaction is that the main mechanism by which the UHI effect influences the indoor environment and energy use is the outdoor air entering the building, which is usually caused by infiltration but also by natural or mechanical ventilation; on the other hand, the waste heat emissions from HVAC systems and (sometimes) exfiltration heat fluxes are the main force by which the building energy performance can affect outdoor thermal conditions. These observations highlight the significance of incorporating the UHI effect in the analysis of building and urban energy systems, which the reader should keep in mind.

12.4.2 Singapore

The case of Singapore, supported by the Singapore National Research Foundation, gave a test of the UWG model in a humid tropical climate (Bueno et al. 2014). Unlike European-type cities, Singapore includes a wide range of different urban configurations and entails a more sophisticated evaluation process.

The field data was measured in urban areas of Singapore in which the use of HVAC systems is extensive. Despite the heterogeneous sites in Singapore, the refined UWG is able to produce good estimates of urban air temperatures and anthropogenic heat fluxes at a neighbourhood resolution when compared with the data from observation and literature, respectively. In addition, the simplifications and assumptions in the UWG model prevent it from capturing some very site-specific microclimate phenomena, but make it sufficient to explore various urban morphology. If one is particularly interested in highly detailed microclimate effects or spatial UHI distributions within a neighbourhood, more advanced computational simulations would be needed.

Perhaps the most interesting and also surprising discovery from the Singapore case is that the location of the rural weather station has a minimal impact on the simulation results and the dependence of the UHI magnitude on city size is quite small. Bueno et al. (2014) suggested an explanation that the energy balances of the urban canopy layer and urban boundary layer are weakly influenced by the advection effect. This argument, if proven to be true in a general context, would make the UWG fairly robust so that the choice of nearby rural weather station is, well, not that crucial for the analysis. Since this is the only time in all the tests that such discussion occurred, statistically speaking, we should revisit this issue in any future cases if sufficient resources are available.

12.4.3 Abu Dhabi

The case of Abu Dhabi (UAE), in partnership with the Masdar Institute of Science and Technology, extended the validation to tropical desert climate zone with highly heterogeneous urban sites (Mao 2018). It was also used to test some more advanced simulation-based techniques (Mao et al. 2017, 2018; Santos et al. 2018).

The comparison between the measurements and predictions indicates that the newest UWG (as of 2019) can roughly capture the UHI pattern and can robustly approximate the thermal behaviour of the urban microclimate system in Abu Dhabi for different seasons. Again, due to simplified parameterizations, the model is not able to significantly improve beyond the spatially and functionally averaged results when applied to a particular urban site. A numerical solution of the momentum equation would be required if the decision is associated with very detailed spatio-temporal microclimate effects. Nevertheless, the climate-mapping capability of the UWG is still useful since it helps to define the morphological parameters of the distinct neighbourhoods within a city and allows physics-based analysis of the building energy performance distribution in spatially heterogeneous cities. Together with previous studies in Europe and Singapore, the UWG can be applied to different climate zones and urban configurations to yield an estimation of the UHI effect for design purpose.

Given the promise of UWG's satisfactory performance with exceptionally low computational requirement, perhaps for the first time, we attempted to enjoy some sophisticated simulation-based techniques in the urban microclimate realm. Specifically, the Monte Carlo-based sensitivity analysis (Mao et al. 2017) and heuristic optimization-based model calibration (Santos et al. 2018; Mao et al. 2018) have been empirically demonstrated to automate the UHI modelling procedure in the Abu Dhabi case. Preliminary results show that the UWG has a catalytic effect in obtaining good solutions even without parallel computing, and hence opens up a potentially broad spectrum of new engineering applications. We hope to witness the power of UWG in simulation-based analysis among interested communities in the near future.

Finally, one appealing takeaway arising from the sensitivity analysis is that the most critical parameters that impact the urban microclimate behaviours in Abu Dhabi are the reference height of the VDM, the UCL-UBL exchange coefficient, the fraction of waste heat into canyon and the night-time urban boundary layer height. Ironically, these parameters remain the most uncertain among all the input parameters of UWG. Their values in specific urban site are often obtained from mesoscale atmospheric simulations, existing literature and/or well-educated guesses since exact observations are hardly available. To see this as a trade-off, the accuracy of the UWG is sacrificed to some extent by simplified parameterizations, which are usually detailed in numerical simulations, with a goal to significantly improve computational efficiency. In order to reduce the uncertainty in modelling the UHI effect with UWG, some particular attention should be paid on these parameters.

12.5 Conclusion and Beyond: Maybe We Are Wrong

During the past decade, *Homo sapiens* has evolved into *Homo urbanus*. The vast bulk of resources that drive modern life are consumed within the buildings constructed in dense urban settlements. As decision makers associate with performance-oriented building design and urban planning, a complex energy interaction between buildings and urban climate needs to be considered carefully given the growing concern of urbanization. One common problem in practice is that there is no weather station near the project site. In order to address this, a compelling case for the use of computer simulations is devised toward comprehensible design strategies.

Bridging the cultures that are often distant, the Urban Weather Generator (UWG) combines expertise from the diverse fields of building physics and urban climatology to create a new physics-based paradigm for energy and environmental computing in an urban context. UWG is configured to be effective and efficient, and questions the notions generally held to be the "physical laws" by design policy practitioners. On a personal note, this has reminded us of the importance of seeking out and appreciating work in other disciplines, even if it seems to lie outside our normal purview. It turns out that many others wanted exactly the same thing and picked up the UWG to pool their own collective knowledge.

Built on four physics-based models with simplified parameterizations, the UWG is able to quickly estimate the microclimate conditions at the urban street level by processing the meteorological information from public rural weather files. It can also be used as a bottom-up model to evaluate the major energy consumption within the specific urban site by aggregating the building stocks. With continuous refinements on the underlying mechanism and validations among different cities, the UWG shows that one can achieve acceptable simulation performance without incurring expensive machine computation. Some recent development of simulation tools further improves the user convenience and interactivity for scientific modelling.

Of course, UWG is still under development. Perhaps one potential direction would be to understand the latent heat exchange within the UWG mechanism. The humidity calculation requires a full consideration of the moisture from vegetation, soil, nearby bodies of water, etc., and these latent components have not been precisely characterized in the current UWG. Thus, in the UWG simulation (as of 2019), Yang (2016) assumed that the absolute humidity in the urban area is the same as in the rural site. It is also worth noting that the latent heat balance has never been empirically validated in our tests. In the current treatment, only a prescribed latent energy fraction for specific component is considered. We believe that formalizing a holistic latent heat network within the four models is an important open question.

In addition, the performance of building and urban systems, which is central to the built environment, is a critical issue in which the UWG actually reinforces a lack of appreciation of complexity. That is, it decodes complexity through simplification, rather than interpretation. To see this, the current UWG assumes that the urban weather interacts with the rural weather via the VDM-UBLM scheme, while the air in the UBL is simplified to be well mixed within the control volume. Furthermore, the energy exchange between the UBL and UCL is also simplified to be essentially adiabatic but joined with an exchange velocity (or exchange coefficient). Such simplified parameterization is indeed useful during an early design stage if one is not too finicky, but certainly will not be enough for advanced analysis to deeply comprehend the system. Given the trade-off between interpretation and computation, a hierarchical or parallel model synthesis might be entertained in the UWG architecture for different purposes in the future.

Finally, just because one has found a compelling mechanism to capture the UHI effect does not mean that there are no other, maybe better, paradigms. For example, the VCWG (Moradi et al. 2019) revises the original VDM-UBLM-UCM scheme and couples two VDMs to simulate the interactions between rural and urban weather. This treatment makes more physical sense and allows some more spatiotemporal estimations within the urban canyon. Given the close relationship between these two simulation paradigms, it is not clear at this moment which one is better; it might be that in seemingly similar situations slightly different perspectives prevail. In practice, it seems reasonable to use whatever makes it easier to obtain results, but this runs the risk of being physically inaccurate. Nevertheless, the UWG offers a hypothesis of simulating the urban microclimate. Some treatment might be inaccurate and, well, even wrong. Interested reader is welcomed to examine the mechanism and maybe create a new prototype.

Nomenclature

A	Area, m^2
A_f	Lateral heat exchange area, m^2
C	Capacitance/conductance, W s K^{-1}/W m^{-2} K^{-1}
C_k	Model parameter related to the diffusion coefficient
c_p	Specific heat at constant pressure, J Kg^{-1} K^{-1}
c_v	Specific heat at constant volume, J Kg^{-1} K^{-1}
d	Thickness, m
E	Turbulent kinetic energy, m^2 s^{-2}
g	Gravity acceleration, m s^{-2}
h	Heat transfer coefficient, W m^{-2} K^{-1}
H_{rur}	Rural sensible heat flux, W m^{-2}
H_{urb}	Urban sensible heat flux, W m^{-2}
K_d	Diffusion coefficient, m^2 s^{-1}
l_k	Length scale, m
Q	Heat flux, W m^{-2}
Q_{surf}	Sum of net radiation, sensible, and latent heat fluxes at the surface, W m^{-2}
R	Resistance, K W^{-1}

t	Time, s
T	Temperature, K
T_{deep}	Annually average ambient air temperature of the site, K
U	Window U-factor, W m^{-2} K^{-1}
u_{circ}	Circulation velocity, m s^{-1}
u_{exc}	Exchange velocity, m s^{-1}
u_{ref}	Reference air velocity, m s^{-1}
V	Volume, m^3
$\dot{V}$	Volume flow rate, m^3 s^{-1}
z	Vertical space component, m
z_i	Boundary layer height, m
z_r	Blending height, m
z_{ref}	Reference height, m
θ	Potential temperature, K
θ_{ref}	Reference potential temperature outside the control volume, K
ρ	Density, kg m^{-3}
$(\rho c)_l$	Volumetric heat capacity of the layer l, J m^{-3} K^{-1}

Abbreviations

AEC	Architecture, engineering and construction
API	Application programming interface
CHTC	Convective heat transfer coefficient
DOE	Department of Energy
EPW	EnergyPlus Weather
HVAC	Heating, ventilation and air conditioning
MIT	Massachusetts Institute of Technology
RC	Resistance-capacitance
RSM	Rural station model
TEB	Town Energy Balance
UBLM	Urban boundary layer model
UC-BEM	Urban canopy and building energy model
UCL/M	Urban canopy layer/model
UHI	Urban heat island
UWG	Urban Weather Generator
VCWG	Vertical City Weather Generator
VDM	Vertical diffusion model

Acknowledgements Over the years, this work has received financial support from the Singapore National Research Foundation through the Singapore-MIT Alliance for Research and Technology (SMART) Center for Environmental Sensing and Modeling (CENSAM), the MIT and Masdar Institute Cooperative Program, the Energy Efficient Buildings Hub of the US Department of Energy (DOE), the MIT Energy Initiative, the MIT Presidential Fellowship, the Leon Hyzen Fellowship and the George Macomber Chair Scholarship.

A special thank you goes to Chris Mackey (at Ladybug Tools LLC) for actively leading the development of the UWG Python version and Dragonfly over the years, as well as for his brilliant comments and fruitful supports during the writing of this chapter.

Finally, we would like to thank all the people who have helped us reach this far. The UWG would not have been possible without your enthusiasm and participations, which allow for potential technical updates and also make it more fun.

References

Augenbroe, G. (2012). The role of simulation in performance-based building. In *Building performance simulation for design and operation* (pp. 15–36). London: Routledge.

Awino, H. R. A. (2019). *Design-integrated Urban Heat Island analysis tool and workflow: Development and application*. Master's thesis. Massachusetts Institute of Technology.

Bueno, B. (2012). *Study and prediction of the energy interactions between buildings and the urban climate*. PhD thesis. Massachusetts Institute of Technology.

Bueno, B, Pigeon, G, et al. (2012a). Development and evaluation of a building energy model integrated in the TEB scheme. Geoscientific Model Development 5(2), pp. 433–448.

Bueno, B., Norford, L. K., et al. (2012b). A resistance-capacitance network model for the analysis of the interactions between the energy performance of buildings and the urban climate. *Building and Environment, 54*, 116–125.

Bueno, B., Hidalgo, J., et al. (2013a). Calculation of air temperatures above the urban canopy layer from measurements at a rural operational weather station. *Journal of Applied Meteorology and Climatology, 52*(2), 472–483.

Bueno, B., Norford, L. K., et al. (2013b). The urban weather generator. *Journal of Building Performance Simulation, 6*(4), 269–281.

Bueno, B., et al. (2014). Computationally efficient prediction of canopy level urban air temperature at the neighbourhood scale. *Urban Climate, 9*, 35–53.

Deru, M. et al. (2011). *US Department of Energy commercial reference building models of the national building stock*. Technical report. National Renewable Energy Laboratory.

Grimmond, C. S. B., Roth, M., Oke, T. R., et al. (2010). Climate and more sustainable cities: Climate information for improved planning and management of cities (producers/capabilities perspective). *Procedia Environmental Sciences, 1*, 247–274.

Mao, J. (2018). *Automatic calibration of an urban microclimate model under uncertainty*. Master's thesis. Massachusetts Institute of Technology.

Mao, J., Yang, J. H., et al. (2017). Global sensitivity analysis of an urban microclimate system under uncertainty: Design and case study. *Building and Environment, 124*, 153–170.

Mao, J., Fu, Y., et al. (2018). Optimization-aided calibration of an urban microclimate model under uncertainty. *Building and Environment, 143*, 390–403.

Masson, V. (2000). A physically-based scheme for the urban energy budget in atmospheric models. *Boundary-Layer Meteorology, 94*(3), 357–397.

Moradi, M., et al. (2019). The vertical city weather generator (VCWG v1.0.0). *Geoscientific Model Development Discussion*. https://doi.org/10.5194/gmd-2019-176.

Nakano, A. (2015). *Urban weather generator user interface development: Towards a usable tool for integrating urban heat island effect within urban design process*. Master's thesis. Massachusetts Institute of Technology.

Oke, T. R. (1973). City size and the urban heat island. *Atmospheric Environment, 7*(8), 769–779.

Sailor, D. J. (2011). A review of methods for estimating anthropogenic heat and moisture emissions in the urban environment. *International Journal of Climatology, 31*(2), 189–199.

Santos, L. G. R., et al. (2018). Evaluating approaches for district-wide energy model calibration considering the urban heat island effect. *Applied Energy, 215*, 31–40.

Stull, R. B. (2012). *An introduction to boundary layer meteorology* (Vol. 13). New York: Springer Science & Business Media.

Wetter, M. (2012). A view on future building system modelling and simulation. In *Building performance simulation for design and operation* (pp. 481–504). London: Routledge.

Yang, J. H. (2016). *The curious case of urban heat island: A systems analysis*. Master's thesis. Massachusetts Institute of Technology.

Chapter 13
The SOLENE-Microclimat Model: Potentiality for Comfort and Energy Studies

Marjorie Musy, Marie-Hélène Azam, Sihem Guernouti, Benjamin Morille, and Auline Rodler

13.1 Introduction

Conducting full-scale experiments in the actual urban context is a difficult task. This is why simulation is needed to test planning and urban design proposals and to predict the impacts of urban developments on local climate and building thermal behavior.

It is necessary therefore to build models that are robust enough and capable of simulating realistic urban settings, including buildings. When examining urban comfort, the scale considered is that of the immediate environment, i.e., street and square. Urban form is the most important parameter that acts on the local climate. It is the reason why its explicit description (including buildings, trees, etc.) is a necessary input into microclimate models. The primary outputs of the models are in return their impacts on sunlight, wind, air temperature, and humidity as well as surface temperature.

SOLENE-microclimat was developed to assess the impact of urban planning on both outdoor and indoor comfort taking into account the interactions between a building and its environment. This requires resolving scale compatibility issues

M. Musy (✉) · S. Guernouti · A. Rodler
Buildings Performance in their Environment (BPE) Research Team, Cerema Ouest & Institut de Recherche en Sciences et Techniques de la Ville (IRSTV), FR CNRS 2488, Nantes, France
e-mail: marjorie.musy@cerema.fr; sihem.guernouti@cerema.fr; auline.rodler@cerema.fr

M.-H. Azam
UMR CNRS 6£183, GeM, Univeristé de Nantes & Institut de Recherche en Sciences et Techniques de la Ville (IRSTV), FR CNRS 2488, Nantes, France
e-mail: marie-helene.azam@univ-nantes.fr

B. Morille
Soleneos & Institut de Recherche en Sciences et Techniques de la Ville FR CNRS 2488, Nantes, France
e-mail: benjamin.morille@soleneos.fr

M. Palme, A. Salvati (eds.), *Urban Microclimate Modelling for Comfort and Energy Studies*, https://doi.org/10.1007/978-3-030-65421-4_13

given the difficulty to accurately represent the urban environment as well as the buildings and their uses within the same tool.

Only a full coupling between building energy models and urban microclimate models allows to:

- Evaluate the reciprocal impact between the urban environment and the thermal behavior of buildings
- Obtain direct and indirect impacts of the selected adaptation techniques
- Better estimate the heat released in the city due to the indoor space thermal control

13.2 The SOLENE-Microclimat Model: Principle

SOLENE-microclimat was first developed for outdoor comfort assessment (Vinet et al. 1999). The addition of new sub-models makes it possible to take into account (1) radiative transfers, including long-wave radiation; (2) conduction and storage in walls and soils; (3) airflow and convective exchanges; (4) evapotranspiration from natural surfaces like vegetation and water ponds or watering systems; and (5) energy balance (energy demand or indoor temperature) for a building in the simulated area. Point 1 corresponds to the historical SOLENE radiative model and can be run alone. Points 1+2 correspond to the thermo-radiative model based on SOLENE (addition of the thermal modeling of different kinds of surfaces). It can also be run independently. Points 1+2+3+4+5 correspond to the so-called SOLENE-microclimat model (Fig. 13.1). For the moment, only the anthropogenic flux due to the heating or cooling of buildings is taken into account.

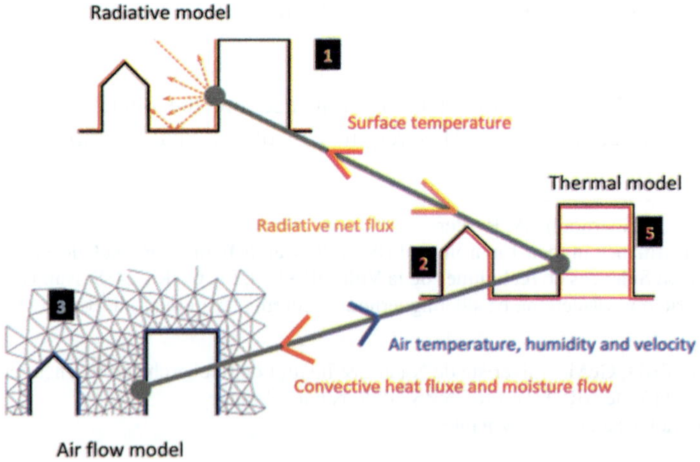

Fig. 13.1 The coupling of the three different models (radiative model = SOLENE, thermal model for urban surfaces and buildings, airflow model = Code_Saturne). Results that are passed from a model to another are represented by colored arrows. Point 4 is not represented; evapotranspiration is taken into account in the energy balance equation used to link all fluxes

This complete coupling of models has not yet been validated, although some comparisons have been attempted by Athamena (2012) on a street canyon composed of containers, showing good agreement for both air temperature and velocity in the street.

In this chapter, we present the modules that compose SOLENE-microclimat and how they can be used to study the direct and indirect impacts of adaptation and mitigation solutions. SOLENE-microclimat consists of the following modules:

- A thermo-radiative module that calculates the thermal behavior and the radiative exchange of urban impervious surfaces (roofs, wall, and pavement), considering their albedo and emissivity properties. This model allows for assessing the impact of cool materials.
- Modules that compute the evaporation and heat fluxes for water ponds, vegetation (trees, lawns, green roofs, green walls), and pavement watering.
- Modules that are used to assess the impacts (thermal comfort assessment and building energy simulation).

For each sub-module, the model it is based on is first described. Then, its validation process when performed is presented. The confidence and credibility of a physical model lie on its validation. However, validation is a complex task when trying to do it for the whole model. It is why a systemic approach was adopted. Most modules implemented in SOLENE-microclimat have been assessed separately instead of assessing the whole model on a variable that integrates all the phenomena such as the atmosphere temperature, as often done. We have considered that this last method would not allow us to conclude on the model's ability to accurately represent each phenomenon. Surface temperatures and temperatures in materials can be accurately measured and their variability in an urban scene allows to assess the model for different conditions (radiative characteristics, materials …) and then to conclude on the submodels' accuracy. This chapter describes the equations and assumptions that are made, their verification, and the validation steps. This valuable information is intended to support the work of researchers and practitioners interested in this model to fully understand its architecture and functioning as well as developers of other tools.

13.3 The Thermo-Radiative Model

13.3.1 Model Description

The *SOLENE-microclimat* simulation tool calculates the solar and luminous effects, due to urban forms and materials and long-wave radiation, conduction, and storage in built surfaces. The model is based on a 3D representation of urban geometry (Fig. 13.2).

Fig. 13.2 3D model of small district in SOLENE (with sun projections over a 1-day period)

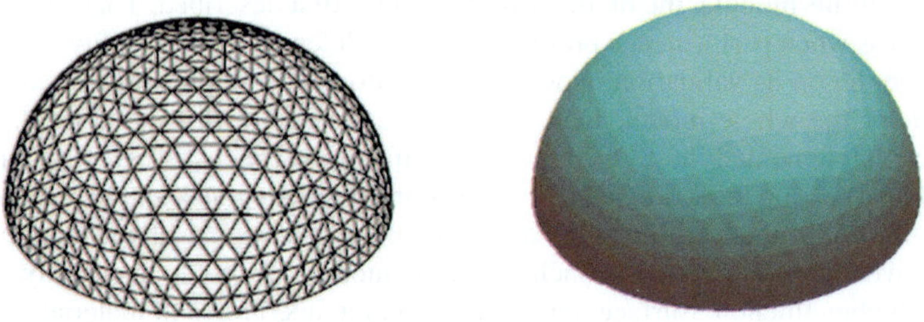

Fig. 13.3 Sky vault representation

The simulated scenes are modeled by the meshing of the 3D developed surface with rectangular or triangular elements, whose area depends on the size of the simulated scene. An additional hemispherical geometry represents the sky vault, which is meshed to take into account the anisotropy of the diffuse solar radiation (Fig. 13.3). The radiation spectrum is divided into two bands corresponding to solar radiation (short wave, SW) and infrared thermal radiation (long wave, LW). The direct solar radiation is emitted by the sky mesh in the sun direction, while the diffuse solar radiation emission is distributed over all sky vault facets by a sky radiance model (Perez et al. 1987). The sky thermal radiation is assumed to be isotropic. These data can be replaced by measurement data if available.

At each time step, the SW radiation absorbed by each facet (R_{n_SW}) is computed as the result of the direct and diffuse SW radiation received directly from the sun and sky (resp. R_{direct_SW} and $R_{diffuse\_SW}$), plus reflections from all other facets in view (R_{IR_SW}), weighted by the absorption coefficient of the surface (Eq. 13.1):

$$R_{n-SW} = \alpha_i \left(R_{direct-SW} + R_{diffuse-SW} + R_{IR-SW} \right) \tag{13.1}$$

These multiple reflections are computed according to a radiosity method that requires (1) computing geometric form factors between all facets of the built surface and the sky vault using contour integral technique (Miguet and Groleau 2002; Vinet 2000), and (2) assigning radiative properties (reflection, transmission, and absorption) to all surfaces of the scene.

The form factors are also used to calculate the net LW radiative with the surrounding surface (noted $R_{\text{n-LW}}$) (Eq. 13.2) and with the sky (noted $R_{\text{n-LW,sky}}$):

$$R_{\text{n-LW}} = \sum_{p}^{j=1} \sigma F_{ij} \left(\varepsilon_j T s_j^4 - \varepsilon_i T s_i^4 \right) \tag{13.2}$$

Then the net radiative flux is the sum of SW and LW net fluxes (Eq. 13.3):

$$R_{\text{n}} = R_{\text{n-SW}} + R_{\text{n LW}} + R_{\text{n-LW,sky}} \tag{13.3}$$

These fluxes are integrated into the energy budget that is calculated for all facets, writing the balance of net radiative flux, convective heat flux, and heat flux by conduction through materials behind the facet (Eq. 13.4):

$$Q_{\text{cond}} = R_{\text{n}} + H - \text{LE} \tag{13.4}$$

The convective heat flux to this atmosphere is computed using surface-to-air temperature difference and a convective heat transfer coefficient (CHTC), which can be a constant value or dependent on wind velocity at a reference point using the law proposed by Jayamaha et al. (1996) (Eq. 13.5). The 1D conductive heat flux through the walls is computed by means of a multilayer model including heat storage in the layers, conduction through the layers (as a function of the material thermal properties), and convection between the wall and an air node inside the building. This temperature is assigned in relation to indoor control temperature (Fig. 13.4). Heat transfer through ground layers and into the deep ground is computed similarly:

$$h_{\text{a}\infty} = 5.85 + 1.7 \cdot V_{\text{a}} \tag{13.5}$$

For applications focusing on soil surface temperature, for example, to assess the impact of local solutions such as watering, it appeared that the rough soil model that is usually used in SOLENE-microclimat lacked accuracy. An alternative detailed soil model has then been designed. Both of these soil models calculate the heat transfer through the layers but neglect the moisture transfer. The difference is that the detailed model is solved by a finite difference method with a nonuniform mesh, finer at the surface than deep in the ground (Fig. 13.5). At the surface, the upper boundary condition is defined by the surface energy balance. In this model, the convective heat flux method has been modified, to take into account mixed convection. Indeed, we have established that it was necessary when wanting to achieve a sufficient accuracy for surface temperature calculation whatever the wind regime. A detailed description of this soil model can be found in Azam et al. (2018b).

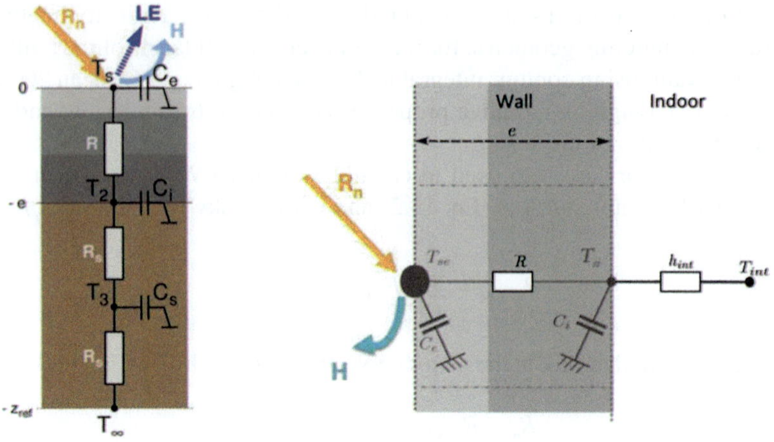

Fig. 13.4 Wall model (right) and ground models (left)

Fig. 13.5 Detailed soil model. Each color from gray to brown represents the soil layers' material variation. The capacity is estimated for each half volume above and under a capacity node

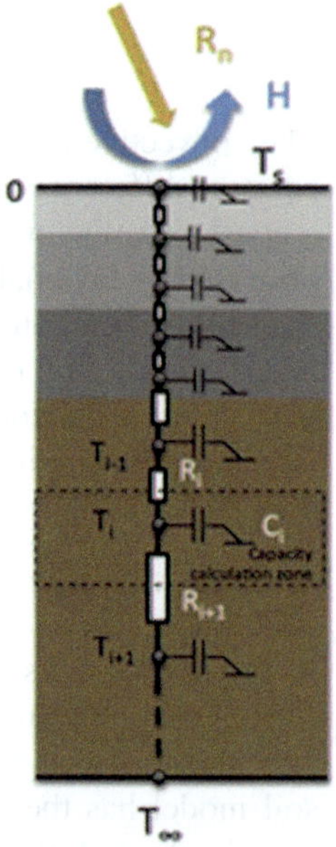

The external surface temperature of each facet (soils, buildings, etc.) is tied to that of all the other facets via long-wave radiation calculation. It results in a nonlinear system of equations that is solved by a Newton-Raphson iterative scheme at each time step.

13.3.2 Model Assessment

Hénon et al. compared simulation results carried out with this thermo-radiative model to temperatures obtained from an airborne thermal camera at facet and pixel scales (Hénon et al. 2012a, b). Measurements were acquired during two intense observation periods in summer and winter over the center of Toulouse (France). The spatial resolution (pixel size at ground) of the thermal images ranging between 1.5 and 3.0 m for zenith viewing angles between 48° and 62° allowed a manual point-to-point comparison with the temperatures calculated for the triangular mesh elements of the model, the dimension of which was about 2 m^2.

Two-thirds of pixels were simulated to within 5 °C and the other third to within 2.5 °C from actual measurements. Differences were mainly attributed to the fact that some surface types were not present in the model. Indeed, the detailed data of all surfaces in the scene were not readily available, in particular inner layers of walls and indoor temperature. Moreover, numerous smaller details cannot be represented.

This second validation step used selected measurement data acquired during the FluxSap campaign (Mestayer et al. 2011). The campaign was carried out over 3 months at a large district scale and we focused on the data acquired on a five-story building built at the beginning of the 1960s and on surrounding lawns (Fig. 13.6). The kind of measurement will be given for each surface in the following. After a survey of the surface materials, Malys (2012) compared the measured and simulated temperatures for a group of points (on a roof, on a wall, and in the ground) over several days.

For simulations, we used the weather data acquired on the roof of the building (air temperature, velocity, and humidity, SW and LW radiations). For SW radiation, as we only have the global value (diffuse and direct), when the total flux is higher than 50% of the theoretical total flux calculated for a clear sky, the flux is distributed between diffuse and direct according to Perez model for a clear sky. When it is under 50%, the flux is split into two equal parts.

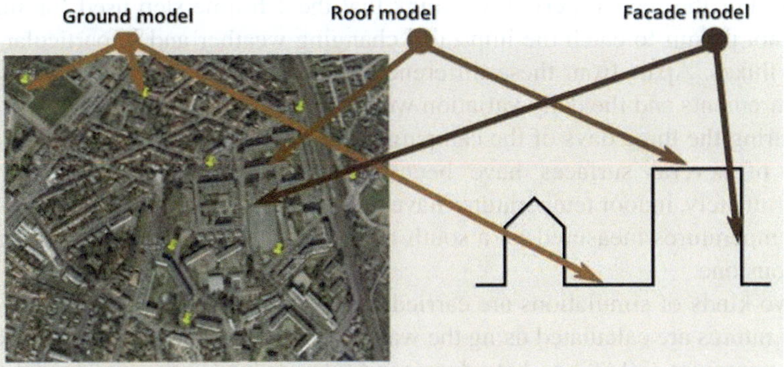

Fig. 13.6 Measurement points and their use for validating the SOLENE-microclimat model surfaces and soil temperature prediction. (Adapted from Malys 2012)

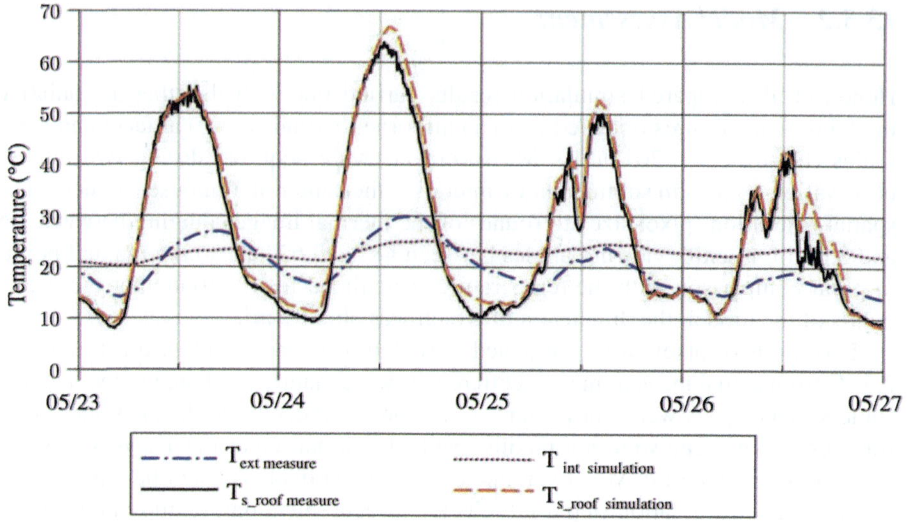

Fig. 13.7 Comparison of measured and simulated roof surface temperature. (Adapted from Malys 2012)

Eight thermocouples were set on this building's roof measuring its temperature over 1 week. The amplitude of the surface temperature daytime variation reached 50 °C during sunny days. For the comparisons, an average of the eight measurements (which were very similar) was used. During the night, the surface temperature was generally lower than the air temperature.

In Fig. 13.7, we compare simulation and measurement. One can verify that there are two short periods during which the model does not well represent the measured values. During the second night, there is a 4 °C difference and at the end of the fourth afternoon there is a 10 °C difference. For the first period, the wind is very weak and the CHTC calculated considering forced regime is certainly underestimated. For the second period, we think that the 1-h time step used for simulation does not permit to catch the impact of changing weather and in particular varying solar fluxes. Apart from these differences, the simulation results are close to the measurements and the daily variation well reproduced.

During the three days of the campaign (May 21st to 24th), the surface temperatures of several surfaces have been manually acquired using a radiometer. Unfortunately, indoor temperatures have not been recorded. For comparison, we use the temperatures measured on a south-facing facade of a building identical to the previous one.

Two kinds of simulations are carried out. For the first one, the building surface temperatures are calculated using the wall model presented above with fixed indoor air temperature (20 °C) as boundary condition of the 1D thermal model for each mesh element. For the second one, we apply the simplified building energy model

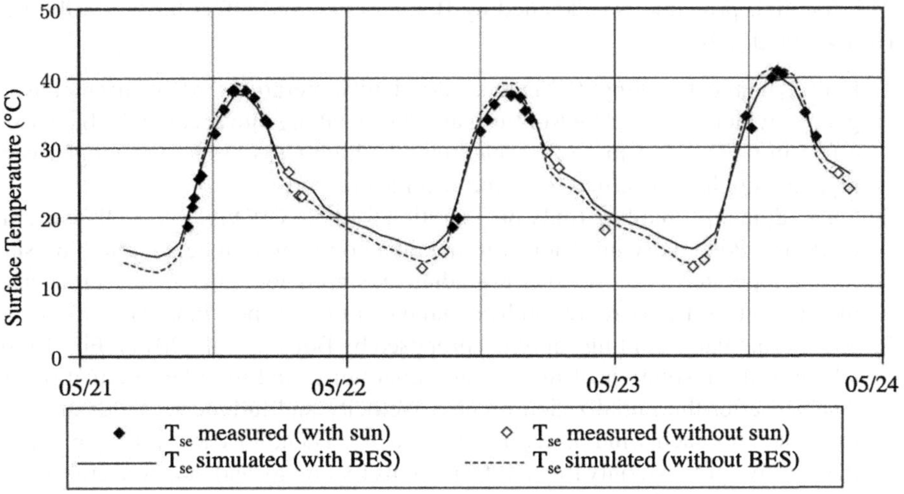

Fig. 13.8 Comparison of measured and simulated facade surface temperature. (Adapted from Malys 2012)

(BEM) that will be presented later (Sect. 13.6.1) and calculate the indoor temperature for each floor (free-running mode). In Fig. 13.8, one can notice that facade's temperatures are lower than the roof's. When applying the BEM, the surface temperature is well assessed when the wall is insolated, but the assessment is worse otherwise. It is the contrary when we do not use the BEM. One explanation could be the simplification that has been done in the BEM when gathering all the internal surface nodes in a unique one for each floor (Bouyer et al. 2011) as it will be shown later in Sect. 13.6.1. However, the model represents well the surfaces' temperature variations.

Finally, the detailed soil model performance has been assessed in an open space (a car park). This case study has been selected to avoid the influence of the surrounding surface on the module evaluation. The simulated temperatures have been compared to the measurements in situ. The mean daily RMSE between the estimated and the observed surface temperature is 0.86, 0.72, 0.58, 0.26, and 0.13 °C, respectively, at the surface, 5, 10, 34, and 50 cm depths. Details can be found in the work of Azam et al. (2018b).

13.4 CFD Coupling

The coupling of the SOLENE model with a CFD model was first performed by Vinet using the N3S CFD code, which has been replaced by FLUENT (Robitu 2005) and later by Code_Saturne (http://code-saturne.org/cms/) (Qu et al. 2012).

The coupling principle was detailed by Bouyer et al. and offers three possibilities (Bouyer et al. 2011):

1. Full dynamic coupling (CFD/thermo-radiative iterations until strict convergence), which was used by Robitu et al. (2006), taking into account the buoyancy effect in CFD: After these first attempts, the high CPU usage was judged very high and could be justified for some specific comfort issues.
2. Quasi-dynamic coupling (only one iteration for both CFD/thermo-radiative simulations): Potentially insufficient to obtain a complete retroactive effect on surface temperatures in the thermo-radiative simulation, this coupling is also insufficient to strictly represent heat and moisture transport in CFD.
3. An intermediate coupling approach proposed by Bouyer et al. (2011) (Fig. 13.9), whereby the resolution of momentum, continuity, and turbulence equations is disabled after the initialization process: Velocity and turbulence fields are preprocessed for each wind direction and velocity. Then, for each time step during the iterative process, only transport equations for energy and moisture are solved. The computational cost is considerably reduced. This intermediate approach is tantamount to assuming that airflow is not disturbed by heat transfer at the wall

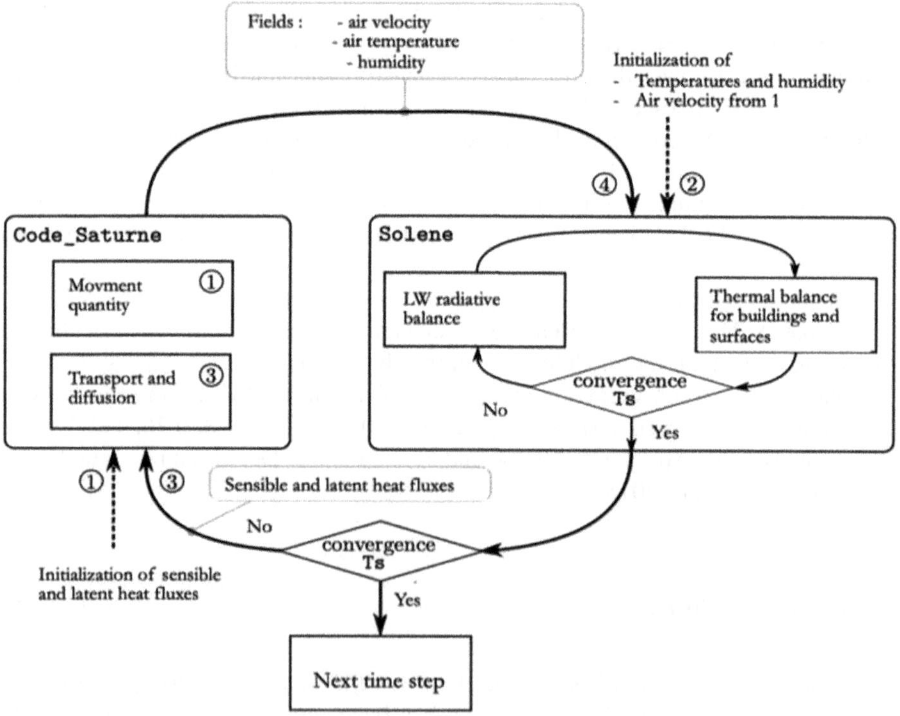

Fig. 13.9 Flowchart of a time step coupling SOLENE and Code_Saturne

surface. Doing that, the impact of insolated walls on airflow will be inaccurate and local recirculation can cause locally high temperatures that would not happen due to buoyancy effect. This will not have a significant effect on buildings' thermal behavior but will affect local comfort assessment.

CFD coupling is also employed to compute the moisture transfer and diffusion of natural surfaces.

In practice, the variables exchanged between the CFD and the thermo-radiative model are as follows:

- From CFD to *the thermal model*:
  - Near-surface air velocity used to calculate the CHTC required for the surface energy balance equation: For this purpose, the linear law derived by Jayamaha et al. has been chosen (Jayamaha et al. 1996) (Eq. 13.5).
  - Near-surface air temperature, which is also required to calculate convective heat flux at the wall surface (in addition to evaluating the sensitive ventilation load).
- From the thermal model to CFD:
  - The convective heat fluxes affecting the energy transport equation and the moisture mass transfer.

Starting from the third coupling method, the impact of coupling with CFD on the assessment of the energy demand of one building in an urban environment has been studied by Malys et al. (2015). In that paper, a sensitivity analysis has been carried out to determine whether the use of local temperature and wind speed was necessary to calculate the convective heat flux. It concludes that for building energy simulation purposes and under winter conditions, thermal coupling with the aerodynamic model in order to represent neighborhood effects can be avoided by using mean values for wind speed and local air temperature. This is less the case during summer since the solar irradiance on surfaces and convective phenomena modify the local temperature in higher proportions.

For winter and energy purposes, a method has been proposed to take into account the impact of the surfaces' temperature on air temperature, and vice versa, in a simple manner. It consists of considering a control volume in which it is assumed that the air temperature is homogeneous. Then, a thermal balance to the air mass contained in this volume is done. The temperature of entering air is considered to be the meteorological temperature, whereas the temperature of exiting air is the mean temperature of the control volume. The airflow entering the volume is calculated by integrating the wind profile (Malys et al. 2015). In this method, the thermo-radiative model is used, without the coupling to the CFD, but a local air temperature can be taken into account.

13.5 Climate Adaptation Techniques

13.5.1 Vegetation

Several green devices have been modeled: trees, lawns, green roofs, and green walls.

Trees are considered as a surface in the radiative model (the tree crown envelop) and as a volume in the CFD model (Fig. 13.10) (Robitu et al. 2006). The impact of trees on the radiative environment is considered taking into account their shading and sun radiation absorption. The tree surface has a solar transmission, reflection, and absorption. It is opaque to long-wave radiation and emits long-wave radiation according to its surface temperature.

In the CFD model, the tree volume contains an aerodynamic resistance. The cells it contains are considered as mixed air/foliage volumes with only one mean temperature for the two entities. This one is calculated as the result of an energy balance equation applied to each tree cell, taking into account radiative fluxes coming from the radiative model (only for cells adjacent to the external surface of the tree), and latent and convective energy exchanges (Bouyer 2009).

Malys introduced green roofs and walls as well as lawns. Unlike trees, these models do not require the addition of a specific entity but are complementary equations added to the wall, roof, and soil models so as to represent the role of foliage in the surface-to-air interface (Fig. 13.11) (Malys et al. 2014). Foliage acts as a partial barrier to radiation. It exchanges with both the surface that supports it (the substrate when present or directly the wall in case of climbing plants) and the surrounding surfaces. It also exchanges convective and latent heat fluxes. For these kinds of surface, the latent heat flux of the substrate is also considered when there is a substrate. As SOLENE-microclimat does not include a hydrological model, this one is

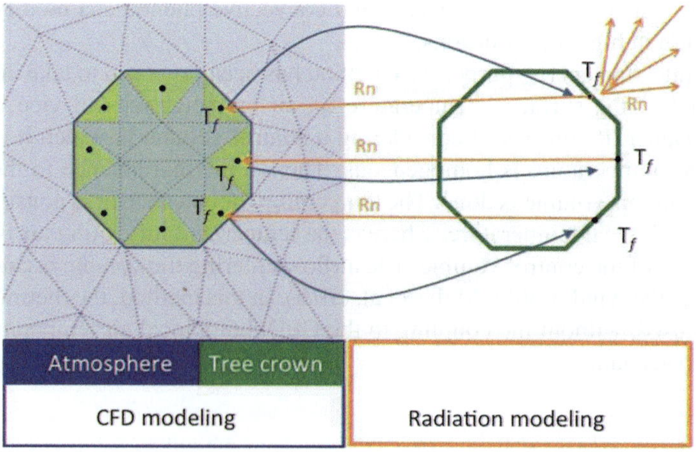

Fig. 13.10 Model used to represent the thermal behavior of trees. (Adapted from Robitu 2005)

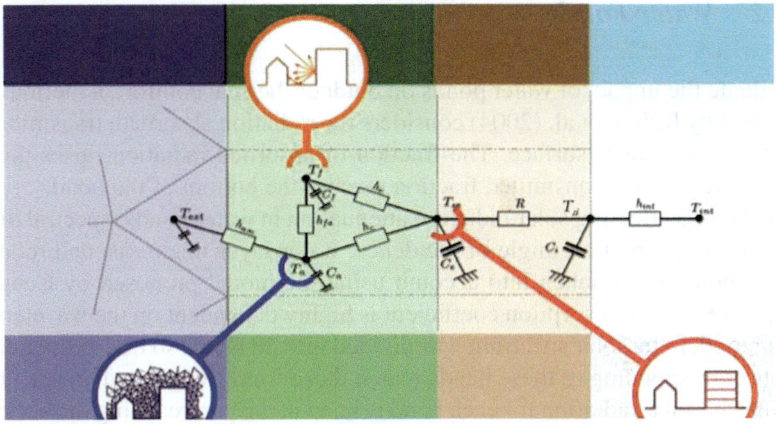

Fig. 13.11 Nodal model used to represent the thermal behavior of green walls and roofs. (Adapted from Malys 2012)

controlled by an evapotranspiration rate (f). It is introduced to consider the occurrence of water stress within the substrate (Eqs. 13.6 and 13.7). The distribution of this flux between the foliage and the substrate is realized through the coefficient α_{lat}:

$$LE_f = \alpha_{lat} f ETP \tag{13.6}$$

$$LE_s = (1 - \alpha_{lat}) f ETP \tag{13.7}$$

The method adopted to calculate evapotranspiration (ETP) is based on the Penman-Monteith equation, as advised by the Food and Agriculture Organization (FAO).

Latent fluxes from vegetation and its substrate are used to calculate the mass rate of moisture released into the air and taken into account in the CFD model.

Malys (2012) drew an initial comparison between the results obtained for green walls and the experimental data recorded by HEPIA, which showed a good agreement.

A comparison has also been carried out for soils covered by lawn as a part of VegDUD project (Musy et al. 2015). The measurement was conducted in the Pin Sec district (in Nantes, France) using Taupe recorders with three sensors at the depths of 0, −5, and −35 cm and at two different locations. The comparison between simulation and measures showed that the simulation reproduces well the surface temperature on sunny days, but also that the model seems to overestimate the nocturnal cooling.

The models used for trees and green roofs have not yet been validated, although temperatures obtained for tree volumes are close to air temperature, as found in the literature.

13.5.2 Water Ponds

To evaluate the impact of water ponds on outdoor thermal comfort, a thermal model developed by Robitu et al. (2004) considers the radiation absorbed, transmitted, and reflected at the water surface. The fraction of absorbed radiation raises the water temperature and the transmitted fraction reaches the bottom of the pond.

The factors affecting solar radiation attenuation in water ponds (spectral distribution of water properties, angle of incidence, water-layer thickness, and reflectivity of pond bottom) are taken into account using the model proposed by Cengel and Özişik (1984). The absorption coefficient is highly dependent on the wavelength; in this submodel, the solar spectrum was divided into 20 bands so that extinction coefficients corresponding to these bands match those from Cengel and Özişik (1984).

Once the solar radiation for each band is known, the part reaching the bottom and the part at various pond depths are summed to obtain the total flux absorbed at each depth in water. Conduction and heat storage in water are solved by using finite difference method (Fig. 13.12).

The water pond model has not been assessed yet, but the impact of such a device has been simulated in an existing place by Robitu et al. (2004). It has been shown that the effect of water ponds is mainly local. The water surface has a local impact on air humidity and temperature. But for comfort assessment, as the radiation budget is not significantly changed for an individual exposed to sun, this impact is very low.

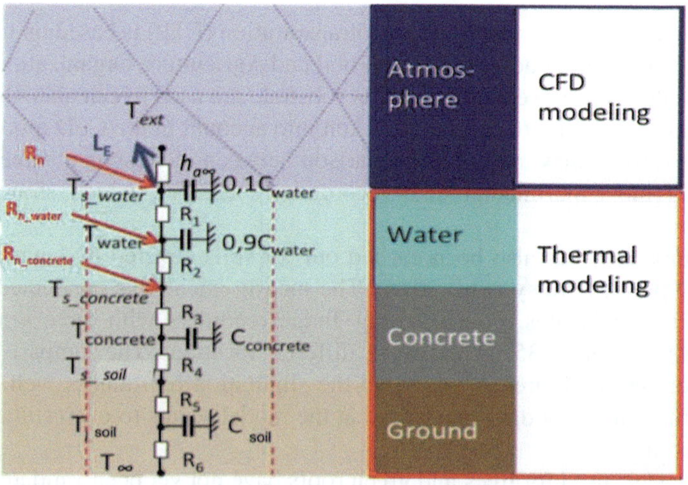

Fig. 13.12 Water pond model illustrated with one intermediate water node

13.5.3 Pavement Watering

A module to simulate the behavior of watered surfaces has been developed. Watering events are modeled through a runoff convective heat flux noted $Q_{\text{wat - pav}}$ and a latent heat flux noted LE (Fig. 13.13). Equation (13.4) is then modified to include these two heat fluxes (Eq. 13.8):

$$Q_{\text{cond}} = R_{\text{n}} + H - \text{LE} - Q_{\text{wat-pav}} \quad (13.8)$$

$$Q_{\text{wat-pav}} = \frac{\rho_{\text{w}} C_{\text{p,w}} \cdot h_{\text{w}}}{\Delta t} \left(T_{\text{s}} - T_{\text{water}}\right) \quad (13.9)$$

The runoff convective heat flux represents the sensible heat exchanged between the soil surface and the water layer (Eq. 13.9). The latent heat flux can then be calculated from the available water height at the surface and the potential latent heat flux (noted LE_{th}). If the potential evaporation water height (noted $h_{\text{evap, th}}$, given by Eq. 13.12) is higher than the available water height (noted h_{w}), all the water present at the surface evaporates. Then Eq. (13.10) is used. However, if the potential evaporation water height is smaller than the available water height, only a part of the water layer evaporates. Then Eq. (13.11) is used:

$$\text{LE} = \text{LE}_{\text{th}} \frac{h_{\text{w}}}{h_{\text{evap,th}}} \quad (13.10)$$

$$\text{LE} = \text{LE}_{\text{th}} \quad (13.11)$$

$$h_{\text{evap,th}} = \frac{\text{LE}_{\text{th}} \Delta t}{L \cdot \rho_{\text{w}}} \quad (13.12)$$

If the simulation time step is small (i.e., 15 min), both heat fluxes are added one at a time as presented in Fig. 13.13. However, for larger time steps (i.e., 1 h), both latent and runoff convective heat fluxes are added at the same time. At each time

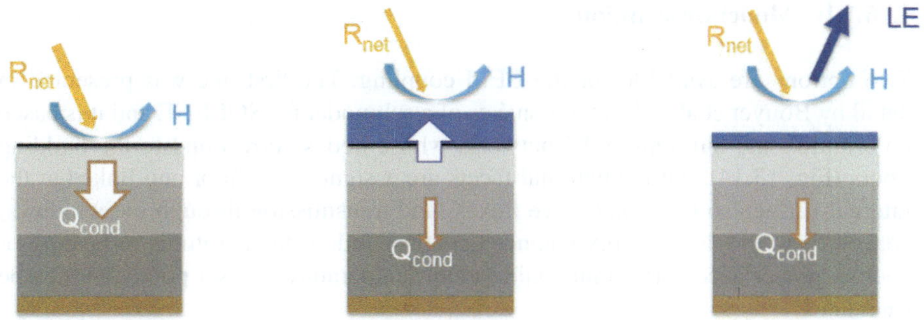

Fig. 13.13 Pavement watering heat flux decomposition

step, the water layer available at the soil surface is monitored for each piece of the mesh.

This watering model has been evaluated using the data from a measurement campaign performed on an asphalt car park during warm days. The measurement campaign reveals that the surface cooling is mainly due to evaporation (80%). However, under warm conditions, the heat flux exchanged between the runoff water and the surface should also be modeled. The mean daily RMSE between estimated and observed surface temperature is 1.04, 0.86, 0.66, 0.35, and 0.21 °C, respectively, at the surface, 5, 10, 34, and 50 cm depths. Details on the model and its assessment can be found in the work of Azam et al. (2018a).

This watering model has also been applied in an urban context, to model pavement watering of a street in the La Part-Dieu district in the city center of Lyon. The model performances are assessed on two hot, clear, and sunny days. Simulated and measured in situ surface temperatures as well as radiation measurements are compared. A RMSE of 2.99 °C is observed, without calibration of the thermal properties of the urban scene. With this model, the effect of pavement watering on soil temperature can be evaluated. In this specific case, an evening watering scenario (17–22 h GMT) decreases the surface temperature by 4.09 °C. Details on this practical example can be found in the work of Azam et al. (2019).

13.6 Assessment Modules

SOLENE-microclimat allows assessing the impact of urban forms, materials, or climate adaptation solutions on both indoor and outdoor conditions. The study of indoor impacts (building's energy consumption or indoor thermal comfort) uses a building energy model while the study of outdoor thermal comfort can be carried out thanks to different outdoor comfort models.

13.6.1 Building Energy Model (BEM)

13.6.1.1 Model Description

Two options are available for the BEM coupling. The first one was presented in detail by Bouyer et al. (2011); it consists of a submodel for SOLENE and it is based on a multizonal building nodal network, whose nodes correspond to the building floors (Fig. 13.14). All the external facets are assigned to a floor and linked at the internal surface so that conductive fluxes (and transmission through the windows) contribute to the floor energy balance equation. Indoor temperatures or heating or cooling power necessary to maintain indoor temperature at a set point can then be calculated.

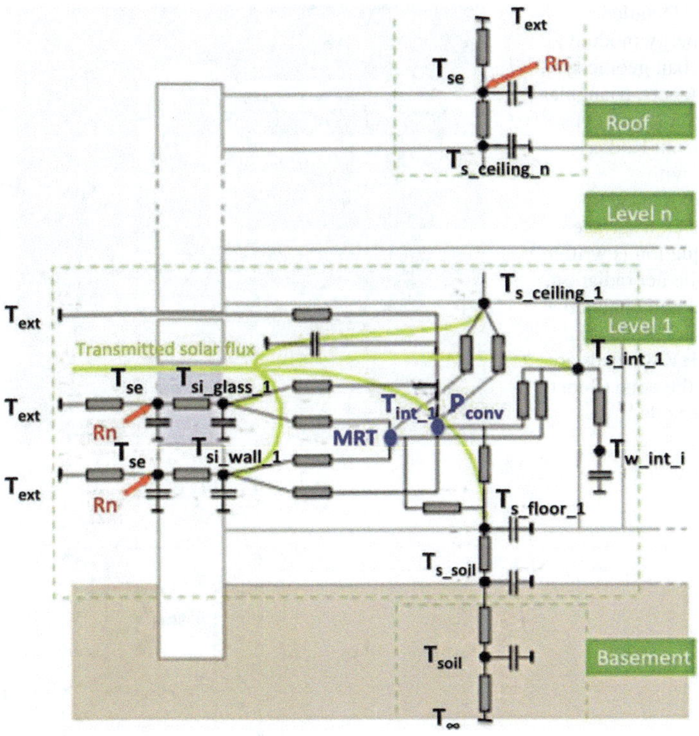

Fig. 13.14 Simplified BEM in SOLENE-microclimat (scheme adapted from Bouyer 2009) resistances and capacities have not been noted in the interest of clarity. It must also be noted that the external nodes are duplicated as many times as the number of external nodes per external surfaces

The second model, called DANA (DetAiled buildiNg simulAtion), has been presented by Rodler et al. (2018). Here, the building's indoor surfaces are differentiated so that the internal surface temperature is calculated for different orientations (Fig. 13.15). In this model, the urban geometry is composed of facets, triangular mesh elements, and cells. To illustrate these terms, some of the facets of the geometry are surrounded in brown in Fig. 13.15. DANA model can represent the dynamic thermal behavior of a single building of any shape using a multizone approach, where each level is a zone. Internal gains and the air leakage through the external envelope are considered. Heating and cooling needs are calculated as well as the air and radiant temperatures for each floor. The external surface temperatures are calculated, with SOLENE-microclimat, for each mesh element of the urban geometry, as we consider the discretized boundary conditions. The thermal conduction exchanges in the walls are still considered unidimensional but the spatial discretization within the walls is refined.

This model can be used as a stand-alone BEM model or can be coupled to SOLENE-microclimat by a weak coupling. The latter consists of, for a time step, launching first the thermo-radiative model whose facade temperatures are used as

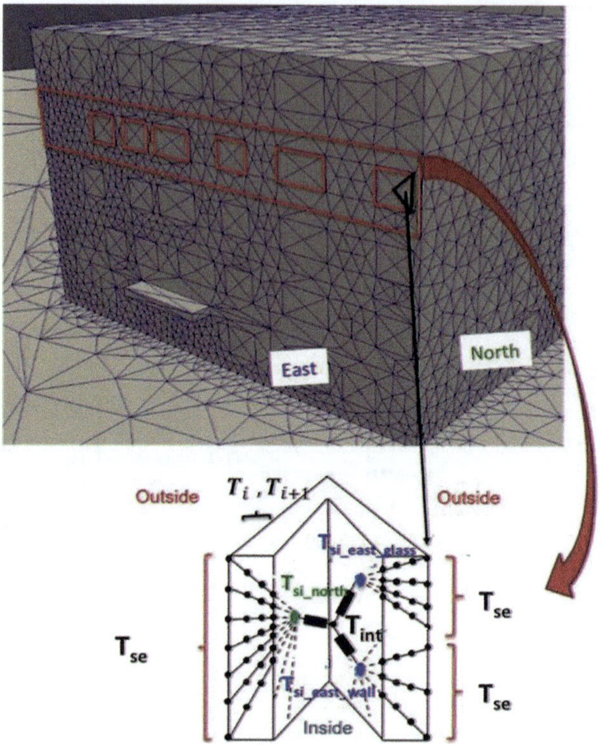

Fig. 13.15 Detailed building energy module DANA: Urban geometry elements: facets, triangular mesh elements, and cells. (Adapted from Rodler et al. 2018 with permission). On each external facet, a surface balance equation is written implying the net radiation, convective, and conductive fluxes, while the internal gains and HVAC loads are applied to the zone (floor) internal air node

initialization to the building energy model DANA. Second, once DANA has finished the resolution of the energy balance, the new calculated buildings outside surface temperatures are used as input to the thermo-radiative function. Then, the resolution can move to the next time step. CFD can be added to the coupling process if the user needs it. If it is the case, the modeler should launch the CFD with the thermo-radiative calculation. In any case, DANA will use all the other fluxes coming from the CFD and the thermo-radiative function, such as long-wave, short-wave, and sensible heat fluxes.

13.6.1.2 Model Assessment

The first approach has not been validated but the walls' and roofs' external surface temperatures were compared to measurements made during the FluxSap campaign as explained above, which led to revising of the model and proposing of a more detailed version.

The results obtained with the second one have been compared to data acquired in the long-term monitoring of a building and the ones that surround it. The comparison was focused on the outside surface temperatures of several buildings within the district, inside air temperatures, and surface temperatures of four flats of the last floor of the central building (which was more intensely equipped with sensors), and its energy demand during winter. They were compared to measures and showed a good agreement (Rodler et al. 2018).

13.6.2 Thermal Comfort Modeling

Several thermal comfort models have been implemented. First, they require calculating the mean radiant temperature (MRT) in the scene. It was performed by Vinet (2000) through defining the so-called bonhomme confort (Fig. 13.16). It is a human representative geometry on which MRT is calculated. For that purpose, all long-wave exchanges between the cylinder representing the human body and all the surfaces, including the sky, are calculated as well as short-wave flux directly from the sun or from the surfaces after multiple reflections with the radiative model presented above. This leads to the calculation of MRT described in the next equation:

$$\mathrm{MRT}^4 = T_1^4 F_{p-1} + T_2^4 F_{p-2} + \cdots + T_n^4 F_{p-n} \tag{13.13}$$

where MRT is the mean radiant temperature of the body, T_n the temperature of the surfaces seen by the body, and F_{p-n} the view angle between the body and the surface seen by the body. This value in turn leads to evaluating various comfort indices: predictive mean vote, PMV* (Robitu 2005); physiological equivalent temperature, PET (Athamena 2012); and universal thermal climate index, UTCI (Weihs et al. 2012).

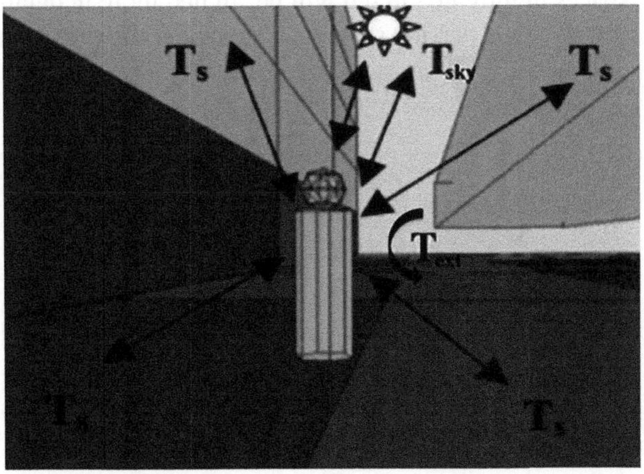

Fig. 13.16 "Bonhomme confort" and the exchanges occurring with its environment

13.7 Applications

Hereafter, the application of the model to the assessment of adaptation strategies' efficiency with regard to outdoor or indoor comfort is described.

13.7.1 Comparison of Different Urban Cooling Strategies Regarding Pedestrian Comfort

The objectives of the EVA project were to compare three different urban cooling strategies: the use of water aspersion systems, vegetation, and high-albedo materials. We present here the results focusing on outdoor thermal comfort.

The assessment of the different cooling strategies was carried out through a set of simulations designed to explore different configurations:

- Cooling strategies were first applied individually to each district so as to compare their impact.
- A composition of the three strategies is then considered trying to optimize their effect (optimized scenario).
- The contribution of each solution to the optimized scenario is finally investigated by deducing it from the optimized configuration (subtractive scenario). This also allows evaluating the cumulative effect of solutions.

The study was applied to three districts in Lyon:

- The Frankfort Square, a multimodal square in front of the La Part-Dieu train station
- The Moncey buildings' block (part of a residential district built in the 1960s)
- The Buire street (part of a recent residential district)

The three districts and the optimized scenarios are presented in Fig. 13.17. The spaces highlighted in red will be studied more in detail regarding pedestrians' comfort. Water aspersion systems are the blue surfaces, vegetation (trees, green walls, or

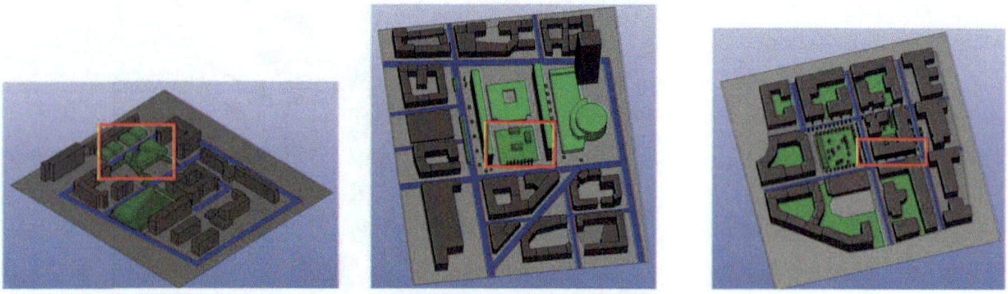

Fig. 13.17 Representation of the optimized scenario applied in the Frankfort district (left), the Moncey district (center), and the Buire district (right)

roofs), and the green ones. The materials are classified based on three albedo levels: low albedo: 0.15; medium albedo: 0.4; and high albedo: 0.8. To produce the high-albedo scenario, for each surface family (roads, buildings, roofs), the albedo value is modified from its actual estimated level to the immediately higher albedo value category.

The simulations are performed using the meteorological data from the heat wave that occurred in July 2003 (26th).

13.7.1.1 Influence on Air Temperature

The results show a reduction of the air temperature at pedestrian height (1.5 m) for each cooling strategy applied individually (Fig. 13.18). High-albedo materials slightly reduce the air temperature but the effect is more widely spread. The reduction induced by the water aspersion systems is higher but concentrated near the space where they are applied. Vegetation also significantly reduces the air temperature in the space where it is applied. It can also induce a slight air temperature increase because of the reduction of the wind speed, which modifies convective exchanges and thus surface temperature. Similar results were observed for all the districts.

13.7.1.2 Influence on Thermal Comfort

The influence of the cooling strategies on thermal comfort is investigated by analyzing the optimized scenario and the subtractive ones (Fig. 13.19). In Fig. 13.19, the dark line with dark points presents the evolution over the day of the average UTCI values for each studied area.

The bars give the contribution of each cooling solution. For each cooling solution, it has been calculated by comparing the UTCI from the optimized scenario with the one from the scenario in which this solution has been deleted. The dark crosses represent the sum of the contributions.

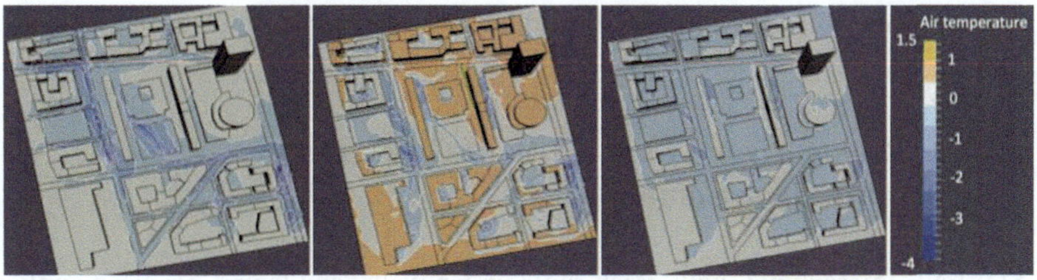

Fig. 13.18 Reduction of air temperature for each cooling strategy applied in the Moncey district at midday. Water aspersion (left), vegetation (center), high-albedo materials (right)

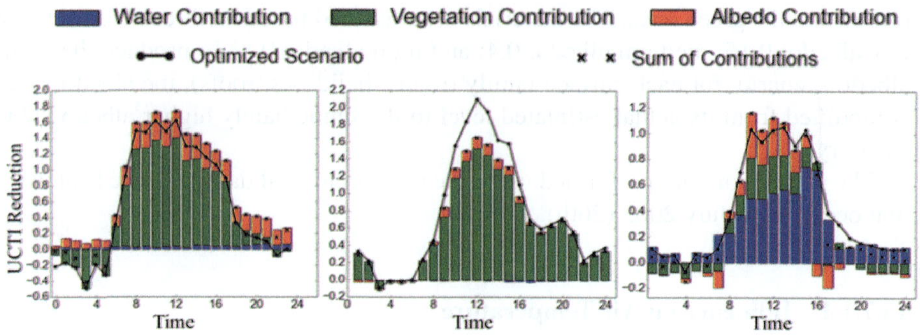

Fig. 13.19 Reduction of UTCI mean values for the studied area in each district: Frankfort (left), Moncey (center), and Buire (right)

The crosses do not match with the points. This means that the total UTCI reduction cannot be considered as the sum of every single effect: the effects are not completely additive. Secondly, the contribution of each cooling strategy as well as its efficiency is different depending on the district. The vegetation has a higher contribution to the optimized solution, especially in the Frankfort place and the Moncey district. As previously presented, vegetation (and in particular trees) can induce a negative influence. Even if they have a positive influence on average where they are applied, in the three cases, trees induce negative effects at nighttime (not visible in Moncey district where few trees were implemented) because they reduce the cooling by infrared radiation exchanges with the sky. Thirdly, the water contribution is low in two of the three cases because it has been implemented on a few surfaces and at such a distance from the studied place that one can only observe the indirect impacts. Lastly, albedo value increase can have a positive or a negative influence, depending on the hour of the day and even more on the urban configuration.

The results of these investigations show that the efficiency of the cooling strategies is very different depending on the district configuration, on the way they are applied, and on the interactions between them. Effects are not additive; they also vary over the day, and can be negative at some times and places. One must also consider that effects on air temperature must be distinguished from those on thermal comfort that encompass several phenomena (air temperature, humidity, radiation, and wind speed). This confirms the difficulty to produce general rules and the need for tools to specifically study each new configuration.

13.7.2 Comparison of the Effect of Green Devices on Indoor Comfort

SOLENE-microclimat, with green roofs, green walls, and lawns, has been applied to the analysis of summer conditions for Pin Sec district of Nantes city (France) (Malys 2012). The effects of green roofs, green walls, and lawns on indoor comfort

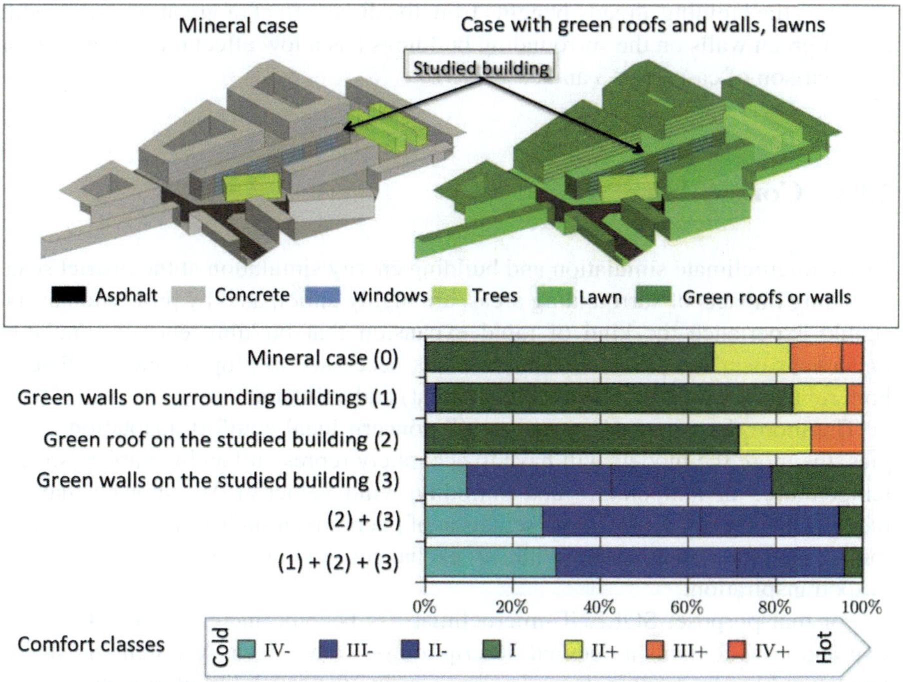

Fig. 13.20 Top: the two extreme cases, case (0) and case (1)+(2)+(3); bottom: the results in terms of percentage of time in each adaptive comfort class over a day

in two kinds of buildings (i.e., insulated or not) taken individually have been compared. First, it has been verified, as often underlined in literature, that green solutions mainly affect non-insulated buildings (Malys et al. 2016). Secondly, it has been demonstrated that the model allowed separating direct and indirect effects of green roofs and walls by simulating successively the implementation of vegetation on the studied building (cases 2 and 3) and on the surrounding (case 1) (Fig. 13.20).

The results are hereafter analyzed for the second floor and in terms of adaptive comfort (Malys et al. 2016). They show that green roofs mainly have a direct effect on the buildings where they are installed, with this effect essentially being verified on the top floor due to the direct shading of the foliage. Comparing reference case to case 2, in Fig. 13.19, one can notice that for the second floor, the comfort is slightly improved. Not represented here, the simulations showed that the indirect effect was very small as the result of a limited effect on air temperature (it has also been verified by de Munck 2013) and due to the fact that buildings are of equal height, so that their roofs have no long-wave radiative effect on the other buildings.

Comparison of reference case, case 1, and case 3 shows that green walls have direct and indirect effects. The direct one is greater due to the effect of shading foliage, while indirect effect is due to the lower radiant flux exchanged with the surrounding buildings.

When the building already benefits from the direct effect of green walls, the addition of green walls on the surrounding buildings has a low effect on indoor comfort (comparison of case 1+2+3 and case 2+3).

13.8 Conclusion

Urban microclimate simulation and building energy simulation at the district scale, including impacts of surrounding areas, are highly anticipated by professionals and should experience the kind of rapid expansion that building energy simulation underwent over the past decades. It is thus necessary to propose models that are both robust (in terms of the cases it can treat) and validated.

The more the expected expertise will concern local comfort adaptation strategies, the more the models will have to accurately represent local variables (surface temperature, air temperature and humidity, wind velocity). We showed that this requires having an accurate representation of phenomena such as mixed convection, long-wave and short-wave radiative exchanges, conduction in materials, and evapotranspiration.

For that purpose, SOLENE-microclimat has become increasingly comprehensive and now contains a detailed description of urban areas. A systematic step of validation has also been instanced so as to ensure the reliability of models.

However, for applications at a larger scale (for instance, the comparison of different master plans) a compromise must be found that considers modeling accuracy, computing capabilities, and data availability. It is the reason why we consider two development strategies:

- The first one consists of continuing to develop accurate submodels that we validate and that will serve as reference to qualify the simpler ones we keep for rougher applications.
- The second one consists of working on models' reduction as already done for the soil model (Azam et al. 2018c) so as to reduce simulation duration while keeping a good accuracy of phenomena.

For the moment, as most of the developers of urban climate models, one difficulty we have to overcome is the lack of reliable input data, at the city scale as well as at the building scale. With the increasing concern about urban climate, the effort put on the coordination between simulation tools and urban databases will provide the documentation necessary to represent urban microclimate characteristics in greater detail. This breakthrough will certainly be very valuable in terms of understanding and knowledge of underlying phenomena, in addition to highlighting the various impacts of urban parameters (the type of buildings, the land use, and the behavior) on climate at different scales.

Nomenclature

$C_a/C_f/C_s/C_e/C_i$	Thermal capacity of air/foliage/soil/external layer/internal layer (J K^{-1})
$C_{water}/C_{concrete}/C_{soil}$	Thermal capacity of water/concrete/soil layer (J K^{-1})
e	Wall thickness (m)
ETP	Reference evapotranspiration (kg m^{-3} s^{-1})
F_{ij}	Form factor of surface j from surface i (–)
f	Evapotranspiration rate (–)
H	Convective heat flux density (W m^{-2})
$h_{a\infty}/h_{int}$	External/internal convective heat transfer coefficient (W m^{-2} K^{-1})
$h_{fa}/h_f/h_c$	Heat transfer coefficient within a green roof or wall, between foliage and air/foliage and wall/air and wall
LE	Latent heat flux density (W m^{-2})
MRT	Mean radiant temperature (K)
P_{conv}	Heating load (W)
$R_{direct\_SW}/R_{diffuse\_SW}/R_{IR\_SW}$	Direct/diffuse/from interreflection, incoming SW radiative heat flux
R_n	Radiative net heat flux density (W m^{-2})
$R_{n\_water}/R_{n\_concrete}$	Radiative heat flux absorbed by the water layer/concrete layer
T_a	Temperature of air within the tree canopy (K)
T_{ext}/T_{int}	External/internal air temperature (K)
T_f	Foliage temperature (K)
$T_{s\_water}/T_{s\_concrete}/T_{s\_soil}/$ $T_{s\_roof}/T_{s\_ceiling}/T_{s\_floor}/T_{s\_wall}$	Temperature of the upper surface of water/concrete/soil/roof/ceiling/floor/wall (K)
$T_s/T_2/T_3/T_\infty$	Temperature of ground from external surface to reference point (K)
T_{se}/T_{si}	Temperature of wall external/internal surface (K)
$T_{water}/T_{concrete}/T_{soil}$	Water/concrete/soil layer temperature (K)
V_a	Air velocity (m s^{-1})
α_i	SW absorption coefficient of the surface (–)
ε_i	Emissivity of the surface (–)
σ	Stefan-Boltzmann constant (W m^{-2} K^{-4})
$Q_{wat\text{-}pav}$	Runoff sensible heat flux density (W m^{-2})
ρ_w	Water density (kg m^{-3})
$C_{p,w}$	Water specific heat (J kg^{-1} K^{-1})
h_w	Water height over the surface (m)
Δt	Simulation time step (h)
L	Latent heat of the vaporization of water (J kg^{-1})

Acknowledgments The development of SOLENE-microclimat has been conducted within the framework of several research projects and Ph.D. theses. We would like to thank ADEME and Région des Pays de la Loire (L. Malys Ph.D., EVA project—contract XXX), Agence Nationale de la Recherche (VegDUD project—contract ANR-09-VILL-0007), Université de Nantes (M. Robitu Ph.D.), CNRS (J. Bouyer Ph.D.), European Community (Built2Spec Project—GA No. 637221-H2020-EeB-2014-2015), and Ecole Nationale Supérieure d'Architecture de Nantes that supported and hosted most of the developments until 2016.

References

Athamena, K. (2012). *Modélisation et simulation des microclimats urbains: Étude de l'impact de la morphologie urbaine sur le confort dans les espaces extérieurs. Cas des éco-quartiers*, Thèse de Doctorat, Ecole Centrale de Nantes & CSTB. Retrieved from https://tel.archives-ouvertes.fr/tel-00811583

Azam, M.-H., Bernard, J., Morille, B., Musy, M., & Andrieu, H. (2018a). A pavement-watering thermal model for SOLENE-microclimat: Development and evaluation. *Urban Climate, 25*, 22–36. https://doi.org/10.1016/j.uclim.2018.04.005.

Azam, M.-H., Morille, B., Bernard, J., Musy, M., & Rodriguez, F. (2018b). A new urban soil model for SOLENE-microclimat: Review, sensitivity analysis and validation on a car park. *Urban Climate, 24*, 728–746. https://doi.org/10.1016/j.uclim.2017.08.010.

Azam, M.-H., Guernouti, S., Musy, M., Berger, J., Poullain, P., & Rodler, A. (2018c). A mixed POD–PGD approach to parametric thermal impervious soil modeling: Application to canyon streets. *Sustainable Cities and Society 42* (1 octobre 2018): 444–61. https://doi.org/10.1016/j.scs.2018.08.010

Azam, M.-H., Rodler, A., Musy, M., Morille, B., & Guernouti, S. (2019). A pavement-watering thermal model validation for SOLENE-microclimate: application to the Buire street. 5th International Conference on Countermeasures to Urban Heat Islands (IC2UHI), Hyderabad, India.

Bouyer, J. (2009). *Modélisation et simulation des microclimats urbains—Étude de l'impact de l'aménagement urbain sur les consommations énergétiques des bâtiments*, PhD thesis. Ecole Polytechnique de l'Université de Nantes.

Bouyer, J., Inard, C., & Musy, M. (2011). Microclimatic coupling as a solution to improve building energy simulation in an urban context. *Energy and Buildings, 43*(7), 1549–1559. https://doi.org/10.1016/j.enbuild.2011.02.010.

Cengel, Y. A., & Özişik, M. N. (1984). Solar radiation absorption in solar ponds. *Solar Energy, 33*(6), 581–591. https://doi.org/10.1016/0038-092X(84)90014-8.

de Munck, C. (2013). *Modélisation de la végétation urbaine et stratégies d'adaptation pour l'amélioration du confort climatique et de la demande énergétique en ville*, PhD thesis. Université de Toulouse, Toulouse.

Hénon, A., Mestayer, P., Lagouarde, J.-P., & Voogt, J. (2012a). An urban neighborhood temperature and energy study from the CAPITOUL experiment with the SOLENE model—Part 1: Analysis of flux contributions. *Theoretical and Applied Climatology, 110*, 177–196. https://doi.org/10.1007/s00704-012-0615-0.

Hénon, A., Mestayer, P., Lagouarde, J.-P., & Voogt, J. (2012b). An urban neighborhood temperature and energy study from the CAPITOUL experiment with the SOLENE model—Part 2: Influence of building surface heterogeneities. *Theoretical and Applied Climatology, 110*, 197–208. https://doi.org/10.1007/s00704-012-0616-z.

Jayamaha, S. E. G., Wijeysundera, N. E., & Chou, S. K. (1996). Measurement of the heat transfer coefficient for walls. *Building and Environment, 31*(5), 399–407. https://doi.org/10.1016/0360-1323(96)00014-5.

Malys, L. (2012). *Évaluation des impacts directs et indirects des façades et des toitures végétales sur le comportement thermique des bâtiments*, PhD thesis. Ecole Centrale de Nantes, Nantes, France.

Malys, L., Musy, M., & Inard, C. (2014). A hydrothermal model to assess the impact of green walls on urban microclimate and building energy consumption. *Building and Environment, 73*(0), 187–197. https://doi.org/10.1016/j.buildenv.2013.12.012.

Malys, L., Musy, M., & Inard, C. (2015). Microclimate and building energy consumption: Study of different coupling methods. *Advances in Building Energy Research*. https://doi.org/10.1080/17512549.2015.1043643.

Malys, L., Musy, M., & Inard, C. (2016). Direct and indirect impacts of vegetation on building comfort: A comparative study of lawns, green walls and green roofs. *Energies, 9*(1), 32. https://doi.org/10.3390/en9010032.

Mestayer, P. G., Rosant, J.-M., Rodriguez, F., & Rouaud, J.-M. (2011). La campagne expérimentale FluxSAP 2010: Mesures de climatologie en zone urbaine hétérogène. *Météorologie (Paris), 1925*(73), 34–44.

Miguet, F., & Groleau, D. (2002). A daylight simulation tool for urban and architectural spaces: Application to transmitted direct and diffuse light through glazing. *Building and Environment, 37*(8–9), 833–843. https://doi.org/10.1016/S0360-1323(02)00049-5.

Musy, M., Malys, L., Morille, B., & Inard, C. (2015). The use of SOLENE-microclimat model to assess adaptation strategies at the district scale. *Urban Climate, 14*(Part 2), 213–223. https://doi.org/10.1016/j.uclim.2015.07.004.

Perez, R., Seals, R., Ineichen, P., Stewart, R., & Menicucci, D. (1987). A new simplified version of the Perez diffuse irradiance model for tilted surfaces. *Solar Energy, 39*(3), 221–231. https://doi.org/10.1016/S0038-092X(87)80031-2.

Qu, Y., Milliez, M., Musson-Genon, L., & Carissimo, B. (2012). Numerical study of the thermal effects of buildings on low-speed airflow taking into account 3D atmospheric radiation in urban canopy. *Journal of Wind Engineering and Industrial Aerodynamics, 104–106*, 474–483. https://doi.org/10.1016/j.jweia.2012.03.008.

Robitu, M. (2005). *Etude de l'interaction entre le bâtiment et son environnement urbain: Influence sur les conditions de confort en espaces extérieurs*, PhD thesis. Ecole Polytechnique de l'Université de Nantes.

Robitu, M., Inard, C., Groleau, D., & Musy, M. (2004). Energy balance study of water ponds and its influence on building energy consumption. *Building Services Engineering Research and Technology, 25*(3), 171–182.

Robitu, M., Musy, M., Inard, C., & Groleau, D. (2006). Modeling the influence of vegetation and water pond on urban microclimate. *Solar Energy, 80*(4), 435–447. https://doi.org/10.1016/j.solener.2005.06.015.

Rodler, A., Guernouti, S., Musy, M., & Bouyer, J. (2018). Thermal behaviour of a building in its environment: Modelling, experimentation, and comparison. *Energy and Buildings, 168*, 19–34. https://doi.org/10.1016/j.enbuild.2018.03.008.

Vinet, J., Antoine, M. J., Raymond, F., & Inard, C. (1999). Modelling the impact of urban vegetation to analyse urban microclimate and outdoor thermal comfort. In *ICUC 1999*.

Vinet, J. (2000). Contribution à la modélisation thermo-aéraulique du microclimat urbain. *Caractérisation de l'impact de l'eau et de la végétation sur les conditions de confort en espaces extérieurs*, PhD thesis. Université de Nantes Ecole Polytchnique de l'Univeristé de Nantes, Nantes, France.

Weihs, P., Staiger, H., Tinz, B., Batchvarova, E., Rieder, H., Vuilleumier, L., et al. (2012). The uncertainty of UTCI due to uncertainties in the determination of radiation fluxes derived from measured and observed meteorological data. *International Journal of Biometeorology, 56*(3), 537–555. https://doi.org/10.1007/s00484-011-0416-7.

Chapter 14
Comparing ENVI-Met and Grasshopper Modelling Strategies to Assess Local Thermal Stress and Urban Heat Island Effects

Martina Pacifici and Matías Nieto-Tolosa

14.1 Introduction

Numerical models are emerging as key tools in urban climate studies due to their capability to simulate complex fluid dynamic interactions, to highlight key processes and to provide effective predictions in support of decision-making (Organization for Economic Co-operation and Development 2011). A model is a simplified representation of the complex real world (Anderson et al. 2015, p. 5), a set of assumptions expressing the nature of a system and its behaviour (Bear 2018, p. 89). In the urban field, modelling approaches can be applied to different city problems, such as land-use policy, mobility network, population distribution as well as slum growth and climate change (Wegener 1994).

Models can function as *forecasting* tools to investigate the effects of certain actions on a selected domain; otherwise, they can be needed to virtually recreate past or present conditions (Anderson et al. 2015, p. 9). Urban models enable making predictions on unbuilt conditions and testing hypotheses of future urban interventions (Batty 2009). Under a physical perspective, modelling techniques are powerful tools to explore urban morphology and related climate processes, enhancing sustainable design practices.

Due to technology advancements, faster computers and availability of free-access databases, modellers can take advantage of increasingly sophisticated codes

M. Pacifici (✉)
Civil Engineering Department, Polytechnic School of São Paulo University—USP (Escola Politécnica da Universidade de São Paulo—USP), São Paulo, Brazil

Allford Hall Monaghan Morris, London, UK

M. Nieto-Tolosa
Urban and Regional Planning Department (departamento de Urbanística y Ordenación Territorial), Technical University of Madrid (Universidad Politécnica de Madrid, UPM), Madrid, Spain
e-mail: matias.nieto.tolosa@alumnos.upm.es

M. Palme, A. Salvati (eds.), *Urban Microclimate Modelling for Comfort and Energy Studies*, https://doi.org/10.1007/978-3-030-65421-4_14

to shed new light on the complex urban dynamics at different scales of analysis. Nonetheless, the range of model solutions which practitioners can address is broad and choosing the proper device could be hard. Among the current district-scale modelling tools capable to simulate urban morphology-climate interactions (Naboni et al. 2017) the following stand out: ENVI-met (Bruse and Fleer 1998), Grasshopper (Rutten 2007), CitySim Pro (LESO-PB), Rayman (Matzarakis et al. 2005), Autodesk CFD (AUTODESK 2009), IMM (DABC 2010) and Urban Weather Generator (Bueno et al. 2013).

To shed light on this complex scene, the chapter develops an in-depth analysis of a standard modelling case study, highlighting how the use of different codes may strongly influence the computing of physical processes and therefore accuracy of results. A conceptual model, including the climate phenomena to be represented into the urban scene, is defined as starting point and associated to an urban domain. On this shared domain, two modelling strategies have been designed, entailing the application of ENVI-met and Grasshopper codes.

ENVI-met and Grasshopper are applied with the goal of exploring microclimate effects involving an urban domain sited at the core of a global south megalopolis. To highlight limits and potential of these two modelling choices, several technical aspects have been discussed. Results explain what strengths and weaknesses characterise both modelling paths, suggesting the more suitable application depending on domain scale, required accuracy, data field availability and modeller's purposes.

Along the chapter, authors will refer to the 'modeller role' in a full sense, indicating who is engaged in the conceptualisation of the urban climate problem, the understanding of its mathematical representation and the computational simulation. Modellers should be able to translate the understanding of the reality in a set of assumptions that emphasise the most important phenomena and disregard the secondary effects. They need to scale data from data sets if necessary and interfere with the computational process to optimise the engine efforts. Unlike code designers or simple users, the role of modeller entails a full awareness of the physics principles driving the climate phenomena, as well as the understanding of their mathematical formulation.

Essentially, this chapter would support urban modellers who are approaching the modelling task, tackling with important modelling choices including the software selection and parameterisation. At the same time, the chapter would help those who are wondering about which modelling strategy is more appropriate for a specific urban case. Contents could also be useful for users and modellers of other software.

14.2 Method

14.2.1 Designing Conceptual Model

According to Anderson et al. (2015, p. 17), once purposes are determined, the modelling process may start following a rigorous workflow including the definition of *conceptual model*. In the *conceptual model*, the modeller has to define a system of interactions and variables characterising the field phenomena. Purposes and conceptual model are essential steps to choose the most appropriate simulation code.

Model purposes should be defined before starting a modelling process. They deal with what the modeller wants to achieve through the modelling application, as well as which assumptions and simplifications are feasible. At this stage, minimum confidence levels in the expected results should be defined.

For this study, the modelling process aims to understand the climate processes affecting an urban fabric embedded in a subtropical megalopolis. Variables as air temperatures (T_a), mean radiant temperatures (MRT) and thermal comfort index will be simulated to discover which specific urban features are affecting their local variation.

The conceptual model developed (Fig. 14.1) summarises the main physical processes occurring in the urban system under analysis, including heat transfer, sun access and building's shading. The selection of the main processes is based on field observations and implies a simplification of the real world, a compromise between aims, available information and resources. This simplification is essential to overcome software limitations and reduce the complexity of model, although it should not undermine the achievement of modeller purposes.

The conceptual model is located on subtropical latitude, characterised by dry winters and rainy wet summers. Being embedded in a metropolitan area (São Paulo, Brazil), the environment carries on urban heat island (UHI) effects. Urban morphology is hybrid, including high-rise and low-rise urban canyons, 10 km far from the Tropic of Capricorn. Due to the low latitude, solar radiation is the most significant driver of urban heating and buildings' shading is the main cause producing climatic

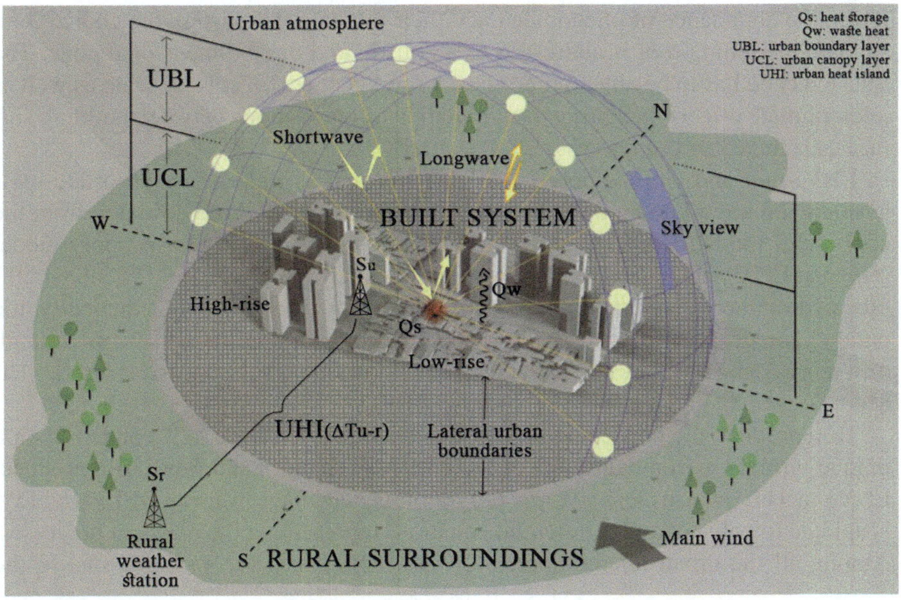

Fig. 14.1 Conceptual model adopted

differences in the area. During the day, buildings filter the sun access, creating exposed or shaded spaces that grant or prevent the heat gain of streets and walls. Since daily excursions occur, temporal modelling discretisation should be hourly at least.

Attention should be drawn to the different microclimate environments between buildings, especially around the tallest ones, with temperature contrasts at opposite exposures. Albedo differences and pavement discontinuities could also be decisive to induce gaps on the temperature field. As consequences, the phenomena scale extends from the urban fabric to the building interstitial spaces, where surfaces exchange heat with atmosphere and neighbouring buildings. Coarse meshes should be avoided since heat exchanges within too large cells would suppress the complexity of the ground surface-atmosphere interactions.

According to field experiments, wind flow is a less relevant force due to the dense urban environment hindering airflows. Scarce vegetation coverage should not be relevant in the T_a modelling but could provide variations to surface temperatures (T_s) and MRT distributions because of the trees' shadow cone. More details on the conceptual model are available in Pacifici et al.'s study (2017, 2019).

14.2.2 Comparing ENVI-Met and Grasshopper Modelling Strategies

The conceptual model highlights the complexity of processes interrelating urban forms and climate variables. It stresses the importance of shadowing and buildings' heights, impact of sun, influence of materials, façade exposures and vegetation. Given the abundance of phenomena and their mutual relationship, the choice for a suitable modelling tool is addressed to a holistic and three-dimensional code. The selected code has to be able to reproduce hybrid geometries at local scale as well as microclimate effects, including the simulation of interactions between building surfaces, air fluxes and sun access.

ENVI-met and Grasshopper have been selected as the most appropriate tools available on the urban modelling scene. Two similar models are designed from the conceptual model described above and then compared. Both Grasshopper model (*Gh*) and ENVI-met model (*En*) adopt a similar study area, even if different hypotheses and mathematic formulations were implemented to model it. Practically, the parallel use of these tools allows investigating domain thermal behaviour through different driving processes. *En* and *Gh* parameterisation is summarised in Table 14.1. More details are given in Pacifici's study (2019).

To reproduce the conceptual model behaviour, *En* 'version 4.3' condenses in a unique virtual environment most of the energy exchanges and flows to reproduce. ENVI-met is coded in *Object Pascal* programming language by using DELPHI. FDM is applied to solve partial differential equations representing the physical processes. Among all the urban microclimate software, ENVI-met is considered one of the

Table 14.1 Setting of input parameter and assumptions for *En* and *Gh* models

	ENVI-met	Grasshopper [DF + HB + LB][a]
Algorithms		
Dimensionality	3D	3D
Problem class	Transient (non-hydrostatic) flow	Average energy transfer function [DF]
Numerical method	Fdm	RSM (Fdm) + VDM (heat-diffusion equation) + UBL (energy balance) + UC-BEM (energy balance) [DF]
Fluid flow calculation	Full 3D CFD	NA
Mathematical model	Process based	Data driven (black box)
Radiative fluxes	Ray-tracing algorithm (Bruse 1995)	Radiance script (Ward 2016) [HB]
Geography characterisation		
City	São Paulo	São Paulo
Lat. (+N, −S), long. (−W, +E)	−23.53; −46.59	−23.53; −46.59
Reference time zone	GMT-3	GMT-3
Spatial and temporal discretisation		
Core model dimension	258 m × 430 m	258 m × 430 m
Border model dimension	111 m (*y*-axis), 119 m (*x*-axis) for each side	NA
Height model dimension	174 m (3D model), 2500 m (1D model)	NA (3D model), 1000 m (boundary layer)
Mesh type	Square	Square
Horizontal mesh grid	3 m × 3 m	3 m × 3 m (MRT); 6 m × 6 m (T_s); NA (T_a)
Vertical mesh grid	2 m + 15% telescoping factor from 10 m	HB zones; NA
Highest building in the domain	84 m	84 m
Model rotation out of grid north	−13.5°	0°
Nested grids	5	NA
Time step	2, 2, 1 s	From 300 to 600 s
Initial conditions		
Start date and time	December 4th at 21:00 h (UTC)	December 9th at 1:00 h (local time)
Initial temperature of atmosphere	22.14 °C	NA
Relative humidity in 2 m	81.57%	NA
Wind direction and speed at 10 m	163°; 2.5 m/s	150°; 2.0 m/s
Roughness length	0.1 (at measurement site)	0.1 (obstacle height at weather station)

(continued)

Table 14.1 (continued)

	ENVI-met	Grasshopper [DF + HB + LB][a]
Specific humidity at 2500 m	10.6 g/kg	NA
Initial soil temp. (0–20–50, 50–200 cm)	297.65; 298.65; 299.65 K	NA
Initial soil wetness (0–20–50–200 cm)	58%; 64%; 69%	NA
Input parameters (urban climate)		
Source of climate data	Six weather stations around domain	INMET Mirante Santana EPW
Temp-height, wind-height at source	2, 10 m (fixed)	2, 10 m (settable) [DF]
Vegetation coverage at source	NA	0.9 [DF]
Scalability of climate data to domain	Direct scalability (simple forcing)	UWG model computation [DF]
LBC for temperature and humidity	Forced	UWG model computation [DF]
LBC for turbulence	Forced	NA
Turbulence model	Standard TKE model	NA
Solar radiation	Estimated but adjusted by field data	INMET Mirante Santana values
Cover of low, medium, high clouds	0; 0; 0 octas	0; 0; 0 octas
UBL, urban boundary layer	1D model (174–2500 m high)	UBL height at daytime and night-time [DF]
Input parameters (urban morphology)		
Tree coverage	Insertion of deciduous plants	Average tree coverage [0–1]: 0.165 [DF]
Wall/roof material [high-rise]	w1 (30 cm) and properties	w: SHGC, 0.6; wall/roof albedo, 0.4 [DF]
Windows surface [high-rise]	NA	Average frac. 0.35 [DF]
Wall/roof material [low-rise]	w2 (30 cm) and properties	w2: SHGC, 0.4, wall/roof albedo: 0.2 [DF]
Window surface [low-rise]	NA	Average frac. 0.1 [DF]
Pavement typologies	Tower, house, road, tree pavements	Average city pavement, tower/house pavement, road, grass [DF]
Building program	NA	Mid-rise apartment [DF]
Building age	NA	Pre-1980s (LR), 1980s–present (HR) [DF]
West heat from air conditioning	NA	Average frac. 0.7 [DF]
Anthropogenic heat from automobiles	NA	8 W/m^2 [DF]
Hourly traffic pattern	NA	Average frac. (0–1), 24 values [DF]

[a]In *Gh* model, values refer to all plug-ins when [DF/HB/LB] are not specified

most complete engines nowadays available, able to simulate surface-vegetation-atmosphere microscale interactions. To configure the ENVI-met model, three virtual environments were set up: the 3D model space, the 1D model and the soil model. The 3D model space or *core model* allows including in the three spatial dimension discretised space (x, y, z) all the features representing the urban area under study, such as buildings, open spaces, vegetation and surfaces. On the surroundings of the 3D model, the 1D model provides the boundary layer conditions. The vertical structure of the 1D model is split into 14 layers made up of grids whose size increases with the height. Finally, the soil model guarantees the computation of the heat transfer through the ground, setting soil temperature and humidity that affect tree evapotranspiration (Bruse and Fleer 1998; Simon 2016; Pacifici 2019).

On the other hand, *Gh* is an open-source graphical algorithm editor running into the proprietary 3D modelling package Rhinoceros 3D or Rhino (developed by McNeel & Associate). It hosts a collection of plug-ins compiled for environmental simulation, some of which include other specific routines such as Urban Weather Generator, EnergyPlus and OpenStudio simulation engines. To design the model, three plug-ins have been interoperated: Dragonfly (DF) 'version 0.0.03, jul_11_2018', Ladybug (LB) 'version 0.0.67, nov_20_2018' and Honeybee (HB) 'version 0.0.64, dec_05_2018'. Each plug-in takes charge of specific tasks in a proper sequence, once their multiple custom components are linked in a logical connectivity (Khabazi 2012). Dragonfly allows the computation of T_a and relative humidity (RH), by implementing Urban Weather Generator (UWG), a validated Python application originally developed by Bueno Unzeta (2010) to estimate UHI effects. Honeybee estimates T_s and MRT by applying EnergyPlus-based OpenStudio as an open-source energy simulation engine from the morphed EPW file and a high-resolution thermal zoning. Finally, Ladybug enables the estimation of the universal thermal comfort index (UTCI) from all the previous inputs.

14.3 Results

The comparative analysis highlights limits and potential of the two modelling choices, considering several technical aspects gathered in four categories (Table 14.2).

14.3.1 Running

14.3.1.1 Algorithms

Although sharing a common conceptual model, *Gh* and *En* models entail different assumptions and computational strategies. To reach their purposes and provide a reliable explanation of the real system, they include some physical processes while

Table 14.2 Comparison categories

Running
Algorithms
Budget computation
Computational effort
Usability
Access
User interface
Customisability
Assumptions and parameterisation
Initial conditions
Input parameters
Spatial and temporal discretisation
Boundary conditions
Practical issues
Scalability
History of use and testing
Application scale

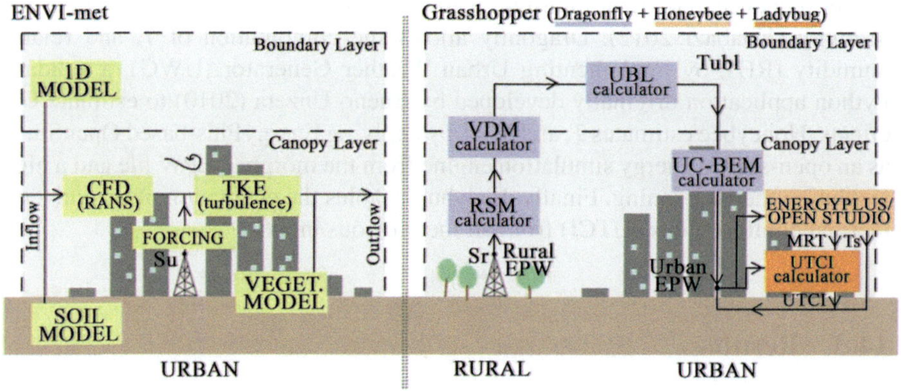

Fig. 14.2 ENVI-met and Grasshopper model workflows

disregarding other ones. To see clearly such differences, the engine architecture of models is better examined.

ENVI-met is a 3D computational fluid dynamics (CFD) model (Fig. 14.2, left). Its architecture may be defined as 'process based' using principles of physics to represent the airflow within the domain. The spatial and temporal evolution of the three-dimensional turbulent airflow is based on the computation of Reynolds-averaged Navier–Stokes (RANS) and continuity equations. The differential equations in the model are solved using the finite difference method. In order to keep the flow field mass conserving, continuity equation has to be satisfied for each time step and each grid cell. Turbulent energy distribution is calculated using turbulent energy production and dissipation equations. The distribution of temperature and specific humidity in the atmosphere is given by the combined advection–diffusion equation with internal source and sinks of sensible heat and vapour (Bruse and Fleer 1998; Bruse 2004; Huttner 2012; Simon 2016).

Gh model is centred on the implementation of UWG via Dragonfly, in combination with Honeybee and Ladybug plug-ins (Fig. 14.2, right). UWG[1] algorithm is a physically based urban canopy and boundary layer model based on the urban energy balance, operating as generator of urban temperatures from rural meteorological station (Bueno et al. 2013, 2014). UWG interoperates four coupled modules, estimating heat fluxes by solving finite difference method, energy balance and heat diffusion equations. The rural station model (RSM) is based on an energy balance at the soil surface and computes sensible heat fluxes by finite difference method (1). The vertical diffusion model (VDM) calculates the vertical profile of air temperature above the weather station (2). The urban boundary layer model (UBL) is based on an energy balance in the urban boundary layer and estimates air temperature above the urban canopy layer (3). The urban canopy + building energy model (UC-BEM) is based on the town energy balance and estimates urban canyon air temperature and humidity (4). Since UWG works on average values, the interconnection with Ladybug and Honeybee results to be very useful to simulate the thermal behaviour of all domain cells, as well as to take into account more 'punctual' phenomena among which shading is once again dominant. Ladybug UTCI computation is based on a Python version of the original UTCI source code written in Fortran (www.utci.org). Similarly, Honeybee runs rely on the *Radiance* ray-tracing script (Ward 2016) and on the EnergyPlus-based OpenStudio validated simulation energy engine.

14.3.1.2 Budget Computation

The energy balance should provide total inflow-total outflow equality. *Gh* and *En* approach the energy balance and heads computation differently in space and time.

En model solves the thermodynamic problem within a closed system that reproduces a virtual simplified microclimate environment crossed by heat and air fluxes. Unless computational errors, the energy budget computation finds numerical solution once initial conditions are specified. Modeller challenge lies in getting closer this solution to the field reality as much as possible by introducing appropriate parameters and initial conditions. The overall budget computation includes the modelling of heat transfer in the air governed by advection and diffusion equations, wind and turbulence air fluxes, pollution transport, incoming radiative shortwave fluxes absorbed by plant and building walls, building and plant shadow projection, long-wave fluxes coming from plants and building walls, heat and vapor exchanges close to plants, interaction between leaves and air by evaporation and transpiration,

[1] The model is described in detail in Chap. 12 by J. Mao and L. Nordford.

underground vertical heat fluxes and moisture content variation, and heat and vapor fluxes at the ground surfaces and walls (Table 14.3).

Gh model is built from scratch, including the main processes affecting the domain. It is an open system, suitable to be integrated by increasingly complex energy exchanges. Its simplicity entails a lower physical consistency. Some phenomena are overlooked since they are not essential to the computation of modelling purposes. For example, computational fluid dynamics operations are not included in the budget computation, giving away the air fluxes and turbulence modelling (Table 14.3). UWG application provides the energy balance at soil surface and urban boundary layer, the vertical heat transfer computation between the urban boundary layer and urban canopy layer, and the energy balance of an average urban canyon. Ladybug and Honeybee integrate the incoming radiative short-wave fluxes absorbed by building walls, shadow projection calculation, interior and exterior light reflections, windows and walls transmissivity.

Table 14.3 Main phenomena reproduced

Phenomenon	*En* model	*Gh* model (DF + HB + LB)	
Heat transfer	Described in the 3D model by the combined advection–diffusion equations based on the conservation law of energy. The computation relies on initial 24-h forced values of T_a and RH	DF	RSM computes sensible heat fluxes by solving a transient heat diffusion equation by FD method. VDM computes vertical profile of air temperature by solving a heat diffusion equation
Wind and turbulence	Mathematically described in the 3D model by turbulence (k) and dissipation (ε) equations based on Yamada and Mellor (1975)	NA	NA in our case
Short-wave and long-wave fluxes	Computed at the model top using a two-stream radiative flux approximation (long wave) and a set of empirical equations (short wave) based on Taesler and Anderson (1984) and Gross (1991). A solar factor allows adjusting short-wave calculation to field radiation data. Within the 3D model, plants and buildings are able to modify radiative fluxes	HB	The solar energy calculation includes interior and exterior light reflections. EPW file is used to run EnergyPlus model in an OpenStudio file. EnergyPlus computes the surface percentage free to see the sun at different time steps and factors them in a dynamic energy balance calculation. Sky temperature is computed from long-wave EPW data
Surface heat conductivity	Wall and roof temperature is estimated based on seven prognostic calculation nodes. Outside node temperature depends on atmospheric variables at the facade and the thermal environment within the view range of the wall/roof. Inner node temperature is based on Fourier's law of heat conduction	HB	Building mass is divided into *HB zones* (smaller volumes) corresponding to how EnergyPlus model should be zoned. Each building floor is split into core and perimetral zones. A window/wall ratio is set. EnergyPlus computes interior/exterior surface temperatures for each individual surface of the zone

14.3.1.3 Computational Effort

Time simulation depends on the amount of mathematical processes included in the algorithms the model is running. The amount of mathematical processes is affected by time and space discretisation (Fig. 14.3), as well as the number of physical exchanges reproduced. The cost of executing a slow numerical model weighs on the modeller time and costs.

Gh model is faster to run in comparison with *En* model. While ENVI-met solves a fixed number of equations, Grasshopper modular scheme allows selecting only the main processes characterising the problem domain, reducing computational effort (Fig. 14.3). Moreover, a simplification of grid discretisation is applied without numerical errors. Grasshopper demands a great clarity of the conceptual model, since the choice of processes to be simulated is set by the user. Given the lower number of simulated processes, Grasshopper allows more complex morphologies and larger domains. In addition, some simplifications have been tested to enhance model performance. As an example, low-rise housings were initially modelled individually; afterwards, they were unified in few blocks and backyards were cancelled. Indeed, as *En* model, Grasshopper code implements a FDM to solve MRT and T_s;

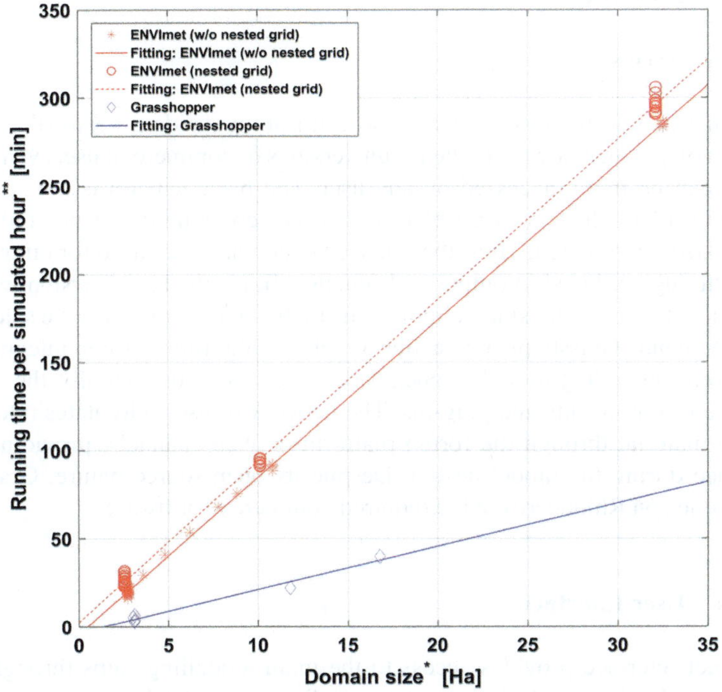

Fig. 14.3 ENVI-met and Grasshopper running times. *For ENVI-met data, domain size is defined based on the assumption that a 3 × 3 m is the maximum cell size to guarantee acceptable solution of turbulent equations. **Minutes needed to simulate for 1 h, obtained dividing the total simulation time for the simulated hours

as a consequence, a simplification of geometries is desired to well discretise the modelling space.

En model simulates many interconnected processes at once, requiring a very high computational effort to solve equations. Practically, since most of the processes are simulated, a full understanding of the study area is not strictly necessary to the model design. In contrast, due to such an accurate atmospheric reproduction, ENVI-met requires a simplification of built morphology and is not suitable for large urban domains. A significant increase is observed when new cells and nested grids are added to the model spacing, or larger domains are tested. Figure 14.3 shows the high correlation (R^2 = 0.99 ENVI-met, 0.96 Grasshopper) between running time [min] and domain size [Ha]. Small and large domain points are scattered on a linear trend and running time variation is largely explained by extension. The addition of nested grids, which improves model stability by reinforcing boundary conditions, requires a greater simulation time for the same domain, shifting upward the regression line. Due to its simplified algorithm and grid design, *Gh* (blue fitting line) entails shorter running times compared to *En*.

14.3.2 Usability

14.3.2.1 Access

ENVI-met is a proprietary software and can run on standard personal computers. Licenses on payment are available for universities or commercial use, even if a basic version can be freely accessed by any user. The basic version is not suitable for modelling urban clusters (large blocks, districts, communities) since only limited domain sizes are granted. Thus, the science version has been used for our modelling case. Although reduced in output and functionalities, the basic version well fits to sensitivity studies where small clippings from the main domain can be selected. On the other hand, Grasshopper is a free graphical algorithm editor integrated with Rhinoceros modelling tool. The open access regards its use but also the chance to create new components and plug-ins. The network of users circulates tutorials and training material through the forum platform or Web channels, providing mutual assistance during the model design. Despite its open-source nature, Grasshopper access relies on Rhinoceros, a fee required commercial software.

14.3.2.2 User Interface

ENVI-met interface provides access to the main modelling steps through pop-up windows, allowing mesh design, setup configuration, database setting (materials and plants) and result checking. These interface windows are fast entry points to model box, even if the whole modelling process could be managed from editable files. ENVI-met interface does not require hard computational skills and it was designed to be used by users with no programming experience. The interface presents an 'assigned attribute-based' structure in which simulation properties are defined linking numerical values or attributes with model equations (Chasznar 2011). Besides the interface, data as saved can be displayed across 'physical' and tabular views and can be read by the model to run accordingly the user's setting. The 3D visualisation allows the user to assign specific materials to every cell through the Open GL Application Programming Interface (Huttner 2012).

On the other side, Grasshopper adopts a 'geometric content-based' structure through the application programming interface (API) that structures data exchange and parameterisation using collections of basic objects called "components", containing inner rules for data processing and input-output requests (Lagios et al. 2010; Davis and Peters 2013). Input-output requests are linkable with geometric primitives, numbers, textual information, formulae and matrices, among others. Users can build up models by dragging the components on a canvas and linking them in a sequence of input-output multiple connections. This set of rules and connections constitutes an editable algorithm that allows automated calculation of physical processes and variables. Grasshopper interface is more demanding than ENVI-met one although a code programming knowledge is not really needed (Fig. 14.4).

14.3.2.3 Customisability

En and *Gh* provide a quite opposite ability to be customised. ENVI-met is a defined microclimate model reproducing mathematically the main atmospheric exchanges occurring at the urban climate layer (UCL). Within this virtual box, the user has to fit spatial configurations (buildings, materials, plants) to the mesh grid, as well as define initial parameters and boundary conditions. *En* workflow exposes the user to minor risks, since the whole model running is ensured at least by default options; on the other side, few customisability choices are available. *En* model is an

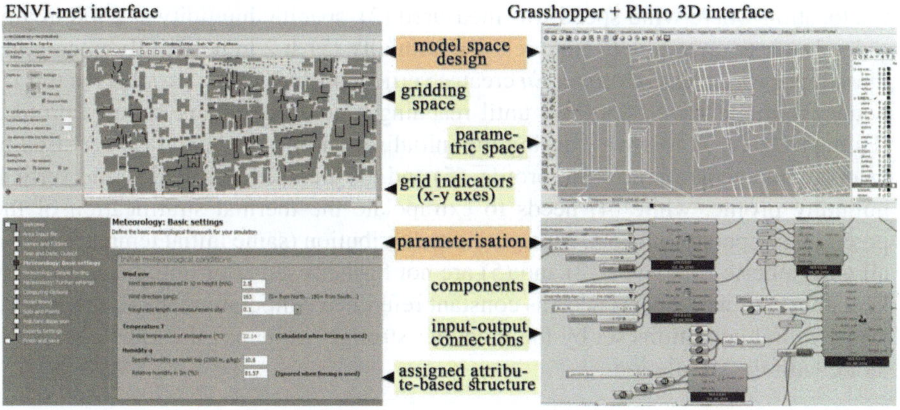

Fig. 14.4 ENVI-met and Grasshopper user interfaces

independent box and does not need external tools to fulfil the modelling purposes, since the majority of processes are simulated.

Conversely, in Grasshopper no initial modelling box is provided to the user. According to the modelling purposes, users can customise their own algorithmic connections. In the case of *Gh* model, a greater understanding of conceptual model is needed to model plausible microclimates without neglecting decisive phenomena. In this way, *Gh* modeller has more risks but a higher chance to build a model suitable to his/her case study. For instance, a set of components related to the volumetric inputs of the 3D Rhino urban scene and thermal zoning can be combined in order to get higher spatial resolution for further calculations. This inclination to customisability says why Grasshopper code is suitable to model a wide range of built environment problems. This applies to the display of charts too, allowing different formats, colours and graphic styles for a better legibility of results. Generally, Grasshopper plug-ins are developed by community users to overcome specific problems related to architectural everyday practice. This fact confers great utility to the software advancements, in contrast with "monolithic" CAD environments. Today, an add-on for Ladybug (not available at the moment of this modelling experiment) enables users to convert Rhinoceros 3D domain to ENVI-met model and run ENVI-met simulations.

14.3.3 Assumptions and Parameterisation

14.3.3.1 Initial Conditions

Initial conditions describe the initial state of all points of a considered domain and could be intended as a boundary condition in time (Anderson et al. 2015; Bear 2018). As a transient model, ENVI-met requires the modeller to specify initial conditions at the beginning of the simulation, usually considered as $t = 0$. *En* initial conditions require horizontal wind speed at 10 m height (1), surface roughness at the location where wind speed was measured (2), specific humidity at 2500 m altitude (3), air temperature (4) and relative humidity (5) at 2 m height. By integrating these input variables over time, *En* creates vertical profiles of the variables in the 1D boundary model up to 2500 m, until reaching a quasi-stationary state of profiles (Huttner 2012). In this way, during the initialisation phase, (1) and (2) are used to calculate the geostrophic wind profile, (5) and (3) are used to calculate the vertical humidity profile, while (4) needs to extrapolate the thermal stratification of the potential temperature from a zero-gradient distribution (same initial temperature for all atmospheric cells). (1), (2) and (5) are not further used after initialisation, (3) is kept constant, while (4) is used as constant reference temperature at 2500 m height. These initial conditions set by the modeller strongly influence the transient results along with the simulation.

In *Gh* model no proper initial conditions are provided. Only input parameters are set but no stress conditions have been introduced at the start of the running on the head distribution. This simplification is a limitation in the estimation of heat transfer.

14.3.3.2 Input Parameters

By setting parameters, modeller brings the model virtual box closer to the real domain. Input parameters should rely on field measurements, official database and good knowledge of the site. Great accurateness should be addressed to data collection, since it enables the model calibration and subsequent predictions. At this phase, uncertainty and too heterogeneous information could require simplification. Urban morphology and climate inputs have been needed to build up *En* and *Gh* (Table 14.1).

For *En* model, morphology parameters include the definition of building forms and tree shapes, façades and ground surfaces, underground soil layers and tree characterisation. Surface layers may have different thickness and base materials characterised by specific absorption, transmission, reflection, emissivity, specific heat, density and thermal conductivity. Climatically, after the initial conditions, 24-h T_a and RH trends have to be provided as *forcing*, together with cloud cover input. A climate campaign on the urban field under study is necessary to collect *forcing* values and validate results. Global solar radiation is estimated by ENVI-met code, although a corrective factor may be applied after comparison with field data.

In *Gh* model, morphology input data are provided by setting specific and general parameters. Specific parameters are directly associated with building surfaces and include building age and typology, average distance between floors, heat waste from air conditioning and envelope characterisation. Envelopes are characterised by the averages of window surface, solar gain heat coefficient and surface albedo. General parameters are related to the urban environment where buildings are placed. Such average parameters include traffic intensity, anthropogenic heat from motor vehicles, vegetation coverage, main pavement materials and tree canopies. Additionally, a district GIS shapefile from municipality database is introduced as input data to provide an influence area to the urban fabric under analysis. Climatically, *Gh* requires the EnergyPlus Weather (EPW) file as a source of input meteorological data. Overall, two types of EPW files are used: the rural EPW and the urban EPW. Rural EPW gathers climate data measured at an operational rural station of reference. Urban EPW is obtained through the application of UWG to reflect the climate conditions within an urban environment. Urban EPW is a UWG outcome and gathers climate data to enable Honeybee computing MRT and T_s. Moreover, Ladybug algorithm uses the rural and the urban EPW files to compute the comfort index in urban and rural conditions. The integration of EPW information within the UWG application regards only some variables and just two of the four UWG models making up the UWG architecture. In particular, radiation heat fluxes from rural

EPW are read by the 'rural station model (RSM)' in order to estimate surface sensible heat fluxes at the rural station. Meantime, the 'vertical diffusion model (VDM)' processes air temperature and wind speed from rural EPW, together with the sensible heat fluxes RSM computed, to estimate the vertical profile of T_a above the rural site (Bueno et al. 2014).

14.3.3.3 Spatial and Temporal Discretisation

In FDM models, numerical solutions are not spatially and temporally continuous; the domain is partitioned and differential equations are solved at points in the discrete space and time (approximated solution). Spatial and temporal discretisations are key factors the modeller can manipulate to optimise the use of computer resources. Fine discretisation allows getting approximated solution closer to continuous solution preventing numerical errors, while coarse discretisation keeps low run times (Lal 1998). The compromise between these two approaches strongly influences model outputs.

Spatially, *En* implements the Arakawa C-grid system to discretise the domain space in a set of orthogonal cells with uniform nodal spacing. Buildings and trees are arranged along *x*, *y* and *z* axes. Inclined or curved building surfaces cannot be reproduced but can be rounded by grid points (Fig. 14.5). At different locations of the cell (face centres, corners), ENVI-met algorithms place quantities (masses, velocities) to be computed. Temporally, ENVI-met splits the simulation time in a sequence of time steps. Time step parameter is very sensitive and its value should be kept low ($\leq$2 s) to avoid troubles in the form of numerical instabilities. Time step intervals vary according to the sun height; the higher the sun, the smaller the interval should be. Overall, too fine meshes and/or too short time steps cause excessive run times in the model, since discretisations are proportional to the number of cells and time steps. Salata et al. (2016) investigate the model mesh sensitivity using grid cells ranging between 1 and 3 m. As expected, the study shows more accurate numerical results using finer mesh, while coarser meshes result in values further from field measurements. Notwithstanding, the finer mesh is not suggested due to the higher computational effort.

Since in *Gh* model different plug-ins are combined, distinct space and time discretisations are implemented by the algorithm. UWG computation does not require the domain discretisation since the solution is continuous on the space and only average values are computed. Buildings, walls, roads and windows integrate the computational equations as input volumes and areas. Average results are provided as valid for the representative domain. Temporally, the model allows runs of large time steps, reducing considerably the entire run time. A time step of 300 s (5 min) is accepted without numerical errors. Conversely, for the computation of surface and mean radiant temperatures, Honeybee implements a numerical scheme to solve heat transfer equations in a discretised space adopting six time steps *per* hour. Two

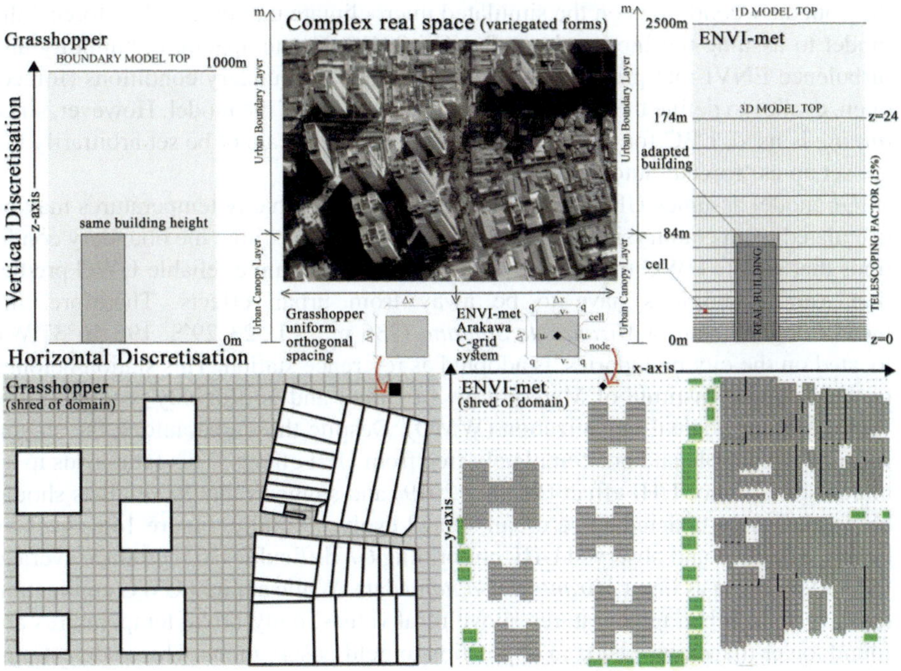

Fig. 14.5 Discretisation for *Gh* and *En* models

different meshes are adopted for MRT and T_s computation. MRT mesh is a 3 × 3 m grid (Fig. 14.5), while T_s mesh varies in function of the surface type and the volume: for roofs and windows, mesh coincides with the entire surface; for façades, mesh has the wall length and inter-floor height; for ground floor, a grid discretisation 6 × 6 m is applied.

14.3.3.4 Boundary Conditions

Boundary conditions define values or flows along the domain borders. In *En*, 1D model allows controlling the boundary layer processes and providing the inflow boundary data to the 3D model. The 1D model entails the extension of the 3D model up to 2500 m. To provide boundary conditions to be applied on the 3D model, 1D model computes one-dimensional vertical profiles of T_a, specific humidity, wind vectors, kinetic energy and turbulent exchanges. Within the 3D model, the conditions of air temperature and humidity are forced by using the *simple forcing* tool. *Simple forcing* defines the diurnal variation of the atmospheric boundary conditions. Its use sounds contradictory to the authors. Without *forcing* the urban climate conditions are far to be reproduced; on the other hand it causes an excessive influence of the boundary conditions on the simulated microclimate that practically ‘forces’ the model to assume the input values (Pacifici 2019). For temperature, humidity and turbulence ENVI-met provides three types of lateral boundary conditions (forced, open, cyclic) to define the flux exchange on the border of the model. However, since *forcing* is used, LBC for air temperature and humidity cannot be set arbitrarily but are set up as ‘forced’ automatically.

Gh model assumes UHI as the driving force by which city temperatures may be calculated from countryside. Rural features and data constitute the boundary conditions that enable UWG to estimate urban values. To obtain a reliable UWG prediction, rural conditions have to be away from urban effects. Therefore, the meteorological station *Mirante de Santana* (786 m a.s.l., 23°29′S, 196 46°37′W), located on the city boundaries, is adopted as reference station. This station belongs to the National Institute of Meteorology of Brazil and is officially recognised by World Meteorological Organization (WMO). Despite these guarantees, the station is found unsuitable and not completely free from UHI effects. This fault leads to an underestimation of UHI effect (Pacifici 2019) and suggests that WMO lists should be continuously updated since urban sprawl evolves rapidly. Figure 14.6 displays hourly modelled and observed UHI, confirming *Gh* difficulties to reproduce average UHI effect using as input *Mirante de Santana* data. While at night UWG application tends to keep urban temperatures above rural values, daily urban temperatures are found lower than countryside in opposition to field observations. This underestimation proves that the use of reference rural station influenced by urbanisation does not guarantee confident results in agreement with Bueno et al. (2013). Otherwise, ENVI-met shows a better accuracy.

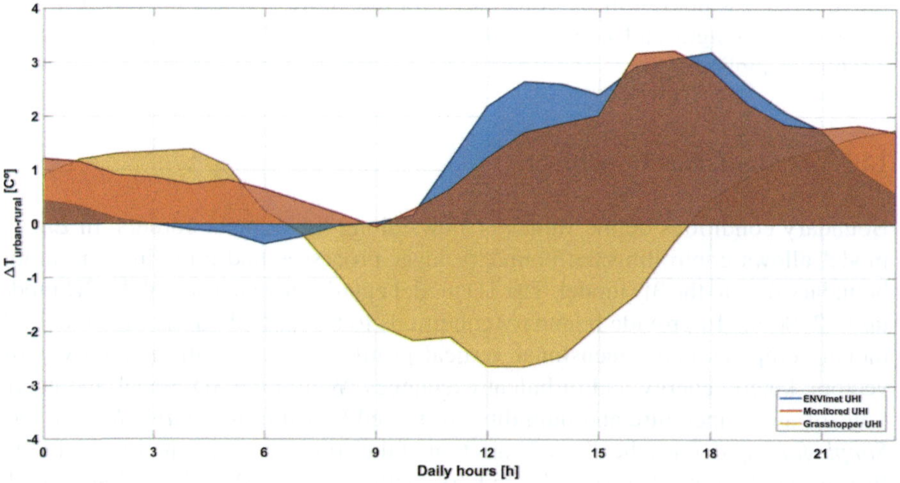

Fig. 14.6 Observed and modelled UHI effect, obtained by subtracting rural from urban temperatures

14.3.4 Practical Issues

14.3.4.1 Scalability

Scalability is intended as the model capable to keep its performance high across expanding domains and workloads, as well as its accuracy to reproduce physical processes at different scales. A good scalability means the model is maintaining its efficiency even under larger operational demands; its suitability to model real phenomena ranges across a variety of scales including the urban area and the overhead boundary layer.

En model provides major accuracy but undeniable rigidity. The code fits well to building/block-scale problems rather than to district/urban-scale ones. Indeed, large-scale domains require larger meshes, increasing the number of cells and mathematical processes within them. Meantime, in order to accurately reproduce the turbulent fluxes, the spatial discretisation should not be too large since air velocity vectors averaged on large cells do not guarantee stable numerical solutions. ENVI-met requires input parameters from field surveys and urban conditions occurring in the reality at different scales and intervals. As an example, to include the influence of boundary layer on the 3D domain, ENVI-met implements scaling operations to transfer atmospheric phenomena (i.e. specific humidity at 2500 m) down to the 3D domain boundaries (i.e. through logarithmic functions). However, the adjustment between field reality and input parameters can also be left to the modeller. Examples regard the solar radiation adjustment, the initial soil temperatures and the forcing inputs. In these cases, the modeller needs to consider the cyclic tendency and variability of phenomena, as well as their role in model equations, until to choose representative input values (not necessarily field values found at the simulation date).

Gh model scheme is more flexible but less accurate in some aspects. The intersection of plug-ins and different computation methods allows considering distinct domains or surfaces of computation. As an example, a reduced cell number is selected from the entire domain to compute MRT values in certain cells without excessive computational efforts (Fig. 14.7). In this way, *Gh* model may support larger domains, up to district scales. The UWG application requires the input of several averaged parameters describing the urban context in which the domain is inserted. Parameters as traffic intensity, vegetation coverage and building typologies are set for an average urban canyon, in contrast with the heterogeneity of global south urban morphologies growing up fast in the megalopolises. This approach has led to mediated UWG results, not always representative. The integration of weather station data via EPW files allows scaling information from countryside until the urban context under analysis by means of coupled algorithms. EPW values include past consolidated data sets, easily editable (see EnergyPlus Statistic and Conversion) if more recent data want to be used.

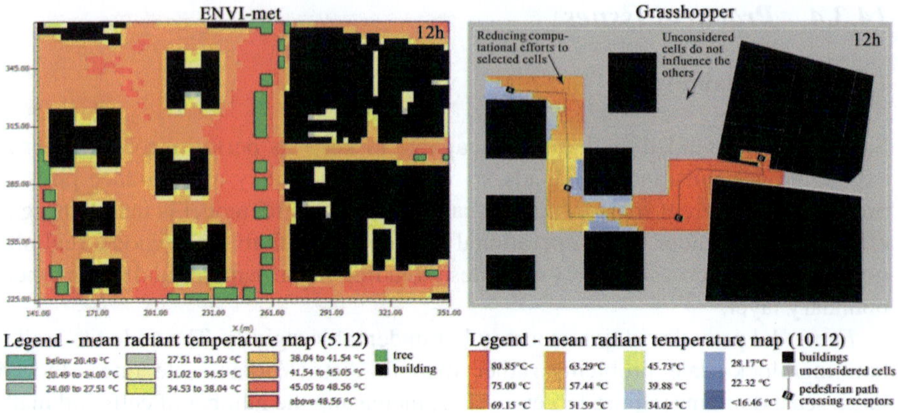

Fig. 14.7 MRT map computed by *G* and *E* models. Two different computational strategies are evidenced

14.3.4.2 History of Use and Testing

The history of applications and uses of ENVI-met, Grasshopper and UWG models concerns a collection of urban studies dealing with urban morphology and environmental problems, parametric experiments, energy consumption and daylighting analyses.

ENVI-met code has a long history of use in the field of microclimate modelling, including thermal comfort investigations (Fazia and Helmut 2006), urban heat island analyses (Emmanuel and Fernando 2007), vegetation studies (Woo-Sik et al. 2009), energy-efficient building forms (Okeil 2010) and modelling experiments. Michael Bruse started to develop ENVI-met in 1995 at the Ruhr-Universität, Bochum (Germany). Such modelling experiments fill the literature with successful findings but also highlight issues and limitations. For instance, in a tropical warm-humid context (Bangladesh), Sharmin et al. (2017) modelled in ENVI-met two residential urban areas in which climate differences were measured on-site. From findings, no relevant variations are simulated in terms of T_a, in agreement with Pacifici's (2019) results obtained at lower latitudes (Brazil). According to the authors, ENVI-met underestimates local air temperature differences resulting from the variation of urban geometry.

Even if the algorithm code was developed by David Rutten (2007), the evolution of Grasshopper is made up by a vast community of users that freely create and share plug-ins. Bueno et al. (2013) evaluated UWG tool against field data extracted from Basel (Switzerland) and Toulouse (France). Nakano et al. (2015) proposed a workflow for the application of UWG as a design tool to climatically orientate the urban geometry and land use in Cambridge, USA. Roudsar et al. (2013) showed the Ladybug potentiality as a design plug-in able to support a broad range of environmental analyses in a single parametric platform. Qingsong and Fukuda (2016) showed in detail the application of Honeybee for building energy and daylighting simulation. Mackey et al. (2017) simulated the outdoor thermal comfort and the buildings surface temperatures in a three-block urban cluster in Singapore, for different wind patterns (CFD studies) and under urban heat island effects. Overall, few studies discuss the simplification of the climate assumptions adopted.

14.3.4.3 Application Scale

The relationship between the numerical code and the scale of application is of critical importance. Models maximise their efficiency within certain ranges of application.

This chapter recommends addressing ENVI-met code to small domains in which the interactions between single urban elements, materials and vegetation at close distance have to be investigated. For the same computational effort, ENVI-met is more adequate to deal with little clusters in which all the physical processes are important for the modeller and a detail scale is required. Other suitable applications could be addressed to large urban domains in which simplified parametric experiments are performed to test sharpened variations. Conversely, to model complex urban fabrics, developers should overcome software limitations and reduce the impact of boundary conditions. The damping of *forcing* should allow the emergence of other important climate exchanges otherwise hidden. Similarly, the interference of wind flux on the heat loss process should also be mitigated.

Conversely, Grasshopper is found more suitable to investigate domains in which only same physical processes are found predominant, allowing more complex geometries and long-term simulation tests. The tool is really appropriate for parametric studies and complex studies requiring the intersection of different environmental analyses. Although in a simpler way, Grasshopper plug-ins are able to reproduce a wide range of phenomena including urban heat island, T_s and MRT distribution, average T_a, outdoor thermal comfort and shadowing. The use of UWG is a convenient choice when modeller is not able to carry out a climate measurement campaign in the studied urban area. For applications whereby a high-detailed scale is needed, ENVI-met is preferred.

14.4 Conclusion

As expected, the application of Grasshopper (*Gh*) and ENVI-met (*En*) models to a standard case of microclimate modelling leads to different advantages and disadvantages depending on the aspects under analysis. *Gh* and *En* are used starting from the same conceptual model and sharing the same modelling purposes. The use of two similar domains guarantees a coherent parallel. The technical comparison involves four key categories and several underlying items. Each item proposes an in-depth analysis of the most important modelling steps, aiming to shed light on these two modelling strategies. In doing so, we hope for supporting users who are tackling with modelling tasks and are wondering about which modelling strategy is more appropriate to their context.

Overall, ENVI-met model is a closed-licence software, whose free basic version is not suitable for urban-scale modelling. Its graphical interface presents an 'assigned attribute-based' structure, easy to be accessed by not programming-proficient users. The all-in-one box structure guarantees full simulations but little customisability. Initial conditions and input parameters require in-depth climate field surveys on the domain under analysis. Temporal discretisation allows short time steps while it does not favour large time steps which would lighten simulation time. Spatial discretisation is a uniform nodal spacing in which irregular buildings need to be rounded. Fine meshes are permitted but with big simulation times. Boundary conditions are imposed by setting the 1D model and forcing air temperature and humidity. Forcing causes excessive influence on the simulated microclimate, hiding other physical processes. The 3D computational fluid dynamics (CFD) allows simulating the evolution of airflow and turbulent exchanges. *En* model simulates many interconnected processes at once, requiring a high computational effort. Longer runs are observed increasing domain size. The code fits well in the modelling of building/block-scale problems rather than district/urban-scale problems. We recommend this tool with simplified geometries and little scales in which all the physical processes are significant and a high-detailed scale is required.

Grasshopper model is freely built up, even if embedded in Rhinoceros, a proprietary 3D modelling software. Interface is more demanding than ENVI-met one but more customisable, involving the user in the development of the algorithm. No initial modelling box is provided to the user. Users have the chance to develop models really close to their study case but with more risks of inconsistency. No proper initial conditions are provided since CFD has not been computed for this modelling case. By means of Urban Weather Generator (UWG), climate input data are obtained from meteorological station, without requiring field investigations. Distinct space and time discretisations are implemented by plug-ins. UWG does not require domain discretisation since average solutions are computed. Temporal discretisation allows large time steps, reducing the entire run time. The urban heat island phenomenon is assumed as a driving force. To obtain reliable UWG predictions, rural conditions have to be away from UHI effects. Grasshopper modular scheme allows selecting the main processes characterising the problem domain, reducing simulation time and enabling larger domains. The interoperating of plug-ins and different computation methods allows considering distinct computation areas. Therefore this model is suggested to investigate domains in which only same physical processes are found predominant, with parametric studies and complex studies requiring the intersection of different environmental analyses.

Summarising, the proper use of modelling tools improves our ability to understand, forecast and plan cities, making possible a more effective design activity, suitable for site-specific contexts and purposes. As shown, the selection of the most appropriate modelling tool is sensitive to a myriad of considerations varying accordingly to specific modeller's purposes, resources and available data. The best compromise should advance our understanding of urban climate systems in the time and cost forecasted.

Acknowledgements I thank the Civil Engineering (POLI) and Architecture (FAU) Faculties of São Paulo University (USP) for supporting my PhD work that inspired this chapter.

References

Anderson, M. P., Woessner, W. W., & Hunt, R. (2015). *Applied groundwater modeling* (2nd ed.). London: Elsevier.

Batty, M. (2009). Cities as complex systems: Scaling, interaction, networks, dynamics and urban morphologies. In *Encyclopedia of complexity and systems science* (pp. 1041–1071). Retrieved from http://link.springer.com/10.1007/978-0-387-30440-3_69

Bear, J. (2018). *Modeling phenomena of flow and transport in porous media* (Vol. 31). Cham: Springer International Publishing.

Bruse, M. (1995). Development of a microscale model for the calculation of surface temperatures in structured terrain (Doctoral dissertation, MSc Thesis, Inst. for Geography, Univ. Bochum).

Bruse, M. (2004). *ENVI-met 3.0: Updated model overview*. Retrieved from http://www.envi-met.com

Bruse, M., & Fleer, H. (1998). Simulating surface–plant–air interactions inside urban environments with a three dimensional numerical model. *Environmental Modelling & Software, 13*(3–4), 373–384. https://doi.org/10.1016/S1364-8152(98)00042-5.

Bueno Unzeta, B. (2010). An urban weather generator coupling a building simulation program with an urban canopy model (Doctoral dissertation, Massachusetts Institute of Technology).

Bueno, B., Norford, L., Hidalgo, J., & Pigeon, G. (2013). The urban weather generator. *Journal of Building Performance Simulation, 6*(4), 269–281. https://doi.org/10.1080/19401493.2012.718797.

Bueno, B., Roth, M., Norford, L., & Li, R. (2014). Computationally efficient prediction of canopy level urban air temperature at the neighbourhood scale. *Urban Climate, 9*, 35–53. https://doi.org/10.1016/j.uclim.2014.05.005.

Chasznar, A. (2011). Navigating complex models in collaborative work for integrated (and sustainable) design. In P. Leclercq, A. Heylighen, & G. Martin (Eds.), *CAAD futures 2011: Designing together* (p. 17). Liege: Les Editions de l'Universite de Liege.

Davis, D., & Peters, B. (2013). Design ecosystems: Customising the architectural design environment with software plug-ins. *Architectural Design, 83*(2), 124–131. https://doi.org/10.1002/ad.1567.

Emmanuel, R., & Fernando, H. (2007). Urban heat islands in humid and arid climates: Role of urban form and thermal properties in Colombo, Sri Lanka and Phoenix, USA. *Climate Research, 34*, 241–251. https://doi.org/10.3354/cr00694.

Fazia, A.-T., & Helmut, M. (2006). Numerical study on the effects of aspect ratio and orientation of an urban street canyon on outdoor thermal comfort in hot and dry climate. *Building and Environment, 41*, 94–108.

Gross, G. (1991). Anwendungsmöglichkeiten mesoskaliger Simulationsmodelle dargestellt am Beispiel Darmstadt, Meteorol. *Rundschau, 43*, 97–112.

Huttner, S. (2012). *Further development and application of the 3D microclimate simulation ENVI-met*. Mainz: Johannes Gutenberg University Mainz.

Khabazi, Z. (2012). *Generative algorithms (using Grasshopper)*. Available online: morphogenesism.

Lagios, K., Niemasz, J., & Reinhart, C. F. (2010). *Animated building performance simulation (ABPS)—Linking rhinoceros/grasshopper with radiance/DAYSIM, New York City, NY* (pp. 321–327).

Lal, A. M. W. (1998). Selection of spatial and temporal discretization in wetland modeling. In *Proceeding Paper, Water Resources Engineering'98, August 7, Memphis, Tennessee, United States*. Reston, VA: American Society of Civil Engineers.

Mackey, C., Galanos, T., Norford, L., Roudsari, M. S., & Architects, P. (2017). *Wind, sun, surface temperature, and heat island: Critical variables for high-resolution outdoor thermal comfort, August 9, San Francisco, USA*.

Matzarakis, A., Rutz, F., & Braunschweig, T. (2005). Application of Rayman for tourism and climate investigations. *Annalen der Meteorologie, 2*, 4.

Naboni, E., Meloni, M., Coccolo, S., Kaempf, J., & Scartezzini, J.-L. (2017). An overview of simulation tools for predicting the mean radiant temperature in an outdoor space. *Energy Procedia, 122*, 1111–1116. https://doi.org/10.1016/j.egypro.2017.07.471.

Nakano, A., Bueno, B., Norford, L., & Reinhart, C. F. (2015). Urban weather generator—A novel workflow for integrating urban heat island effect within urban design process. In *BS2015, December 9, Hyderabad, India*.

Okeil, A. (2010). A holistic approach to energy efficient building forms. *Energy and Buildings, 42*(9), 1437–1444. https://doi.org/10.1016/j.enbuild.2010.03.013.

Organisation for Economic Co-operation and Development (OECD). (2011). Effective modelling of urban systems to address the challenges of climate change and sustainability. *Global Science Forum*.

Pacifici, M. (2019). *Urban morphology and climate: Field assessment and numerical modeling of interactions*. PhD dissertation, USP University, São Paulo. https://doi.org/10.11606/T.3.2019.tde-13082019-150128.

Pacifici, M., de Castro Marins, K. R., de Mello Catto, V., Rama, F., & Lamour, Q. (2017). Morphological and climate balance: Proposal for a method to analyze neighborhood urban forms by way of densification. *Sustainable Cities and Society, 35*, 145–156. https://doi.org/10.1016/j.scs.2017.07.023.

Pacifici, M., Rama, F., & Marins, K. R. C. (2019). Analysis of temperature variability within outdoor urban spaces at multiple scales. *Urban Climate, 27*, 90–104. https://doi.org/10.1016/j.uclim.2018.11.003.

Qingsong, M., & Fukuda, H. (2016). Parametric office building for daylight and energy analysis in the early design stages. *Social and Behavioral Sciences, 216*, 818–828.

Roudsar, M. S., Pak, M., & Smith, A. (2013). *Ladybug: A parametric environmental plugin for grasshopper to help designers create an environmentally-conscious design*. Presented at the 13th International IBPSA Conference, Chambéry, France.

Rutten, D. (2007). *Grasshopper & Galapagos* (1st ed.). Seattle, WA: Robert McNeel& Associates.

Salata, F., Golasi, I., de Lieto Vollaro, R., & de Lieto Vollaro, A. (2016). Urban microclimate and outdoor thermal comfort. A proper procedure to fit ENVI-met simulation outputs to experimental data. *Sustainable Cities and Society, 26*, 318–343. https://doi.org/10.1016/j.scs.2016.07.005.

Sharmin, T., Steemers, K., & Matzarakis, A. (2017). Microclimatic modelling in assessing the impact of urban geometry on urban thermal environment. *Sustainable Cities and Society, 34*, 293–308. https://doi.org/10.1016/j.scs.2017.07.006.

Simon, H. (2016). *Modeling urban microclimate. Development, implementation and evaluation of new and improved calculation methods for the urban microclimate model ENVI-met*. Mainz: Johannes Gutenberg University Mainz.

Taesler, R., & Andersson, C. (1984). Method for solar radiation computations using routine meteorological observations. *Energy and Buildings, 7*(4), 341–352.

Ward, G. (2016). Anyhere software. Photosphere v1, 8.

Wegener, M. (1994). Operational urban models state of the art. *Journal of the American Planning Association, 60*(1), 17–29. https://doi.org/10.1080/01944369408975547.

Woo-Sik, J., Jong-Kil, P., Hwa, W. L., Eun-Byul, K., & Hyo-Jin, C. (2009). Wind speed variation over the leeward region according to vegetation under the strong wind. *Berichte Des Meteorologischen Instituts Der Albert-Ludwigs-Universität Freiburg, 19*, 255–262. https://doi.org/10.1007/s00501-007-0331-8.

Yamada, T., & Mellor, G. (1975). A simulation of the Wangara atmospheric boundary layer data. *Journal of the Atmospheric Sciences, 32*(12), 2309–2329.

Chapter 15
Urban Microclimate and Building Energy Simulation Coupling Techniques

Auline Rodler, Nicolas Lauzet, Marjorie Musy, Marie-Hélène Azam, Sihem Guernouti, Dasaraden Mauree, and Thibaut Colinart

15.1 Introduction

The urban climate is the result of many factors that modify the climatic exchanges in the city: the shape and density of urban fabric, the thermo-physical properties of urban surface and the generation of heat and pollutants from anthropogenic sources. Indeed, many climate phenomena alter the energy balance of urban systems, such as the trapping of solar radiation, the enhancement of thermal storage and the decrease of latent heat transfer due to lack of water and vegetation (Oke 2002).

The resulting urban microclimatic conditions are often characterised by higher temperatures and reduced mean wind velocities than in the surrounding rural area.

A. Rodler (✉)
Cerema, Equipe-Projet BPE, Nantes, France
e-mail: auline.rodler@cerema.fr

N. Lauzet
Université de Bretagne Sud, UMR CNRS 6027, IRDL, Lorient, France

TRIBU Environmental and Sustainable Development Consulting Agency, Lyon, France

M. Musy · M.-H. Azam · S. Guernouti
Cerema, Equipe-Projet BPE, Nantes, France

Université Nantes, CNRS UMR 6183, GeM, Nantes, France
e-mail: marjorie.musy@cerema.fr; marie-helene.azam@univ-nantes.fr; sihem.guernouti@cerema.fr

D. Mauree
Solar Energy and Building Physics Laboratory, Ecole Polytechnique Fédérale de Lausanne, Lausanne, Switzerland

BG Ingénieurs Conseil SA, Vernier, Switzerland

T. Colinart
Université de Bretagne Sud, UMR CNRS 6027, IRDL, Lorient, France
e-mail: thibaut.colinart@univ-ubs.fr

M. Palme, A. Salvati (eds.), *Urban Microclimate Modelling for Comfort and Energy Studies*, https://doi.org/10.1007/978-3-030-65421-4_15

Nevertheless, the weather input data usually used in BEM are not those of the specific site where the building is built and are hence not representative of the local microclimatic conditions. This uncertainty is partially responsible for energy performance discrepancies between simulations and actual measured performances (Bouyer et al. 2011; Johnston et al. 2015; Malys et al. 2015; Mauree et al. 2017; Santamouris 2014). In several works, it has been highlighted that the urban microclimate has a significant impact on the calculation of energy performance and comfort conditions in buildings (Goffart et al. 2017; Rabouille et al. 2013). In particular:

- The solar energy potential of a city can be very heterogeneous due to shadows and solar protection of buildings and multi-reflections between urban surfaces. This variation has an impact on building cooling and heating demand.
- The sky view factor can be reduced in streets. This has a direct impact on net long-wave exchanges (modulating the part that occurs with other built or non-built surfaces and the part that occurs with the sky).
- The modification of the local wind velocity, pressure and air temperature affects convective exchanges.
- The increase of the air temperature can lead to less heating demand but higher cooling demand.
- The nature of surfaces (imperviousness and water content, inertia, optical characteristics) affects the urban surface temperature as well as the air temperature and long-wave exchanges.

These elements explain why the urban climate can differ from the rural one, leading to temperatures that are often higher in urban districts than in rural areas; this is commonly called the urban heat island (UHI) effect. One of the reasons why the UHI is usually not taken into account in BEM is that it needs to be assessed at a scale larger than that of buildings (Mirzaei and Haghighat 2010; Salvati et al. 2017a, b; Santamouris et al. 2001).

The heat fluxes through building envelopes may be substantially modified in an urban context; these are:

- Convective heat flux at the external surfaces and the heat losses due to ventilation and air infiltration.
- Solar radiation exchanges.
- Long-wave radiation exchange with the sky and surrounding surfaces.

Many approaches have been proposed to account for urban microclimate in building performance simulation. However, in most of them, not all the urban climate and energy phenomena are considered. For example, the surrounding urban forms are considered for the calculation of shadows, but they are often not considered in the thermal long-wave exchanges occurring between the building and the environment (Miller et al. 2018). Yet, some authors have demonstrated that in an urban context, calculating the thermal interactions and solar inter-reflections is necessary to achieve an accurate energy performance assessment (Lauzet et al. 2017; Vallati et al. 2018). Additionally, the local wind intensity and direction can vary strongly between different sites (Allegrini et al. 2015), thereby modifying significantly the convective heat exchanges. This chapter aims to present the different coupling strategies between BEMs and urban climate models for a better estimation of building's performance and comfort.

First, we present different UCMs that can be used for this purpose. Then we show how BEM can handle climate boundary conditions produced by UCM. Finally, we explain the strategies used to implement the interaction between BEM and UCM models.

15.2 Overview of Urban Climatic Models

A set of computational models were developed to characterise urban microclimates from the largest to the smallest scale. These approaches include urban canopy models and city energy models, district models (either parametric or with explicit geometry), zonal models or canyon street models. These models involve different spatial and temporal resolutions and modelling assumptions. The list of tools presented below is not meant to be exhaustive; we have only highlighted here tools that are used more frequently and are more relevant to the current discussions.

- The city models are large-scale models. The local characteristics are parametrised at low resolution and do not represent the real 3D urban forms of the city. The dimension of the cells is larger than 200 m. The models can study the interaction between the urban surface and the urban boundary layer. In this category, we can find tools like MESO-NH (Lac et al. 2017) or WRF (Skamarock et al. 2008) where urban parameterisation like TEB (Masson 2000) and BEP (Martilli et al. 2002) is, respectively, integrated. TEB (Masson 2000) will be further presented in the coupling projects. TEB is a two-dimensional urban canyon model, composed of a road and two buildings of equal height. The surfaces of the urban canyon—roof, walls and road—are characterised in terms of materials (i.e. albedo, emissivity and thermal capacity) and vegetation. WRF-BEP has a similar approach except that it has multiple layers representing the building in the urban canopy where it computes local fluxes while MESO-NH-TEB performs this operation in one single layer. In WRF-BEP, the canopy is divided into several vertical layers where the model computes each variable. This allows better prediction of the canopy physics as it can include building height distributions and detailed prediction of street-level climate. However, compared to single-layer models, this approach is more computationally consuming.
- Regarding the district-scale models, we can find two types of models. The first does not represent explicitly the urban forms but uses parameters to translate their impact on UWG (Bueno et al. 2013a, b), CAT (Erell and Williamson 2006) and CIM (Mauree et al. 2017). The second type of models relies on the real 3D geometry of the district (ENVI-met (Bruse and Fleer 1998), SOLENE-Microclimat (Musy et al. 2015)). Among the first family of models, we can find UWG. It is based on energy conservation principles and is a bottom-up building

stock model. It considers parameters representative of the urban morphology, geometry and surface materials. UWG has several coupled modules. It comprises the rural station model, the vertical diffusion model, the urban boundary layer model and the urban canopy. This model translates rural temperatures into urban canyon temperatures. CIM is a column model that resolves the flow using the Navier-Stokes equations reduced to one dimension. It explicitly resolves the turbulent kinetic energy taking into account the presence of obstacles using a modified 1.5-order turbulence closure. Among the second type of models, we can find ENVI-met. The district geometry is explicitly represented in 3D. It is a widely used tool that includes airflows, short-wave radiation and heat and moisture transfer and considers vegetation. Another tool belonging to this second type of models is SOLENE-Microclimat.[1] It is a coupling of a CFD tool (Saturne) with radiation models and thermal urban surface models. The advantage of the second group of models is that there is an explicit representation of the buildings and hence they resolve the flow with a higher resolution. This however means that they require more computational resources and cannot be used for simulations longer than a few days. Longer temporal simulation can thus only be performed with the first group of models, although they lack a detailed urban geometry representation.
- The street-scale model is the lowest scale of urban climate models. It is a street bordered by walls in which solar trapping and airflow are calculated. The canyon model TEB can be used to study a particular street for example. Other models exist, such as the one developed by Bozonnet (2006): these zonal approaches mainly calculate airflows on the street canyon scale and can involve thermo-radiative modelling. Solene tool is used for the solar radiation and inter-reflections (resolutions and methods are given in Chap. 13). Their spatial resolution and physical bases enable local microclimatic conditions.

15.3 Building Energy Models

The BEMs presented here are physics-based models used to estimate building energy consumption and thermal comfort based on the thermal transfer principles. They require a weather file (ambient temperature, humidity, radiation, etc.) and information about relevant physical characteristics of a building, such as building geometry, HVAC systems, usage patterns, building materials, thermostat set points, occupancy rates and schedules, and internal loads. These tools are usually adapted to the simulation of a single building with a high temporal resolution (some minutes) and are used for long-period simulation (typically a year).

[1]The model physics is described in detail in Chap. 13 by Musy, Azam, Guernouti, Morille and Rodler.

In this section, we focus only on the description of the heat fluxes affecting the buildings' energy performance in an urban context that can potentially be simulated using UCM models and be used by the BEM as input. Solar radiation affecting the outside of the building envelope including the short-wave and the long-wave radiation is going to be detailed. Finally, the aeraulic exchanges and ventilation are discussed (Fig. 15.1). Thermal conduction through walls will not be explored in this chapter, as this heat flux is often considered in the same way in the BEMs.

15.3.1 Short-Wave Radiation

Direct and diffuse radiations are always considered in the BEMs. The shadows from distant obstructions can be considered as well as the shadow from closer obstructions (neighbouring obstacles).

The calculation of the inter-reflections between building surfaces and the surrounding urban surfaces or buildings requires geometrical models and the knowledge of the properties such as the albedo of the surfaces (Fig. 15.1). The inter-reflections are often only considered with the soil and not between the buildings (Table 15.1). Some tools like EnergyPlus (Crawley et al. 2001) have a 'reflection option' to consider the building surface reflection. This allows to consider the

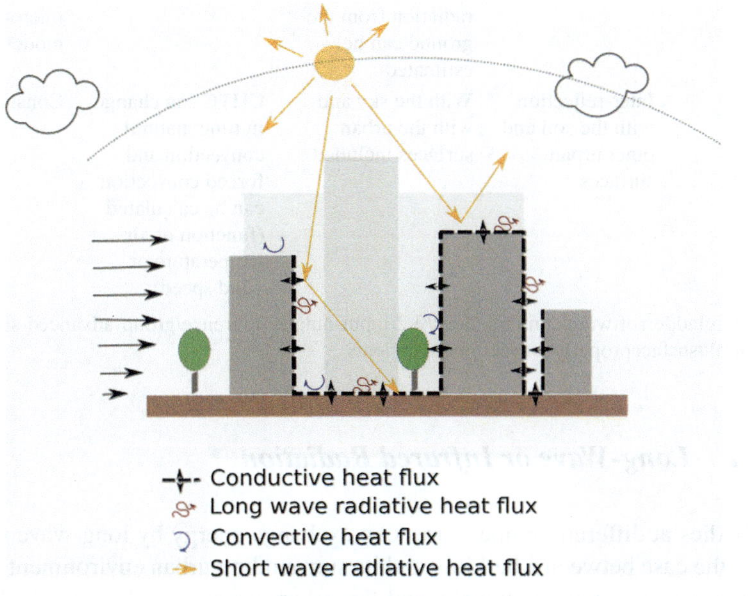

Fig. 15.1 Heat exchanges occurring around the building

reflections from the surrounding buildings' facades on the calculation of the building solar gains. However multiple reflections are not included.

Table 15.1 Radiation, convection and ventilation heat fluxes for several BEMs

BEM	Short-wave heat flux (solar radiation)	Long-wave heat flux (infrared radiation)	Convective heat flux (aeraulic)	Ventilation and infiltration flux
TRNSYS	Inter-reflection only with the soil	With the sky and ground	Constant CHTC	Constant or TRNFLOW model
EnergyPlus (design builder)	Reflection with the soil in the standard version and one bounce with the building surfaces when taking the option	Only with the sky when it is the default setting. The ground and surrounding surfaces can be included using 'local environment objects'	Usually constant CHTC but can be changed by the user[a]	Constant or natural ventilation can be modelled with the 'airflow network'
PLEIADES	Inter-reflection only with the soil	Only with the sky included in the CHTC	Constant CHTC	Constant or flow rate based on CONTAM model
BuildSysPro	Inter-reflection with the soil and can include other urban surfaces by external calculation	With the sky and with the urban surfaces included (an additional module is required)	Constant CHTC	Constant or needs to be implemented
IES-VE	Inter-reflection only with the soil	Only with the sky and long-wave radiation from the ground can be estimated	Constant CHTC	Constant or macro-flow or micro-flow model
CitySim	Inter-reflection with the soil and other urban surfaces	With the sky and with the urban surfaces included	CHTC can change in time: natural convection and forced convection can be calculated (function of air temperature or wind speed)	Constant

[a]https://bigladdersoftware.com/epx/docs/9-2/input-output-reference/group-advanced-surface-concepts.html#surfacepropertyconvectioncoefficients

15.3.2 Long-Wave or Infrared Radiation

Two bodies at different temperatures can exchange energy by long-wave radiation. This is the case between a building and its surrounding urban environment. The surrounding surfaces include the sky and the urban surfaces such as roads, buildings or trees.

Usually, only exchanges with the sky are considered in BEM. The sky temperature is calculated as a function of the air temperature and cloud cover. Therefore, it is possible to consider this transfer as aggregate with the external convection by modifying the convective heat transfer coefficient. This is done in PLEIADES (Peuportier and Sommereux 1990) and IES-VE. In EnergyPlus (Crawley et al. 2001), the infrared exchange is usually only calculated with the sky by using view factors between surfaces and the sky. However, these tools neglect the thermal exchanges between built surfaces (ground, buildings) though it is possible now with EnergyPlus.[2]

Only few studies consider the long-wave exchanges between urban surfaces, which is possible now. BuildSysPro (Plessis et al. 2014) and CitySim (Robinson et al. 2009) consider explicitly the exchanges between the sky and the built surfaces (Table 15.1) (Walter and Kämpf 2015).

It should be noted that TRNSYS (Klein et al. 2017) and EnergyPlus have a high modularity meaning that the infrared calculation could be extended and changed by an expert user (Salvati et al. 2020) (Table 15.1).

15.3.3 Convection

The calculation of the convective heat flux is based on the computation of the convective heat transfer coefficients (CHTC). The CHTCs depend on the wind direction and intensity as well as the local thermal air gradient. The CHTCs are commonly assumed to be constant in the BEM. If implemented, they can be a function of the wind speed when the CHTC is for the outside walls and/or a function of surface-air temperature for the inside walls. As the BEM does not calculate near-wall values of wind speed or outside-air temperature, these values generally come directly from the weather files and neglect the local microclimatic conditions.

Some tools are more flexible than others so that the CHTC can be changed. In TRNSYS and BuildSysPro, for example, natural and forced convection can be implemented. In EnergyPlus, the CHTC can be changed by using the adaptive convection algorithm. In IES-VE forced convection can be chosen (Table 15.1).

BEMs use pressure coefficients to calculate wind pressure and infiltration airflows. Those coefficients depend upon the exposure of the surface to wind. The wind speed can be corrected by considering the surface and site rugosity. The pressure coefficients are usually considered constant in BEMs. The heat gains or losses due to ventilation airflows are calculated in BEMs using the outside-air temperature from the weather file.

[2] https://bigladdersoftware.com/epx/docs/8-8/input-output-reference/group-advanced-surface-concepts.html#surfacePropertysurroundingSurfaces.

15.3.4 Ventilation and Infiltration

BEMs use pressure coefficients that vary depending on the exposure of the surface to wind (upwind or downwind) to calculate the wind pressure. The wind speed is often calculated based on the surface height and the site rugosity. Infiltration flows can be assessed and different models exist (Table 15.1).

15.3.5 Other Heat Fluxes

The moisture transfer and corresponding heat fluxes in the materials which can impact the calculation of the internal air temperature can be considered in tools such as EnergyPlus, but not in all the existing BEMs. Often, relative humidity of the air is given in the weather files and it is used directly as input.

Evapotranspiration of the surrounding vegetation is often neglected in BEMs. These physical phenomena can influence the surface temperature of a building due to solar absorption of the vegetation or due to latent energy consumed for evapotranspiration by the vegetation. Among the tools presented only CitySim has an evapotranspiration model (Coccolo et al. 2018).

As BEMs have been developed before UCM, they are adapted to integrate external climatic conditions at mesoscale only. This means that BEMs can be forced with only one weather file.

15.4 Co-simulation Strategies: Coupling or Chaining

Co-simulation in the area of building physics has already been addressed whereas this has hardly been the case for UCM. In the building performance simulation area, authors have shown that there is a need for interoperable simulation environments (Marija and Jan 2006; Vanpachtenbeke et al. 2015; Wetter 2011).

Different methods of coupling tools exist (Hensen 1999) and three different strategies are identified:

- Strong coupling (de Sturler et al. 2000) strategy or full dynamic coupling (Zhai 2004) is defined as the combination of two models that iterate at the same time within each time step until the error estimate falls within a predefined tolerance (Fig. 15.2). This method is also named as onion coupling in building performance simulation field (Hensen 1999).
- Weak coupling: The models are run in sequence, and each model uses as boundary condition the known output values (of the previous time step) of the other model (Fig. 15.2). This method is also named the ping-pong coupling (Hensen 1999).

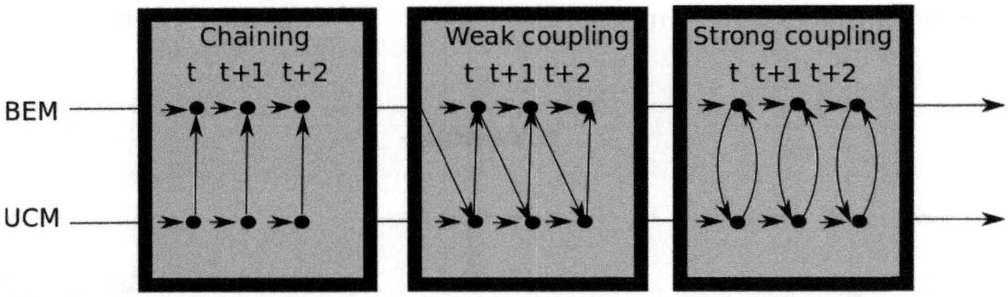

Fig. 15.2 Chaining, weak coupling or strong coupling between BEM and UCM

– Chaining methods: The first model is solved and then the outputs are used directly as input to the second model to be solved for the same time step: it is a one-way communication (Fig. 15.2).

Most of the time the weak coupling or chaining is used, while the strong coupling is rarely used. In the next sections of this chapter, we will address the coupling strategies between UCM and BEM.

15.4.1 Coupling or Chaining UCM and BEM

UCM and BEM tools can be linked through physical variables or heat fluxes. The coupling strategies allow feedbacks of buildings' use on the local climate and reciprocal. For instance, the use of air-conditioning systems releases heat calories directly to the outside environment and can increase the outdoor-air temperature. At the opposite heat waves occurring outside the building can affect the comfort of occupants in the building and the building's thermal behaviour (Fig. 15.3).

In strong and weak coupling methods, UCM and BEM simulations interact with each other. Finally, the chaining method consists of making a whole UCM simulation and providing the results into weather files used in a BEM tool (Fig. 15.3).

Most of the BEM tools need a single year of typical data where all the weather variables are defined, similar to a typical meteorological year. The hourly weather data given are the global horizontal solar radiation, dry bulb temperature, humidity ratio and wind speed. However, the local heat fluxes are influenced not only by these variables but also by the solar inter-reflections, thermal long-wave exchanges with the urban surfaces and the sky, etc. Therefore, tools with explicit geometry are needed to calculate the shadows, and the short-wave and long-wave exchanges between buildings.

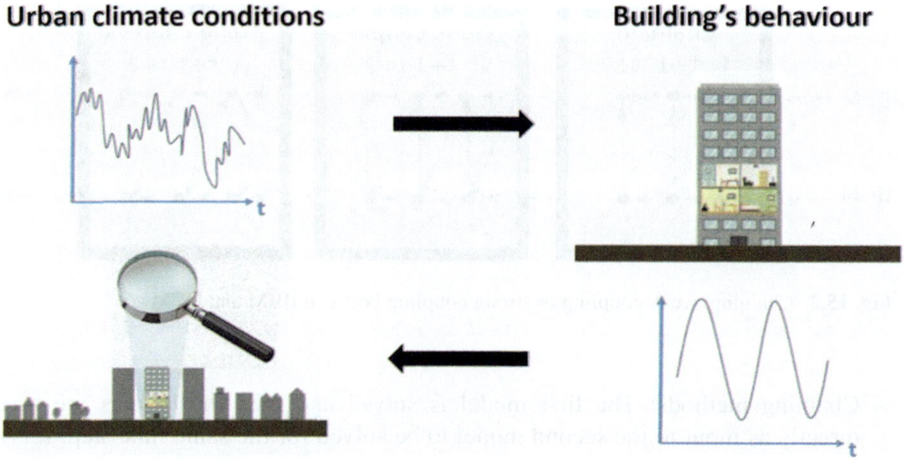

Fig. 15.3 Interaction between UCM and BEM

15.4.2 Chaining Strategies Between UCM and BEM

The chaining strategies use the outputs of a UCM directly as inputs to the BEM but with no feedback to the UCM. In the presented projects (Fig. 15.4), the weather inputs can be directly modified, such as air temperature and humidity, or the heat fluxes, like the long-wave or convective fluxes.

Palme et al. (2017) presented a methodology to link a modified local weather file to a TRNSYS building simulation. In this case, chaining is quite simple as it only consists of modifying part of the input weather files without changing particular parameters of the model: the local air temperature and the relative humidity are modified by UWG (Fig. 15.4). Salvati et al. (2017a, b) also used UWG to calculate hourly values of urban air temperature given the weather data measured at an operational weather station located outside the city. This calculated urban air temperature is then used as an input in the BEM (EnergyPlus).

The energy needs of a typical building located in the urban centre of Rome are estimated by chaining WRF to EnergyPlus (Ciancio et al. 2018) on a yearly simulation. WRF is used to produce the input file for EnergyPlus. In this study, WRF gave the dry bulb temperature, dew-point temperature, relative humidity, atmospheric pressure, and wind speed and direction.

Pappaccogli et al. (2018) used WRF coupled with the building effect parameterisation (BEP) to provide the meteorological file which is used as input data for the building energy model implemented with TRNSYS. Then WRF is launched to provide meteorological fields at the urban scale and surface temperature of the external building envelope (400 m spatial resolution). The WRF/urban output is provided according to the height and orientation of the façade of the simulated building in the TRNSYS model. The weather parameters (i.e. beam and diffuse irradiation on a horizontal surface, air temperature and humidity, wind velocity) calculated by

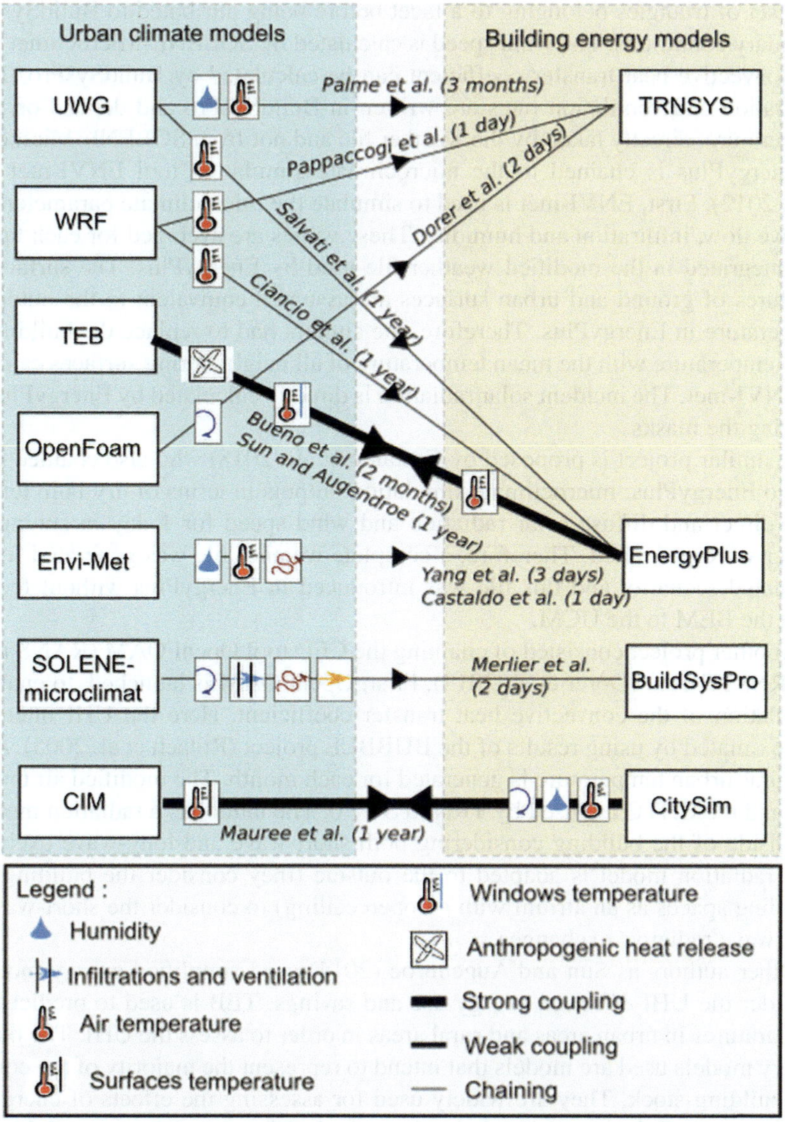

Fig. 15.4 Co-simulation projects between BEM and UCM. (Adapted from Lauzet et al. 2019)

WRF/urban are finally chained and then used in a daily TRNSYS simulation for only 1 day in July. External convective heat transfer coefficients, CHTCs, were provided as inputs to the TYPE56 of TRNSYS and they were correlated to the wind velocity.

Other authors chained the BEM BuildSysPro to SOLENE-Microclimat (Merlier et al. 2019a, b). The long-wave exchanges with the sky and urban scene are considered and are directly calculated by SOLENE-Microclimat. They are averaged for each set of triangles belonging to a facet before being attributed to BuildSysPro as boundary conditions. The wind speed is calculated by SOLENE-Microclimat so that the convective heat transfer coefficient can be calculated by BuildSysPro. The air infiltration and ventilation rates are written in BuildSysPro and depend on the air temperature, directly taken by the weather file and not from SOLENE-Microclimat.

EnergyPlus is chained to the microclimate simulation tool ENVI-met (Yang et al. 2012). First, ENVI-met is used to simulate the microclimate parameters, convective flow, infiltration and humidity. These values are averaged for each face and are integrated in the modified weather file used by EnergyPlus. The surface temperatures of ground and urban surfaces are assumed equivalent to the outdoor-air temperature in EnergyPlus. Therefore, the authors had to replace the building surface temperature with the mean temperature of all neighbouring surfaces calculated by ENVI-met. The incident solar radiation is directly calculated by EnergyPlus considering the masks.

A similar project is proposed by Castaldo et al. (2018) who also chained ENVI-met to EnergyPlus: microclimate simulation outputs in terms of dry bulb temperature, direct and diffuse solar radiation and wind speed for 1 day in summer and winter were generated. Therefore, a complete weather file was generated for each simulated scenario, and this file was introduced in EnergyPlus without feedback from the BEM to the UCM.

Another project consisted of chaining the CFD tool OpenFOAM (RANS model) to TRNSYS 17.0 (Dorer et al. 2013). First, OpenFOAM is launched, to enable the calculation of the convective heat transfer coefficient. Here the UHI intensity is approximated by using results of the BUBBLE project (Rotach et al. 2005). A diurnal rural-urban temperature is generated for each month. The modified air temperature and the CHTC are used by TRNSYS 17.0. The latter has a radiation model for the inside of the building considering both short-wave and long-wave exchanges. This radiation model is adapted to the outside (they consider the building's surrounding spaces as an atrium with an open ceiling) to consider the short-wave and long-wave radiative exchanges.

Other authors as Sun and Augenbroe (2014) used simplified energy models to consider the UHI effect on energy use and savings. TEB is used to predict the air temperatures in urban areas and rural areas in order to assess the UHI. The building energy models used are models that intend to represent the majority of the commercial building stock. They are widely used for assessing the effects of energy efficiency, new building energy codes and standards. These models are available in the form of EnergyPlus input files (Deru et al. 2011).

15.4.3 Coupling Strategies Between UCM and BEM

EnergyPlus and TEB are coupled with a two-way interaction (Bueno et al. 2011). The algorithm corresponds to a weak coupling aiming to analyse how the outdoor-air temperature is impacted by heat emissions of a HVAC system. First, the solar radiation received by the building surfaces and the long-wave radiation are calculated by the average-oriented canyon in TEB. The convective heat flux is calculated in TEB using wall CHTC. Then, the external surface temperature is imposed in EnergyPlus as boundary conditions. It is not clear whether canyon air temperature is taken into account as an inlet temperature for the ventilation and the air infiltration heat flux.

Another project consists of coupling the canopy interface model (CIM) and CitySim with the aim to provide an annual estimation of the energy demand at the district scale (Mauree et al. 2017). First, a simulation with CitySim is performed with the standard typical meteorological year data and the surface temperatures are obtained. CIM is then forced with the surface temperature from CitySim. Long-wave radiation from buildings and ground surfaces is considered. The vertical profile of meteorological variables is recalculated in CIM, such as the air temperature and the wind speed. Finally, CitySim runs with the generated localised meteorological variables and simulates thus the district energy demand. The variable modification influences the computation of the convection as well as the calculation of the long-wave exchanges.

15.5 Discussion

15.5.1 Impact of Making UCM and BEM Co-simulation on the BEM Behaviour

In this section, we will quantify the impact of coupling or chaining a UCM to a BEM on the calculation of the heating or cooling demand. Warmer outdoor temperatures with respect to those in standard weather data files were shown to lead to an average increase of total building energy demand in a year from 10% to 35% (Salvati et al. 2017a, b). These results can be extended to the Mediterranean climate but not to all climates.

An evident observation is that the inclusion of the UHI effect leads to lower heating demand and higher cooling demand estimations for building in urban environment (Guattari et al. 2018; Sun and Augenbroe 2014), whatever the co-simulation method employed. For example, an average annual increase in air temperature of 2 °C corresponding to the UHI amplitude leads to an average increase of 17.25% in office building cooling needs when building heating energy use is affected by an average of 17.04% decrease. This is a mean value given over 15 climate zones of the USA (Sun and Augenbroe 2014).

Concerning the cooling load, the impact of coupling or chaining a UCM to a BEM has been extracted from the literature studied in this chapter and is shown in Fig. 15.5. When the coupling to a UCM is neglected, we can observe a median building cooling energy consumption difference of 50%. We can observe a large variation between studies, ranging from 10% to 200% globally (Fig. 15.5) (Ciancio

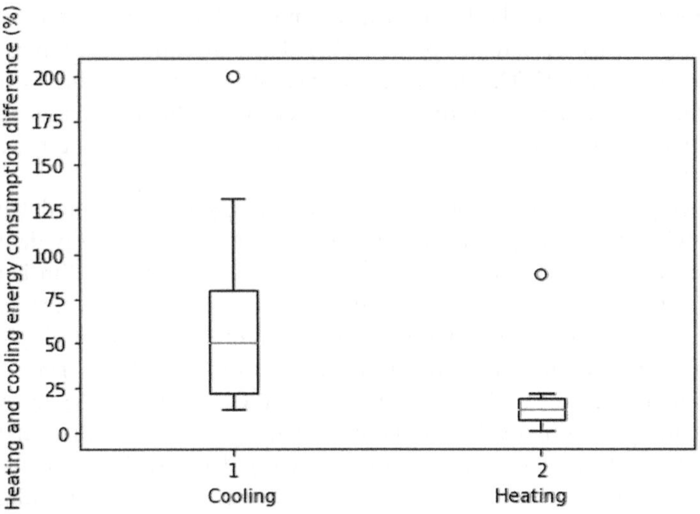

Fig. 15.5 Impact of co-simulation between BEM and UCM on heating and cooling consumptions

et al. 2018; Kolokotroni et al. 2007; Merlier et al. 2019a; Palme et al. 2017; Pappaccogli et al. 2018; Salvati et al. 2017a, b; Sun and Augenbroe 2014; Yang et al. 2017).

There are less papers focusing on the impact of coupling or chaining a UCM to a BEM to consider the UHI on building heating demand. When the local urban climate is considered, a median decrease in heating demand of 10% is observed, with variations between 3% and 89% (Fig. 15.5) (Ciancio et al. 2018; Kolokotroni et al. 2007; Merlier et al. 2019a; Palme et al. 2017; Salvati et al. 2017a, b; Sun and Augenbroe 2014; Yang et al. 2017).

The differences due to the UHI on heating and cooling demand are very different from one paper to another, probably due also to the chaining or coupling method accuracy, and the accuracy of the inputs used. Authors have tried to show if these discrepancies are dependent on the region's climate. It is difficult to conclude, as Li et al. (2019) conclude in their review that the discrepancies do not seem to be linked to a geographical pattern. On the other hand, Salvati et al. (2020) show that urban climate varies significantly across a city depending on the density of urban texture and its impact on the annual energy demand depends on the region's climate.

However, these differences can be due to intercity variations of the UHI. For example, Sun and Augenbroe (2014) showed that there is a clear intercity variation of the UHI, which has an impact on the energy consumption. For different urban areas, the same authors present results based on case studies of office buildings in 15 representative cities across different climate zones in the USA. Urban areas are divided into categories according to urban density: tall-building districts, and high-, medium- and low-density cities. The building models represent 16 types willing to represent the majority of the building stock (small, medium and large offices). UHI leads to an average of 12.5% reduction in heating for those in urban and suburban areas, and an average of 18.9% reduction for those in large city centres (Sun and Augenbroe 2014).

Also, the differences observed between the studies in this field can be explained by the different building uses analysed: for offices or residential buildings the occupancy presence and the internal thermal loads are different so the impact of the UHI on the energy demand can vary. For example, Street et al. (2013) reported stronger UHI impacts on building cooling energy consumption for office buildings than those for residential buildings, both located in the same city and district.

There are few papers focusing on the impact of coupling or chaining a UCM to BEM and highlighting the contribution of the different boundary conditions (microclimatic factors).

It was found that the contribution of long-wave radiative heat transfer mostly explains the differences in the cooling needs observed when considering local boundary conditions derived from the microclimatic tools instead of the default approach (no coupling or chaining between BEM and UCM) (Merlier et al. 2019a). The cooling needs increase by 130% when considering the long-wave radiative heat transfer. The simultaneous modifications of all boundary fluxes increase cooling needs by 100% (Merlier et al. 2019a). Yang et al. (2012) found that the total cooling load reduces by 10.6% after taking into account the full microclimatic environment and by 1.1% when considering only the long-wave exchanges.

The second most important change in cooling needs shown by some authors is induced by the modification of short-wave radiation fluxes in urban context, representing a reduction of 25% of energy (Dorer et al. 2013; Merlier et al. 2019a). Here the short-wave radiation considers the solar masks and the inter-reflections and is calculated by the UCM, but could have been modelled by the BEM directly. In some studies, this is found to be the most important contribution of urban context, reducing cooling loads and increasing heating loads, respectively, by 18.8% and 0.8% (Yang et al. 2012). The reduction of incident radiation due to urban obstruction can be desirable in Mediterranean climate; it allows a reduction of annual energy demand up to 19% compared to an unobstructed environment (Salvati et al. 2017a, b).

Considering the increased CHTC derived from the microclimatic boundary conditions induces a decrease in cooling needs by 25% (Merlier et al. 2017). Finally, the aeraulic-induced heat transfer increases cooling needs by 0.8% and infiltration by 9.6% (Yang et al. 2012). However, it is the UHI impact represented by the air temperature variable which seems to be more important than the radiation or the CHTC coefficient (Dorer et al. 2013).

In the presented approaches, authors tend to consider only the UHI (Ciancio et al. 2018; Palme et al. 2017; Salvati et al. 2017a, b) but do not correct all the local boundary conditions. On the other hand, few authors correct the local heat fluxes but do not consider the UHI effect on the air temperature (Merlier et al. 2019a; Yang et al. 2012).

15.5.2 *The Remaining Technical Locks for BEM and UCM to Co-simulate*

Most of the authors do not consider all urban local fluxes (boundary conditions) simultaneously. The heat fluxes are only partially corrected in the BEM. For example, in some projects, authors alter the convective exchanges only and the ventilation by introducing a local air temperature (Salvati et al. 2017a, b). But for the same authors the wind velocity is not locally calculated so the convective heat flux is only partially corrected using the local air temperature. A key limitation is the assumption that only the air temperature varies in the urban context compared to the rural one. A change in humidity and wind velocity in the urban area is expected to produce a change in the natural ventilation rate, convective heat transfer and latent loads.

If for some authors all the local urban heat fluxes are taken as input in the BEM, they do not introduce a feedback from the BEM to the UCM as they do only a chaining as seen before (Merlier et al. 2019a, b). This can lead to an error due to the difference in assumed and actual surface temperatures for example.

Some BEMs are not adapted to co-simulate with other tools as they do not offer the possibility to modify some of their variables of equations. For example, Yang et al. (2012) needed to modify the equation of the CHTC so as to consider the local outside-air temperature variations. It cannot be modified directly in the used version of EnergyPlus.

It can be difficult to match the geometries of both UCM and BEM. For example, Merlier et al. (2017) calculated the radiation with SOLENE-Microclimat for each small triangle of the urban scene (less than 1 m^2). This heat flux is then averaged for each wall (more than 20 m^2) orientation in Dymola. It leads to a 'loss' in accuracy between tools. Pappaccogli et al. (2018) have the same problem as TRNSYS represents a component of the simulated building, while the WRF/urban model is a representative surface of the 400 × 400 m grid cell. These geometrical simplifications will not allow to calculate the impact of a local adaptation solution which would be placed especially to optimise its effect. In this context, Mauree et al. (2018) have proposed a new coupling that provides improved coupling between WRF and BEP to create high-resolution profiles that can then be used directly in standard BEMs.

Building energy simulation is usually performed by using as an input the weather conditions of an entire year. Urban microclimate models can be launched in relatively small periods because of the computational time, so making UCM and BEM co-simulate is only possible for a period of some days, not more. On the opposite, when using the tool UWG, monthly and annual simulations can be launched, but the disadvantage is that only the air temperature and humidity are corrected and not all the urban heat fluxes.

15.6 Conclusion

In this chapter, an overview of the different ways to consider the urban microclimate in building design simulations is presented. The UHI shows to have a clear impact on the heating and cooling demand of buildings with great variation, ranging from 10% to 200% for cooling demand and 3% to 89% for heating demand. The impact of the local microclimate boundary conditions on the BEM is less studied. However, they show to have an important impact on the heating and cooling demand mainly due to the long-wave and short-wave heat fluxes, reducing cooling loads, respectively, by 130% and 25%.

Secondly, the remaining technological locks for BEM and UCM to co-simulate are given. Workarounds are often considered by the authors to make BEM and UCM co-simulate. It can be difficult to match the geometries of both UCM and BEM. Also, some equations of the BEM cannot be changed, so the heat fluxes coming from the UCM cannot be directly used as a boundary condition to the BEM. Finally, in the coupling or chaining projects, authors do not correct all microclimatic boundary conditions. All these drawbacks can lead to a loss in accuracy between both UCM and BEM. Finally, making UCM and BEM co-simulate is today only possible for a period of some days, but not for annual simulations, unless parametric UCM tools are used, as UWG or WRF.

Future work should be about how to validate with experimental measurements the simulation results obtained when making UCM and BEM co-simulate. Some authors have started validating UCM simulations. For example, Tsoka et al. (2019) show that their microclimate results can capture microclimate characteristics, where dry bulb temperatures were reported compared to the corresponding values at the reference location, with indicative mean daily deviations up to 1.0 and 0.75 °C in February and July, respectively. Pappaccogli et al. (2018) compared the air temperature given by WRF in an urban context to measures of the city of Bolzano: WRF performs satisfactorily, with most bias values around or below 1 °C.

If the authors tend to validate the UCM results, few tend to validate both the UCM and BEM results, which should be done in the future. With the upcoming developments of UCM, a part of the urban microclimate parameters should be more easily transferred to the BEM through a new local climatic file (local wind velocity, air temperature or humidity). However, taking into account the fluxes that depend on surfaces that are directly surrounding the buildings, such as radiative fluxes, requires access to these fluxes' calculation within the BEM.

For BEM issues when considering sufficiently small time steps, the weak and strong coupling strategies give similar results (Marija et al. 2007). However, for hourly time steps, the errors could become important, but this has not been shown in the presented coupling projects. Finally, for BEM authors show that weak coupling can lead to shorter execution time than strong coupling, but might not be as accurate. Also, regarding the time step, modellers should be able to specify the simulation time step, and it is sufficiently small to assure the stability of the results (Marija and Jan 2006). The algorithms to chain or couple BEM and UCM (weak or strong coupling and chaining) have not been compared yet. In future, a benchmark between each technique should be done in the context of co-simulation between UCM and BEM, to know if they have the same level of accuracy.

References

Allegrini, J., Orehounig, K., Mavromatidis, G., Ruesch, F., Dorer, V., & Evins, R. (2015). A review of modelling approaches and tools for the simulation of district-scale energy systems. *Renewable and Sustainable Energy Reviews, 52*, 1391–1404. https://doi.org/10.1016/j.rser.2015.07.123.

Bouyer, J., Inard, C., & Musy, M. (2011). Microclimatic coupling as a solution to improve building energy simulation in an urban context. *Energy and Buildings, 43*, 1549–1559. https://doi.org/10.1016/j.enbuild.2011.02.010.

Bozonnet, E. (2006). *Impact des microclimats urbains sur la demande énergétique des bâtiments—Cas de la rue canyon*. PhD.

Bruse, M., & Fleer, H. (1998). Simulating surface–plant–air interactions inside urban environments with a three dimensional numerical model. *Environmental Modelling & Software, 13*, 373–384. https://doi.org/10.1016/S1364-8152(98)00042-5.

Bueno, B., Norford, L., Pigeon, G., & Britter, R. (2011). Combining a detailed building energy model with a physically-based urban canopy model. *Boundary-Layer Meteorology, 140*, 471–489. https://doi.org/10.1007/s10546-011-9620-6.

Bueno, B., Hidalgo, J., Pigeon, G., Norford, L., & Masson, V. (2013a). Calculation of air temperatures above the urban canopy layer from measurements at a rural operational weather station. *Journal of Applied Meteorology and Climatology, 52*, 472–483. https://doi.org/10.1175/JAMC-D-12-083.1.

Bueno, B., Norford, L., Hidalgo, J., & Pigeon, G. (2013b). The urban weather generator. *Journal of Building Performance Simulation, 6*, 269–281. https://doi.org/10.1080/19401493.2012.718797.

Castaldo, V. L., Pisello, A. L., Piselli, C., Fabiani, C., Cotana, F., & Santamouris, M. (2018). How outdoor microclimate mitigation affects building thermal-energy performance: A new design-stage method for energy saving in residential near-zero energy settlements in Italy. *Renewable Energy, 127*, 920–935. https://doi.org/10.1016/j.renene.2018.04.090.

Ciancio, V., Falasca, S., Golasi, I., Curci, G., Coppi, M., & Salata, F. (2018). Influence of input climatic data on simulations of annual energy needs of a building: EnergyPlus and WRF modeling for a case study in Rome (Italy). *Energies, 11*, 2835. https://doi.org/10.3390/en11102835.

Coccolo, S., Kämpf, J., Mauree, D., & Scartezzini, J.-L. (2018). Cooling potential of greening in the urban environment, a step further towards practice. *Sustainable Cities and Society, 38*, 543–559. https://doi.org/10.1016/j.scs.2018.01.019.

Crawley, D. B., Lawrie, L. K., Winkelmann, F. C., Buhl, W. F., Huang, Y. J., Pedersen, C. O., Strand, R. K., Liesen, R. J., Fisher, D. E., & Witte, M. J. (2001). EnergyPlus: Creating a new generation building energy simulation program. *Energy and Buildings, 33*, 319–331.

de Sturler, E., Hoeflinger, J., Kalé, L., & Bhandarkar, M. (2000). A new approach to software integration frameworks for multi-physics simulation codes. In *Conference Paper in IFIP Advances in Information and Communication Technology, Ottawa, Canada* (pp. 87–104). https://doi.org/10.1007/978-0-387-35407-1_6.

Deru, M., Field, K., Studer, D., Benne, K., Griffith, B., Torcellini, P., Liu, B., Halverson, M., Winiarski, D., Rosenberg, M., Yazdanian, M., Huang, J., & Huang, J. (2011). *U.S. Department of Energy commercial reference building models of the national building stock*. Technical report, TP-5500-46861, National Renewable Energy Laboratory, Golden, CO.

Dorer, V., Allegrini, J., Orehounig, K., Moonen, P., Upadhyay, J., Kaempf, J., & Carmeliet, J. (2013). Modelling the urban microclimate and its impact on the energy demand of buildings and building clusters. In *IBPSA* (pp. 3483–3489).

Erell, E., & Williamson, T. (2006). Simulating air temperature in an urban street canyon in all weather conditions using measured data at a reference meteorological station. *International Journal of Climatology, 26*, 1671–1694. https://doi.org/10.1002/joc.1328.

Goffart, J., Mara, T., & Wurtz, E. (2017). Generation of stochastic weather data for uncertainty and sensitivity analysis of a low-energy building. *Journal of Building Physics, 41*, 41–57. https://doi.org/10.1177/1744259116668598.

Guattari, C., Evangelisti, L., & Balaras, C. A. (2018). On the assessment of urban heat island phenomenon and its effects on building energy performance: A case study of Rome (Italy). *Energy and Buildings, 158*, 605–615. https://doi.org/10.1016/j.enbuild.2017.10.050.

Hensen, J. L. M. (1999). A comparison of coupled and de-coupled solutions for temperature and air flow in a building. *ASHRAE Transactions, 105*(2), 962–969.

Johnston, D., Miles-Shenton, D., & Farmer, D. (2015). Quantifying the domestic building fabric 'performance gap'. *Building Services Engineering Research and Technology, 36*, 614–627. https://doi.org/10.1177/0143624415570344.

Klein, S., Duffie, J., Mitchell, J., Kummer, J., Thornton, J., Bradley, D., & Arias, D. (2017). *TRNSYS 17. A transient system simulation program* (Mathematical reference) (Vol. 4). Madison, WI: Solar Energy Laboratory, University of Wisconsin.

Kolokotroni, M., Zhang, Y., & Watkins, R. (2007). The London heat island and building cooling design. *Solar Energy, 81*(1), 102–110.

Lac, C., Chaboureau, P., Masson, V., Pinty, P., Tulet, P., Escobar, J., Leriche, M., Barthe, C., Aouizerats, B., Augros, C., et al. (2017). Overview of the Meso-NH model version 5.4 and its applications. Geoscientific Model Development Discussions (pp. 1929–1969).

Lauzet, N., Morille, B., Leduc, T., & Musy, M. (2017). What is the required level of details to represent the impact of the built environment on energy demand? *Procedia Environmental Sciences, Sustainable Synergies from Buildings to the Urban Scale, 38*, 611–618. https://doi.org/10.1016/j.proenv.2017.03.140.

Lauzet, N., Rodler, A., Musy, M., Azam, M.-H., Guernouti, S., Mauree, D., & Colinart, T. (2019). How building energy models take the local climate into account in an urban context—A review. *Renewable and Sustainable Energy Reviews, 116*, 109390. https://doi.org/10.1016/j.rser.2019.109390.

Li, X., Zhou, Y., Yu, S., Jia, G., Li, H., & Li, W. (2019). Urban heat island impacts on building energy consumption: A review of approaches and findings. *Energy, 174*, 407–419. https://doi.org/10.1016/j.energy.2019.02.183.

Malys, L., Musy, M., & Inard, C. (2015). Microclimate and building energy consumption: Study of different coupling methods. *Advances in Building Energy Research, 9*, 151–174. https://doi.org/10.1080/17512549.2015.1043643.

Marija, T., & Jan, H. (2006). Model and tool requirements for co-simulation of buildings performance. In *Proceedings of the 15th IASTED International Conference on Applied Simulation and Modelling 7*.

Marija, T., Michael, W., & Jan, H. (2007). Comparison of co-simulation approaches for building and HVAC/R system simulation. In *Proceedings of the 10th IBPSA Building Simulation Conference*, (pp. 1418–1425).

Martilli, A., Clappier, A., & Rotach, M. W. (2002). An urban surface exchange parameterisation for mesoscale models. *Boundary-Layer Meteorology, 104*, 261–304. https://doi.org/10.1023/A:1016099921195.

Masson, V. (2000). A physically-based scheme for the urban energy budget in atmospheric models. *Boundary-Layer Meteorology, 94*, 357–397. https://doi.org/10.1023/A:1002463829265.

Mauree, D., Blond, N., Kohler, M., & Clappier, A. (2017). On the coherence in the boundary layer: Development of a canopy interface model. *Frontiers in Earth Science, 4*. https://doi.org/10.3389/feart.2016.00109.

Mauree, D., Blond, N., & Clappier, A. (2018). Multi-scale modeling of the urban meteorology: Integration of a new canopy model in the WRF model. *Urban Climate, 26*, 60–75. https://doi.org/10.1016/j.uclim.2018.08.002.

Merlier, L., Frayssinet, L., Kuznik, F., Rusaouen, G., Johannes, K., Hubert, J.-L., & Milliez, M. (2017). Analysis of the (urban) microclimate effects on the building energy behaviour. In *Proceedings of the 15th IBPSA Conference, San Francisco, CA, USA*.

Merlier, L., Frayssinet, L., Johannes, K., & Kuznik, F. (2019a). On the impact of local microclimate on building performance simulation. Part II: Effect of external conditions on the dynamic thermal behavior of buildings. *Build Simul, 12*, 747–757.

Merlier, L., Frayssinet, L., Johannes, K., & Kuznik, F. (2019b). On the impact of local microclimate on building performance simulation. Part I: Prediction of building external conditions. *Build Simul, 12*, 735–746.

Miller, C., Thomas, D., Kämpf, J., & Schlueter, A. (2018). Urban and building multiscale co-simulation: Case study implementations on two university campuses. *Journal of Building Performance Simulation, 11*, 309–321. https://doi.org/10.1080/19401493.2017.1354070.

Mirzaei, P. A., & Haghighat, F. (2010). Approaches to study urban heat island—Abilities and limitations. *Building and Environment, 45*, 2192–2201. https://doi.org/10.1016/j.buildenv.2010.04.001.

Musy, M., Malys, L., Morille, B., & Inard, C. (2015). The use of SOLENE-microclimate model to assess adaptation strategies at the district scale. *Urban Climate, 14*, 213–223.

Oke, T. R. (2002). *Boundary layer climates* (2nd ed.). London: Routledge.

Palme, M., Inostroza, L., Villacreses, G., Lobato, A., & Carrasco, C. (2017). Urban weather data and building models for the inclusion of the urban heat island effect in building performance simulation. *Data in Brief, 14*, 671–675. https://doi.org/10.1016/j.dib.2017.08.035.

Pappaccogli, G., Giovannini, L., Cappelletti, F., & Zardi, D. (2018). Challenges in the application of a WRF/urban-TRNSYS model chain for estimating the cooling demand of buildings: A case study in Bolzano (Italy). *Science and Technology for the Built Environment, 24*, 529–544. https://doi.org/10.1080/23744731.2018.1447214.

Peuportier, B., & Sommereux, I. B. (1990). Simulation tool with its expert interface for the thermal design of multizone buildings. *International Journal of Solar Energy, 8*, 109–120.

Plessis, G., Kaemmerlen, A., & Lindsay, A. (2014). BuildSysPro: A Modelica library for modelling buildings and energy systems. In *Proceedings of the 10th International Modelica Conference; March 10–12, 2014, Lund, Sweden* (pp. 1161–1169). Linköping: Linköping University Electronic Press.

Rabouille, M., Wurtz, E., & Perrotin, P. (2013). Analysis of dynamic thermal simulation for refurbishment. In *IBPSA* (pp. 358–365).

Robinson, D., Haldi, F., Kämpf, J., Leroux, P., Perez, D., Rasheed, A., & Wilke, U. (2009). CitySim: Comprehensive micro-simulation of resource flows for sustainable urban planning. In *Proc. Building Simulation* (pp. 1614–1627).

Rotach, M. W., Vogt, R., Bernhofer, C., Batchvarova, E., Christen, A., Clappier, A., Feddersen, B., Gryning, S.-E., Martucci, G., Mayer, H., Mitev, V., Oke, T. R., Parlow, E., Richner, H., Roth, M., Roulet, Y.-A., Ruffieux, D., Salmond, J. A., Schatzmann, M., & Voogt, J. A. (2005). BUBBLE—An urban boundary layer meteorology project. *Theoretical and Applied Climatology, 81*, 231–261. https://doi.org/10.1007/s00704-004-0117-9.

Salvati, A., Coch Roura, H., & Cecere, C. (2017a). Assessing the urban heat island and its energy impact on residential buildings in Mediterranean climate: Barcelona case study. *Energy and Buildings, 146*, 38–54. https://doi.org/10.1016/j.enbuild.2017.04.025.

Salvati, A., Coch, H., & Morganti, M. (2017b). Effects of urban compactness on the building energy performance in Mediterranean climate. *Energy Procedia, 122*, 499–504. https://doi.org/10.1016/j.egypro.2017.07.303.

Salvati, A., Palme, M., Chiesa, G., & Kolokotroni, M. (2020). Built form, urban climate and building energy modelling: Case-studies in Rome and Antofagasta. *Journal of Building Performance Simulation, 13*, 209–225. https://doi.org/10.1080/19401493.2019.1707876.

Santamouris, M. (2014). On the energy impact of urban heat island and global warming on buildings. *Energy and Buildings, 82*, 100–113. https://doi.org/10.1016/j.enbuild.2014.07.022.

Santamouris, M., Papanikolaou, N., Livada, I., Koronakis, I., Georgakis, C., Argiriou, A., & Assimakopoulos, D. N. (2001). On the impact of urban climate on the energy consumption of buildings. *Solar Energy, 70*, 201–216. https://doi.org/10.1016/S0038-092X(00)00095-5.

Skamarock, W. C., Klemp, J. B., Dudhia, J., Gill, D. O., Barker, D. M., Duda, M. G., Huang, X.-H., Wang, W., & Powers, J. G. (2008). *A description of the advanced research WRF version 2*. DTIC Document.

Street, M., Reinhart, C., Norford, L., & Ochsendorf, J. (2013). Urban heat island in Boston: An evaluation of urban air temperature models for predicting building energy use. In *Proceedings of BS2013: 13th Conference of International Building Performance Simulation Association in Chambéry, France* (p. 1022).

Sun, Y., & Augenbroe, G. (2014). Urban heat island effect on energy application studies of office buildings. *Energy and Buildings, 77*, 171–179. https://doi.org/10.1016/j.enbuild.2014.03.055.

Tsoka, S., Tsikaloudaki, K., & Theodosiou, T. (2019). Coupling a building energy simulation tool with a microclimate model to assess the impact of cool pavements on the building's energy performance. application in a dense residential area. *Sustainability, 11*, 2519. https://doi.org/10.3390/su11092519.

Vallati, A., Mauri, L., & Colucci, C. (2018). Impact of shortwave multiple reflections in an urban street canyon on building thermal energy demands. *Energy and Buildings, 174*, 77–84.

Vanpachtenbeke, M., Van De Walle, W., Janssen, H., & Roels, S. (2015). Analysis of coupling strategies for building simulation programs. In *Energy Procedia, 6th International Building Physics Conference, IBPC 2015* (Vol. 78, pp. 2554–2559). https://doi.org/10.1016/j.egypro.2015.11.276.

Walter, E., & Kämpf, J. H. (2015). A verification of CitySim results using the BESTEST and monitored consumption values. In *Proceedings of the 2nd Building Simulation Applications Conference* (pp. 215–222).

Wetter, M. (2011). Co-simulation of building energy and control systems with the building controls virtual test bed. *Journal of Building Performance Simulation, 4*, 185–203. https://doi.org/10.1080/19401493.2010.518631.

Yang, X., Zhao, L., Bruse, M., & Meng, Q. (2012). An integrated simulation method for building energy performance assessment in urban environments. *Energy and Buildings, 54*, 243–251. https://doi.org/10.1016/j.enbuild.2012.07.042.

Yang, X., Jin, T., Yao, L., Zhu, C., & Peng, L. L. (2017). Assessing the impact of urban heat island effect on building cooling load based on the local climate zone scheme. *Procedia Engineering, 205*, 2839–2846. https://doi.org/10.1016/j.proeng.2017.09.904.

Zhai, Z. (2004). *Developing an integrated building design tool by coupling building energy simulation and computational fluid dynamics program*. Massachusetts Institute of Technology.

Chapter 16
RayMan and SkyHelios Model

Andreas Matzarakis, Marcel Gangwisch, and Dominik Fröhlich

16.1 Introduction

Complex environments make it hard to assess meteorological parameters in a comprehensive and representative way (Matzarakis et al. 2010). One of the most complex environment types is the street level of urban areas (Hwang et al. 2011). The volume ranging from the ground to the roof level, the urban canopy layer (Oke 1987, p. 274), is the most relevant and at the same time most people are affected here. Measurements can only provide very limited insights, as the urban canopy layer is highly heterogeneous in time and space (Mirzaei and Haghighat 2010). Additionally, for most purposes spatial distributions of many meteorological and human-biometeorological parameters are required. Setting up lots of measuring stations and interpolating their readings to obtain spatial information for a desired area is mostly not an option due to being expensive and error-prone (Mirzaei and Haghighat 2010). Thus, for urban environments the application of urban microscale models is the most promising alternative (Hwang et al. 2011; Matzarakis et al. 2018), especially for the purpose of future planning and quantification of adaptation and mitigation measures (Herrmann and Matzarakis 2012; Ketterer and Matzarakis 2014).

This becomes even more complicated when thermal comfort or stress of humans is to be assessed. Information about thermal bioclimate is mostly generated by the calculation of thermal indices, e.g., predicted mean vote (PMV, Fanger 1972), physiologically equivalent temperature (PET, Höppe 1993, 1999; Matzarakis et al. 1999; Mayer and Höppe 1987), perceived temperature (PT, Staiger et al. 2012), universal thermal climate index (UTCI, Jendritzky et al. 2012), standard effective

A. Matzarakis (✉) · M. Gangwisch · D. Fröhlich
Research Centre Human Biometeorology, German Meteorological Service, Freiburg, Germany
e-mail: andreas.matzarakis@dwd.de; marcel.gangwisch@dwd.de; dominik.froehlich@mailbox.org

M. Palme, A. Salvati (eds.), *Urban Microclimate Modelling for Comfort and Energy Studies*, https://doi.org/10.1007/978-3-030-65421-4_16

temperature (SET*, American Society of Heating 2005; Gagge et al. 1986; Gonzalez et al. 1974), or modified physiologically equivalent temperature (mPET, Chen and Matzarakis 2018). These thermal indices combine meteorological and thermo-physiological aspects to approximate the thermal perception of a sample human being (Höppe 1993). Most thermal indices do allow for the setting of the physiological parameters: age, weight, height, metabolic rate and activity, as well as posture and sex (e.g., PMV, PET, mPET, and PT). The meteorological variables for all the indices mentioned above are air temperature (T_a), vapor pressure (VP), wind speed (v), and different radiation fluxes (e.g., Fanger 1972; Höppe 1999; Jendritzky et al. 2012; Staiger et al. 2012). The short- and long-wave radiation fluxes from and to the sample person are usually summarized as the mean radiant temperature (T_{mrt}). T_{mrt} is defined as the surface temperature of a perfect black and equal surrounding environment, which leads to the same energy balance as the current environment (Fanger 1972; VDI 1988, 2008).

16.2 Methods and Data

16.2.1 Thermal Indices

Human beings have no senses to "feel" individual meteorological parameters. In particular, humans are unable to distinguish between heat stress by actinic and thermal processes (e.g., radiation and air temperature). They rather feel the integral effect of several parameters in terms of modification to their skin as well as to their blood temperature (Staiger et al. 2018, 2019). The thermal sensation of humans, as well as the physiological strain, can be assessed by thermal indices.

16.2.1.1 Perceived Temperature

The perceived temperature (PT) is an equivalent temperature for the assessment of human thermal comfort based on the human energy balance model "Klima-Michel model" (Staiger et al. 2012), which is designed for outdoor use. Thermal assessment in PT is based on a modification of the (indoor) thermal index predicted mean vote (PMV, Fanger 1972) after Gagge et al. (1986). PT is defined as "the air temperature of a reference environment in which the thermal perception would be the same as in the actual environment" (Staiger et al. 2012, 2019). The perceived temperature does consider a self-adapting clothing model that will automatically try to achieve thermally comfortable conditions. In case it fails to do so, thermal stress is occurring.

16.2.1.2 Universal Thermal Climate Index

The universal thermal climate index (UTCI), like PT, is an equivalent temperature. In UTCI the meteorological conditions are compared to a reference environment with 50% relative humidity, calm air, and T_{mrt} being equal to T_a (Jendritzky et al. 2012). UTCI is defined as "the isothermal air temperature of the reference condition that would elicit the same dynamic response (strain) of the physiological model" (Jendritzky et al. 2012). UTCI calculates the current heat load based on a heat transfer model (Fiala et al. 2012) considering a fully automatic clothing model (Havenith et al. 2012) adapting to the current meteorological conditions. For being too complex to be determined numerically, UTCI is usually estimated based on a regression equation (Bröde et al. 2012). While the regression speeds up the calculation, it limits the possible input to the meteorological parameters: air temperature, vapor pressure, wind speed (at a height of 10 m), as well as mean radiant temperature. In addition, the range of parameters is restricted by the regression. For example, T_a may only vary from −50 to +50 °C, and wind speed may range from 0.5 to 17.0 m/s. If the conditions are exceeding the limits, there are workarounds proposed allowing to calculate UTCI anyway, but with some imprecision (Bröde et al. 2012).

16.2.1.3 Physiologically Equivalent Temperature

A widely applied thermal index for all kinds of studies about human thermal comfort is the physiologically equivalent temperature (PET, Höppe 1999; Mayer and Höppe 1987). It is defined as "the air temperature at which, in a typical indoor setting (without wind and solar radiation), the energy budget of the human body is balanced with the same core and skin temperature as under the complex outdoor conditions to be assessed" (Höppe 1999). PET is based on a simplified version of the human energy balance model "Munich energy balance model for individuals" (MEMI, Höppe 1984). In contrast to PT and UTCI, the clothing model in PET is not self-adapting. This allows for the setting of a clothing insulation in terms of the clothing index clo.

16.2.1.4 Standard Effective Temperature

The standard effective temperature (SET*, American Society of Heating 2005; Gagge et al. 1986; Gonzalez et al. 1974) is originally defined as the equivalent dry bulb temperature of an isothermal environment. The environment is at 50% relative humidity and 0.25–0.3 m/s wind speed in which a human being would have the same heat stress (skin temperature) and thermoregulatory strain (skin wetness) as in the actual environment.

16.2.1.5 Modified Physiologically Equivalent Temperature

The modified physiologically equivalent temperature (mPET, Chen and Matzarakis 2018) is a further development of PET adding a self-adapting or manual clothing model, as well as an improved consideration of humidity.

16.2.2 Human Biometeorological Models

16.2.2.1 RayMan

The RayMan model is a microscale model developed at the Chair for Environmental Meteorology of the University of Freiburg, calculating radiation fluxes in simple and complex urban environments (Matzarakis et al. 2007, 2010). This allows for the estimation of T_{mrt}, which is an important input parameter for the calculation of thermal indices like PT, UTCI, and PET.

RayMan is one-dimensional in space (all calculations are performed for a single point in space). It is a diagnostic model and therefore fully time independent. RayMan was developed with performance and usability in mind. All calculations and settings can be set up and controlled through the graphical user interface (Fig. 16.1). The short runtime is a precondition to run calculations for long datasets covering several years in high temporal resolution (e.g., Fröhlich and Matzarakis 2013).

Another idea behind RayMan is to require only a minimal number of meteorological parameters as input and all of them being parameters that are typically recorded at normal climate stations (Matzarakis et al. 2010). Due to the one-dimensionality of the model, the output of the model is only valid for the given meteorological input conditions. For a spatial output and spatially varying conditions consider the SkyHelios model.

A key feature of RayMan is the calculation of the sky view factor (SVF, the fraction of free sky within the upper hemisphere) from either a fisheye image or a spatial input in terms of an obstacle file. RayMan obstacle files include the surrounding urban morphology of the point of interest. The local, spherical SVF can be used to estimate the mean radiant temperature for the current location.

The obstacle file's format is defined as a vector-based ASCII file format, which can be created manually using the RayMan obstacle editor or exported from other spatial geodata using the plug-in "Shp to Obs" developed for Quantum GIS (QGIS, Open Source Geospatial Foundation 2018).

RayMan is capable of considering resolved objects of the urban environment (buildings and trees). These are saved in obstacle files (obs) in a specific format. Obstacle files are raw plain text files containing a header and each individual obstacle:

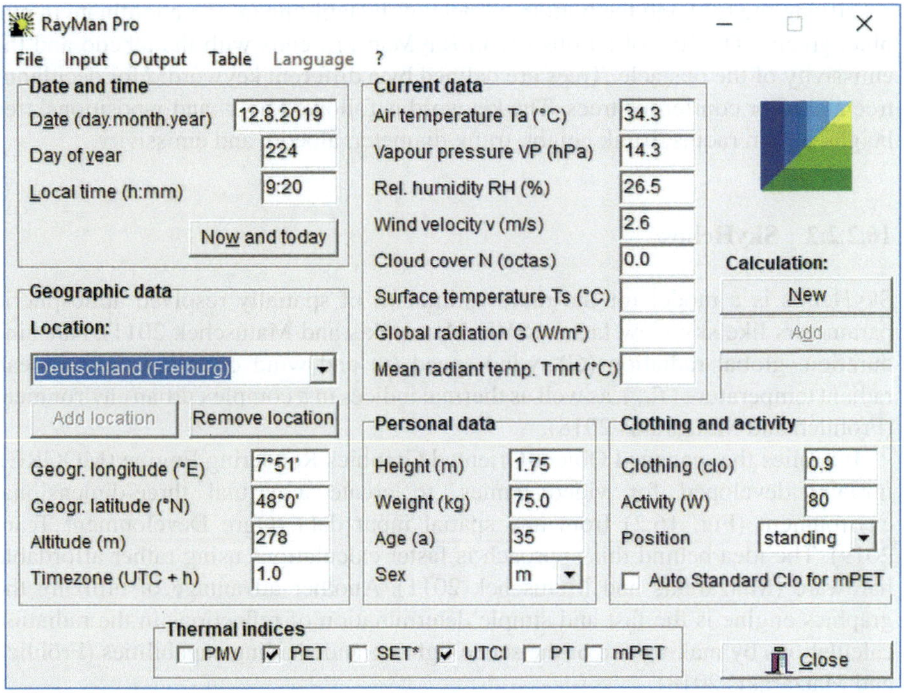

Fig. 16.1 RayMan main window, showing the graphical user interface of RayMan. All meteorological variables required to estimate thermal sensation by thermal indices

```
# RayMan Pro obstacle file #
# Building with roof and bottom coordinates
g    6.25 17.38 10.00 6.75 7.75 10.00 15.13 7.13 10.00
     14.75 17.25 10.00 6.25 17.38 0.00 6.75 7.75 0.00
     15.13 7.13 0.00 14.75 17.25 0.00
     0.30 0.95     # albedo and emissivity

# Broadleaf tree with x, y coordinate, tree height,
# crown radius, trunk height, trunk diameter,
# albedo and emissivity
l    3.88 12.38 10.00 3.33 4.00 0.33 0.30 0.95

# Coniferous tree
n    12.88 1.75 10.00 5.00 4.00 0.00 0.25 0.96
```

The obstacle files can be displayed and modified using a plain raw text editor (e.g., Vim, Notepad++). Thereby one obstacle is defined in one individual row. A building obstacle is represented in one line by the *g* keyword. The coordinates of the building are given in a subsequent way, defining all corners of the building, by specifying the roof and the bottom corners of the building (x, y, z coordinate in red, blue, green). The line of an obs file in RayMan Pro ends with the albedo and the emissivity of the obstacle. Trees are defined by a different keyword: *l* for deciduous tree and *n* for coniferous trees. The keyword is followed by x- and y-positions, tree height, crown radius, trunk height, trunk diameter, albedo, and emissivity.

16.2.2.2 SkyHelios

SkyHelios is a model for the rapid estimation of spatially resolved atmospheric parameters like sky view factor (SVF, Matzarakis and Matuschek 2011), sunshine duration, global radiation (G), wind speed (v) and wind direction (WD), mean radiant temperature (T_{mrt}), as well as thermal indices in a complex urban environment (Fröhlich and Matzarakis 2018).

It applies the managed Object-Oriented Graphics Rendering Engine (MOGRE), initially developed for video games, to create a virtual three-dimensional environment (Fig. 16.2) from any spatial input data (Ogre Development Team 2019). The idea behind this approach is faster calculations using rather affordable hardware (Matzarakis and Matuschek 2011). Another advantage of utilizing the graphics engine is the fast and simple determination of reflections in the radiation calculations by making use of the scene lighting and shading capabilities (Fröhlich and Matzarakis 2018).

Fig. 16.2 SkyHelios main window, showing the 3D study area with buildings (gray) and urban trees (green). The red dot specifies a point of interest and determines the calculation of sky view factor for single points (top, right). The diagram (center, right) gives an overview of the sun path throughout the day and year

Like RayMan, SkyHelios is a fully diagnostic model (Fröhlich and Matzarakis 2018). Therefore, the steady-state model is time independent allowing for running the model for specific points in time without having to consider any spin-up period. The model, e.g., can calculate for 20.03.2019 14:00 UTC, 23.09.2019 14:00 UTC, and 22.12.2019 14:00 UTC the proposed parameters. Meteorological input data for each time step (date and time) desired can be provided by a delimited text file (Fröhlich and Matzarakis 2018).

Another big advantage of the SkyHelios model is the various spatial input options. The SkyHelios model accepts a wide range of well-known spatial formats. These are to be divided into raster and vector formats. Raster formats consist of equidistant grids of parameters (e.g., elevation) for every grid cell. SkyHelios is capable of importing the most common raster file formats by including the Geospatial Data Abstraction Library (GDAL, GDAL/OGR Contributors 2019). Vector formats, on the contrary, do specify the position of corners and points in the outline of obstacles (vertices). Several vertices together can form polygons representing, e.g., buildings. It is necessary that input files incorporate three-dimensional data either by providing an additional height field or by polygons based on three-dimensional vertices. A very common example for vector file formats is the ESRI shapefile format. SkyHelios can read a number of vector formats by including the OpenGIS Simple Features Reference Implementation library (OGR, GDAL/OGR Contributors 2019). RayMan obstacle files can be imported as well. Starting from the passed urban geometry data, a three-dimensional model of the city is rendered by the MOGRE engine.

Based on the provided input, several astronomical, meteorological, as well as biometeorological quantities can be estimated both spatially (area of interest) and for individual preselected points in space (point of interest). The parameter sky view factor (planar and spherical, Hämmerle et al. 2011) can be estimated from rendered fisheye imagery within the provided model domain. SVF can be further utilized to derive the radiation properties in the urban radiational regime (short- and long-wave radiation fluxes from the upper and lower hemisphere).

The mean radiant temperature (T_{mrt}) summarizes all radiation fluxes and is the most important parameter for human thermal comfort assessment. T_{mrt} is dependent on astronomic conditions (e.g., solar altitude and azimuth angle), atmospheric conditions (e.g., relative humidity, cloud cover, air temperature, surface temperature of the neighborhood, global radiation (divided into direct and diffuse radiation)), and surrounding urban morphology (with different albedo and emissivity coefficients).

T_{mrt} can be determined by solving the Stefan-Boltzmann law for all surrounding surfaces i of the neighborhood (Matzarakis et al. 2007):

$$T_{mrt} = \sqrt[4]{\sum_i \left(\epsilon_{lw,i} \cdot T_{s,i}^4 + \frac{\alpha_{abs,s,i} \cdot D_{s,i}}{\epsilon_{lw,p} \cdot \sigma} \right) \cdot \mathrm{Pr}_{p,i}} \left[{}^\circ\mathrm{C} \right] \quad (16.1)$$

T_{mrt} is dependent on the surface's long-wave emissivity ($\epsilon_{lw,i}$), surface's temperature ($T_{s,i}$), surface's short-wave absorption coefficient (1.0 – albedo, $\alpha_{abs,s,i}$), surface's absorbed scattered reflected global radiation ($D_{s,i}$), surface's projection factor ($Pr_{p,i}$), and Stefan-Boltzmann constant $\left(\sigma = 5.67 \times 10^{-8}\,\mathrm{W} \cdot \mathrm{m}^{-2} \cdot T_s^{-4}\right)$. An emissivity of 0.97 is applied to the human body (Fanger 1972).

The calculation pipeline for the mean radiant temperature in SkyHelios is depicted in Figs. 16.3, 16.4, and 16.5:

T_{mrt} in SkyHelios is calculated based on the radiational impact of the lower and upper hemispheres for long- and short-wave radiation fluxes. Direct and scattered short-wave radiation fluxes (including single reflections by the neighborhood) as well as long-wave emittance of the ground and the surroundings are considered (Fig. 16.3). In order to obtain better approximations for radiation fluxes, short- and long-wave radiation fluxes are considered separately:

The calculation of direct short-wave radiation flux (I) is assumed to be independent of SVF for unshaded locations and zero for shaded conditions (Fig. 16.4). According to Jendritzky (1990), I can be estimated as a function of the solar constant (I_0), solar zenith angle based on Bouguer-Lambert's equation in combination with the relative optical air mass (r_{opt}), optical depth of the atmosphere (δ_{opt}), and Linke turbidity factor (T_L):

$$I = I_0 \cdot \exp\left(-\delta_{opt} \cdot T_L \cdot r_{opt} \cdot \frac{p}{p_0}\right) \cdot \cos(\zeta) \cdot \left(1 - \frac{cc}{8}\right) \tag{16.2}$$

The relative optical air mass can be estimated after (Kasten and Young 1989) as a function of solar azimuth angle (ζ):

$$r_{opt} = \left(\sin(90^\circ - \zeta) + 0.50572 \cdot \left((90^\circ - \zeta) + 6.07995^\circ\right)^{-1.6364}\right)^{-1} \tag{16.3}$$

The optical depth is described by the optical air mass (Kasten 1980):

$$\delta_{opt} = \frac{1}{0.6 \cdot r_{opt} + 9.4} \tag{16.4}$$

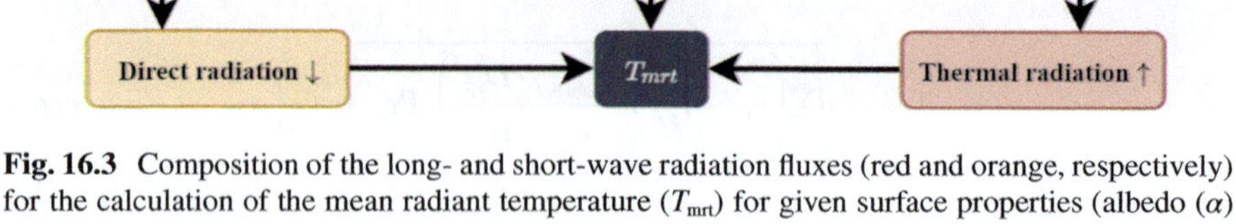

Fig. 16.3 Composition of the long- and short-wave radiation fluxes (red and orange, respectively) for the calculation of the mean radiant temperature (T_{mrt}) for given surface properties (albedo (α) and emissivity (ϵ))

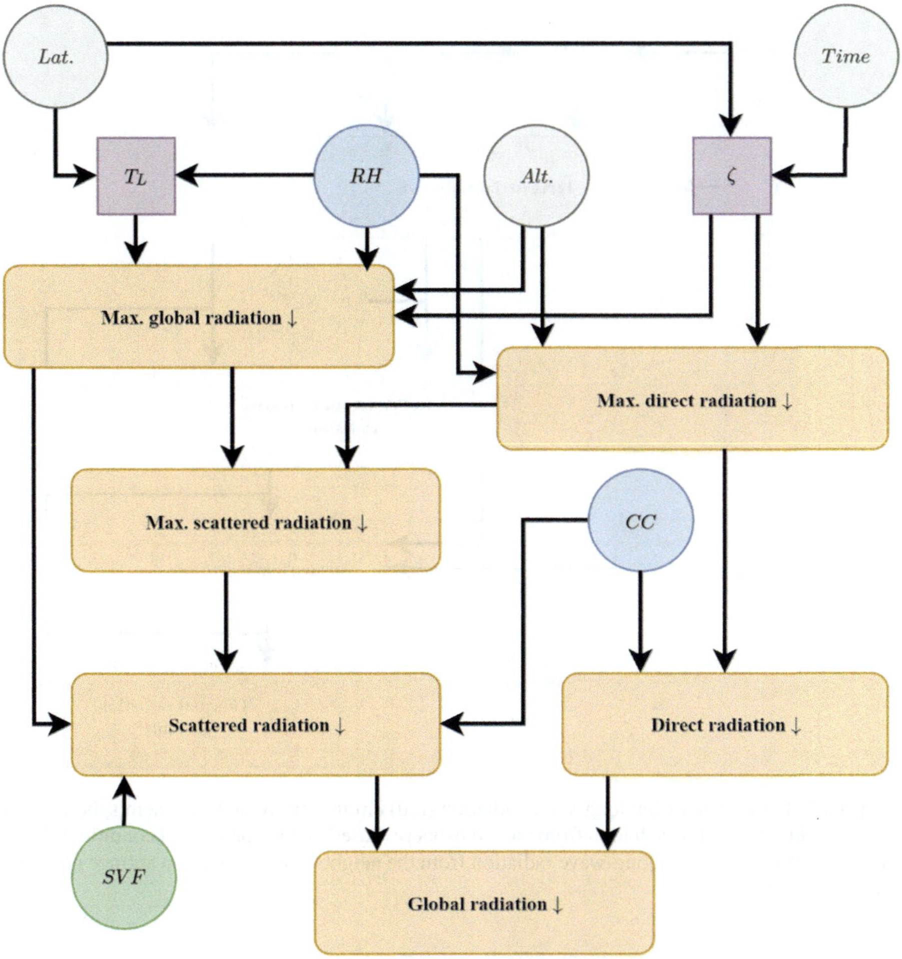

Fig. 16.4 Calculation of short-wave radiation fluxes for diffuse (based on SVF) and direct solar radiation (based on the visibility of the sun) for given cloud cover (CC) and relative humidity (RH), depending on the geographical parameters for location (latitude (lat) and altitude (alt)) and time

The diffuse short-wave irradiation is calculated from the direct solar irradiation as the sum of isotropic and anisotropic scattered radiation ($D = D_{\text{iso}} + D_{\text{aniso}}$) (Valko 1966). The isotropic part is calculated by the following equation:

$$D_{\text{iso}} = \left(G_0 - I_{\text{clear}}\right) \cdot \left(1 - \frac{I_{\text{clear}}}{I_0 \cdot \cos(\zeta)}\right) \cdot \text{SVF} \tag{16.5}$$

The direct solar irradiation for clear sky conditions (without clouds and horizon limitations) is calculated based on Eq. (16.2) (for CC = 0). The anisotropic fracture of the scattered radiation can be approximated if the sun is visible (Valko 1966):

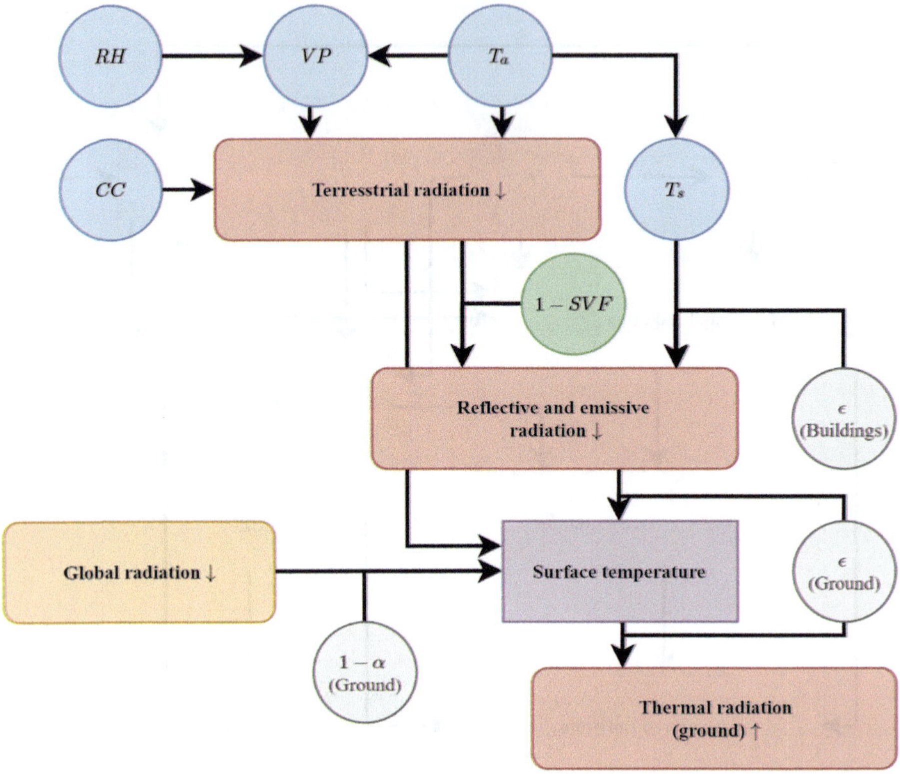

Fig. 16.5 Estimation of the long-wave radiation (red) emitted from the lower hemisphere for the T_{mrt} calculus. Terrestrial radiation from the atmosphere (based on meteorological conditions (blue)) and reflected and emitted long-wave radiation from the neighborhood (based on surface properties (gray)) are considered

$$D_{aniso} = \left(G_0 - I_{clear}\right) \cdot \frac{I_{clear}}{I_0 \cdot \cos(\zeta)} \tag{16.6}$$

The global radiation (G_0) for clear sky conditions can be directly calculated as proposed by Jendritzky (1990) and VDI (1994, 2008) for given air pressure (pr), T_L and ζ:

$$G_0 = 0.84 \cdot I_0 \cdot \cos(\zeta) \cdot \exp\left(\frac{-0.027 \cdot \frac{pr}{pr_0} \cdot T_L}{\cos(\zeta)}\right) \tag{16.7}$$

These fluxes are calculated based on given short-wave albedo and long-wave emission coefficients for each urban obstacle in the neighborhood. The estimation of the long-wave radiation fluxes as depicted in Fig. 16.5 is mainly based on the Stefan-Boltzmann law for a gray body (e.g., non-perfectly black surface) (Fanger 1972; VDI 2008):

$$P_{\mathrm{lw}} = \epsilon \cdot \sigma \cdot A_{\mathrm{s}} \cdot T_{\mathrm{s}}^{4} \tag{16.8}$$

Based on this law, SkyHelios calculates the surface temperature (T_s) and emitted long-wave radiation flux density (P_{lw}), and vice versa. The emitted long-wave radiation depends on ϵ of the surface as well as on σ.

Besides the calculation of T_{mrt}, SkyHelios is capable of sunshine duration, wind speed and direction (Fröhlich 2017; Fröhlich et al. 2019; Fröhlich and Matzarakis 2018), and aerodynamic roughness (Ketterer et al. 2017). Based on these, the thermal indices PT, UTCI, and PET can be calculated for any location within the model domain.

16.2.3 Area of Interest

A study area in the West of Freiburg (South-West Germany), the urban quarter *Rieselfeld* is selected for this study. The district has an extent of 1015 m on 916 m (0.92 km^2) and is located at 47.9991 N, 7.7921 E, 232 m a.s.l. The whole area can be divided into 0.15 km^2 built-up area and 0.77 km^2 free space. Detailed spatial input of the buildings in terms of a level of detail (LOD) 1, as well as a LOD 2 city model (based on the CityGML data format (Gröger et al. 2012)) and an urban tree cadaster, was provided by the municipality of Freiburg. The model is applied for the specified spatial resolution of 1 m, resulting in a discrete model domain of 1054 on 916 grid cells. All results are assessed for the target height of 1 m.

16.2.4 Meteorological Input Data

As a sample dataset, records of the official weather station 01443 (Freiburg airport) of the German Meteorological Service were selected, providing the data via the Climate Data Center (CDC). The meteorological input comprises the parameters air temperature (°C), relative humidity (%), air pressure (hPa), wind speed (m/s) and direction (°), as well as global radiation (W/m^2) for the whole day in 10-min temporal resolution.

From the dataset, July 25, 2019, was selected for the analysis in this study. The day is selected for being part of a heat wave and for providing clear-sky conditions. The maximum of the solar altitude angle is 64.6° (11:36 UTC), while solar azimuth angle ranges from 58.5° during sunrise (05:55 UTC) to 301.0° during sunset (21:14 UTC).

The prevailing air temperature rises from 17.5 °C in the morning (04:20 UTC) to 37.2 °C in the afternoon (15:00 UTC). Together with the low wind speed of 0.0–4.7 m/s in 10 m height and the global radiation of up to 855 W/m^2, thermal stress is to be expected.

16.3 Results and Discussion

Human thermal sensation is assessed for the prevailing weather conditions of July 25, 2019. Besides the final output, intermediate results, which are required to quantify human thermal comfort in terms of thermal indices, are analyzed. The spatial resolution for all calculations is 1 m per default, but may be specified on demand in SkyHelios.

16.3.1 Intermediate Results

16.3.1.1 Obstruction of the Upper Hemisphere: Analysis by the Spherical Sky View Factor

The obstruction of the upper hemisphere by the urban morphology (e.g., buildings and trees), in terms of the spherical and planar sky view factor (SVF) (Hämmerle et al. 2011), can be calculated by SkyHelios (Fig. 16.6). SVF is the fraction of the visible sky, as seen from a certain point (Oke 1987, p. 353). Spherical SVF is shown here, because it is more suitable for issues about objects with a vertical extension (e.g., human body), which can be represented by a cylinder, instead of a plane. Planar SVF can be applied for flat surfaces (e.g., the calculation of material heating and reflection) (Hämmerle et al. 2011). The urban environment is built up by vertical surfaces. Urban areas with wide spaces and small obstacles are represented by high SVF, whereas areas with narrow street canyons and high obstacles are more obstructed, resulting in a low SVF.

The spatial mean of spherical SVF for the study area is 0.67, while SVF ranges from 0.99 at the borders to 0.05 in the center of the study area beside buildings. Planar SVF is in the range of 0.07–0.99 with an average of 0.78. SVF is calculated in 1 m spatial resolution, so that 965,464 values for the land area of 0.93 km^2 are calculated.

The spatial average of SVF can be compared to SVF footprints of other cities (Middel et al. 2018) in order to characterize the city in terms of superelevation of the horizon. According to Middel et al. (2018, 2019), SVF can be assessed from Google Street View Imagery, based on four cardinal images and one image facing upwards. The study area in Freiburg (average planar SVF: 0.78, spherical SVF: 0.67) is comparable to Bonn (average planar SVF: 0.74). Nevertheless, SVF by Google Street View is only available for specific points along streets. "Google Street View

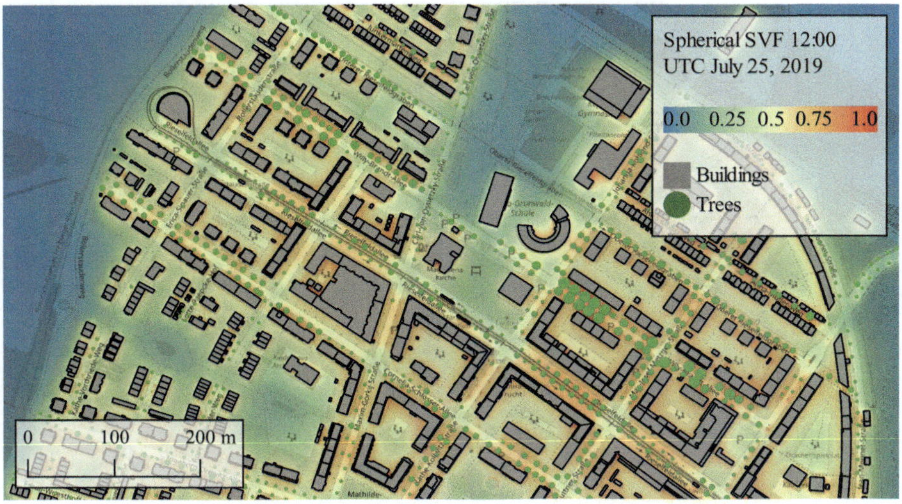

Fig. 16.6 The obstruction of the upper hemisphere by the urban morphology (e.g., buildings and trees) can be calculated by SkyHelios in terms of spherical and planar sky view factor (SVF). SVF can be computed spatially resolved with variable resolution for different urban environments due to preexisting urban city models. This SVF is calculated for 12:00 UTC on July 25, 2019, in 1 m height

images are inherently biased towards street locations and therefore do not provide continuous spatial coverage of the urban environment" (Middel et al. 2018). For the land area of 141.1 km^2 only 93,188 locations have been used for the estimation of SVF in Bonn. This leads to a spatial density of 0.0006 SVF estimation per m^2. Therefore, the spatial resolution and the density of SkyHelios (in this case 1 m) cannot be reached. This leads to the assumption that the spatial density of Google Street View imagery is not sufficient in order to assess spatially resolved SVF. The estimation of SVF by Google Street View imagery is also limited to the fixed target height of 2 m, while the user of SkyHelios can determine the target height freely.

16.3.1.2 Limited Direct and Diffuse Short-Wave Radiation by Shading: Analysis by the Mean Radiant Temperature

The sky view factor (SVF) has a direct impact on the incoming short- and long-wave radiation fluxes within the urban radiation regime. The effects of all different short- and long-wave radiation fluxes in the study area are shown in Fig. 16.7 as T_{mrt}. Shading of direct and diffuse short-wave radiation by urban geometries (e.g., buildings) as well as urban vegetation is clearly observable in T_{mrt} (blue areas with reduced T_{mrt}). SkyHelios is able to model full shading as well as partial shading. Therefore, it is possible to distinguish between shading by buildings and shading by vegetation. The mean T_{mrt} in shaded areas by buildings is 48.9 °C (ranging from 45.2 to 54.6 °C) compared to the mean T_{mrt} in shaded areas by vegetation 49.9 °C (ranging

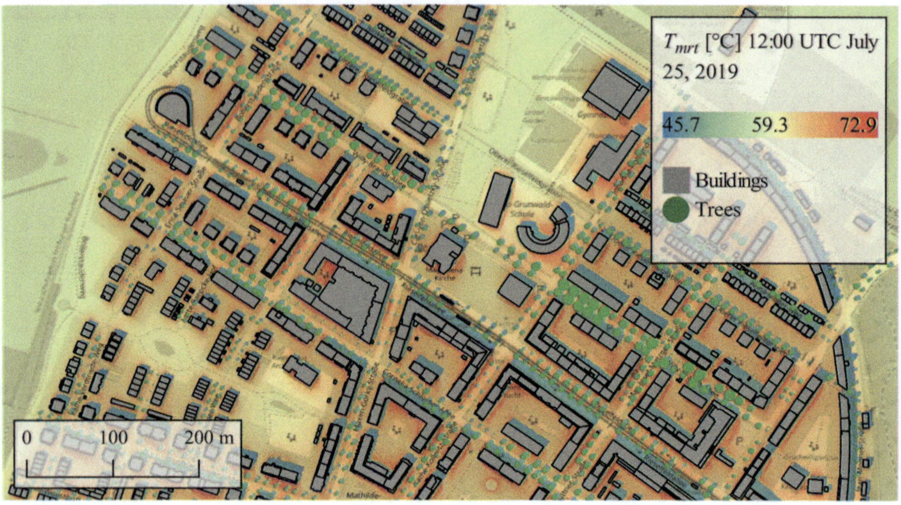

Fig. 16.7 The effect of all different short- and long-wave radiation fluxes in the urban environment can be calculated by SkyHelios in terms of the mean radiant temperature (T_{mrt}). T_{mrt} is calculated for 12:00 UTC on July 25, 2019

from 46.6 to 53.6 °C). The mean T_{mrt} in sunlit areas is 62.2 °C (ranging from 57.7 to 73.7 °C). Therefore, shade by buildings reduces T_{mrt} by 13.3 °C, while shade by vegetation reduces T_{mrt} by 12.3 °C compared to sunlit areas.

Different types of shade can explain this difference. While buildings block solar irradiance completely, vegetation can be considered as porous media, allowing for partial transmission of solar radiation. For the study area this makes a difference of 1 °C in terms of T_{mrt}. The effect of porosity has to be further discussed under the aspect of wind speed and direction. The size of the shaded area of buildings and vegetation is dependent on solar altitude and solar azimuth angle and therefore on time of the day, day of the year, and geographical latitude.

The shaded area by buildings in the study area is 0.08 km^2 (10.38% of the area with free space), while 0.03 km^2 (3.9% of the area with free space) of shade is provided by vegetation.

16.3.1.3 Diagnostic Wind Speed and Direction

SkyHelios is capable of running a diagnostic steady-state wind model to simulate a time-independent wind field (Fröhlich et al. 2019). Modelled wind speed and direction are further incorporated into the assessment of human thermal comfort, to consider the effect of wind chill. The sensitivity of thermal indices with high wind speed has been described by Fröhlich and Matzarakis (2016). The incident wind (3 m/s from 350°) is modified by the urban morphology, resulting in wind speeds between 0.0 and 6.2 m/s. The magnitude of the mean wind speed is 1.2 m/s.

Sheltering effects near buildings, which are orthogonal to the prevailing wind direction, are observable with reduced wind speed. Wind speed is significantly increased in narrow-street canyons, which are aligned to incident wind, due to channeling effect (Fig. 16.8).

16.3.2 Spatial Assessment of the Outdoor Thermal Sensation

Thermal sensation and physiological strain due to heat load and radiation interception are quantified in terms of the physiologically equivalent temperature (PET). PET varies between 40.4 and 61.2 °C. The mean PET is 47.5 °C approving the assumption of heat stress for July 25, 2019, in Rieselfeld, Germany. According to the scale for thermal perception (Matzarakis and Mayer 1996), physical strain is equivalent to *extreme heat stress* (Table 16.1). Impact of decreased wind speed (sheltering effect) at the lee side of buildings, increased wind speed (channeling effect) in narrow-street canyons, as well as shading by buildings and vegetation is clearly visible (Fig. 16.9). Spatial mean of PET in shaded areas by urban obstacles (e.g., shading by buildings) is 42.9 °C (ranging from 40.37 to 46.1 °C), compared to the spatial mean of PET in shaded areas by vegetation 42.8 °C (ranging from 41.1 to 46.4 °C). PET in sunlit areas varies between 45.5 and 61.2 °C with an average of 48.4 °C. Therefore, building's shade reduces mean thermal strain on average by 5.5 °C, while vegetation's shade reduces heat stress on average by 5.6 °C in terms of PET. The air temperature and humidity in the model are assumed to be constant

Fig. 16.8 Diagnostic wind speed under prevailing boundary conditions of 3 m/s and a wind direction of 350°, calculated by SkyHelios. Wind speed and wind direction are calculated for 12:00 UTC on July 25, 2019

Table 16.1 Thermal sensation classes for human beings in Central Europe (based on an internal heat production of 80 W and a heat transfer resistance of the clothing of 0.9 Clo) by Matzarakis and Mayer (1996)

PET (°C)	Thermal perception	Level of physiological stress
≥41	Very hot	Extreme heat stress
35–41	Hot	Strong heat stress
29–35	Warm	Moderate heat stress
23–29	Slightly warm	Slight heat stress
18–23	Comfortable	No thermal stress
13–18	Slightly cool	Slight cold stress
8–13	Cool	Moderate cold stress
4–8	Cold	Strong cold stress
<4	Very cold	Extreme cold stress

Fig. 16.9 SkyHelios can assess human thermal comfort in terms of the physiologically equivalent temperature (PET). PET can be computed spatially resolved with variable spatial resolution for different urban environments by SkyHelios. PET is calculated for 12:00 UTC on July 25, 2019

throughout the study area. The temperature of different surfaces in the model domain is not coupled with the air temperature by now. In addition to that, cooling by biophysiological processes (e.g., transpiration and photosynthesis) is also not included in the model, but can reduce air temperature in green parks by 0.94 °C according to literature (Bowler et al. 2010) or even up to 2.0 °C (Zardo et al. 2017). This would also reduce thermal heat stress.

16.3.3 Temporal Analysis of the Thermal Sensation

SkyHelios can also be used to run simulations for long time series. The calculation is conducted locally for one point of interest, instead of the complete study area. The time series is calculated in accordance to the predefined meteo input file (10-min temporal resolution). It is possible to select all named human biometeorological parameters.

The diurnal cycle of the thermal indices physiologically equivalent temperature (PET) and Universal Thermal Climate Index (UTCI), recorded wind speed, modelled wind speed, mean radiant temperature, and global radiation throughout the day (July 25, 2019) for a predefined location of interest (easting: 409863, northing: 5316954—EPSG:32632) are presented in Fig. 16.10. Both indices show the same diurnal cycle but vary in magnitude. UTCI is systematically of higher magnitude compared to PET during night, while PET results in higher values during the day. In both cases, thermal heat load is reduced with increasing wind velocity and decreasing mean radiant temperature.

PET varies between 30.9 and 49.8 °C with an average of 38.7 °C, while UTCI is in the range of 34.2–43.2 °C with an average of 37.4 °C.

Thermal heat stress is reduced with increasing wind speed in the afternoon. The modelled diagnostic wind speed is of same shape compared to recorded wind velocity at the station, but is systematically reduced by increased aerodynamic roughness of the built-up urban neighborhood in the proximity to the point of interest. The modelled wind velocity varies between 0.0 and 2.7 m/s, while the recorded dataset ranges from 0.0 to 4.7 m/s. Mean wind speed is reduced due to the neighborhood by 0.8 m/s.

Mean radiant temperature follows a similar diurnal cycle compared to global radiation but is damped during midday and in the late afternoon. The mean radiant temperature curve corresponds to a Gaussian bell curve, with a minimum of 26.4 °C, maximum of 62.7 °C, and an average value of 40.3 °C.

T_{mrt} is dependent on the local, spherical sky view factor (SVF). Spherical SVF for the location of interest is 0.77 (planar SVF: 0.94). The corresponding fisheye image is shown in Fig. 16.11. It also shows the sun path at the selected day and location.

Thermal sensation is assessed by UTCI and PET, while thermal perception differs between these indices. PET evaluates the prevailing meteorological conditions from *warm* in the night (from 00:00 to 07:00 UTC and from 17:00 to 23:50 UTC) to *very hot* (during noon at 11:20 UTC), associated with *strong* to *extreme heat stress*. The range of UTCI values is denser, resulting in lower level of physiological stress. The conditions assessed by UTCI vary between *warm* and *hot*, associated with mainly *strong* and *very strong heat stress* according to the UTCI classification of thermal stress (Table 16.2) (Błażejczyk et al. 2013).

UTCI classification is a classification for heat stress while PET classes are suitable for the assessment of thermal comfort and discomfort. Both indices are

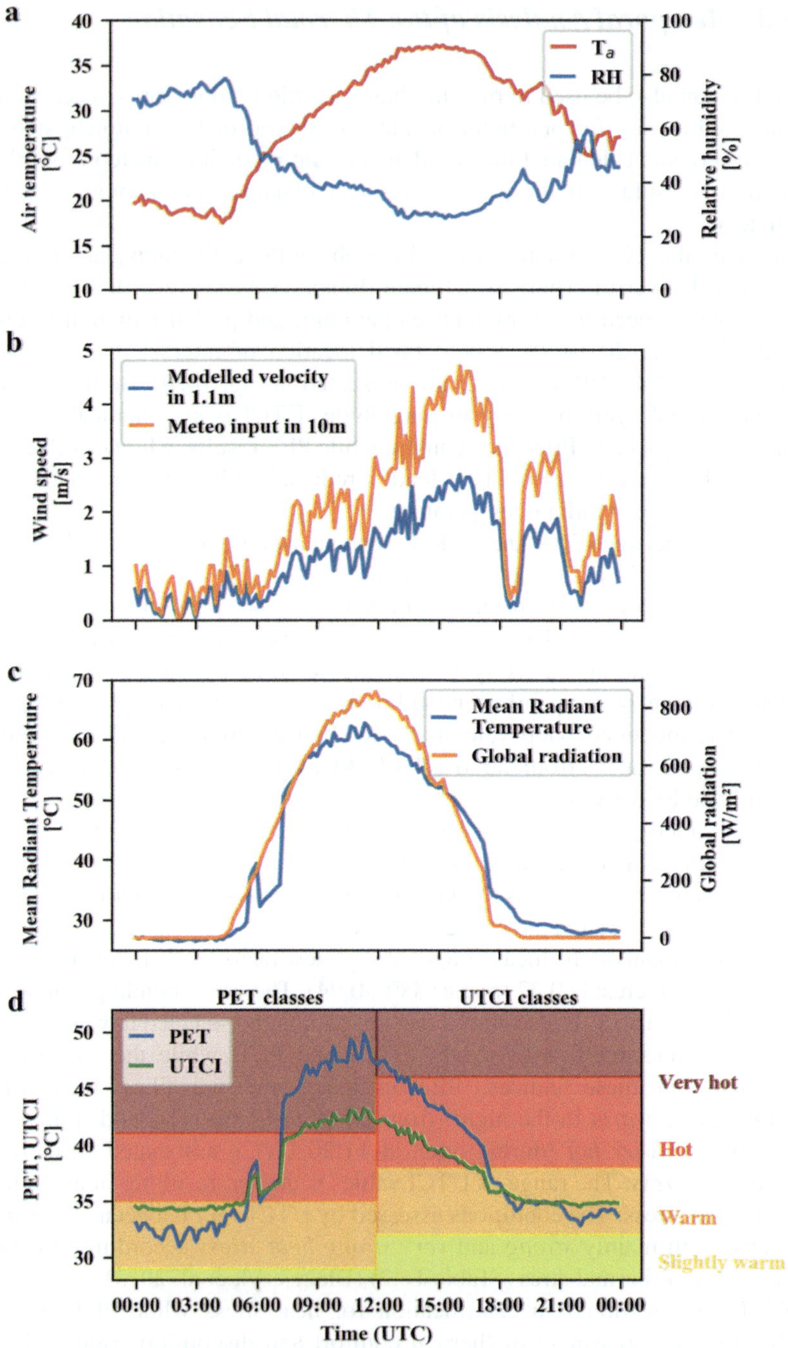

Fig. 16.10 Time series representing the center of the study area. The dependency of the thermal indices (**d**) on air temperature (**a**), relative humidity (**a**), wind speed (**b**), and global radiation (**c**) as well as mean radiant temperature (**c**) is shown

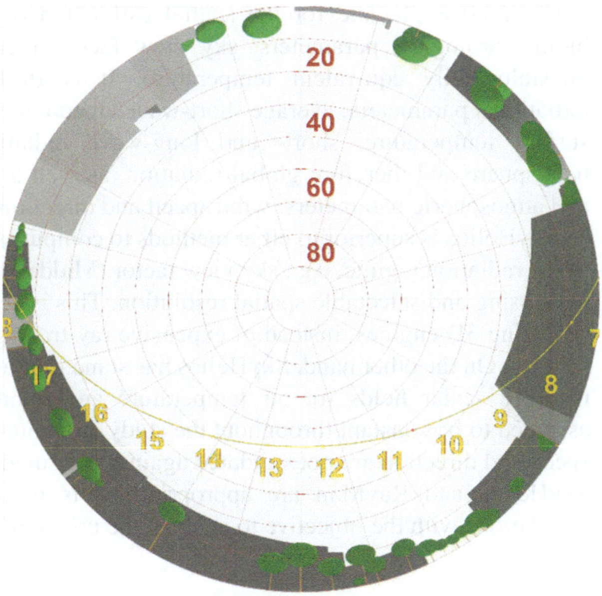

Fig. 16.11 Fisheye image showing the urban morphology in the proximity and sun path (yellow) for the location of interest. The sun is hidden by buildings and vegetation in the morning and the evening

Table 16.2 Classification of UTCI equivalent temperature in terms of thermal stress (Błażejczyk et al. 2013)

UTCI (°C)	Level of physiological stress
≥46	Extreme heat stress
38–46	Very strong heat stress
32–38	Strong heat stress
26–32	Moderate heat stress
9–26	No thermal stress
0–9	Slight cold stress
−13 to 0	Moderate cold stress
−27 to −13	Strong cold stress
−40 to −27	Very strong cold stress
<40	Extreme cold stress

appropriate for thermal assessment in human biometeorological studies in the outdoor area (Staiger et al. 2019). Based on these results, the former assumption of heat stress for given meteorological parameters is approved.

16.4 Conclusions

The time-independent RayMan and SkyHelios models are suitable models for human biometeorological analysis in urban areas. RayMan is advantageous over SkyHelios for selective point analysis, whereas the advanced SkyHelios model is much more applicable for the spatial estimation of the named, derived human biometeorological parameters: sky view factor, mean radiant temperature, and physiologically equivalent temperature. It is further applicable to compute radiational parameters: average short-wave albedo, long-wave emission coefficient, surface temperature, short- and long-wave radiation fluxes from the upper hemisphere and therefore global radiation, as well as sunshine duration, shading, and atmospheric parameters: wind speed and direction.

SkyHelios is superior to other methods to compute spatially resolved parameters of the radiation regime, e.g., sky view factor (Middel et al. 2018), due to fast spatial processing and selectable spatial resolution. This is achieved by utilization of fast-rendering 3D engines, instead of expensive ray-tracing approaches or manual field surveys. On the other hand, SkyHelios has some limitations due to missing spatially resolved scalar fields for air temperature and humidity. These parameters are assumed to be constant throughout the study area. Further, the calculations of wind speed and direction are not validated against measured data. Based on these results, SkyHelios and RayMan are appropriate tools to conduct urban microclimate simulations with the objective to analyze the urban bioclimate.

References

American Society of Heating, Refrigerating and Air-Conditioning Engineers (2005). *2005 ASHRAE handbook: Fundamentals, inch-pound edition*. Norwood, MA: Books24x7.com. Retrieved September 19, 2019 from http://www.books24x7.com/marc.asp?bookid=18504

Błażejczyk, K., Jendritzky, G., Bröde, P., Fiala, D., Havenith, G., Epstein, Y., et al. (2013). An introduction to the Universal Thermal Climate Index (UTCI). *Geographia Polonica, 86*(1), 5–10. https://doi.org/10.7163/GPol.2013.1.

Bowler, D. E., Buyung-Ali, L., Knight, T. M., & Pullin, A. S. (2010). Urban greening to cool towns and cities: A systematic review of the empirical evidence. *Landscape and Urban Planning, 97*(3), 147–155. https://doi.org/10.1016/j.landurbplan.2010.05.006.

Bröde, P., Fiala, D., Błażejczyk, K., Holmér, I., Jendritzky, G., Kampmann, B., et al. (2012). Deriving the operational procedure for the Universal Thermal Climate Index (UTCI). *International Journal of Biometeorology, 56*(3), 481–494. https://doi.org/10.1007/s00484-011-0454-1.

Chen, Y.-C., & Matzarakis, A. (2018). Modified physiologically equivalent temperature—Basics and applications for western European climate. *Theoretical and Applied Climatology, 132*(3–4), 1275–1289. https://doi.org/10.1007/s00704-017-2158-x.

Fanger, P. O. (1972). *Thermal comfort: Analysis and applications in environmental engineering*. New York: McGraw-Hill.

Fiala, D., Havenith, G., Bröde, P., Kampmann, B., & Jendritzky, G. (2012). UTCI-Fiala multi-node model of human heat transfer and temperature regulation. *International Journal of Biometeorology, 56*(3), 429–441. https://doi.org/10.1007/s00484-011-0424-7.

Fröhlich, D. (2017). *Development of a microscale model for the thermal environment in complex areas*. Freiburg: Albert-Ludwigs-Universität Freiburg. https://doi.org/10.6094/UNIFR/11614.

Fröhlich, D., & Matzarakis, A. (2013). Modeling of changes in thermal bioclimate: Examples based on urban spaces in Freiburg, Germany. *Theoretical and Applied Climatology, 111*(3–4), 547–558. https://doi.org/10.1007/s00704-012-0678-y.

Fröhlich, D., & Matzarakis, A. (2016). A quantitative sensitivity analysis on the behaviour of common thermal indices under hot and windy conditions in Doha, Qatar. *Theoretical and Applied Climatology, 124*(1–2), 179–187. https://doi.org/10.1007/s00704-015-1410-5.

Fröhlich, D., & Matzarakis, A. (2018). Spatial estimation of thermal indices in urban areas—Basics of the SkyHelios model. *Atmosphere, 9*(6), 209. https://doi.org/10.3390/atmos9060209.

Fröhlich, D., Gangwisch, M., & Matzarakis, A. (2019). Effect of radiation and wind on thermal comfort in urban environments—Application of the RayMan and SkyHelios model. *Urban Climate, 27*, 1–7. https://doi.org/10.1016/j.uclim.2018.10.006.

Gagge, A. P., Fobelets, A. P., & Berglund, L. G. (1986). A standard predictive index of human response to the thermal environment. *ASHRAE Transactions (United States), 92*, 2B.

GDAL/OGR Contributors. (2019). *GDAL/OGR geospatial data abstraction software library*. Retrieved March 15, 2019 from http://gdal.org

Gonzalez, R. R., Nishi, Y., & Gagge, A. P. (1974). Experimental evaluation of standard effective temperature a new biometeorological index of man's thermal discomfort. *International Journal of Biometeorology, 18*(1), 1–15. https://doi.org/10.1007/BF01450660.

Gröger, G., Kolbe, T. H., Nagel, C., & Häfele, K.-H. (2012). In G. Gröger, T. H. Kolbe, C. Nagel, & K.-H. Häfele (Eds.), *OGC city geography markup language (CityGML) encoding standard*. Wayland, MA: 2.0.0. Open Geospatial Consortium.

Hämmerle, M., Gál, T., Unger, J., & Matzarakis, A. (2011). Comparison of models calculating the sky view factor used for urban climate investigations. *Theoretical and Applied Climatology, 105*(3–4), 521–527. https://doi.org/10.1007/s00704-011-0402-3.

Havenith, G., Fiala, D., Błażejczyk, K., Richards, M., Bröde, P., Holmér, I., et al. (2012). The UTCI-clothing model. *International Journal of Biometeorology, 56*(3), 461–470. https://doi.org/10.1007/s00484-011-0451-4.

Herrmann, J., & Matzarakis, A. (2012). Mean radiant temperature in idealised urban canyons—Examples from Freiburg, Germany. *International Journal of Biometeorology, 56*(1), 199–203. https://doi.org/10.1007/s00484-010-0394-1.

Höppe, P. (1984). *Die Energiebilanz des Menschen* (Vol. 49). München: Universität München, Meteorologisches Institut.

Höppe, P. (1993). Heat balance modelling. *Experientia, 49*(9), 741–746. https://doi.org/10.1007/BF01923542.

Höppe, P. (1999). The physiological equivalent temperature—A universal index for the biometeorological assessment of the thermal environment. *International Journal of Biometeorology, 43*(2), 71–75. https://doi.org/10.1007/s004840050118.

Hwang, R.-L., Lin, T.-P., & Matzarakis, A. (2011). Seasonal effects of urban street shading on long-term outdoor thermal comfort. *Building and Environment, 46*(4), 863–870. https://doi.org/10.1016/j.buildenv.2010.10.017.

Jendritzky, G. (1990). *Methodik zur räumlichen Bewertung der thermischen Komponente im Bioklima des Menschen: fortgeschriebenes Klima-Michel-Modell*. Hannover: Akademie für Raumforschung und Landesplanung.

Jendritzky, G., de Dear, R., & Havenith, G. (2012). UTCI—Why another thermal index? *International Journal of Biometeorology, 56*(3), 421–428. https://doi.org/10.1007/s00484-011-0513-7.

Kasten, F. (1980). A simple parameterization of the pyrheliometric formula for determining the Linke turbidity factor. *Meteorologische Rundschau, 33*, 124–127.

Kasten, F., & Young, A. T. (1989). Revised optical air mass tables and approximation formula. *Applied Optics, 28*(22), 4735.

Ketterer, C., & Matzarakis, A. (2014). Human-biometeorological assessment of heat stress reduction by replanning measures in Stuttgart, Germany. *Landscape and Urban Planning, 122*, 78–88. https://doi.org/10.1016/j.landurbplan.2013.11.003.

Ketterer, C., Gangwisch, M., Fröhlich, D., & Matzarakis, A. (2017). Comparison of selected approaches for urban roughness determination based on Voronoi cells. *International Journal of Biometeorology, 61*(1), 189–198. https://doi.org/10.1007/s00484-016-1203-2.

Matzarakis, A., & Matuschek, O. (2011). Sky view factor as a parameter in applied climatology rapid estimation by the SkyHelios model. *Meteorologische Zeitschrift, 20*, 39–45. https://doi.org/10.1127/0941-2948/2011/0499.

Matzarakis, A., & Mayer, H. (1996). Another kind of environmental stress: Thermal stress. *WHO Newsletter, 18*, 7–10.

Matzarakis, A., Mayer, H., & Iziomon, M. G. (1999). Applications of a universal thermal index: Physiological equivalent temperature. *International Journal of Biometeorology, 43*(2), 76–84. https://doi.org/10.1007/s004840050119.

Matzarakis, A., Rutz, F., & Mayer, H. (2007). Modelling radiation fluxes in simple and complex environments—Application of the RayMan model. *International Journal of Biometeorology, 51*(4), 323–334. https://doi.org/10.1007/s00484-006-0061-8.

Matzarakis, A., Rutz, F., & Mayer, H. (2010). Modelling radiation fluxes in simple and complex environments: Basics of the RayMan model. *International Journal of Biometeorology, 54*(2), 131–139. https://doi.org/10.1007/s00484-009-0261-0.

Matzarakis, A., Fröhlich, D., Ketterer, C., & Martinelli, L. (2018). Urban bioclimate and micro climate: How to construct cities in the era of climate change. In *Climate change and sustainable heritage* (pp. 38–61). Newcastle upon Tyne: Cambridge Scholars Publishing.

Mayer, H., & Höppe, P. (1987). Thermal comfort of man in different urban environments. *Theoretical and Applied Climatology, 38*(1), 43–49. https://doi.org/10.1007/BF00866252.

Middel, A., Lukasczyk, J., Maciejewski, R., Demuzere, M., & Roth, M. (2018). Sky view factor footprints for urban climate modeling. *Urban Climate, 25*, 120–134. https://doi.org/10.1016/j.uclim.2018.05.004.

Middel, A., Lukasczyk, J., Zakrzewski, S., Arnold, M., & Maciejewski, R. (2019). Urban form and composition of street canyons: A human-centric big data and deep learning approach. *Landscape and Urban Planning, 183*, 122–132. https://doi.org/10.1016/j.landurbplan.2018.12.001.

Mirzaei, P. A., & Haghighat, F. (2010). Approaches to study urban heat island—Abilities and limitations. *Building and Environment, 45*(10), 2192–2201. https://doi.org/10.1016/j.buildenv.2010.04.001.

Ogre Development Team. (2019). *OGRE—Open source 3D graphics engine*. Retrieved March 15, 2019 from https://www.ogre3d.org/

Oke, T. R. (1987). *Boundary layer climates* (2 new ed.). London: Routledge.

Open Source Geospatial Foundation. (2018). *QGIS geographic information system*. Retrieved from http://qgis.osgeo.org

Staiger, H., Laschewski, G., & Grätz, A. (2012). The perceived temperature—A versatile index for the assessment of the human thermal environment. Part A: Scientific basics. *International Journal of Biometeorology, 56*(1), 165–176. https://doi.org/10.1007/s00484-011-0409-6.

Staiger, H., Laschewski, G., & Matzarakis, A. (2018). A short note on the inclusion of sultriness issues in perceived temperature in mild climates. *Theoretical and Applied Climatology, 131*(1–2), 819–826. https://doi.org/10.1007/s00704-016-2013-5.

Staiger, H., Laschewski, G., & Matzarakis, A. (2019). Selection of appropriate thermal indices for applications in human biometeorological studies. *Atmosphere, 10*(1), 18. https://doi.org/10.3390/atmos10010018.

Valko, P. (1966). Die Himmelsstrahlung in ihrer Beziehung zu verschiedenen Parametern. *Archiv für Meteorologie, Geophysik und Bioklimatologie, Serie B, 14*(3–4), 336–359.

VDI. (1988). *Stadtklima und Luftreinhaltung*. Berlin: Springer. https://doi.org/10.1007/978-3-662-10001-1.

VDI. (1994). *3789 Part 2: Environmental meteorology. Interactions between atmosphere and surfaces, calculation of short-wave and long-wave radiation* (p. 52). Düsseldorf: VDI.

VDI. (2008). *VDI 3787, Part 2: Environmental meteorology, Methods for the human biometeorological evaluation of climate and air quality for urban and regional planning at regional level.*

Zardo, L., Geneletti, D., Pérez-Soba, M., & Van Eupen, M. (2017). Estimating the cooling capacity of green infrastructures to support urban planning. *Ecosystem Services, 26*, 225–235. https://doi.org/10.1016/j.ecoser.2017.06.016.

Chapter 17
A Methodology for Assessing the Impact of Climate Change on Building Energy Consumption

Carolina Ganem Karlen and Gustavo Javier Barea Paci

17.1 Introduction

The latest report from the World Meteorological Organization (WMO 2017) confirms that 2016 was the warmest year of all recorded ones: there was a surprising increase of 1.1 °C above pre-industrial levels and 0.06 °C with respect to the previous record of 2015. As greenhouse gas (GHG) concentrations continue to rise, future changes to the climate system greater than those already observed are expected.

The warming of the climate system is unequivocal according to the fifth and final Evaluation Report of the Intergovernmental Panel on Climate Change (IPCC 2014), with human influence being the dominant cause. Despite efforts to reduce emissions to the atmosphere, the global climate is undergoing a consistent and inevitable change. The level of GHG in different scenarios projected by IPCC demonstrated dramatic rise in the coming future (Shen 2017).

In Fig. 17.1, the upper panel shows the Model for the Assessment of Greenhouse-gas Induced Climate Change (MAGICC), and the lower panel shows in solid line the multi-model mean global surface temperature, and in shaded line and spectra of all models time series of global annual mean surface-air temperature anomalies (relative to 1986–2005) from CMIP5 simulations. The numbers indicate how many models are available for the corresponding time period and scenario.

In the Fifth IPCC Report (2014) four new emission scenarios have been defined, the so-called representative concentration paths (RCP). These are characterized by their total radiative forcing (RF) for the year 2100, which ranges between 2.6 and 8.5 W/m^2. The four CPR trajectories comprise a scenario in which mitigation efforts lead to a very low level of forcing (RCP 2.6), two stabilization scenarios (RCP 4.5 and RCP 6.0) and a scenario with a very high level of GHG emissions (RCP 8.5).

C. Ganem Karlen (✉) · G. J. Barea Paci
INAHE, CCT-CONICET, Mendoza, Argentina
e-mail: cganem@mendoza-conicet.gov.ar

M. Palme, A. Salvati (eds.), *Urban Microclimate Modelling for Comfort and Energy Studies*, https://doi.org/10.1007/978-3-030-65421-4_17

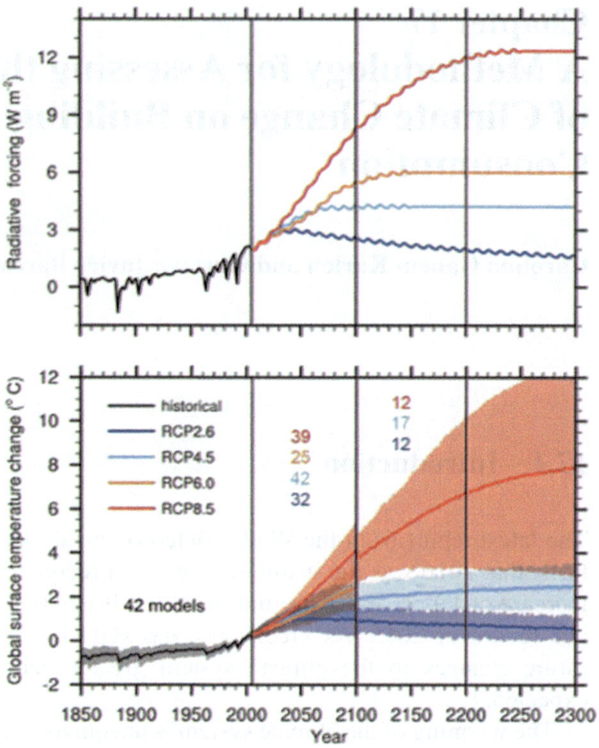

Fig. 17.1 Global mean radiative forcing and global surface temperature change for the representative concentration pathway (RCP) scenarios. (Source: Stocker et al. 2013)

Table 17.1 Representative Concentration Paths (RCP) scenarios defined in the Fifth IPCC Report (2014). Source: Own elaboration

Scenarios	Radiative forcing (RF) (W/m^2)	RF tendency	CO_2 in 2100 (ppm)
RCP 2.6	2.6	Decreasing	421
RCP 4.5	4.5	Stabilizing	538
RCP 6.0	6.0	Stabilizing	670
RCP 8.5	8.5	Increasing	936

The new CPRs can contemplate the effects of policies aimed at limiting climate change in the twentieth century against the emission scenarios used in the IPCC Fourth Assessment Report: Climate Change 2007 (AR4) (called SRES) that did not contemplate the effects of possible international policies or agreements aimed at mitigating emissions. See Table 17.1.

The combination of the projected population and economic growth together with climate change results in placing greater stress on vital resources in the future if there is a continuation of the business-as-usual scenario. The energy sector in urban areas could thus play an important role in tackling climate change and in decreasing the carbon/energy footprint of urban areas (Mauree et al. 2019). According to the Intergovernmental Panel on Climate Change (IPCC 2015) and the International Energy Agency (IEA 2018) around two-thirds of global primary energy demand is attributed to urban areas, inducing 75% of global direct energy-related GHG emissions. In the near future urban areas will be the main sources and emissions of intensive heat energy and all kinds of environmental pollution (including air pollution), sources which have already today, and will increase in the future, an important impact on micro-, meso-, and maybe even global climatological changes which will result in a reduction in the quality of life.

One hundred years ago, only 14% of the population lived in cities and in 1950, less than 30% of the world population was urban (Bitan 1992). Nowadays, around 3.5 billion people live in urban areas around the world and by 2050 more than two-thirds of the urban population will live in cities (DOE 2019).

Today, at least 170 cities support more than one million inhabitants each. The situation is even more dramatic in developing countries. Already 23 of the 34 cities with more than 5 million inhabitants are in developing countries, and 11 of these cities have populations of between 20 and 30 million. Estimations show that urban populations will occupy 80% of the total world population in 2100 (Santamouris et al. 2001).

Since the 1950s many of the changes observed have been unprecedented. Global carbon dioxide (CO_2) emissions from residential, commercial, and public buildings are expected to grow from 1.9–2.9 to 1.9–3.3 GtC/year in 2020, and 1.9–5.3 GtC/year in 2050. As a result of this IPCC projection, and under the simulation of different emission scenarios, the average temperature could increase between 2.6 and 4.8 °C by the year 2100 (IPCC 2018).

17.2 Local Context

17.2.1 Argentina

In Argentina a comparison of temperatures between 1961 and 2016, displayed in Fig. 17.2, indicates that the annual mean temperature increased throughout the country between 0.3 and 2 °C, with the higher increments in West Patagonia. The exception is the central zone, where the annual mean temperature remained constant, probably due to higher cloudiness and precipitation levels. In summer (December to February), the mean maximum temperature increased between 0.3 and 2 °C. As in the annual average, higher increments were in the West Patagonia region and the exception is the central zone where it decreases between −0.3 and −1 °C. In winter (June to August), average maximum temperature increased between 0.5 and 1 °C. The most affected regions were the Patagonia (South), the Andean region (West), and the Littoral (Northeast). Future projections for Argentina indicate that the observed tendencies described below will deepen in the future (Cabré et al. 2016).

Projections by the end of the twenty-first century show both an increase of mean temperature ranging from 1.5 to 5.5 °C (depending on the season and region, with

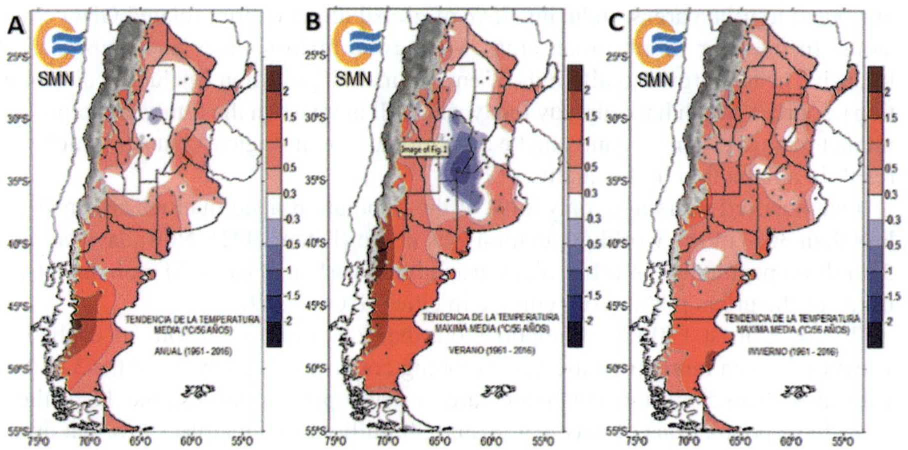

Fig. 17.2 Temperature differences (°C/56 years) comparing the year 1961 with the year 2016 for Argentina: (**a**) annual mean temperature (left), (**b**) average maximum in summer (center), and (**c**) average maximum in winter (right). (Source: National Meteorological Service (SMN) 2018)

the highest increases over subtropical latitudes) and an increase of precipitation in central Argentina. This regional model also projects a maximum temperature increase (between 4 and 5 °C) over subtropical latitudes mainly during spring (Agosta et al. 2015; Flores Larsen et al. 2018).

In the middle of December 2013, an intense heat wave began that lasted, with few interruptions, until almost mid-January. I covered the center of Argentina: from Buenos Aires to Córdoba and Mendoza, with maximum temperatures above 40 °C and minimums greater than 24 °C. It was the longest and most intense recorded in the region. The distribution of electricity collapsed in many sectors of the metropolitan area of Buenos Aires due to the record consumption due to the intense use of air-conditioning equipment.

Climate projections indicate that there will be an increase in days with heat waves in most regions of the country. The projected increase in the number of days with heat waves would be greater in the North, and especially in the Northwest of the country, where it would increase by more than 60 days in the near future. As the North of the country is the region with the greatest social vulnerability to disasters, it would be the region with the greatest risks of social impacts due to heat waves (IPCC 2018).

The increase in temperatures will cause changes in the energy demand associated with weather. Without additional mitigation policies, global energy demand from air-conditioning systems is estimated to increase from approximately 300 TWh, in 2000, to about 4000 TWh by 2050 (Chalmers 2015). From the consumption of energy in residential buildings, it follows that 42% of residential consumption corresponds to the use of electricity and natural gas (IEA 2014). Therefore, it is a sector of great interest to investigate, develop, and innovate in the rational use of energy in order to reduce these consumptions.

17.2.2 Mendoza

In Mendoza, the climate presents high heliophany and abundant solar radiation. Based on the climate classification by Roig et al. (1991), Mendoza is identified as a continental Mediterranean province with an arid to semiarid climate located in western central Argentina. It is exposed to the action of the Atlantic anticyclone movements with foothills, depressions, and plains on the Atlantic side and with high mountains and some volcanic regions on the Pacific side. It is noted that in the Köppen classification (Kottek et al. 2006), Mendoza belongs to both BWh, warm desert, and BWk, cold desert, categories.

Air temperatures: Both maximum and minimum temperatures demonstrate severe winter periods with temperatures below 5 °C. Summers are rigorous, as temperatures exceed 24 °C during most of the summer cycle. The average annual temperature is 11.9 °C. The annual heating degree days (base temperature 18 °C) is 1384 °C day/year.

Rainfall is scarce and most intense in the summer. Precipitation reaches an annual average of 218 mm, reaffirming the semiarid characteristics of the land. Prevailing winds are from the South throughout the year. The fastest southern winds are recorded in the summer. In winter, the speeds are usually low, with 35% of the time characterized as calm.

The city of Mendoza has a significant incidence of global solar radiation. In the summer, GSR between 25 and 27 MJ/m^2 day is usually registered, while in the winter, the lowest values are between 9 and 10 MJ/m^2 day on the horizontal plane.

17.3 The Building Sector

Climate change is having and will have great impacts on the energy consumption of buildings, and the design of future buildings should consider this situation. Specific microclimate conditions, which occur in urban areas, are expected to change with almost certainly disadvantageous effects on the energy consumption of buildings and the quality of life at outdoor spaces. The severity of heat waves has an impact on the design of buildings, indicating a need to review current architectural designs and adopt different strategies for new constructions, adapted to the climate change. Most buildings have a life span of several decades, during which the urban microclimate will continue to change gradually.

The building sector, including residential and commercial sectors, is responsible for approximately 30–40% of total global energy demand and emits a third of global GHG emissions (Bhamare et al. 2019). However, energy use and related emissions are expected to double, or even triple, by mid-century due to several key trends, especially as the standards of living standards of the population increase. The use of energy for cooling in buildings has doubled from 2000 to the present, which makes it the fastest growing end use in buildings. The International Energy Agency in the report "Future of Cooling" (IEA 2017) projects that the demand for space cooling energy will triple by the year 2050.

This dangerous trend could be partially reversed if today's best practices and cost-effective technologies are widely disseminated. In this context, buildings represent a critical piece for a future low carbon with low emissions and a global challenge for integration with sustainable development (Lucon et al. 2014; Flores Larsen et al. 2018). As can be seen in Fig. 17.3, buildings have significant potential for reducing emissions, the most important compared to other sectors, after implementing a rational use of energy.

The potential for energy savings in both new and existing buildings ranges from 50% to 90% (Chalmers 2015). Many researchers worldwide have put their focus on the influence of climate change in buildings. Shen (2017), after studying different cases in four representative cities of the USA, shows that the climate change will have a major impact on energy use in residential and office buildings between 2040 and 2069. A subsequent study by the same author (Shen et al. 2019) proposes a methodological framework to assess the impacts of improvements on existing buildings under climate change, based on the fact that the current energy use for building operation will increase by 1.7% annually until 2025. The authors agree that existing buildings have great potential for energy reduction.

Roshan et al. (2019) believe that to ensure thermal comfort in new and existing residential buildings, it is necessary to review and adapt the design criteria of our buildings to climate change. The authors reviewed the bioclimatic design recommendations, with climatic data for two periods: 1986–2015 and 2020–2050. A modified Givoni bioclimatic chart was used to visualize the climatic variation and inform designers about precise bioclimatic design recommendations that are resistant to climate change. The effects of climate change for different cooling and heating strategies were examined, using three concentration representative pathways (CRP)

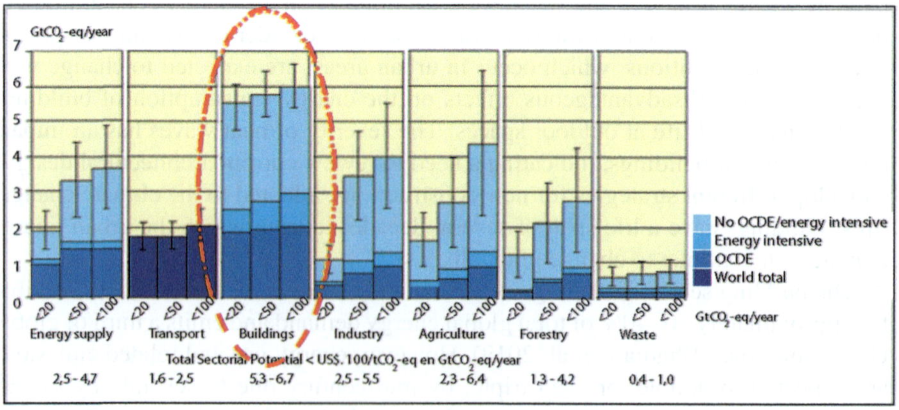

Fig. 17.3 Potential for economic mitigation estimated in 2030, by sector, compared to the respective reference values. These potentials do not include nontechnical options, such as altering lifestyles. (Source: Plan Innovative Argentina 2020)

2.6, 4.5, and 8.5, proposed in the fifth and final Evaluation IPCC Report (see Fig. 17.1). The results showed that, for all the climates studied, the trend in increasing temperature in the coming decades will be imminent, and therefore the use of heating strategies will decrease, and the use of cooling strategies will increase. Andric et al. (2017) also agree that heating demand will decrease in the future due to climate change.

Zhai and Helman (2019) analyzed the influence of 23 climate models and 56 future climate change scenarios on building consumption. They identified four representative climate models that cover the full range of all scenarios, for seven climate zones in the USA. The results are in accordance with the conclusions of other authors: the energy for cooling increases by 5%, 28%, 20%, and 52%, respectively, for the four models: low CPR (2.6), low-medium CPR (4.5), medium-high CPR (6), and high CPR (8.5). The authors also state that the energy to heat buildings decreases. Shen and Lior (2016) used global climate models to predict the future performance of renewable energy using low-energy residential buildings. The results showed that buildings with current renewable energy configurations are losing their ability to meet the zero-energy (ZEB) criteria, in half of the climatic zones considered. It was discovered that renewable energy generation systems, for a future ZEB, should be resized and reconfigured to adapt to the impacts of climate change.

It is thus evident that there is a need to increase the comfort and the design of buildings to adapt in the most comprehensive way to the negative impacts of climate change. A few cities are already taking adaptive actions and cities are seen as the "first responders" to climate change. However, most of the time, the climatic challenges are assessed individually (Bai et al. 2018). The current practice of building design was, for years, shaped by building codes with little use of such knowledge and modeling capabilities that have been marginalized from the design. A network approach, based on the modeling of the climate, city, buildings, outdoor spaces, and human variables, is necessary to understand their interdependencies (Bouyer et al. 2011).

Building energy simulation (BES) programs are capable of predicting building energy performance in detail in a dynamic model. These models should ensure that new buildings will adapt to future conditions. Nevertheless, the meteorological data used as input, even if it is in situ-measured urban microclimatic data, are generally based on current or past weather conditions and do not attend future scenarios.

The objective of this chapter is to present a parameterization of the impact of urban microclimatic conditions on energy performance of buildings in four tentative scenarios for the city of Mendoza, Argentina. The scenarios combine two time horizons with three climatic scenarios: Time horizons are the near-future climate (2015–2039) of interest for adaptation policies and the far-future climate (2075–2099) informative in the long term. The climatic scenarios are the current situation (2020), CPR 4.5 which corresponds to moderate emissions that stabilize around 2075, and RCP 8.5 which corresponds to the extreme case in which emissions will continue to grow with current trends until the end of the century to stabilize in 2200. The exact periods were chosen based on the outputs of the GCMs and

MCRs available for the region. Taking all of the exposed into account, the defined scenarios for this study for the city of Mendoza, Argentina, are:

1. The current situation (2020)
2. CRP 4.5 (2015–2039)
3. CRP 8.5 (2015–2039)
4. CRP 8.5 (2075–2099)

17.4 Methodology

A global climate model (GCM) is a complex system of computer programs that numerically represent the physical processes, and to a lesser extent chemical and biological, that occur in the atmosphere, oceans, cryosphere, and the earth's surface and in a very simplified form of the biosphere. It is the most reliable tool currently available to simulate the climate system and its variations. Currently, the GCMs simulate the weather considering a three-dimensional grid on earth with a general horizontal resolution of between 60 and 200 km, and up to 60 vertical levels in the atmosphere and similarly in the ocean. It has been verified that these models are capable of simulating the global characteristics of the climate and its changes recorded in the recent past.

One of the biggest limitations of the MCGs is that they do not always make an adequate representation of the climate at the regional level due to their low horizontal resolution. This particularly affects regions that have strong topographical accidents or thermal contrasts. In the case of Mendoza, Argentina, this is a source of errors in the simulation of the climate in the entire Andean area and its surroundings due to the impossibility of adequately representing the Los Andes mountain range in models with low spatial resolution.

17.4.1 Morphing Strategy

The problem of low resolution of GCMs can be addressed with the use of higher resolution regional climate models (RCM) that, due to their computational demand, are usually limited to a certain region. As the regional climate is in permanent connection with that of the rest of the planet, this approach makes sense if the simulation of the regional climate that the RCM performs incorporates information from a GCM. The strategy used is to feed the boundary conditions of the RCM cross-linking, every certain time step, with the GCM outputs, a process known as "nesting" (Nuñez et al. 2005; Solman and Pessacg 2012).

In order to include in the projections data observed and simulated by climate models on the Argentine Republic, the 3CN Database was used (CIMA 2020). For the elaboration of these projections, 14 experiments of 42 MCGs that presented a horizontal resolution of 2° or greater, of the CMIP5 base, were selected. A validation process was then applied to generate nested RCMs to the selected GCMs. These models are available through the Model Diagnostic and Inter-Comparison Program (MDICP) and were used in the Fifth Assessment Report of the Intergovernmental Panel on Climate Change (IPCC). For this chapter we selected the MRI-CGCM3 model developed by the Meteorological Research Institute (Taylor et al. 2012) because it has a low error index and narrow horizontal spatial resolution: 1.1° latitude per 1.2° longitude. It was built with two RCMs, MRI/JMA and ETA-HADCM3. See Table 17.2.

The nesting researchers (CIMA) developed a metric to identify which of the climate models exhibited highly erroneous behavior and in which regions. For each model, maps of the percentage errors for precipitation and mean temperature in the southern region of South America, including all of continental Argentina, were prepared. These errors were calculated with respect to the observed data from the Climatic Research Unit of East Anglia (CRU) (New et al. 1999, 2000) base for the annual average and the quarters of summer (DEF) and winter (JJA) and were averaged in each of the four regions. The metric, in the case of the average temperature of each GCM, includes the mean error defined as

$$\text{Error } T^{\circ}\text{C} = T_{\text{MC}} - T_{\text{CRU}} \tag{17.1}$$

where T_{MC} and T_{CRU} represent the average of each region (summer, winter), and the annual temperature of each climate model (MC). Likewise, the linear correlation coefficients between the annual cycle of the regional average temperature of CRU

Table 17.2 Selection of 14 experiments of 42 MCGs that presented a horizontal resolution of about 2°, of the CMIP5. Source: Own elaboration

Modelo	Error T_{anual} (°C)	Error T_{verano}(°C)	Error $T_{inv.}$ (°C)	$R_{ciclo\ anual}$	$\sigma T_{MCG}/\sigma T_{CRU}$
CCSM4	0,1	-0,1	0,2	1,00	1,01
CMCC-CM	-1,2	-0,5	-2,0	0,99	0,93
CNRM-CM5	-2,4	-1,2	-3,3	0,99	0,83
CSIRO-Mk3-6-0	-0,9	0,0	-1,3	1,00	0,93
GFDL-ESM2G	1,1	1,1	1,5	0,99	1,09
HadGEM2-CC	-1,3	-1,3	-0,5	0,92	0,90
HadGEM2-ES	-1,0	-0,7	-0,2	0,93	0,93
INMCM4	-3,3	-1,2	-5,6	0,99	0,75
IPSL-CM5A-MR	-0,1	0,4	-0,4	1,00	1,00
MIROC5	2,1	1,9	2,6	1,00	1,17
MPI-ESM-LR	-0,3	0,3	-0,4	0,99	1,01
MPI-ESM-MR	-0,5	0,1	-0,7	0,98	0,97
MRI-CGCM3	0,0	0,0	0,3	0,99	1,01
NorESM1-M	0,1	0,0	0,6	1,00	1,01

and that derived from each MC (*R* annual cycle) were calculated. To assess the inter-annual variability of temperature represented by the GCMs, the quotient between the standard deviations of the annual temperature of each model (σT_{MC}) and the reference (σT_{CRU}) was calculated.

For the creation of future climates, point #50 of the regional model grid belonging to the Metropolitan Area of Mendoza (MAM) was taken. Figure 17.4 shows the complete grid for the Province of Mendoza, Argentina, where it can be observed point #50 located in the North-Center. Average climatic data including maximum and minimum average temperatures was used to develop local CPR scenarios 4.5 and 8.5, for the years 2039 and 2099. See Table 17.3 for the climatic information. With these data and following the ASHRAE Handbook Fundamentals 2017—Chap. 14 methodology, the design days were set up with 99.6% percentile for winter and 0.4% for summer, for future scenarios.

The term "designday" is used very frequently in the building simulation and HVAC industry. Designdays are used to describe a climatic period with the maximum conditions under which HVAC systems should respond to desired indoor temperature and humidity conditions.

Warm seasons are based on annual percentiles of 0.4% or 1.0%, while cold season conditions are based on annual percentiles of 99.6% or 99%. In this work we will use 0.4% and 99.6%.

In other words, 0.4% of the dry bulb temperature means that summer temperatures will exceed this design value 0.4% of the time, for a period of 25 years, and

Fig. 17.4 Regional model grid (MAM). See point #50. (Source: Own elaboration)

Table 17.3 Average climatic data including for point #50. Source: Own elaboration

	RCP 4.5				RCP 8.5			
	Min temperature		Max temperature		Min temperature		Max temperature	
	June	December	June	December	June	December	June	December
2015–2039	6.00	18.16	12.06	29.67				
2075–2099	6.48	19.44	12.79	31.29	8.16	25.55	17.83	37.41

99.6% of the design dry bulb temperature. Winter means that the outdoor temperature remains above 99.6% all hours of the year, also according to an average of 25 years.

With the future daily data of 25 years (2015–2039 and 2075–2099), the percentiles of 99.6% for winter and 0.4% for summer were calculated using the RStudio software.

17.4.2 Case Study

The selected case is a west-facing detached house. Due to the environmental conditions, surrounding buildings, and orientation of the land, the house is organized around a central courtyard, which works environmentally as it moderates light and solar radiation, and functionally as it divides the daytime area from the night. Daytime spaces are oriented towards the east main courtyard. The front façade faces west, where vertical parasols sunscreens were designed according to the orientation. See Fig. 17.5 for graphical information of the house under study: plan and section and west façade photography. It also presents the 3D models developed with EnergyPlus.

The construction is traditional, with reinforced concrete structure and brick masonry revoked on both sides. In addition, on the south façade, 5 cm of thermal insulation is incorporated in the outside layer of the wall. The carpentries are made of aluminum with DVH (3 + 3–9–5). The *U* value is 2.1. The thermal properties of the construction layers of walls, roofs, and floors are shown in Table 17.4.

17.4.3 Measurements and Simulation Calibration

The house was simulated with the EnergyPlus V9.2 software. The house was divided into ten thermal zones whose floor areas, volumes, and internal heat gains are shown in Table 17.5. The internal heat gain due to people is described in the hours of occupation and activity level (sleeping: 40 W, cooking: 100 W, reading: 80 W). Air renewals due to infiltrations were estimated considering the total length of the frame, air infiltration for aluminum carpentry (IRAM 11507, 2001), correction factor of the external environment, and wind speed. The infiltration rate was estimated

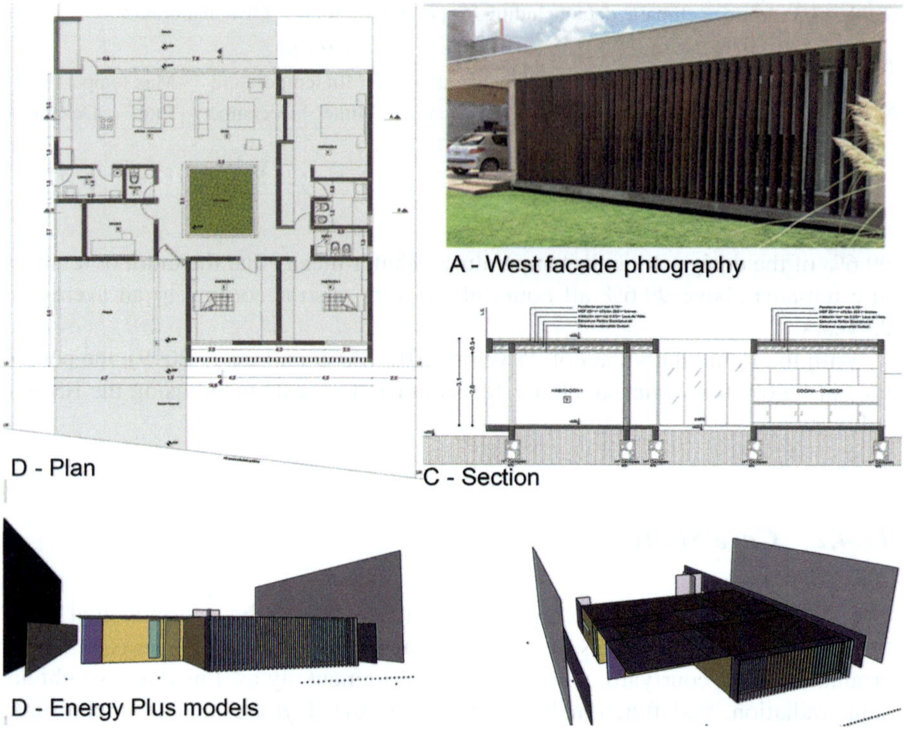

Fig. 17.5 Graphical information of the house under study: (**a**) West façade photography, (**b**, **c**) plan and section of the house under study, and (**d**) 3D models developed for EnergyPlus simulations. (Source: Own elaboration)

at 2.8 m^3/h for the bedrooms and the living room. Total air renewals were estimated at 3 ach/h (Czajkowski and Gómez 1994).

The simulated energy consumption results were compared with historical on-site energy consumption measurements from 2015 to 2019. The averaged measured energy consumption for heating and cooling allowed a reliable reference to calibrate the physical model in EnergyPlus, in order to study the influence of the climate change on the same building.

The average consumption of 5 years, destined to heating with natural gas, was 4369.76 kWh (25.70 kWh/m^2). Daily consumption and daily consumption/m^2 recorded were 11.97 kWh and 0.07 kWh day/m^2, respectively. For cooling, 3123.15 kWh was consumed during the summer months. Daily consumption and daily consumption/m^2 were 8.56 kWh and 0.05 kWh day/m^2, respectively.

Table 17.6 shows the adjustment values achieved with the EnergyPlus software, for a winter thermostat at 19.5 °C and summer set at 22.5 °C (temperatures according to adaptive comfort of ASHRAE 55). The setting has a difference between measurements and simulation of 1.16 kWh/m^2 in winter and of 0.79 kWh/m^2 in summer.

Table 17.4 Thermal properties of the construction layers. Source: Own elaboration

Perimeter (m)	Useful area (m^2)	Volume (m^3)	Envelope (m^2)				Compactness coefficient (FAEP)	Thermal transmittance ($W/m^2\ °C$)				U value
68	170	425	Walls	Roofs	Transparent	Total	2.037	Walls N, E, O	South wall	Windows	Roof	2.1
			93.5	170	79.5	343		2.41	0.64	3.2	0.41	

Table 17.5 Thermal zones regarding geometry and internal heat gains of the house under study. Source: Own elaboration

Geometry			Internal heat gains	
Thermal zone	Floor area (m^2)	Volume (m^3)	Electric appliances (3 W/m^2)	People (W)
Bedroom 1	25.38	68.53	76.14 W	160
Bedroom 2	21.50	58.05	64.50 W	80
Bedroom 3	21.50	58.05	64.50 W	80
Bedroom 4	9.99	26.97	30.00 W	80
Living-kitchen	48.16	130.04	144.50 W	420
Corridor	14.77	39.88	44.31 W	0
Bathroom 1	4.20	11.34	12.60 W	0
Bathroom 2	6.44	17.39	19.32 W	0
Bathroom 3	2.40	6.48	7.20 W	0
Laundry	5.39	14.56	43.68 W	0

Table 17.6 Calibration of EnergyPlus for heating and cooling. Source: Own elaboration

	Measurements		EnergyPlus simulation		Calibration	
	kWh	kWh/m^2	kWh	kWh/m^2	kWh	kWh/m^2
Heating	4369.76	25.70	4173.36	24.55	196.40	1.16
Cooling	3125.15	17.36	3085.35	18.15	39.79	0.79

17.5 Assessing the Impact of Climate Change on Building Energy Consumption

Four simulation sequences were run in the calibrated BSM (EnergyPlus) with the established microclimatic models. Figure 17.6 shows heating and cooling energy consumption results for each scenario. There is a clear tendency to the increment of cooling loads and a reduction of the heating requirement.

- Scenario 1: The current situation (2020). Results are consistent with consumptions registered in the house (heating = 25.7 kWh/m^2 and cooling = 17.36 kWh/m^2). Heating represents 57.5% and cooling 42.5%.
- Scenario 2: CRP 4.5 (2015–2039). This scenario in a time horizon in the near future of interest for adaptation policies corresponds to moderate emissions that stabilize around 2075. Heating represents 40.6% and cooling 59.4%.
- Scenario 3: CRP 8.5 (2015–2039). This scenario in a time horizon in the near future of interest for adaptation policies corresponds to the extreme case in which emissions will continue to grow with current trends until the end of the century to stabilize in 2200. Heating represents 34% and cooling 66%.
- Scenario 4: CRP 8.5 (2075–2099). This scenario is in a time horizon in the far future informative in the long term. Its simulation is justified because it did not stabilize in the near-future time horizon. Heating represents 22.2% and cooling 77.8%.

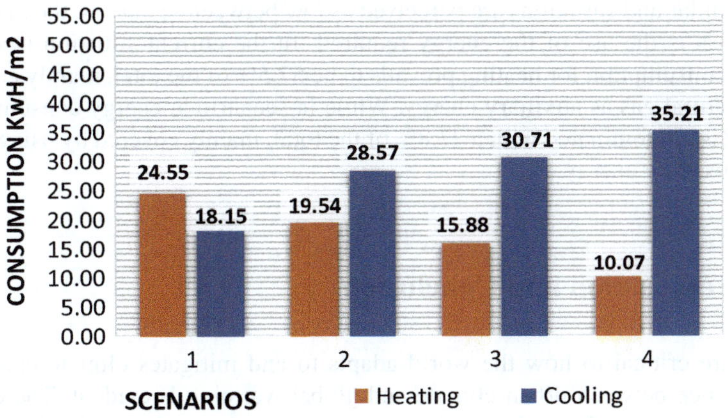

Fig. 17.6 Heating and cooling energy consumption results for each scenario. Scenario 1: The current situation (2020). Scenario 2: CRP 4.5 (2015–2039). Scenario 3: CRP 8.5 (2015–2039). Scenario 4: CRP 8.5 (2075–2099). (Source: Own elaboration)

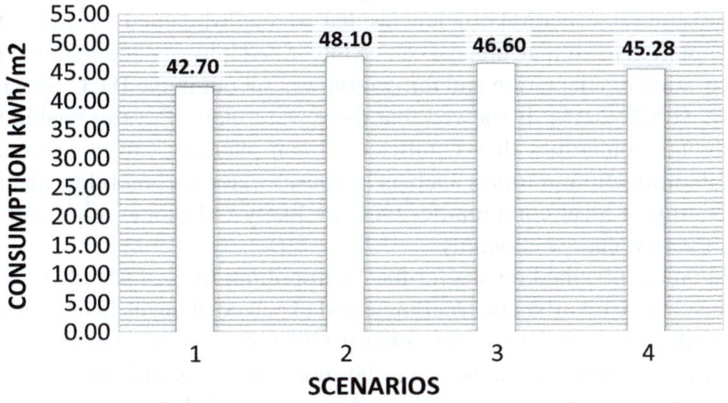

Fig. 17.7 Total energy consumption results for each scenario. Scenario 1: The current situation (2020). Scenario 2: CRP 4.5 (2015–2039). Scenario 3: CRP 8.5 (2015–2039). Scenario 4: CRP 8.5 (2075–2099). (Source: Own elaboration)

Figure 17.7 presents total energy consumption for each scenario. When heating and cooling needs are added in a single figure it is interesting to analyze the whole energy consumption trends. In the case under study, as well as similar constructions in the city of Mendoza, Argentina, total energy consumption presents a maximum variation of 12.6% with a tendency to diminish in the worst-case scenario CRP 8.5 (2075–2099).

Consumptions of 42.70 kWh/m^2 are registered in the current situation (2020) and consumptions of 48.10, 46.6, and 45.28 kWh/m^2 are predicted, respectively, for each scenario. It is important to take into account that even though consumptions seem similar and variations are perceived as low between scenarios, the main aspect to discuss is the use of the energy required. In the current situation (scenario 1) energy consumption for heating prevails to be 57.5% of the total, mainly covered by using natural gas as auxiliary energy. While in scenario 4, energy consumption for cooling is predominated with a 77.8% of the total, mainly covered by using electricity as auxiliary energy.

17.6 Discussion and Conclusion

Cities are critical to how the world adapts to and mitigates climate change. The interference between urban climate and global warming is evident. The combination of the projected population and economic growth together with climate change results in placing greater stress on vital resources in the future if there is a continuation of the business-as-usual scenario.

Climate change action requires both mitigation and adaptation. Mitigation needs cities to demonstrate the decoupling of greenhouse emissions from wealth generation. This requires structural efficiencies that are mainstreamed through urban planning. Adaptation to climate change can be integrated into city and regional planning. The integration of adaptation and mitigation was set out in the IPCC Special Report on 1.5 °C (IPCC 2018). It showed that there were many ways to enable both processes though sometimes they conflicted. Urban planning can manage adaptation and enable mitigation in topics such as land use, water use, mobility planning, and understanding of local communities who are likely to become more and more the focus of the new energy economy.

Based on the obtained results for the Metropolitan Area of Mendoza, the impact of climate change on urban microclimate and expected changes in energy consumption of buildings during the next century will present new challenges. The heat waves with respect to the present will be more extensive and are estimated to be incremented in 115 more days by 2050 (Agosta et al. 2015).

By efficiently preparing for the imminent threats posed by climate change, social and economic stability can be preserved, enabling growth in a changing environment. Therefore, adaptation strategies to climate change must be developed to create more robust urban environments that are resilient to both gradual climate change-driven events and extreme events.

In the current microclimatic situation, main attention is paid to heating systems as energy consumption demand for heating is higher than that for cooling. Therefore, direct gain combined with conservation measures is the prevailing passive strategy recommended to take into account in a new project or in a refurbishment.

Nevertheless, this tendency reverts in every studied scenario. The double relation between buildings and climate sets an almost certain snowball effect of cooling needs for the future. Because of that, special attention should be paid to the design of passive conditioning systems that will adapt to the future hotter microclimate. The use of direct solar gain in winter will be likely to cause overheating in summer.

To counteract the effects of heat waves, the need of passive cooling strategies is evident. Natural crossed ventilation and sun protection are the main strategies recommended in the current scenario. In future scenarios, natural ventilation will be less efficient due to the higher minimum temperatures in summer. Therefore, the use of well-designed sunscreens in the building envelope is of utmost importance, especially in glazed areas that should also be carefully sized.

The increase in the maximum electricity load during the cooling seasons will put more pressure on the electricity network, especially during peak hours. This tendency is consistent with prognostics expressed by Shen et al. (2019) and Filippín and Flores Larsen (2019). As the demand for electricity will increase significantly in the future, this issue should be included in political discussions today in order to be prepared not only to produce electricity in the less aggressive way towards the environment, but also because power line capacity will soon be exceeded with great risk of frequent blackouts.

Users can also incorporate photovoltaic systems connected to the network to reduce pressure to the energy production system. Nevertheless, this measure that eases the production demand also depends on the capacity of the power lines.

In conclusion, in a changing climate, a regional and localized climate analysis is absolutely necessary to design appropriate passive strategies. Regions are affected differently by global climate change, depending on local factors such as topographic relieves of mountains and large water reserves.

If in the twentieth century attention was paid mainly to a better indoor climate and a saving in energy for the single building, in the twenty-first century, assessing the possible impact of urban microclimate on energy consumption of buildings seems mandatory. In this context, buildings represent a critical piece of a low-carbon future and a global challenge for integration with sustainable development.

References

Agosta, E., Araneo, D., Camilloni, I., Carril, A., Doyle, M., Frumento, O., Nuñez, M., Ortiz, M. I., Penalba, O., Rusticucci, M., Saulo, C., & Solman, S. (2015). *Cambio Climático en Argentina: Tendencias y Proyecciones. Tercera Comunicación Nacional sobre Cambio Climático.* Secretaría de Ambiente y Desarrollo Sustentable de la Nación, Argentina.

Andric, I., Pina, A., Ferrão, P., Fournier, J., Lacarrière, B., & Le Corre, O. (2017). The impact of climate change on building heat demand in different climate types. *Energy and Buildings, 149*, 225–234.

Bai, X., Dawson, R. J., Ürge-Vorsatz, D., Delgado, G. C., Barau, A. S., Dhakal, S., Dodman, D., Leonardsen, L., Masson-Delmotte, V., Roberts, D. C., & Schultz, S. (2018). Six research priorities for cities and climate change. *Nature, 555*, 23–25. Retrieved from http://www.nature.com/articles/d41586-018-02409-z.

Bhamare, D. K., Rathod, M. K., & Banerjee, J. (2019). Passive cooling techniques for building and their applicability in different climatic zones—The state of art. *Energy and Buildings, 198*, 467–490.

Bitan, A. (1992). The high climatic quality city of the future. *Atmospheric Environment, 26*(3), 313–329.

Bouyer, J., Inard, J. C., & Musy, M. (2011). Microclimatic coupling as a solution to improve building energy simulation in an urban context. *Energy and Buildings, 43*(7), 1549–1559.

Cabré, M. F., Solman, S., & Núñez, M. (2016). Regional climate change scenarios over southern South America for future climate (2080–2099) using the MM5 model. Mean, interannual variability and uncertainties. *Atmósfera, 29*(1), 35–60.

Chalmers, P. (2015). *Cambio climático: Implicaciones para los Edificios. Hallazgos claves del Quinto informe de evaluación del IPCC.* Cambridge: University of Cambridge.

CIMA - Centro de Investigaciones del Mar y la Atmósfera [Sea and Atmosphere Research Centre] (2020). School of Natural and Exact Sciences. National University of Buenos Aires. Last consulted May 2020. Retrieved from http://3cn.cima.fcen.uba.ar/

Czajkowski, J., & Gómez, A. (1994). Introducción al diseño bioclimático y economía energética edilicia. Fundamentos y métodos. Publicación de la Editorial de la Universidad Nacional de La Plata, República Argentina.

DOE—Department of Energy. (2019). *World population prospects: The 2019 revision, key findings and advance tables.* Retrieved from https://population.un.org/wpp/

Filippín, C., & Flores Larsen, S. (2019). Monitoreo post-ocupación y simulación de un edificio bioclimático no residencial durante la transición primavera-verano en clima de máxima irradiancia. IBPSA LATAM 2019. Mendoza, Argentina.

Flores Larsen, S., Filippín, C., & Barea, G. (2018). Impact of climate change on energy use and bioclimatic design of residential buildings in the 21st century in Argentina. *Energy and Buildings, 184*, 216–229.

IEA—International Energy Agency. (2014). *Programme on energy in buildings and communities. Final report annex 53, November 2014.* Sendai: Graduate School of Engineering, Tohoku University.

IEA—International Energy Agency. (2017). *The future of cooling.* Retrieved from https://www.iea.org/reports/the-future-of-cooling

IEA—International Energy Agency. (2018). *World energy outlook 2018.* Retrieved from https://www.iea.org/weo/

IPCC—Intergovernmental Panel on Climate Change. (2014). *Climate change 2014: Synthesis report. Contribution of working groups I, II and III to the fifth assessment report of the Intergovernmental Panel on Climate Change.* Geneva: IPCC.

IPCC—Intergovernmental Panel on Climate Change. (2015). In O. Edenhofer et al. (Eds.), *Climate change 2014: Mitigation of climate change.* Cambridge: Cambridge University Press.

IPCC—Intergovernmental Panel on Climate Change. (2018). In V. Masson-Delmotte, P. Zhai, H.-O. Pörtner, D. Roberts, J. Skea, P. R. Shukla, A. Pirani, W. Moufouma-Okia, C. Péan, R. Pidcock, S. Connors, J. B. R. Matthews, Y. Chen, X. Zhou, M. I. Gomis, E. Lonnoy, T. Maycock, M. Tignor, & T. Waterfield (Eds.), *Global warming of 1.5°C. An IPCC special report on the impacts of global warming of 1.5°C above pre-industrial levels and related global greenhouse gas emission pathways, in the context of strengthening the global response to the threat of climate change, sustainable development, and efforts to eradicate poverty.* New York: IPCC.

Kottek, M., Grieser, J., Beck, C., Rudolf, B., & Rubel, F. (2006). World Map of the Köppen-Geiger climate classification updated. *Meteorologische Zeitschrift, 15*(3), 259–263.

Lucon, O., Urge-Vorsatz, D., Zain Ahmed, A., Akbari, H., Bertoldi, P., Cabeza, L. F., Eyre, N., Gadgil, A., Harvey, L. D. D., Jiang, Y., Liphoto, E., Mirasgedis, S., Mu-Rakami, S., Parikh, J., Pyke, C., & Vilarino, M. V. (2014). *Buildings, climate change 2014: Mitigation of climate change, contribution of working group III to the fifth assessment report of the Intergovernmental Panel on Climate Change.* New York: Cambridge University Press.

Mauree, D., Naboni, E., Coccolo, S., Perera, A. T. D., Nik, V. M., & Scartezzini, J. L. (2019). A review of assessment methods for the urban environment and its energy sustainability to guarantee climate adaptation of future cities. *Renewable and Sustainable Energy Reviews, 112*, 733–746.

New, M., Hulme, M., & Jones, P. (1999). Representing twentieth-century space–time climate variability. Part I: Development of a 1961–90. Mean monthly terrestrial climatology. *Journal of Climate, 12*, 829–856.

New, M., Lister, D., Hulme, M., & Makin, I. (2000). A high-resolution data set of surface climate over global land areas. *Climate Research, 21*, 1–25.

Nuñez, M. N., Solman, S., Cabré, M., & Rolla, A. (2005). *Estimación de escenarios regionales de cambio climático mediante el uso de modelos climáticos regionales*. Informe Final.

Plan Argentina Innovadora. (2020). *Sector Energía. Desarrollo de envolventes y de sistemas de controles inteligentes para edificaciones energéticamente sustentables*. Retrieved from https://www.argentina.gob.ar/plan-argentina-innovadora-2020/energia-0

Roig, F. A., González Loyarte, M. M., Abraham, E. M., Mendez, E., Roig, V. G., & Martínez Carretero E. (1991). *Maps of desertification hazard of Central Western Argentina (Mendoza Province) study case World Atlas of Thematic Indicators of Desertification*. United Nations Environment Programme (UNEP) (pp. 50–53).

Roshan, G. R., Oji, R., & Attia, S. (2019). Projecting the impact of climate change on design recommendations for residential buildings in Iran. *Building and Environment, 155*, 283–297.

Santamouris, M., Papanikolaou, N., Livad, A. I., Koronakis, I., Georgakis, C., Argiriou, A., & Assimakopoulos, D. N. (2001). On the impact of urban climate on the energy consumption of buildings. *Solar Energy, 70*(3), 201–216.

Shen, P. (2017). Impacts of climate change on U.S. building energy use by using down-scaled hourly future weather data. *Energy and Buildings, 134*, 61–70.

Shen, P., & Lior, N. (2016). Vulnerability to climate change impacts of present renewable energy systems designed for achieving net-zero energy buildings. *Energy, 114*, 1288–1305.

Shen, P., Braham, W., & Yi, Y. (2019). The feasibility and importance of considering climate change impacts in building retrofit analysis. *Applied Energy, 233*, 254–270.

SMN-National Meteorologica lService of Argentina. (2018). Retrieved from https://www.smn.gob.ar/caracterizacion-estadisticas-de-largo-plazo

Solman, S., & Pessacg, N. (2012). Evaluating uncertainties in regional climate simulations over South America at the seasonal scale. *Climate Dynamics, 39*, 59–76.

Stocker, T. F., Qin, D., Plattner, G.-K., Alexander, L. V., Allen, S. K., Bindoff, N. L., Bréon, F.-M., Church, J. A., Cubasch, U., Emori, S., Forster, P., Friedlingstein, P., Gillett, N., Gregory, J. M., Hartmann, D. L., Jansen, E., Kirtman, B., Knutti, R., Krishna Kumar, K., Lemke, P., Marotzke, J., Masson-Delmotte, V., Meehl, G. A., Mokhov, I. I., Piao, S., Ramaswamy, V., Randall, D., Rhein, M., Rojas, M., Sabine, C., Shindell, D., Talley, L. D., Vaughan, D. G., & Xie, S. P. (2013). *Climate change 2013: The physical science basis. Contribution of working group I to the fifth assessment report of the Intergovernmental Panel on Climate Change*. Cambridge: Cambridge University Press.

Taylor, K., Stouffer, R. J., & Meehl, G. A. (2012). An overview of CMIP5 and the experiment design. *Bulletin of the American Meteorological Society, 93*, 485–498.

WMO—World Meteorological Organization. (2017). *Statement on the state of the global climate in 2017*. Provisional release November 2017.

Zhai, Z. J., & Helman, J. M. (2019). Implications of climate changes to building energy and design. *Sustainable Cities and Society, 44*, 511–519.

Part III
Applying Urban Climate Modelling in Policy, Planning and Design: Case Studies

Chapter 18
Spatial Metrics to Investigate the Impact of Urban Form on Microclimate and Building Energy Performance: An Essential Overview

Michele Morganti

18.1 Introduction

In recent years, the topic of cities and climate change has drawn the attention of media, experts as well as researchers from different fields. The increasing frequency of natural disasters raises the question of how to adapt cities and counteract to these challenging conditions in the short-term period. The IPCC recalls for the urgency of urban adaptation and energy transition and determines spatial configuration of cities as a driver of high importance in these issues (IPCC 2014, 2018).

It has been widely demonstrated in the past 50 years that urban form has relevant effects on the urban microclimate, thermal comfort—both indoor and outdoor—and building energy performance (Emmanuel and Steemers 2018). A huge amount of information has been developed in the field of geography, climate, urban ecology, urban physics and resilient design. The characterization of these effects in every field has grown remarkably in the last years. However, the complex nature of the issue produces a rich but intricate body of theoretical knowledge making use of hundreds of spatial metrics that depend on the aim, perspective and tools of each study.

A broad set of urban form attributes and associated metrics with relevance for urban climate and energy performance is available today, but the proposed approaches are often partitioned and limited to few attributes. Moreover, the use in practice made by professionals in regenerative design is very limited. Architects, urban designers, planners and policymakers are generally aware of the importance of the issue, but they often turn out to be frustrating to include this knowledge when making decisions. Multiple metrics, sometimes conflicting, are available and which are the most important with the design objectives is not easy to define. The unintended interaction among urban form, microclimate and energy and the associated

M. Morganti (✉)
Department of Architecture and Urban Studies, Polytechnic of Milan, Milan, Italy
e-mail: michele.morganti@polimi.it

M. Palme, A. Salvati (eds.), *Urban Microclimate Modelling for Comfort and Energy Studies*, https://doi.org/10.1007/978-3-030-65421-4_18

variables together with the difficulties to isolate the urban form from other drivers of microclimate and energy performance requires a more holistic comprehension of the relevance, suitability and effectiveness of metrics in the description of cities' functioning (Silva et al. 2017) (Fig. 18.1).

18.2 A Cross-Disciplinary Framework

This chapter provides an essential overview on spatial attributes of cities with relevance on microclimate and energy efficiency: a general framework on the most appropriate spatial metrics in quantitative design research method concerning urban form is introduced and discussed. On the one side, the overview wants to clarify the existing scientific "babble" on the metrics and promote an effective and suitable use. On the other side, it wants to foster the application of the metrics to climate-responsive regenerative design and planning. Quoting Jacobson, the most urgent challenge in this field is to enable architects and planners to easily understand, control and assess cumulative impacts and effects of their design decisions (Jacobson 2019). Metrics represent a powerful tool, especially for the possibility of integration into parametric design tools, nowadays widely used in professional practice as well as in scientific research.

The overview of the metrics is divided into three subsections. Each section describes the capability of existing methods to investigate the effects of urban form on microclimate and outdoor thermal comfort, solar irradiation and building energy performance, respectively. Recent relevant studies are analysed in terms of two additional factors—the scale of investigation and the climate of reference—as well as in terms of suitability and effectiveness of urban form attributes and the

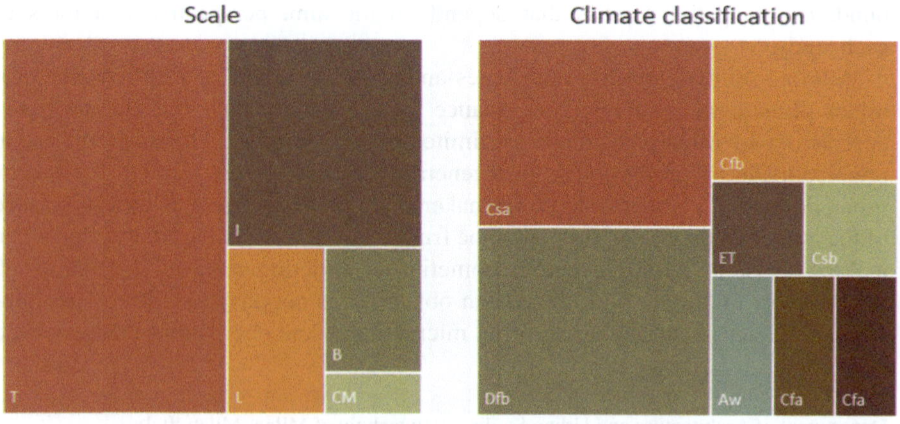

Fig. 18.1 Scale of analysis and Köppen climate classification of recent studies using spatial metrics to investigate urban microclimate and energy efficiency

associated metrics (Fig. 18.2). Taking advantage of metrics' neutrality to wave together three sectorial perspectives in a single framework is the main goal of this chapter.

18.2.1 Microclimate

The effect of physical characteristics of the urban environment on local microclimate is well known in the field. It has been a long time since we analysed and understood the main factors that affect this effect and possible solutions to counteract or mitigate it (Golany 1996; Oke 1982). Due to unprecedent climate change consequences on cities—temperature rise, extreme droughts and heatwaves in summer—recent advancements are focusing on outdoor thermal comfort and urban heat island effects. For this reason, most of the studies deal with warm and hot climates.

The most common metrics used in these research methods describe key morphology attributes of the urban fabric that are correlated with air temperature in the urban canopy layer: compactness, plot ratio, distance between buildings, vertical density and sky view factor—both of open space and façades (Table 18.1). Urban texture "porosity" and "building intensity" are the attributes of interest that play a crucial role in the urban heat energy balance. Both these attributes significantly affect most of the components of urban energy balance: wind flow, incident solar radiation, energy absorption and (re-)emission. The former expresses the amount of open space available in the urban area; the latter, conversely, expresses the amount of buildings that rise in the urban area. Several density metrics—such as porosity, compactness, plot ratio and vertical density—have been defined and used to measure these attributes and have proved to be indicators of outdoor comfort conditions and energy performance of buildings. For example, compactness, plot ratio and building volume density are directly related to summer air temperatures in the case of the Mediterranean climate (Petralli et al. 2014). Moreover, UHI intensity is strongly related to compactness in winter and to vertical density in summer (Salvati et al. 2019).

Stewart and Oke (2012) developed the local climate zones, a spatial metrics-based taxonomy of urban microclimates in order to overcome the inadequacy of simple urban-rural climate division and to determine different magnitudes of UHI in urban textures. The study characterizes built form types and land types using six urban form metrics out of eight. The idea of downscaling climate analysis in cities considering the effects of physical attributes—too often neglected by practitioners—is of interest concerning not only temperature studies but also the associated building energy demand. Through appropriate metrics—such as aspect ratio and compactness—consistency and accuracy of energy performance evaluation can be improved, including differences in UHI intensity (Vuckovic et al. 2017). This relation can be described even more effectively by using a combination of urban density metrics: compactness, vertical density and average building height show good correlation with energy demand for residential heating and cooling (Salvati et al. 2017).

Table 18.1 Urban form indicators investigated in recent studies

Main topic[a]	Ref.	Metric	Scale[b]	Climate[c]	Location	Main findings
E	Martins et al. (2019)	Plot ratio/ floor area ratio Compactness Aspect ratio Shape factor Courtyard/ patio aspect ratio Street width Standard deviation of building heights	T I B	Cfb	Toulouse, FR	For compact city blocks, the courtyard aspect ratio and the standard deviation of the built height account for almost 50% of the overall impact on heating demand of buildings. The relevance of certain factors was strongly linked to their typologies or to their urban environment.
E	Natanian et al. (2019)	Shape factor Plot ratio/ floor area ratio Window-to-wall ratio Distance between buildings	T	Csa	Tel Aviv, IL	The findings demonstrate the correlation between the shape factor and the energy load match index as well as the benefits of the courtyard typology in terms of energy balance, with its challenging daylight performance.
E	Nault et al. (2015)	Plot ratio/ floor area ratio	T I	ET Dfb Csb	Bern, CH Yverdon-les-Bains, CH San Francisco, USA	The study tests the role of simple metrics as performance indicators when applied to neighbourhoods. To do so, the selected metrics, including geometrical parameters (e.g. compactness) and solar exposure levels (e.g. annual irradiation), are compared with simulation results (e.g. heating need), taken as reference values.

(continued)

Table 18.1 (continued)

Main topic[a]	Ref.	Metric	Scale[b]	Climate[c]	Location	Main findings
E	Rodríguez-álvarez (2016)	Compactness Plot ratio/floor area ratio Compactness ratio	T	Cfb Dfb Bsk Csa	London, UK Paris, FR Berlin, D Madrid, ES Barcelona, ES	The UEIB is a tool that has been specifically designed to assess the energy performance of buildings in large urban areas. It is based on the reduction of the urban geometry into a simpler national grid that retains critical information to perform meaningful estimations. Compactness, plot ratio and compactness ratio have been used to model urban form.
E	Rode et al. (2014)	Compactness Plot ratio/floor area ratio Building height Shape factor Open space ratio	T I	Cfb Dfb Csa	London, UK Paris, FR Berlin, D Istanbul, TR	The average building height and building density were found to be good indicators for heat-energy efficiency, each correlating negatively with the heat-energy demand. The shape factor also correlates well but positively with heat-energy demand.

(continued)

Table 18.1 (continued)

Main topic[a]	Ref.	Metric	Scale[b]	Climate[c]	Location	Main findings
E	Steadman et al. (2014)	Building volume density Plan depth Exposed surface area	T I	Cfb	London, UK	A correlation of energy use with the exposed surface areas of buildings has been demonstrated, although the full significance of the correlation cannot be assigned to this parameter. Urban form-induced heat-energy efficiency is significant and can lead to a difference in heat-energy demand of up to a factor of six.
M	Salvati et al. (2017)	Compactness	T B	Csa	Rome, IT	Considering UHI intensity and solar obstruction determined by different urban texture, compact urban textures, with compactness above 0.5, contribute to reduced energy consumption in a Mediterranean climate.
M	Chatzidimitriou and Yannas (2016)	Aspect ratio Shading factor	T	Csa	Thessaloniki, GR	Aspect ratio has a strong effect on PET in summer season.
M	Leo et al. (2018)	Compactness Frontal area density Average building height	T I	Cfa	Oklahoma City, USA	The use of morphometric parameters based on scale-adaptive methods provides better agreement with measured data than those in which scales for their calculations are arbitrarily chosen.

(continued)

Table 18.1 (continued)

Main topic[a]	Ref.	Metric	Scale[b]	Climate[c]	Location	Main findings
M	Perini and Magliocco (2014)	Compactness Building height	T	Cfa Csa	Milan, IT Rome and Genova, IT	Compactness and height of buildings in a city area influence potential temperature and mean radiant temperature. Higher density causes higher potential temperatures. The height of buildings plays an important role in determining potential temperatures at 1.6 m from the ground level: with taller buildings temperatures are lower due to their shading effect.
M	Petralli et al. (2014)	Plot ratio/ floor area ratio Compactness Street surface Building volume density	T	Csa	Florence, IT	The minimum night temperature in summer is positively affected by an increase of street and building indicators.
M	Salvati et al. (2019)	Compactness Vertical density Average building height	T I	Csa	Rome, IT Barcelona, ES	In the Mediterranean climate, an increase of horizontal density of urban textures determines an increase of the UHI intensity in winter, while an increase of vertical density entails an increase of UHI intensity in summer.

(continued)

Table 18.1 (continued)

Main topic[a]	Ref.	Metric	Scale[b]	Climate[c]	Location	Main findings
M	Stewart and Oke (2012)	Sky view factor/sky factor Aspect ratio Compactness Average building height Impervious/pervious surface fraction	L	–	–	The proposed metrics are the basis for the classification of the LCZ as a comprehensive climate-based classification of urban and rural sites for temperature studies.
M	Tsitoura et al. (2016, 2017)	Aspect ratio Sky view factor/sky factor	T	Csa	Crete, GR	During summer, the effects of the aspect ratio on microclimate depend on the orientation of the urban canyon; SVF is used as an indicator of shading but orientation is crucial in thermal comfort (radiant temperature).
M	Vuckovic et al. (2017)	Aspect ratio of urban canyon Compactness Impervious/pervious surface fraction	T I B	Dfb	Wien, AU	A higher impervious surface fraction is noticeably correlated with higher night-time air temperature in the urban canyon. Significant deviations are noted in computed heating and cooling loads, as well as in overheating levels, with regard to standardized climate input data.
S	Martins et al. (2014, 2016)	Porosity Shape factor Aspect ratio Plot ratio/floor area ratio Contiguity Distance between buildings	T I B	Aw	Maceiò, BR	Results indicate a significant impact of the aspect ratio and the distance between buildings.

(continued)

Table 18.1 (continued)

Main topic[a]	Ref.	Metric	Scale[b]	Climate[c]	Location	Main findings
S	Chatzipoulka et al. (2016)	Mean outdoor distance Compactness Directionality Vertical density Standard deviation of building height	T	Cfb	London, UK	Mean outdoor distance, compactness, directionality and complexity were the most influential for the solar performance of open spaces, while building façades were mostly affected by standard deviation of building height and directionality (among others). Direct solar irradiance on ground and façades was found to be influenced by different variables in January and July.
S	Lee et al. (2016)	Compactness Plot ratio/ floor area ratio Number of storeys	T I	Dfb Dfa	Copenhagen, DK NYC, USA	The lower the density, the higher the solar potential. Density has less effect on the active solar potential of roofs, where climate plays a more influential role. Solar accessibility appears to be less a function of density than of building layout and configuration.
S	Morganti et al. (2017)	Compactness Vertical density Sky view factor/sky factor	T I	Csa	Rome, IT Barcelona, ES	Compactness, vertical density and sky factor show very good correlation with solar irradiation and could be used to develop a comparative assessment tool of solar performance at fabric scale.

(continued)

Table 18.1 (continued)

Main topic[a]	Ref.	Metric	Scale[b]	Climate[c]	Location	Main findings
S	Mohajeri et al. (2019)	Street orientation Street width Street length Sky view factor/sky factor	L T	Dfb	Geneve, CH	Street orientation has a strong effect on received annual solar radiation by street surfaces and façades. Received solar radiation, both for street surfaces and façades, shows only moderate correlations with the other measured geometric parameters, namely street width, street length, asymmetric aspect ratio and SVF, the highest coefficient of determination being between received street-surface radiation and SVF.
S	Mohajeri et al. (2016)	Compactness Plot ratio/floor area ratio Volume-area ratio	L T	Dfb	Geneve, CH	Compactness, plot ratio and volume-area ratio are the most effective in describing the solar performance of roofs and façades.
S	Sarralde et al. (2015)	Average building height Compactness Average building perimeter Standard deviation of building heights Plot ratio/floor area ratio Distance between buildings	L T	Cfb	London, UK	Results show that by optimizing combinations of up to eight metrics of urban form the renewable energy potential of solar irradiation of roofs could be increased by ca. 9%, while that of façades could be increased by up to 45%.

[a](M) Microclimate, UHI and outdoor thermal comfort; (E) urban building energy performance; (S) solar irradiation
[b](L) Local > (T) urban texture > (I) urban island > (B) building
[c]Köppen-Geiger climate classification

In addition, spatial metrics have been positively applied to design-oriented research, with special attention to outdoor thermal comfort. Aspect ratio and sky view factor can be used in combination with orientation of the urban texture to predict temperature variation even in the warm climate, where the direct radiation plays a major role in comfort conditions (Chatzidimitriou and Yannas 2016; Tsitoura et al. 2016, 2017).

18.2.2 Solar Energy

Urban solar energy studies investigate and assess passive and active solar energy, and the solar potential of buildings to produce renewable energy, as well as their effects on the thermal gains on building façades and heat in open spaces, the desirability of which differs from winter to summer and depends on the local climate. Based on the latter and the research method, it must be noted that the same metric can be highlighted as relevant or not to solar performance. For example, at tropical latitudes, plot ratio has a minor impact on façade solar radiation, while at continental latitude it is one of the most effective in describing the same performance (Lee et al. 2016; Martins et al. 2016; Mohajeri et al. 2016). Another crucial aspect concerning the research method is the modelling of case studies. In this regard, two main approaches have been developed: dealing with real urban forms or using theoretical archetypes. When solar analyses on façades are conducted on homogeneous urban fabrics, rather than cells of fixed dimension, the relation between form archetypes and solar performance can be investigated. Through a combination of compactness, vertical density and sky factor, a clear trend can be fitted (Morganti et al. 2017). Moreover, the combination of spatial metrics can be easily applied in statistical model aiming at optimizing renewable energy potential of roof and façade. It has been demonstrated that it is possible to increase this potential up to 45% (Sarralde et al. 2015).

Several studies confirm the validity of two additional spatial metrics: sky factor and sky view factor. Considerable insights can be derived by the use of these metrics in investigating the solar performance of open spaces (Chatzipoulka et al. 2016; Chatzipoulka and Nikolopoulou 2018). Moreover, among several physical attributes of our city at the neighbourhood scale, the sky view factor has been effectively used for characterizing the solar access of street pattern and façades of adjacent buildings (Mohajeri et al. 2019).

18.2.3 Building Energy Performance

In recent years, due to the urgent need to meet the energy efficiency target established at international and national policy levels, the evaluation of energy performance has been upscaled from building to the urban environment. Firstly, building

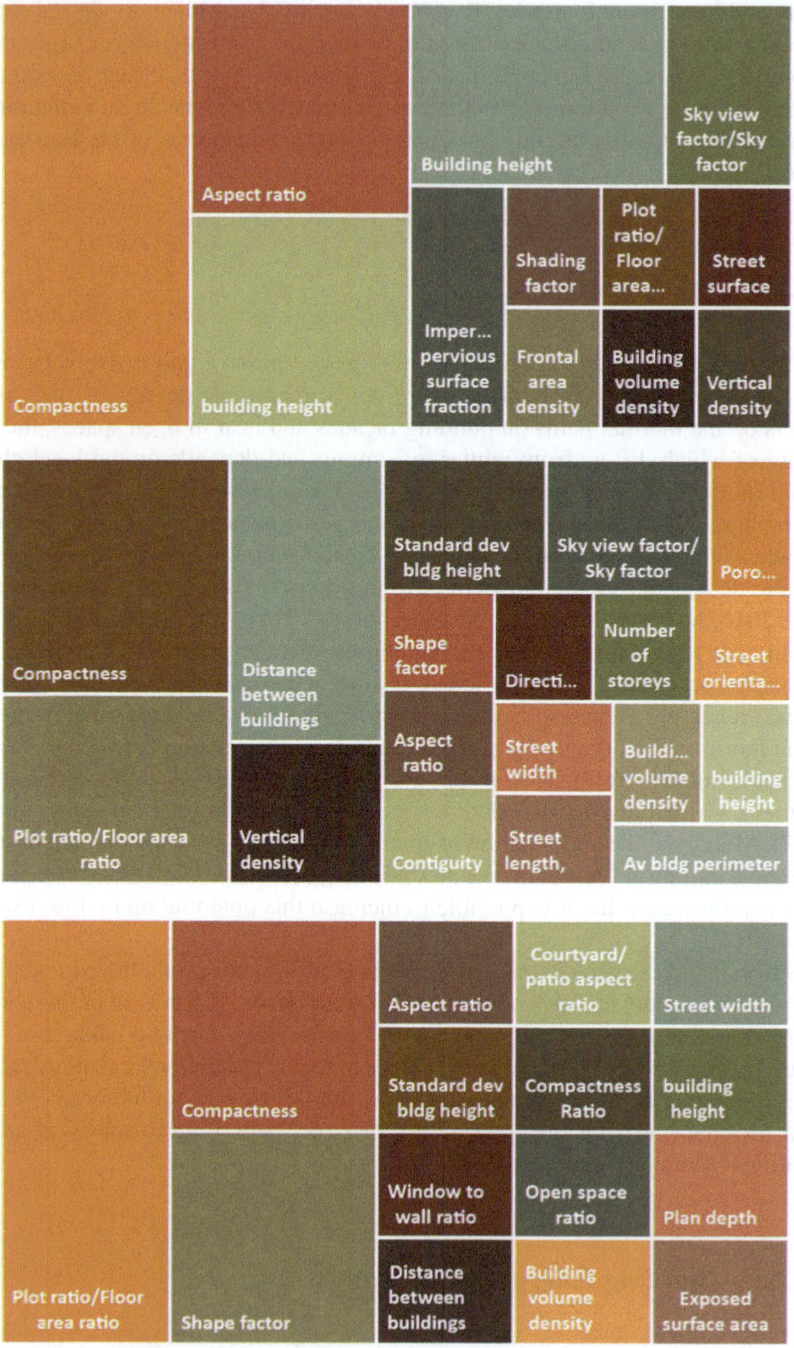

Fig. 18.2 Predominant metrics found in recent studies to correlate urban form attributes with microclimate and energy performance: (**a**) microclimate, and (**b**) solar and (**c**) energy performance

energy efficiency actions and regulations have found very limited application in the renovation process of the building stock. Secondly, researchers point out the importance of addressing the interactions among the building, the urban context and the microclimate to develop reliable building energy performance assessments. Thirdly, the need for analysing this performance for thousands of buildings at once, in order to develop efficient urban energy system, has emerged.

Therefore, new hybrid methods have been developed, merging statistical analysis of the building stock and building energy modelling (Allegrini et al. 2015; Reinhart and Davila 2016). Among these research methods, bottom-up approaches make use of spatial metrics as well as other building variables to generate archetypes for energy analysis of the building stock.

A difference of a factor up to six of the heating demand induced by urban form has been demonstrated (Steadman et al. 2014). In addition, the authors established a correlation between exposed surface area and energy use. Similar findings have been proposed by Rode et al. (2014): they demonstrate that urban density metrics and shape factor are good indicators of heating demand in European compact city model. Besides, spatial metrics related to urban form have been recently used in the development of urban building energy tools, such as Urban Energy Index for Buildings, in which the urban geometry is retained in a national grid through compactness, plot ratio and compactness ratio (Rodríguez-álvarez 2016).

As in the case of solar radiation, different energy demands and climates require different metrics: in the Oceanic climate, standard deviation of building height accounts for almost half of the overall impact on heating, while for cooling its effect is reduced due to the prevalence of other building indicators (Martins et al. 2019). In the Mediterranean climate, shape factor of urban form directly affects the energy load of buildings, demonstrating the effect of courtyard and patio morphology on the global performance (Natanian et al. 2019).

18.3 Reliability and Role of Spatial Metrics in Urban and Building Studies

This chapter presents an essential overview of the relevance and role of spatial metrics in recent research methods on microclimate and building energy performance. We report about the ongoing debate on the most relevant metrics associated with urban form attributes that produces important advancements.

Due to the use made by authors this world of metrics can be a "muddy terrain" for several reasons, especially for professionals without specific knowledge in the field. On the one side, different definitions for the same name have been used. For example, the plot ratio that is one of the most common density indicators has no unique definition: it is calculated by using the total footprint area of a building as well as the total floor area. On the other side, different names are used for the same definition. Intending to overcome this barrier and share knowledge on urban form

Table 18.2 Relevant and reliable metrics for urban microclimate and energy studies

Metric	Units	Definition
Aspect ratio	–	Ratio of building height to the width of the distance between buildings (street width + building setbacks)
Average building perimeter	m	The mean value of the building perimeters included in the urban site area
Building height	m	The value of the building heights included in the urban site area
Building volume density	m	The total volume of buildings in the urban site area
Compactness	–	The ratio of the building footprint to the urban site area
Compactness ratio	–	The proportion of exposed building envelope area (roof and external walls) to the floor area
Contiguity	–	The ratio of total vertical surface adjacent with other building envelopes to the total envelope surface that is exposed to the outside environment within the urban site area
Courtyard/patio aspect ratio	–	The ratio of the building height to the patio width
Directionality	–	The standard deviation of ground's permeability in 36 directions weighted by compactness
Distance between buildings	m	The minimum spacing between adjacent buildings
Exposed surface area	m^2	The total surface of the buildings' envelope in the urban site area
Impervious/pervious surface fraction	–	The ratio of the impervious surface coverage to the urban site area
Number of storeys	–	The mean value of the building storeys included in the urban site area
Plot ratio/floor area ratio	–	The ratio of the total built floor area to the urban site area where the buildings are located
Porosity	–	The ratio of the useful open volume to the total volume of the urban site area
Shape factor	–	The ratio of the non-contiguous building envelope to their built volume, over the urban site area
Sky view factor/ sky factor	–	SF: The ratio of the solid angle of visible sky from each point of the considered urban components to the sky vaultSVF: The cosine-weighted ratio of the solid angle of visible sky from each point of the considered urban components to the sky vault
Standard deviation of building height	–	A quantity expressing by how much the building heights included in the urban site area differ from the mean value for the group
Vertical density	–	The ratio of the building façade area to the urban site area
Window-to-wall ratio	–	Average wall-to-window ratio in all the elevations of the urban site area

metrics the proposed overview draws a general framework on 20 essential metrics for microclimate and energy studies (Table 18.2). These metrics can be easily derived or calculated by means of GIS maps and open data set, and therefore are available for the great majority of cities. It is worth to underline that, due to their definitions and indirect correlations among basic variables, some metrics are equivalent indicators of urban form performances, as exemplified by the relation of sky view factor with compactness and mean outdoor distance variables (Chatzipoulka and Nikolopoulou 2018). Moreover, care should be taken in the selection of metrics in relation to the scale of analysis. Most of these metrics are not suitable to act as reliable performance indicators at all the scales, as highlighted in Table 18.1. Since they measure specific urban attributes, to consider their variation within the study area is crucial.

The most important insights are the following.

- Within the recent studies in the field, we can assume that the most successful metrics are density-based parameters, both for relevance and effectiveness: compactness and plot ratio, as well as shape factor, aspect ratio and building height.
- It should be highlighted that in microclimate studies the main focus of analysis is the relation among physical description and macroclimate of the cities: in particular, warm and hot climates are the most critical due to discomfort, health issues and increasing cooling demand during the summer season.
- When focusing on both the passive and the active performance of solar energy it is important to notice that, depending on the specific built element (ground, façade, roof) and on the season of the year, different metrics will result to be the most reliable to describe the solar performance. Therefore, attention must be reserved in the use of metrics within the research method and the associated design process.
- The characterization of urban form attributes in the definition of different building stock archetypes through urban metrics in building energy performance has the potential to easily control the effect of the urban texture at neighbourhood level on microclimate (UHI intensity) and energy demand of buildings.
- The full significance of the correlation between urban metrics and performance cannot be assigned to one single metric, as the complex interaction of climate-form-energy occurring in the urban environment cannot be represented by one single variable. For this reason, most of the studies discussed in this overview make use of a combination of urban metrics, both for describing the urban textures and testing the effectiveness in the description of a specific urban phenomenon, both for defining trends among metrics and performance, e.g. energy demand, urban heat island intensity and passive solar irradiation. Such kind of approach has proved to be more effective in the description of microclimate and energy performance and can be easily integrated into digital design tools for architects and urban planners. In addition, the approach helps the assessment of cumulative impacts of decisions on regenerative design process. However, attention must be paid to unintended conflicts and redundancies among metrics when used in combination for prediction or design purposes.
- To prioritize certain urban form variables widespread in urban design and architecture is crucial to support design strategies and to help architects understand the above-mentioned impacts. These metrics include plot ratio, compactness, shape factor and aspect ratio that can easily be included in urban climate and energy planning or policies.

The description through metrics of the effects of urban form on microclimate and energy performance is of primary importance for cities' adaptation to climate change and energy transition. In order to distinguish urban form metrics as a reliable indicator and to foster climate-sensitive design, several challenges remain. Nowadays theoretical knowledge and associated research methods are sectorial, confused, dispersed and too often complicated. A systematization of this knowledge on physical characteristics and spatial metrics relevant for urban microclimate and energy efficiency makes it possible to overcome current limitations. Firstly, to investigate unexplored connections among different urban phenomena is needed for climate change and energy transition agenda. Secondly, this is important because discussion on this issue would promote cross-disciplinary approach. Thirdly, a guidance on relevant metrics would allow urban designers and architects to control the impacts of design choices during the entire creative process. The diffusion of such kind of approach is crucial in the regenerative design of cities that plays a big part in the SDG for the near future.

References

Allegrini, J., Orehounig, K., Mavromatidis, G., Ruesch, F., Dorer, V., & Evins, R. (2015). A review of modelling approaches and tools for the simulation of district-scale energy systems. *Renewable and Sustainable Energy Reviews, 52*, 1391–1404. https://doi.org/10.1016/j.rser.2015.07.123.

Chatzidimitriou, A., & Yannas, S. (2016). Microclimate design for open spaces: Ranking urban design effects on pedestrian thermal comfort in summer. *Sustainable Cities and Society, 26*, 27–47. https://doi.org/10.1016/J.SCS.2016.05.004.

Chatzipoulka, C., & Nikolopoulou, M. (2018). Urban geometry, SVF and insolation of open spaces: London and Paris. *Building Research and Information, 46*(8), 881–898. https://doi.org/10.1080/09613218.2018.1463015.

Chatzipoulka, C., Compagnon, R., & Nikolopoulou, M. (2016). Urban geometry and solar availability on façades and ground of real urban forms: Using London as a case study. *Solar Energy, 138*, 53–66. https://doi.org/10.1016/j.solener.2016.09.005.

Emmanuel, R., & Steemers, K. (2018). Connecting the realms of urban form, density and microclimate. *Building Research and Information, 46*(8), 804–808. https://doi.org/10.1080/09613218.2018.1507078.

Golany, G. S. (1996). Urban design morphology and thermal performance. *Atmospheric Environment, 30*(3), 455–465. https://doi.org/10.1016/1352-2310(95)00266-9.

IPCC. (2014). *Climate change 2014: Impacts, adaptation, and vulnerability. Part A: Global and sectoral aspects. Contribution of working group II to the fifth assessment report of the Intergovernmental Panel on Climate Change*. New York: Cambridge University Press. Retrieved from https://www.ipcc.ch/site/assets/uploads/2018/02/WGIIAR5-PartA_FINAL.pdf

IPCC. (2018). Global research and action agenda on cities and climate change science. In *Cities and Climate Science Conference*. Edmonton: IPCC. Retrieved from https://www.ipcc.ch/event/cities-and-climate-change-science-conference/

Jacobson, A. M. (2019). New metrics for digital regenerative design. In E. Naboni & L. Havinga (Eds.), *Regenerative design in digital practice. A handbook for the built environment*. Bolzano: EURAC.

Lee, K. S., Lee, J. W., & Lee, J. S. (2016). Feasibility study on the relation between housing density and solar accessibility and potential uses. *Renewable Energy, 85*, 749–758. https://doi.org/10.1016/j.renene.2015.06.070.

Leo, L. S., Buccolieri, R., & Di Sabatino, S. (2018). Scale-adaptive morphometric analysis for urban air quality and ventilation applications. *Building Research and Information, 46*(8), 931–951. https://doi.org/10.1080/09613218.2018.1501797.

Martins, T. A. L., Adolphe, L., & Bastos, L. E. G. (2014). From solar constraints to urban design opportunities: Optimization of built form typologies in a Brazilian tropical city. *Energy and Buildings, 76*, 43–56. https://doi.org/10.1016/j.enbuild.2014.02.056.

Martins, T. A. d. L., Adolphe, L., Bastos, L. E. G., & Martins, M. A. d. L. (2016). Sensitivity analysis of urban morphology factors regarding solar energy potential of buildings in a Brazilian tropical context. *Solar Energy, 137*, 11–24. https://doi.org/10.1016/j.solener.2016.07.053.

Martins, T. A. d. L., Faraut, S., & Adolphe, L. (2019). Influence of context-sensitive urban and architectural design factors on the energy demand of buildings in Toulouse, France. *Energy and Buildings, 190*, 262–278. https://doi.org/10.1016/j.enbuild.2019.02.019.

Mohajeri, N., Upadhyay, G., Gudmundsson, A., Assouline, D., Kämpf, J., & Scartezzini, J. L. (2016). Effects of urban compactness on solar energy potential. *Renewable Energy, 93*, 469–482. https://doi.org/10.1016/j.renene.2016.02.053.

Mohajeri, N., Gudmundsson, A., Kunckler, T., Upadhyay, G., Assouline, D., & Kämpf, J. H. (2019). A solar-based sustainable urban design: The effects of city-scale street-canyon geometry on solar access in Geneva, Switzerland. *Applied Energy, 240*, 173–190. https://doi.org/10.1016/j.apenergy.2019.02.014.

Morganti, M., Salvati, A., Coch, H., & Cecere, C. (2017). Urban morphology indicators for solar energy analysis. In *9th International Conference on Sustainability in Energy and Buildings, SEB-17, 5–7 July 2017, Chania, Crete, Greece* (Vol. 134, pp. 1–8). Chania: Elsevier Ltd.. https://doi.org/10.1016/j.egypro.2017.09.533.

Natanian, J., Aleksandrowicz, O., & Auer, T. (2019). A parametric approach to optimizing urban form, energy balance and environmental quality: The case of Mediterranean districts. *Applied Energy*. https://doi.org/10.1016/j.apenergy.2019.113637.

Nault, E., Peronato, G., Rey, E., & Andersen, M. (2015). Review and critical analysis of early-design phase evaluation metrics for the solar potential of neighborhood designs. *Building and Environment, 92*, 679–691. https://doi.org/10.1016/j.buildenv.2015.05.012.

Oke, T. R. (1982). The energetic basis of the urban heat island. *Quarterly Journal of the Royal Meteorological Society, 108*(455), 1–24. https://doi.org/10.1002/qj.49710845502.

Perini, K., & Magliocco, A. (2014). Effects of vegetation, urban density, building height, and atmospheric conditions on local temperatures and thermal comfort. *Urban Forestry & Urban Greening, 13*(3), 495–506. Retrieved August 9, 2018 from https://www.sciencedirect.com/science/article/pii/S1618866714000296.

Petralli, M., Massetti, L., Brandani, G., & Orlandini, S. (2014). Urban planning indicators: Useful tools to measure the effect of urbanization and vegetation on summer air temperatures. *International Journal of Climatology, 34*(4), 1236–1244. https://doi.org/10.1002/joc.3760.

Reinhart, C. F., & Davila, C. C. (2016). Urban building energy modeling: A review of a nascent field. *Building and Environment, 97*, 196–202. https://doi.org/10.1016/j.buildenv.2015.12.001.

Rode, P., Keim, C., Robazza, G., Viejo, P., & Schofield, J. (2014). Cities and energy, urban morphology and heat energy demand. *Environment and Planning B, 41*, 138–162. https://doi.org/10.1068/b39065.

Rodríguez-Álvarez, J. (2016). Landscape and urban planning urban energy index for buildings (UEIB): A new method to evaluate the effect of urban form on buildings' energy demand. *Landscape and Urban Planning, 148*, 170–187. https://doi.org/10.1016/j.landurbplan.2016.01.001.

Salvati, A., Coch, H., & Morganti, M. (2017). Effects of urban compactness on the building energy performance in Mediterranean climate. In *Energy Procedia, CISBAT 201(Future Buildings &*

Districts—Energy Efficiency from Nano to Urban Scale, CISBAT 2017, 6–8 September 2017, Lausanne, Switzerland) (pp. 499–504). https://doi.org/10.1016/j.egypro.2017.07.303.

Salvati, A., Monti, P., Coch Roura, H., & Cecere, C. (2019). Climatic performance of urban textures: Analysis tools for a Mediterranean urban context. *Energy and Buildings, 185*, 162–179. https://doi.org/10.1016/j.enbuild.2018.12.024.

Sarralde, J. J., Quinn, D. J., Wiesmann, D., & Steemers, K. (2015). Solar energy and urban morphology: Scenarios for increasing the renewable energy potential of neighbourhoods in London. *Renewable Energy, 73*, 10–17. https://doi.org/10.1016/j.renene.2014.06.028.

Silva, M., Oliveira, V., & Leal, V. (2017). Urban form and energy demand: A review of energy-relevant urban attributes. *Journal of Planning Literature, 32*(4), 346–365. https://doi.org/10.1177/0885412217706900.

Steadman, P., Hamilton, I., & Evans, S. (2014). Energy and urban built form: An empirical and statistical approach. *Building Research and Information, 42*(1), 17–31. https://doi.org/10.1080/09613218.2013.808140.

Stewart, I. D., & Oke, T. R. (2012). Local climate zones for urban temperature studies. *Bulletin of the American Meteorological Society, 93*(12), 1879–1900. https://doi.org/10.1175/BAMS-D-11-00019.1.

Tsitoura, M., Michailidou, M., & Tsoutsos, T. (2016). Achieving sustainability through the management of microclimate parameters in Mediterranean urban environments during summer. *Sustainable Cities and Society, 26*, 48–64. https://doi.org/10.1016/J.SCS.2016.05.006.

Tsitoura, M., Michailidou, M., & Tsoutsos, T. (2017). A bioclimatic outdoor design tool in urban open space design. *Energy and Buildings, 153*, 368–381. https://doi.org/10.1016/J.ENBUILD.2017.07.079.

Vuckovic, M., Kiesel, K., & Mahdavi, A. (2017). The extent and implications of the microclimatic conditions in the urban environment: A Vienna case study. *Sustainability, 9*, 177. https://doi.org/10.3390/su9020177.

Chapter 19
Green Infrastructure to Mitigate Extreme Temperatures in Cities

Francisco de la Barrera and Sonia Reyes-Paecke

19.1 Greening the Urban Climate

This chapter analyses the capacity of urban green infrastructure (GI) to determine urban climates and mitigate extreme temperatures in cities. An exploratory literature review was conducted focused on empirical studies to evaluate the state of the art of GI and urban climate at the metropolitan and urban scales.

Urban climates are the result of the interaction between meteorological conditions, urban structures and human activities. The meteorological variables (e.g. temperature, humidity, radiation, wind patterns) are modified by the physical characteristics of the urban landscape (e.g. density and height of buildings, amount and distribution of open spaces, vegetation cover, construction materials), and by human activities that produce heat and atmospheric pollution (e.g. motorised transport, air conditioning, emissions from industries; Parlow 2011; Nastran et al. 2019). One of the most conspicuous characteristics of urban climate is that city temperatures are higher than temperatures in surrounding rural areas, which is referred to as urban heat islands (UHI). According to Venter et al. (2020), the effect of UHI is to increase average temperature by 6.4 ± 2.3 °C. The magnitude of the temperature difference (UHI intensity) correlates positively with the size of cities, although there are differences between European, North American and Asian cities related to building

F. de la Barrera (✉)
Departamento de Geografía, Facultad de Arquitectura Urbanismo y Geografía, Universidad de Concepción, Concepción, Chile

Centro de Desarrollo Urbano Sustentable, Santiago, Chile

S. Reyes-Paecke
Centro de Desarrollo Urbano Sustentable, Santiago, Chile

Departamento de Ecosistemas y Medio Ambiente, Facultad de Agronomía e Ingeniería Forestal, Pontificia Universidad Católica de Chile, Santiago, Chile
e-mail: sonia.reyes@uc.cl

M. Palme, A. Salvati (eds.), *Urban Microclimate Modelling for Comfort and Energy Studies*, https://doi.org/10.1007/978-3-030-65421-4_19

typology and level of use of air conditioning (Parlow 2011; Venter et al. 2020). The geographic locations of cities also affect the intensity of UHI. To illustrate this point, Fig. 19.1 shows the spatial distribution of an indicator of UHI (land surface temperature) in four Latin American cities: Santiago (Chile), Concepción (Chile), Rosario (Argentina) and Buenos Aires (Argentina). Concepción and Buenos Aires are coastal cities, while Santiago and Rosario are continental and located along rivers, which in the case of Rosario is the wide Paraná River. Differences between urban and rural surface temperature are from +4.0 to +6.9 °C.

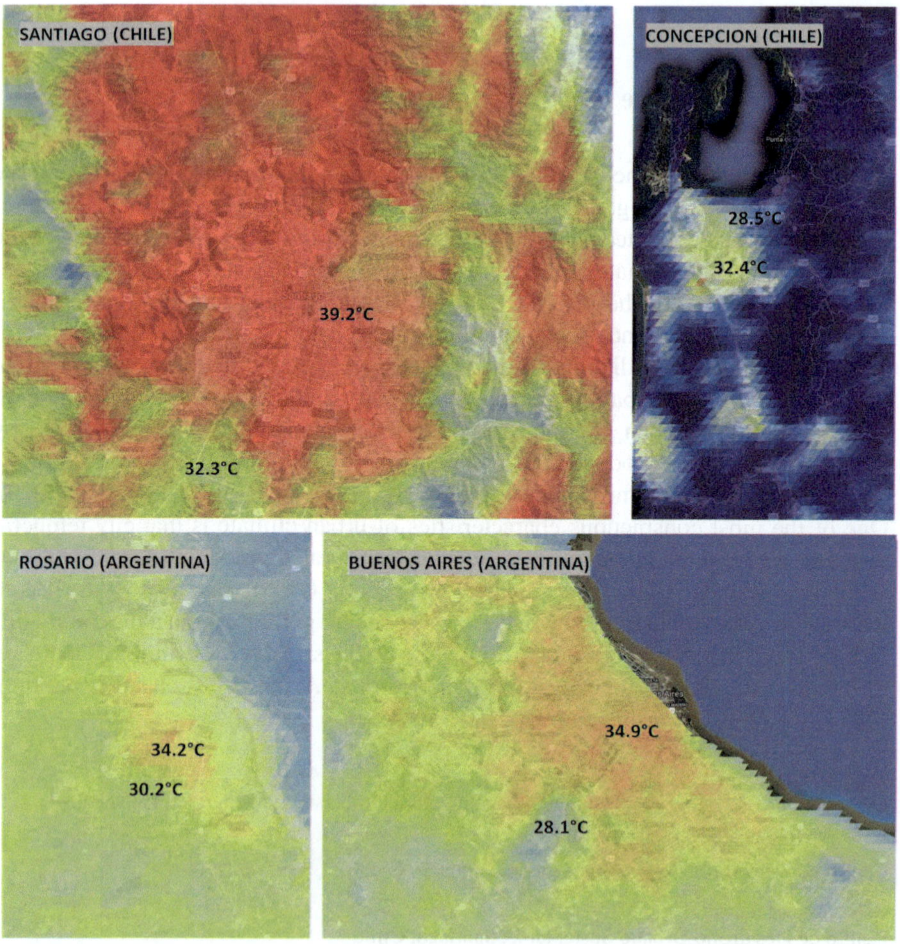

Fig. 19.1 Spatial distribution of land surface temperatures in four Latin American cities: Santiago (Chile), Concepción (Chile), Rosario (Argentina) and Buenos Aires (Argentina). Land surface temperature data represent an average of the austral summer from December 1, 2016, to February 28, 2017, using MODIS data sets managed in the Google Earth Engine platform

Since cities are complex systems in terms of their components and operational dynamics, the occurrence of extremely high or low air temperatures is due to many causes. Thinking of the city in just two dimensions, as a two-dimensional plane, it is possible to visualize a mosaic of surfaces with different capacities to accumulate or reflect solar energy, as well as different speeds in releasing stored heat. In this mosaic, areas with plant cover accumulate less heat and release it more quickly than bare soil or areas covered by asphalt or concrete. Thermal extremes in areas with plant cover are reduced by the humidity generated by evapotranspiration, shade from trees and cool breezes. In contrast, paved surfaces heat up more quickly due to lack of shade, accumulate more heat and release it more slowly than surfaces with vegetation (Whitford et al. 2001; Lobaccaro and Acero 2015; Inostroza et al. 2016).

The greater the amount of green space in a city, the lower the probability of urban heat island forming and the less their impact if they do form (Moreno 1994, 1999). However, in addition to the amount of green space, the typology in terms of size, shape and spatial distribution should be considered, since larger green spaces or widely distributed tree cover are highly effective in regulating city air temperatures. These capacities increase when green spaces and tree cover are connected in a green network that encompasses the entire city (Nastran et al. 2019; Pauleit and Duhme 2000).

Visualizing the city in three dimensions, that is, adding the volume or the vertical plane to the horizontal plane, adds complexity to our understanding of urban infrastructure and its mechanisms. The shape, arrangement, density and height of buildings modify urban temperature. As an example, high-rise buildings contribute more to decreasing surface temperature than low-rise buildings because they cast longer shadows (Song et al. 2020). Spatial clustering of buildings also plays an important role in urban temperature. Wang et al. (2019) reported that clustered paved surfaces elevate urban temperature, and that the warming effect was stronger during the day than at night in cities like Los Angeles, Portland and Chicago, even though they are located in different climatic zones. In relation to vegetation, an area covered by short green grass has a different thermal effect than an area covered by tall grass (0.5 m) or an area with mature trees, since more plant material or biomass has a greater effect on shading, relative humidity and wind speed (Lobaccaro and Acero 2015; Skelhorn et al. 2014).

19.2 Heat-Mitigating Potential of a Green Infrastructure

Green infrastructure is defined as "the strategically planned network of natural and semi-natural areas with other environmental features designed and managed to deliver a wide range of ecosystem services" (European Commission 2013). This network includes street trees, all kinds of green spaces, urban parks, hedgerows, wetlands and riparian areas, as well as man-made elements such as green walls and facades, green roofs and cycle paths.

To analyse the effects of the GI on urban temperature (or on UHI), we discuss four types of elements: tree canopies or tree cover; green spaces (of all types and sizes); green roofs; and vertical greenery systems (green walls and facades), and their spatial scales of influence. These typologies are useful for urban planning because they allow for identifying and ensuring the diversity of components at different scales while recognizing their distinct contributions. Green infrastructure typologies cover a wide range of functions and users, with equitable spatial distribution being essential to ensure the broadest accessibility to green infrastructure benefits, especially to the most disadvantaged groups (Tzoulas et al. 2007).

One of the main attributes of green infrastructure is its multifunctionality, which is understood as the capacity to carry out several functions and provide various benefits in the same spatial area (Ahern 2011). In their review of articles on green infrastructure, Wang and Banzhaf (2018) identified 28 functions of green infrastructure; among them are notably temperature control, regulating of floods, freshwater storage, providing of green spaces and habitats, and biological control. There has been a growing number of publications in recent years that deal with the multifunctionality of green infrastructure, ranging from site-specific studies that analyse the multiple benefits and functions of green areas to landscape ecology approaches that propose concepts like multifunctional networks and landscapes (Wang and Banzhaf 2018).

GI multifunctionality is attractive to decision makers because public agencies usually evaluate their investments with cost-benefit or cost-efficiency criteria, and in both cases, the cost of investment in green infrastructure, ecosystem restoration and construction of new parks is offset by multiple and diverse benefits (Alves et al. 2020). Generating both social and environmental benefits is especially important since city government priorities usually involve these two dimensions. The large number of documents on green infrastructure planning and assessment produced by governmental (e.g. Federal Agency for Nature Conservation, Germany; Green Infrastructure Northwest England, UK; United States Environmental Protection Agency) and international agencies (e.g. European Commission, European Environment Agency) is clear evidence of this interest.

19.3 Cooling Mechanisms

Research has provided rich evidence of the capacity of GI to decrease high urban temperatures and mitigate UHI. Here we present four mechanisms that explain how GI mitigate extreme city temperatures: reducing reflected solar radiation, providing shade, cooling by evapotranspiration and regulating wind speed (Table 19.1).

Table 19.1 Synthesis of the contribution of green infrastructure to regulating air temperature at different spatial scales

	Tree cover	Urban parks	Green roofs	Vertical greenery systems
Mechanisms of air temperature regulation				
Reduction of reflected solar radiation	+	++	+	
Provision of shade	++	+		+
Cooling by evapotranspiration	++	++	+	+
Regulation of wind speed	+	++		+
Spatial scale of influence				
Urban	++	++		
Local	++	++	+	
Building	+		++	++

19.3.1 Reducing Reflected Solar Radiation

Urbanization modifies land surface albedo (the fraction of reflected solar radiation) because of the use of dark construction materials and pavement, and the effect of complex vertical topography of built-up areas (Trlica et al. 2017). Lower albedo implies increased solar radiation absorption by surface areas, which in turn increases surface and air temperature. The albedo is lower in densely built-up areas like urban cores, and increases in areas with more plant cover like green spaces, low-density residential areas and peri-urban agricultural areas. Vegetation reflects the sun's energy better than man-made structures, stores less heat and releases it faster, which decreases surface and air temperature.

19.3.2 Providing Shade

The shade provided by plants keeps the air cooler by simply intercepting solar radiation and preventing energy absorption by urban surfaces and reradiation of heat to the canopy layer atmosphere (Gunawardena et al. 2017; Oke 1989). The effectiveness of shade is determined by leaf size, crown area and leaf area index (LAI) of plant canopies. Trees and shrubs offer more effective shade than other types of plants (Gunawardena et al. 2017). Trees can create microclimates below their canopies, with leafy and wide-crowned trees being more effective in this regard. The percentage of tree (and shrub) cover is directly related to the reduction of air temperature by interacting with the reduction of solar radiation and evapotranspiration, as Chang et al. (2007) reported based on measurements of unshaded areas of urban parks. Shade is one of the most valued attributes of urban trees in cities with Mediterranean climates given that high temperatures and solar radiation on summer days can negatively affect the well-being of urban residents (Avolio et al. 2015; Gatto et al. 2020).

19.3.3 Cooling by Evapotranspiration

The combination of evaporation of soil humidity and plant transpiration is known as evapotranspiration, which constitutes a mechanism of transferring heat from the earth's surface to the atmosphere (Gunawardena et al. 2017). Plant transpiration basically consists of water loss by evaporation through the aerial parts of plants, which has a cooling effect on the surrounding atmosphere. Evapotranspiration is influenced by the availability of moisture, types of plants and wind flow. Soil humidity in urban areas depends on regional climate and irrigation. Watering is vital to maintain soil moisture in periods of drought, especially in cities with Mediterranean and semi-arid climates (Rana et al. 2020; Reyes-Paecke et al. 2019). The vegetation type is relevant because there are two large groups of plants differentiated by their transpiration regulation mechanisms. Most plants from cold and humid climates maintain leaf pores (stomata) open during the day (called C3 photosynthetic metabolism), with the transpiration of significant volumes of water, while plants from hot and dry climates have developed an adaptation to avoid water loss, keeping their stomata closed during the day in what is known as C4 photosynthetic metabolism. Consequently, the latter group of plants contribute less to cooling by transpiration, but make more efficient use of water and have a greater capacity for retaining soil moisture.

19.3.4 Regulating Wind Speed

Ventilation is associated with the dispersion of air pollutants, so the effect of plants on wind speed can be very important (Ng et al. 2012). However, from the point of view of the urban temperature, there are diverse benefits associated with facilitating or reducing the occurrence of wind and its speed. On the one hand, wind control in cold winters can be decisive in preventing air infiltration into buildings and avoiding heat loss from energy transfer. Trees, shrubs and grassy areas located in the immediate vicinity of homes, especially in residential yards, can be decisive. According to Lobaccaro and Acero (2015), the presence of trees can decrease wind speed by 32%, while grass 0.5 m high can decrease winds by 12% by increasing roughness. However, fresh winds are highly valued during hot dry summers (Lafortezza et al. 2009). Given this, the combination of wind breaks and open spaces like parks that allow wind flows is important while providing shade and lower temperatures at the ground level.

19.4 Green Infrastructure to Mitigate Heat

The four mechanisms of heat mitigation are controlled by GI elements (Table 19.1), which contribute to heat mitigation according to how the vegetation is expressed in each element. The percentage of urban area that is covered by GI elements and how these elements are distributed in a city are also important.

19.4.1 Urban Parks

In 1989 T.R. Oke proposed the concept of (urban) park cool island (PCI), which in mid-latitude cities should be 1–2 °C cooler than surrounding areas, and rarely more than 3 °C cooler, although under ideal conditions it could be as much as 5 °C, as Spronken-Smith and Oke (1998) found in Sacramento (5–7 °C), while Barradas (1991) recorded differences of up to 5.6 °C. According to a systematic review by Bowler et al. (2010), parks are around 1 °C cooler than surrounding areas in the day. They also found that size strongly influences the cooling effect of parks. Comparing an urban park in Santiago (Chile) with its immediate surrounding, differences were found of up to 4 °C in surface temperature, with an average of 1 °C, and differences of up to 16 °C in air temperature, with daily averages of around 3 °C (de la Barrera et al. 2016). Based on a field survey of 61 parks in Taipei City, Chang et al. (2007) found that 3 ha is a threshold in this regard, while Barradas (1991) tested five parks ranging from 1.9 to 9.9 ha and reported that differences in air temperature were explained by park size. Grilo et al. (2020) reported that even small green spaces can regulate microclimates, reducing air temperature by 1–3 °C. All these studies report that the magnitude of temperature difference varies because of external variables like the macroscale climate or the use of surrounding land (e.g. homes with green yards or paved surfaces). They also describe the importance of internal characteristics of urban parks, such as morphology, irrigation and tree density, to maximize the advective influence beyond its borders. These characteristics affect shade, albedo and water availability, and consequently the absorption of solar radiation and evaporative cooling.

19.4.2 Tree Cover

Increasing tree cover has positive effects on mitigating extreme temperatures (Skelhorn et al. 2014), with differences of up to 9–10 °C between planted and paved surfaces, as has been registered in Oslo and New York (Venter et al. 2020). The effects on temperature can be measured empirically through satellites using the thermal band and the land surface temperature indicator (e.g. Inostroza et al. 2016; Fig. 19.1), or sensors that register air temperature. However, air temperature depends on other factors that correlate weakly with surface temperature (Sarricolea et al. 2008). It is important to not only increase vegetation cover, but also locate it in suitable places, especially areas more exposed to heat (Zölch et al. 2016).

19.4.3 Green Roofs

Green roofs, which are becoming increasingly popular, have a significant effect on internal building temperature (Jänicke et al. 2014). However, their effect on overall urban temperature is less than that of plant cover (Lobaccaro and Acero 2015). Most research on the topic has used models to calculate the contribution of green roofs on air temperature in buildings, which is mainly measured with air temperature sensors that can be real-time monitored to adjust air-conditioning systems and register the reduction in energy demand. Recent investigations suggest that increasing the area covered by green roofs in a city can increase evapotranspiration in summer, and decrease the heat captured and stored in buildings, and returned to the atmosphere, thus contributing to reducing urban temperatures (Getter et al. 2011; Vaz Monteiro et al. 2017; Peng and Jim 2013). Roofs represent around 30% of urban surfaces and are therefore an important resource for any strategy to reduce urban temperatures (Peng and Jim 2013). The type of plants used in green roofs is important since not all plants have the same cooling and thermal insulation capacity in the warm season. Vaz Monteiro et al. (2017) compared roofs covered by different types of plants (succulent, herbaceous and broadleaved with different leaf colours) and observed differences in outgoing long-wave radiation, heat flux, substrate heat, evapotranspiration rates and other relevant parameters to reduce surrounding temperatures. The selection of appropriate plants is highly important to effectively improve the thermal insulation of building and to contribute to cooling urban environments (Blanusa et al. 2013; Vaz Monteiro et al. 2017).

19.4.4 Vertical Greenery Systems

Vertical greenery systems cover the facades of buildings with vegetation like creepers, climbers, herbs, ferns and shrubs. There are two types of systems: green facades and green walls. Green facades are traditional systems where the vegetation, usually creepers or climbers, is rooted to the ground or in large planters, and grows leaning on structures like trellis, wires or meshes that can be freestanding (detached green facade) or attached to the wall (attached green facade or direct systems) (Bartesaghi Koc et al. 2017; Oldham and Karima 2018). Green walls, which are also called living walls, consist of prefabricated panels where the plants are cultivated prior to installation and the panels are subsequently fixed either to a vertical structural support or directly to a building wall. Hence, plants are not in contact with the ground (Bartesaghi Koc et al. 2017). The detached systems leave a gap between the building wall and the greening system, and this gap acts as a thermal buffer that improves the building's thermal insulation (Pérez et al. 2014). The thermal regulation capacity of green facades has been evaluated in diverse climatic regions using thermal sensors and real-time monitoring. Vox et al. (2018) carried out long-term monitoring of two green facades and a control wall in Thessaloniki, Greece, and registered daytime temperatures up to 9 °C lower on the green facades than on the control during the warm season, and up to 3.5 °C higher at night in the cold season. The effect of mitigating high temperatures was observed throughout the year (Vox et al. 2018). Also in Thessaloniki, the exterior surface temperature of green facades covered with common ivy (*Hedera helix*) was 5.7 °C lower than the surface temperature of bare walls (Eumorfopoulou and Kontoleon 2009). The green walls of house-like cubicles in an experiment in Catalonia, Spain, had summer temperatures 16.5 °C lower than control walls (Coma et al. 2017). In Berlin, Germany, the surface temperature of traditional green facades was 15.5 °C lower than that of bare walls in summer (Hoelscher et al. 2016). In Wuhan (China), which is characterized by a hot humid climate, the exterior temperature of green walls was registered with differences as much as 20.8 °C from that control walls, while interior walls were as much as 7.7 °C lower (Chen et al. 2013). In addition to mitigating UHI, green facades and walls are a sustainable means to improve the energy efficiency of buildings (Vox et al. 2018). Using an experimental approach, Coma et al. (2017) compared the thermal performance of green facades and green walls in experimental house-like cubicles. Their results show that both systems have the potential to mitigate heat and consequently save energy. The best cooling performance was provided by green walls, which reduced temperatures by 58.9%, while green facades reduced temperature by 33.8%, both compared to the reference cubicle with an internal temperature of 24 °C (Coma et al. 2017).

19.5 Scaling Effects in the Mitigation of Extreme Temperatures by GI

Green infrastructure and its natural components like different types of plants, soils and even bodies of water can mitigate extreme temperatures at distinct spatial scales. In this final section, we explore how elements of GI can be incorporated at each scale to strengthen synergy in greening interventions (Bowler et al. 2010).

At the **metropolitan scale**, it is possible to appreciate the urban heat island phenomenon, depicted by differences in maximum and minimum temperatures and in the hourly behaviour of temperatures by comparing two antagonistic situations: the city and the rural area. Basically, city temperatures are higher both during the day and at night because more heat is accumulated and released more slowly than in rural areas. The extensive use of impervious materials that modify the thermal and radiative properties of surface areas increases heat storage and the resulting heat re-emission as long-wave radiation or sensible heat. The heat absorbed, stored and released to the atmosphere can also be intercepted by air pollutants and redirected back to the urban environment, reinforcing the warming effect (Vaz Monteiro et al. 2017). This scale points to the difference between rural and urban environments considering the city as a unit, without distinguishing among its components (e.g. infrastructure, buildings, streets) or neighbourhood types. Analyses at this scale have found reductions in air temperature associated with the percentage of areas with plant cover, and provide evidence that trees are more effective in mitigating high temperatures than other types of plants (Ng et al. 2012; Skelhorn et al. 2014). Urban parks and small green spaces contribute to mitigating extreme temperatures, but their effects can be reduced if land use in their immediate surroundings is dominated by paved surfaces instead of tree-lined streets or planted yards and gardens. Large urban parks have a great capacity to alleviate heatwaves, but they have to be located strategically to take advantage of wind. For example, the borders of the green spaces need to be designed to facilitate the horizontal transfer of the cooling effect.

When the spatial scale is adjusted to smaller areas within cities, the difference in temperatures between greener and less green urban areas is greater than the temperature difference between urban and rural areas. A study in Santiago, Chile, found a strong correlation between plant cover (using NDVI as an indicator) and surface temperature (based on Landsat imagery), with differences among neighbourhoods of up to 1.6 °C (de la Barrera et al. 2019). In the same city, air temperature measured in and outside of parks in various neighbourhoods showed that the temperatures in parks in greener neighbourhoods were 3.5 °C cooler than their surroundings, while in neighbourhoods with little plant cover, the difference was 7.1 °C (de la Barrera et al. 2019). This shows the importance of urban parks in regulating microclimates in neighbourhoods that lack vegetation, and that usually have low-income populations. The vegetation structure in urban parks greatly influences its cooling capacity. For example, the combination of trees and herbs or grass allows to reduce air temperatures by as much as 10 °C (Lobaccaro and Acero 2015). Likewise, park management practices, in particular irrigation, are important in optimizing temperature regulation given that trees subject to water stress do not develop broad and leafy canopies essential to providing shade (Mariani et al. 2016). In this respect, cities in arid and semi-arid regions face a trade-off between using vegetation (which requires irrigation) and regulating temperature or making alternative uses of water as a scarce resource. This trade-off is more significant in the context of climate change, given that one of the main urban adaptation measures is to increase plant cover. However, cities will face greater water shortages in the future (Reyes-Paecke et al. 2019). In this context, selection of plant species adapted to arid climates is essential to optimize GI thermal regulation and, at the same time, efficiently use available water.

As well as the effects of green infrastructure at the aforementioned mentioned scales, there are the effects at the scale of **buildings and their immediate surroundings**. The aspect ratio and orientation of buildings can be determining factors for urban microclimates (Ali-Toudert and Mayer 2006). Further study is urgently needed to develop solutions that spatially combine the attributes of constructed elements, such as building morphology and orientation, with the size, shape and spatial distribution of different kinds of green infrastructure components, like urban parks, street trees, green roofs, green facades and walls, residential yards, urban orchards and other planted spaces. Buildings located near urban parks or small green spaces can be favoured with fresh air, as well as having a dense tree cover nearby. To get this the buildings need to be part of urban projects that include open spaces and offer deep soils for growing plants while do not interrupting the solar radiation. The shade projected by high buildings should not affect urban parks, but rather areas without tree cover.

19.6 Other Urban Variables Influencing UHI

In relation to building typology, some cities, like New York, have an abundance of skyscraper, while compact cities like Barcelona typically have a mix of tall and medium-sized buildings, and yet other cities have many historical and commercial buildings, and even industrial sites. Cities that have spread out as they have grown are dominated by suburban homes in low-density developments. This wide diversity of urban landscape patterns indicates the need for detailed planning of green infrastructure that takes into account site-specific variation in urban climates. At the fine scale, natural and constructed features are interwoven and can even co-occur, like tree canopies over pavement. Latin American, African and Asian cities can be socio-ecologically unique due to the combination of the exuberant nature of their surroundings, intense land-use changes and high levels of social inequality. A study by Inostroza et al. (2016) focused on higher degree of vulnerability of low-income populations in Santiago, Chile, to the risks posed by heat, and the need for specific improvements in housing materials. In this vein, increasing green infrastructure coverage for urban heat control can be a complementary measure, but not a replacement to improving the buildings.

Regulating the urban climate by using vegetation has effects at different scales that are highly dependent on the type, quantity and spatial arrangement of green infrastructure elements. A single tree can offer shade or help decrease solar radiation on a roof or open space, but a myriad of street trees can help to reduce the average air temperature at the urban scale. As noted above, it is not only important to have trees, but they should also be mature and combined with shrubbery and different types of grass. After an extensive review of studies that measured the impact of urban vegetation on different urban climates, Ng et al. (2012) stated that researchers agree in suggesting that greening is important for regulating urban climates.

The use of green infrastructure to regulate urban temperatures has many advantages over man-made cooling systems. Firstly, green infrastructure is linked to urban planning and can be incorporated into municipal regulations, designs of new developments and urban renewal plans. As a planning approach, it links the supply of ecosystem services to population needs and with features of the constructed built environment. For example, the design of urban parks must take into account the amount of vegetation present in the streets and residential yards, because these influence the cooling capacity of parks. Another great advantage of green infrastructure is multifunctionality, given that along with regulating temperature, it also regulates flooding, contributes to rainwater infiltration and pollination, and provides spaces for recreation, contact with nature, social integration and a sense of place, among others, all of which increase people's well-being (Lafortezza et al. 2009).

19.7 Conclusions

Urban climate and thermal experience of people depend on many variables that are subject to intervention, mainly through construction and urban planning. This chapter focused on the contribution of other disciplinary approaches to a better understanding of the role of natural structures introduced into cities. The diversity of heat mitigation strategies is particularly important when we consider trends in climate change that indicate that cities worldwide are going to face even higher temperatures in the future. This can be critical in warm climates where the UHI can elevate urban temperatures by as much as 8 °C (Moreno 1994), representing a significant health threat during heatwaves to vulnerable populations like the elderly, children and people living in poor-quality housing.

There is increasing recognition in urban plans and policies of the ecosystem services provided by GI, as in Madrid (Spain), Mérida (Mexico), New York (USA) and many other cities worldwide (Ayuntamiento de Madrid 2018; Municipio de Mérida 2016; NYC Environmental Protection 2010). However, it is necessary to continue studying how to build the best solutions that spatially combine aspects like the ratio and orientation of buildings, with the size, shape and spatial distribution of green infrastructure elements, such as large parks, urban trees, green roofs and walls, and residential gardens.

Acknowledgement This work was partially funded by ANID/FONDAP/15110020 and ANID/FONDECYT/1202003.

References

Ahern, J. (2011). From fail-safe to safe-to-fail. Sustainability and resilience in the new urban world. *Landscape and Urban Planning, 100*, 341–343. https://doi.org/10.1016/j.landurbplan.2011.02.021.

Ali-Toudert, F., & Mayer, H. (2006). Numerical study on the effects of aspect ratio and orientation of an urban street canyon on outdoor thermal comfort in hot and dry climate. *Building and Environment, 41*(2), 94–108.

Alves, A., Vojinovic, Z., Kapelan, Z., Sanchez, A., & Gersonius, B. (2020). Exploring trade-offs among the multiple benefits of green-blue-grey infrastructure for urban flood mitigation. *Science of the Total Environment, 703*, 134980. https://doi.org/10.1016/j.scitotenv.2019.134980.

Avolio, M. L., Pataki, D. E., Pincetl, S., Gillespie, T. W., Jenerette, G. D., & McCarthy, H. R. (2015). Understanding preferences for tree attributes: The relative effects of socio-economic and local environmental factors. *Urban Ecosystems, 18*, 73–86. https://doi.org/10.1007/s11252-014-0388-6.

Ayuntamiento de Madrid. (2018). *Plan de Infraestructura Verde y Biodiversidad.* Retrieved December 10, 2019 from https://www.madrid.es/UnidadesDescentralizadas/ZonasVerdes/TodoSobre/PlanInfraestructuraVerdeYBiodiversidad/DocumentacionAsociada/Resumen%20Ejecutivo.pdf

Barradas, V. L. (1991). Air temperature and humidity and human comfort index of some city parks of Mexico City. *International Journal of Biometeorology, 35*(1), 24–28.

Bartesaghi Koc, C., Osmond, P., & Peters, A. (2017). Towards a comprehensive green infrastructure typology: A systematic review of approaches, methods and typologies. *Urban Ecosystems, 20*, 15–35. https://doi.org/10.1007/s11252-016-0578-5.

Blanusa, T., Vaz Monteiro, M. M., Fantozzi, F., Vysini, E., Li, Y., & Cameron, R. W. F. (2013). Alternatives to Sedum on green roofs: Can broad leaf perennial plants offer better "cooling service"? *Building and Environment, 59*(2013), 99–106.

Bowler, D. E., Buyung-Ali, L., Knight, T. M., & Pullin, A. S. (2010). Urban greening to cool towns and cities: A systematic review of the empirical evidence. *Landscape and Urban Planning, 97*(3), 147–155.

Chang, C. R., Li, M. H., & Chang, S. D. (2007). A preliminary study on the local cool-island intensity of Taipei city parks. *Landscape and Urban Planning, 80*(4), 386–395.

Chen, Q., Li, B., & Liu, X. (2013). An experimental evaluation of the living wall system in hot and humid climate. *Energy and Buildings, 61*, 298–307. https://doi.org/10.1016/j.enbuild.2013.02.030.

Coma, J., Pérez, G., de Gracia, A., Burés, S., Urrestarazu, M., & Cabeza, L. F. (2017). Vertical greenery systems for energy savings in buildings: A comparative study between green walls and green facades. *Building and Environment, 111*, 228–237. https://doi.org/10.1016/j.buildenv.2016.11.014.

de la Barrera, F., Rubio, P., & Banzhaf, E. (2016). The value of vegetation cover for ecosystem services in the suburban context. *Urban Forestry & Urban Greening, 16*, 110–122.

de la Barrera, F., Henriquez, C., Ruiz, V., & Inostroza, L. (2019). Urban parks and social inequalities in the access to ecosystem services in Santiago, Chile. In *IOP Conference Series: Materials Science and Engineering* (Vol. 471(10), p. 102042). Bristol: IOP Publishing.

Eumorfopoulou, E. A., & Kontoleon, K. J. (2009). Experimental approach to the contribution of plant-covered walls to the thermal behaviour of building envelopes. *Building and Environment, 44*(5), 1024–1038. https://doi.org/10.1016/j.buildenv.2008.07.004.

European Commission. (2013). Communication from the Commission to the European Parliament, The Council, the European Economic and Social Committee and the Committee of the Regions. Green Infrastructure (GI) Enhancing Europe's Natural Capital. COM (2013) 249 Final.

Gatto, E., Buccolieri, R., Aarrevaara, E., Ippolito, F., Emmanuel, R., Perronace, L., & Santiago, J. L. (2020). Impact of urban vegetation on outdoor thermal comfort: Comparison between a Mediterranean City (Lecce, Italy) and a Northern European City (Lahti, Finland). *Forests, 11*(2), 228. https://doi.org/10.3390/f11020228.

Getter, K. L., Rowe, D. B., Andresen, J. A., & Wischman, I. S. (2011). Seasonal heat flux properties of an extensive green roof in a Midwestern U.S. climate. *Energy and Buildings, 43*(12), 3548–3557. https://doi.org/10.1016/j.enbuild.2011.09.018.

Grilo, F., Pinho, P., Aleixo, C., Catita, C., Silva, P., Lopes, N., et al. (2020). Using green to cool the grey: Modelling the cooling effect of green spaces with a high spatial resolution. *Science of the Total Environment, 724*, 138182.

Gunawardena, K. R., Wells, M. J., & Kershaw, T. (2017). Utilising green and bluespace to mitigate urban heat island intensity. *Science of the Total Environment, 584*, 1040–1055.

Hoelscher, M. T., Nehls, T., Jänicke, B., Wessolek, G. (2016). Quantifying cooling effects of façade greening: shading, transpiration and insulation, *Energy and Buildings 114,* 283–290. https://doi.org/10.1016/j.enbuild.2015.06.047.

Inostroza, L., Palme, M., & De la Barrera, F. (2016). A heat vulnerability index: Spatial patterns of exposure, sensitivity and adaptive capacity for Santiago de Chile. *PLoS One, 11*(9), e0162464.

Jänicke, B., Meier, F., Hoelscher, M. T., & Scherer, D. (2014). Evaluating the effects of facade greening on human bioclimate in a complex urban environment. *Advances in Meteorology, 2015*, 15. https://doi.org/10.1155/2015/747259.

Lafortezza, R., Carrus, G., Sanesi, G., & Davies, C. (2009). Benefits and well-being perceived by people visiting green spaces in periods of heat stress. *Urban Forestry & Urban Greening, 8*(2), 97–108.

Lobaccaro, G., & Acero, J. A. (2015). Comparative analysis of green actions to improve outdoor thermal comfort inside typical urban street canyons. *Urban Climate, 14*(Part 2), 251–267. https://doi.org/10.1016/j.uclim.2015.10.002.

Mariani, L., Parisi, S. G., Cola, G., Lafortezza, R., Colangelo, G., & Sanesi, G. (2016). Climatological analysis of the mitigating effect of vegetation on the urban heat island of Milan, Italy. *Science of the Total Environment, 569*, 762–773.

Moreno, M. (1994). Intensity and form of the urban heat island in Barcelona. *International Journal of Climatology, 14*(1994), 705–710.

Moreno, M. (1999). *Climatología urbana*. Barcelona: Edicions de la Universitat de Barcelona.

Municipio de Mérida. (2016). *Plan Municipal de Infraestructura Verde*. http://www.merida.gob.mx/sustentable/contenidos/doc/PlanInfraestructuraVerde.pdf. Accessed 10 December 2019.

Nastran, M., Kobal, M., & Eler, K. (2019). Urban heat islands in relation to green land use in European cities. *Urban Forestry & Urban Greening, 37*, 33–34. https://doi.org/10.1016/j.ufug.2018.01.008.

Ng, E., Chen, L., Wang, Y., & Yuan, C. (2012). A study on the cooling effects of greening in a high-density city: An experience from Hong Kong. *Building and Environment, 47*, 256–271.

NYC Environmental Protection. (2010). *NYC Green Infrastructure Plan*. http://www.nyc.gov/html/dep/pdf/green_infrastructure/NYCGreenInfrastructurePlan_ExecutiveSummary.pdf. Accessed 10 December 2019.

Oke, T. R. (1989). The micrometeorology of the urban forest. *Philosophical Transactions of the Royal Society of London B, Biological Sciences, 324*(1223), 335–349.

Oldham, C., & Karima, A. (2018). *Performance of green walls in Mediterranean climates: A literature review*. The University of Western Australia. Report to Water Corporation of Western Australia. Retrieved from https://api.research-repository.uwa.edu.au/files/43627161/Green_Wall_Literature_Review_final.pdf

Parlow, E. (2011). Urban climate. In J. Niemelä (Ed.), *Urban ecology* (pp. 31–44). New York: Oxford University Press.

Pauleit, S., & Duhme, F. (2000). Assessing the environmental performance of land cover types for urban planning. *Landscape and Urban Planning, 52*(2000), 1–20.

Peng, L. L. H., & Jim, C. Y. (2013). Green-roof effects on neighborhood microclimate and human thermal sensation. *Energies, 6*, 598–618.

Pérez, G., Coma, J., Martorell, I., & Cabeza, L. F. (2014). Vertical Greenery Systems (VGS) for energy saving in buildings: A review. *Renewable & Sustainable Energy Reviews, 39*, 139–165. https://doi.org/10.1016/j.rser.2014.07.055.

Rana, G., De Lorenzi, F., Mazza, G., Martinelli, N., Muschitiello, C., & Ferrara, R. M. (2020). Tree transpiration in a multi-species Mediterranean garden. *Agricultural and Forest Meteorology, 280*, 107767. https://doi.org/10.1016/j.agrformet.2019.107767.

Reyes-Paecke, S., Gironás, J., Melo, O., Vicuña, S., & Herrera, J. (2019). Irrigation of green spaces and residential gardens in a Mediterranean metropolis: Gaps and opportunities for climate change adaptation. *Landscape and Urban Planning, 182*, 34–43. https://doi.org/10.1016/j.landurbplan.2018.10.006.

Sarricolea, P., Aliste, E., Castro, P., & Escobedo, C. (2008). Análisis de la máxima intensidad de la isla de calor urbana nocturna de la ciudad de Rancagua (Chile) y sus factores explicativos. *Revista de Climatología, 8*(2008), 71–84.

Skelhorn, C., Lindley, S., & Levermore, G. (2014). The impact of vegetation types on air and surface temperatures in a temperate city: A fine scale assessment in Manchester, UK. *Landscape and Urban Planning, 121*(2014), 129–140. https://doi.org/10.1016/j.landurbplan.2013.09.012.

Song, J., Chen, W., Zhang, J., Huang, K., Hou, B., & Prishchepov, A. V. (2020). Effects of building density on land surface temperature in China: Spatial patterns and determinants. *Landscape and Urban Planning, 198*, 103794. https://doi.org/10.1016/j.landurbplan.2020.103794.

Spronken-Smith, R. A., & Oke, T. R. (1998). The thermal regime of urban parks in two cities with different summer climates. *International Journal of Remote Sensing, 19*(11), 2085–2104.

Trlica, A., Hutyra, L. R., Schaaf, C. L., Erb, A., & Wang, J. A. (2017). Albedo, land cover, and daytime surface temperature variation across an urbanized landscape. *Earth's Future, 5*, 1084–1101. https://doi.org/10.1002/2017EF000569.

Tzoulas, K., Korpela, K., Venn, S., Yli-Pelkonen, V., Kazmierczak, A., Niemela, J., & James, P. (2007). Promoting ecosystem and human health in urban areas using green infrastructure: A literature review. *Landscape and Urban Planning, 81*(3), 167–178. https://doi.org/10.1016/j.landurbplan.2007.02.001.

Vaz Monteiro, M., Blanusa, T., Verhoef, A., Richardson, M., Hadley, P., & Cameron, R. W. F. (2017). Functional green roofs: Importance of plant choice in maximising summertime environmental cooling and substrate insulation potential. *Energy and Buildings, 141*, 56–68.

Venter, Z. S., Krog, N. H., & Barton, D. N. (2020). Linking green infrastructure to urban heat and human health risk mitigation in Oslo, Norway. *Science of the Total Environment, 709*, 136193. https://doi.org/10.1016/j.scitotenv.2019.136193.

Vox, G., Blanco, I., & Schettini, E. (2018). Green façades to control wall surface temperature in buildings. *Building and Environment, 129*(2018), 154–166. https://doi.org/10.1016/j.buildenv.2017.12.002.

Wang, X., & Banzhaf, E. (2018). Towards a better understanding of green infrastructure: A critical review. *Ecological Indicators, 85*, 758–772. https://doi.org/10.1016/j.ecolind.2017.09.018.

Wang, C., Li, Y., Myint, S. W., Zhao, Q., & Wentz, E. A. (2019). Impacts of spatial clustering of urban land cover on land surface temperature across Köppen climate zones in the contiguous United States. *Landscape and Urban Planning, 192*, 103668. https://doi.org/10.1016/j.landurbplan.2019.103668.

Whitford, V., Ennos, A., & Handley, J. (2001). City form and natural process: Indicators for the ecological performance of urban areas and their application to Merseyside, UK. *Landscape and Urban Planning, 57*, 91–103.

Zölch, T., Maderspacher, J., Wamsler, C., & Pauleit, S. (2016). Using green infrastructure for urban climate-proofing: An evaluation of heat mitigation measures at the micro-scale. *Urban Forestry & Urban Greening, 20*, 305–316.

Chapter 20
Urban Morphology as a Mitigation Strategy of Urban Warming in "Oasis Cities" of Arid Regions

Erica Norma Correa, Maria Belén Sosa, Maria Alicia Cantón, and Maria Angélica Ruiz

20.1 Introduction

Thermal behavior of cities is largely known for being a by-product of urban morphology or, more specifically, the composition and three-dimensional configurations that define the urban frame (McPherson 1994). Therefore, the process of urban planning is closely related to climate, given that the central purpose of planning is to create an environment suited to humans (Zhao et al. 2011).

An appropriate planning can mitigate the negative effects cities have on local climate and improve their environmental quality. In this regard, a comprehensive understanding of the relationship between urban form and energy use is important to formulate climate change mitigation policies at a city level (Li et al. 2018). These topics are of particular importance in developing countries that have experienced an accelerated process of urbanization. The urban sprawl rate is almost three times the population growth observed during the same period of time. Regarding urban morphology, researchers often focus on urban canyons (UCs). In built environments, UCs usually cover more than a quarter of urbanized areas (Shashua-Bar and Hoffman 2003), or in some historically planned cities, such as Manhattan or Barcelona, the urbanized area covers up to 30% of their built surface (UN-Habitat 2013). There are many studies worldwide that discuss how geometry, materiality, and forestation of UCs can mitigate UHI (Sanusi et al. 2016; Alchapar and Correa 2015; Ruiz et al. 2015; Lin 2009; Shashua-Bar and Hoffman 2003). However, there is not enough research that considers how planning strategies related to the morphologies of low-density residential neighborhoods can contribute to cooling urban air temperatures. In this respect, Middel et al. (2014) demonstrated that urban grid

E. N. Correa (✉) · M. B. Sosa · M. A. Cantón · M. A. Ruiz
Institute of Environment, Habitat and Energy (INAHE), National Council of Scientific and Technical Research (CONICET), CCT - Mendoza, Argentina
e-mail: ecorrea@mendoza-conicet.gob.ar; msosa@mendoza-conicet.gob.ar; mcanton@mendoza-conicet.gob.ar; aruiz@mendoza-conicet.gob.ar

M. Palme, A. Salvati (eds.), *Urban Microclimate Modelling for Comfort and Energy Studies*, https://doi.org/10.1007/978-3-030-65421-4_20

forms have a larger impact on daytime temperatures than urban greening. Additionally, Middel et al. (2015) assessed the combined cooling benefits of trees and cool roofs. Middel et al. have demonstrated that a 25% increase in tree canopy in residential neighborhoods resulted in an average decrease of 2.0 °C during the day, whereas cool roofs showed a reduction of only 0.3 °C in air temperature. While there is scientific evidence that urban forms, or some of its attributes, have an impact on the energy demand, the characterization of the effect of urban form per se represents a major challenge (Silva et al. 2017; Lee and Lee 2014). Therefore, it is important to evaluate how the urban form variables affect the building energy consumption (Ko and Radke 2014).

Particularly, in Latin America and the Caribbean, cities are discontinuously dispersed and spread in low-density patterns. It is estimated that, by 2050, the urban population in these cities will grow 87%. This intense anthropization of the environment raises urban temperatures and energy consumption in buildings. The International Energy Agency estimates that between 30% and 40% of the energy used worldwide is consumed by buildings (International Energy Agency, IEA 2017). This building sector contributes to 30% of the global CO2 emissions. Therefore, acquiring knowledge about this subject would be useful to plan social housing, considering the outdoor microclimatic variables as part of the design process. In this regard, a good neighborhood planning in Latin American cities, considering the layout form, street and block proportions and orientations, urban trees, and cooler materials, can help to reduce the impact of urban form on the microclimate (Sosa et al. 2017; Middel et al. 2015; NRDC 2011; Jusuf et al. 2007).

This chapter aims to discuss the impact of urban morphology on the city microclimate, taking into account not only the form of urban canyons but also the neighborhood morphology in which they are inserted. In addition, it presents diverse methods and simulation tools to reach such objectives. First, it analyzes how geometry, materiality, and forestation aspects of urban canyons can be adequately combined to be used as an urban warming mitigation strategy and to enhance thermal comfort. Second, it analyzes how the residential neighborhood morphology can contribute to cool cities. It will evaluate the impact of different forms of urban grids on the urban temperature, and analyze which is the impact of design decisions on energy consumption for cooling.

20.1.1 Oasis Cities of Arid Regions and Thermal Performance

What is the meaning of "oasis city" and how have its features influenced its thermal performance? Mendoza Metropolitan Area (MMA) is located in central western Argentina (32°53°S, 68°51°W, 750 m. a. s. l.). According to the Köppen-Geiger climate classification, it is an arid continental climate: BWh or BWk, depending on the isotherm used (Kottek et al. 2006). It is characterized by cold winters and hot summers with significant daily and seasonal thermal amplitudes. Winds are moderate and infrequent, the amount and intensity of solar radiation is high, and the average annual rainfall is 198 mm (González Loyarte et al. 2009). Arid cities at a global scale usually present one of the two types of streetscapes: compact or open models. The compact model has continuous urban development, consisting of tall buildings and narrow streets where the use of urban forestation is absent or scarce. In contrast, the open model has lower buildings with wide and forested streets (Correa et al. 2010). According to these models, MMA has been an oasis city since its foundation. Its urban grid is defined by wide streets and buildings contained in checkered frames with rows of trees flaking their boundaries. This forest frame is the main strategy of shading and the "green tunnels" form an authentic forest within the city. Since 2003, the INAHE-CONICET has been studying the urban climate of the Mendoza Metropolitan Area (MMA) with a focus on the magnitude, causes, and consequences of the urban heat island (UHI). It was found that, despite its features, the UHI in the city reaches a maximum 10.6 °C (Correa et al. 2008). This phenomenon is perceived with greater intensity during summer nights, because the sky view blocked by the tree canopy diminishes the radiative and convective cooling of outdoor spaces. Previous studies have established some of the impacts caused by this phenomenon: an increase up to 20% of energy consumption by air-conditioning, as well as worse outdoor thermal comfort conditions, with people feeling uncomfortable due to the heat in outdoor spaces during 46–62% of the time (Correa et al. 2008, 2012; Ruiz and Correa 2014). Their work highlights the importance of considering both day and nighttime thermal behavior. In urban canyons, the daytime thermal behavior determines the degree of comfort of spaces and energy consumption of buildings. In addition, nighttime behavior affects the cooling possibilities of the city and the energy consumption of buildings.

20.2 Urban Morphology Variables and Their Impact on Thermal Performance of Oasis Cities

20.2.1 Selection of Case Studied and Methodological Procedure

When Mendoza's features are considered, it is useful to investigate the consequences of implemented design decisions and their influences on outdoor thermal behavior in summer. In this context, this research aims to define and quantify the variables—forest, morphological, and material ones—that determine air temperature during day and nighttime as well as the degree of outdoor thermal comfort to identify the best urban configurations.

Nineteen representative urban canyons were selected according to three axes: tree species, street widths, and building densities. These characteristics reflect the variety of the prevailing urban features of the MMA. Eighteen forested urban canyons were selected. The forested canyons have tree species of first and second magnitude. The classification of forest magnitude is based on the end height that a tree reaches 20 years after plantation. The first magnitude is for species whose end height exceeds 15 m such as "London plane" (*Platanus hispanica*). The second magnitude varies from 8 to 15 m such as "white mulberry" (*Morus alba*) and "European ash" (*Fraxinus excelsior* L.). In addition, one urban canyon had no trees. Pictures of forested urban canyons of Mendoza are shown in Fig. 20.1. According to the street widths, three typical were selected: 16, 20, and 30 m; this is based on the MMA urban regulation policies (Regulatory Law N°4341/1978). In relation to building density, low buildings (from 3 to 6 m heights) and high buildings (from 6 to 24 m height) were chosen. All selected urban canyons are oriented East-West to show the greatest difference in temperatures in summer.

Several types of variables were collected: morphological, forest, and material ones. However, this work only analyzes the variables that can be included in building codes and used by developers and urban planners. Variables selected by category include the following:

- *Urban canyon structure*: building volume (BV), compactness (*C*), urban canyon length (UCL), volume/width (*V*/*W*), urban canyon width (UCW), volume/length (*V*/*L*), mean building height (MBH), height/width (*H*/*W*), building view factor (BVF), and sky view factor (SVF).
- *Urban forest structure*: solar permeability of tree species (SP), number of trees (NT), trees per meter (*T*/m), mean tree height (MTH), tree cover (TC), and tree view factor (TVF).
- Solar radiation permeability of tree species is defined as a characteristic of the tree canopy to transmit the incoming global solar radiation. It is a value between 0 and 1.
- *Optical properties of materials*: horizontal surface albedo (HA) and vertical surface albedo (VA).

In each urban canyon selected, these variables were contrasted with the following microclimatic variables: daytime air temperature (DTair), daytime surface temperature of pavement (DTpav), daytime surface temperature of sidewalk (DTsw), daytime surface temperature of wall (DTwall), solar radiation (SR), relative humidity (RH), wind speed (WS), daytime thermal comfort (COMFA), and nighttime air temperature (NTair). A detailed description and discussion of methodology procedure are given by Ruiz et al. (2015, 2017).

Fig. 20.1 Forested urban canyons of the MMA. (**a**) *Platanus hispanica*, (**b**) *Morus alba*, and (**c**) *Fraxinus excelsior*. (Adapted from Correa et al. 2010)

COMFA index was used to assess thermal comfort based on the results reached by Ruiz and Correa (2015). The researchers presented a comparison between six thermal comfort models, contrasted with 667 subjective reports, in order to identify which of the models could be used to correctly predict thermal comfort outdoor space of MMA. COMFA expresses the energy balance in W/m^2 of an individual in an open environment (Gaitani et al. 2007). When the balance is near zero, it may be expected that an individual feels thermally comfortable. Figure 20.2 describes the sensation of human comfort related to the values of energy balance. The microclimatic variable "daytime thermal comfort" coincides with the average of registered values between 5 and 6 pm. This time frame coincides with the thermal discomfort peak of the city.

In order to detect and weigh the variables that determine the day and nighttime thermal behavior and the thermal comfort, a set of multiple linear regressions were performed in R software (R Development Core Team 2011), and three statistical models were obtained (Ruiz et al. 2015, 2017):

Daytime model is shown in Eq. (20.1):

$$\mathrm{DTair} = 29.18 - 6.20 * {}^{T}\!/_{\mathrm{m}} - 0.05 * \mathrm{MBH} - 0.13 * \mathrm{UCW} - 6.97 * \mathrm{HA} + 0.28 * \mathrm{DTwall} \tag{20.1}$$

where DTair stands for daytime air temperature, *T*/m stands for trees per meter, MBH stands for mean building height, UCW stands for urban canyon width, HA stands for horizontal surface albedo, and DTwall is daytime surface temperature of wall.

Nighttime model is shown in Eq. (20.2):

$$\mathrm{NTair} = 23.50 + 1.797 * {}^{H}\!/_{W} - 2.649 * \mathrm{MBH} - 0.0000001386 * \mathrm{BV} + 2.135 * \mathrm{VA} \tag{20.2}$$

where NTair stands for nighttime air temperature, *H*/*W* stands for height/width ratio, MBH stands for mean building height, BV stands for building volume, and VA stands for vertical surface albedo.

COMFA (W/m^2)	*Sensation*
COMFA < -150	Would prefer to be much warmer
-150 > COMFA < -50	Would prefer to be warmer
-50 > COMFA < 50	Would prefer no change
50 > COMFA < 150	Would prefer to be cooler
COMFA > 150	Would prefer to be much cooler

Fig. 20.2 Sensation of human comfort related to the values of energy balance, COMFA

Note that at nighttime only morphological and optical variables have an influence on air temperature.

Thermal comfort model is shown in Eq. (20.3):

$$\text{COMFA} = -491.59 - 12.90 * \text{SP} - 4.86 * \text{NT} + 313.23 * \text{SVF} + 20.89 * \text{DTwall} \quad (20.3)$$

where COMFA stands for daytime thermal comfort, SP stands for solar permeability, NT stands for number of trees, SVF stands for sky view factor, and DTwall stands for daytime surface temperature of wall.

The goodness of fit and the testing of assumption validation are shown in Table 20.1 for each of the three models. Note that three models meet the assumptions.

Finally, daytime and nighttime models were carried out as morphological expressions.

20.2.2 Results

20.2.2.1 Testing Predesign Tools

A series of possible urban configurations were developed from multivariate regression models, as this research does not want to cover the universe of urban morphology, but it seeks to analyze the most typical cases of the studied city.

From the multivariate regression model, 12 nomograms were developed. A nomogram consists of a set of n scales, one for each variable in an equation. Knowing the values of $n - 1$ variables, the value of the unknown variable can be found, or by fixing the values of some variables, the relationship among the unfixed ones can be studied.

In the case of ***daytime model*** in three of typical UCW in the studied city (16, 20, and 30 m), MBH, HA, and *T*/m values vary. The DTwall is set at 31 °C as the average in monitored cases. The results suggest the following:

- To equal the value of mean building height, horizontal albedo, and number of trees, a greater urban canyon width determines a lower air temperature (1.3 °C for every 10 m width). This would highlight the influence of increasing the convective exchange of the urban canyon surfaces during the heating period.

Table 20.1 Goodness of fit and testing of assumption validation for each developed model

	Goodness of fit		Normality	Homoscedasticity
	Adjusted R^2	RMSE	Shapiro-Wilk test	Studentized Breusch-Pagan test
Model	–	%	–	–
DTair	0.7227[a]	2.51	0.9329	10.8949
NTair	0.6396[a]	0.93	0.9517	3.7689
COMFA	0.8606[a]	13.40	0.9464	3.3406

- A greater density of trees and a fixed value of horizontal albedo, mean building height, and urban canyon width means air temperature is lower (−0.6 °C for every 0.11 trees/m, which means that there is 1 tree every 6.5 m). In this case, the ability of forest would have influenced to provide shade and to cool the air by evapotranspiration.
- For a set number of trees and width and height of the urban canyon, an increase of 0.10 in the horizontal albedo produces an air temperature decrease of 0.7 °C. This indicates that the higher reflective capacity of horizontal surfaces avoids the overheating of buildings, so the air temperature is lower.
- As mean building height increases, and all other variables are fixed, daytime air temperature decreases 0.8 °C for every 15 m (five stories).
- In the graphed cases, the best response is given by using huge density of trees (0.24 trees/m = 8.5 m of tree spacing) and a high horizontal albedo equal to 0.54, in high and wide urban canyons (30 m wide by 30 m high). This configuration offers 6.8 °C less than the worst configuration.
- Urban canyons without trees are, in all cases, warmer.

The behavior of the ***nighttime model*** was assessed in two different building volumes (9000 and 23,000 m ^{3}). These values represent low and high building density for the studied city. In each case, there are two mean building heights: 3 and 6 m, and 15 and 30 m, respectively. The *H*/*W* ratio varies between 0.10 and 1.88, according to these heights and the typical widths of urban canyons. Vertical albedo data are also varied. The results show the following:

- To equal *H*/*W* ratio, BV, and MBH, an increase of 0.25 in the vertical albedo represents an increase of 0.5 °C in air temperatures.
- To equal BV, MBH, and vertical albedo, when the *H*/*W* ratio increases (UCW decreases), the night air temperature increases. These differences are five times more pronounced in high density than in low density. This might be due to the configurations with high *H*/*W* that do not allow a proper convective and radiative cooling.
- Lower BV (9000 m^3), MBH (3 m), *H*/*W* (0.1), and vertical albedo (0.35) is the best configuration in terms of thermal response, among the cases considered (difference of air temperature = 3.5 °C).

For both periods, this analysis determined that daytime air temperature is more sensitive to variations in the forest, morphological, and optical variables than in nighttime.

Thermal comfort model was tested in 36 urban canyons. Taking into account the variables involved in the model, the DTwall value was fixed at 31 °C because it is the resulting average, the median, and the mode in the monitored cases. The SP and NT vary according to tree species. Typical species that exist in the MMA were used (Cantón et al. 2003), one of first magnitude—*P. hispanica* (SP = 0.098, NT = 18)—and one of second magnitude—*F. excelsior* (SP = 0.162, NT = 25). The NT values were calculated according to the tree implantation distance (first magnitude = 11 m and second magnitude = 8 m). The trees were arranged in two sidewalks of a 100 m length street. The SVF values vary according to building heights (3, 6, 15, and 30 m), street widths (16, 20, and 30 m), and tree species.

These values were calculated by using ENVI-met 3.1® (Bruse 2004).

The ***comfort behavior*** corresponding to the observations induces the following:

- According to sensation ranges of COMFA index (Fig. 20.2), urban canyons do not present thermal comfort conditions. This coincides with the empirical results obtained by Correa et al. (2012).
- To equal SVF and SP, an increase in the number of trees represents a -170 W/m^2 decrease in the energy balance.
- When SVF decreases, people seem to feel more comfortable. This causes an increase in shading in the streets.
- People seem to feel more uncomfortable when the number of trees decreases. This observation helps us to confirm the cooling effect by shading and evapotranspiration that the urban forest provides.
- The best tree species, for this universe of study, is *Fraxinus excelsior* (SP = 0.314), as it contributes to decreasing the energy balance of the street.
- In urban configurations without trees the habitability of outdoor spaces notoriously decreases because people feel hotter. This fact confirms the benefits of forests in the cities.

20.2.3 Morphological Expressions

From the model implications for day and nighttime, the study provided bases to generate urban canyon configurations in both dimensions. The UCW (horizontal dimension) and MBH (vertical dimension) were used. *H*/*W* ratio was calculated from the relationship between the two indicators. The thermal behavior of each morphological relationship was expressed in sections of the urban canyon. Each section has a color scale that represents its thermal behavior, four sections in total, due to the possible combinations of height and width. In order to simplify the reading of Fig. 20.3, the thermal differences (ΔT °C) between the possible combinations were incorporated.

In the daytime, the high density of forested and non-forested urban canyons remains 1.3 °C cooler compared to the respective cases of low density. As a consequence, we can assume that the increase in the edification height helps to keep the streets cooler by the shadow effect. In addition, the cooler configurations for both densities are those with a street width equal to 30 m. This would be produced through the convective exchanges by the wind in the analyzed channels. Another fact to take into account is the comparison between the thermal behavior of forested and non-forested configurations, where there is a difference of 1.4 °C at equal conditions of urban morphology. This result demonstrates the important contribution of urban forest in terms of cooling the outside environment through shading and evapotranspiration. For nighttime, low building density presents the best

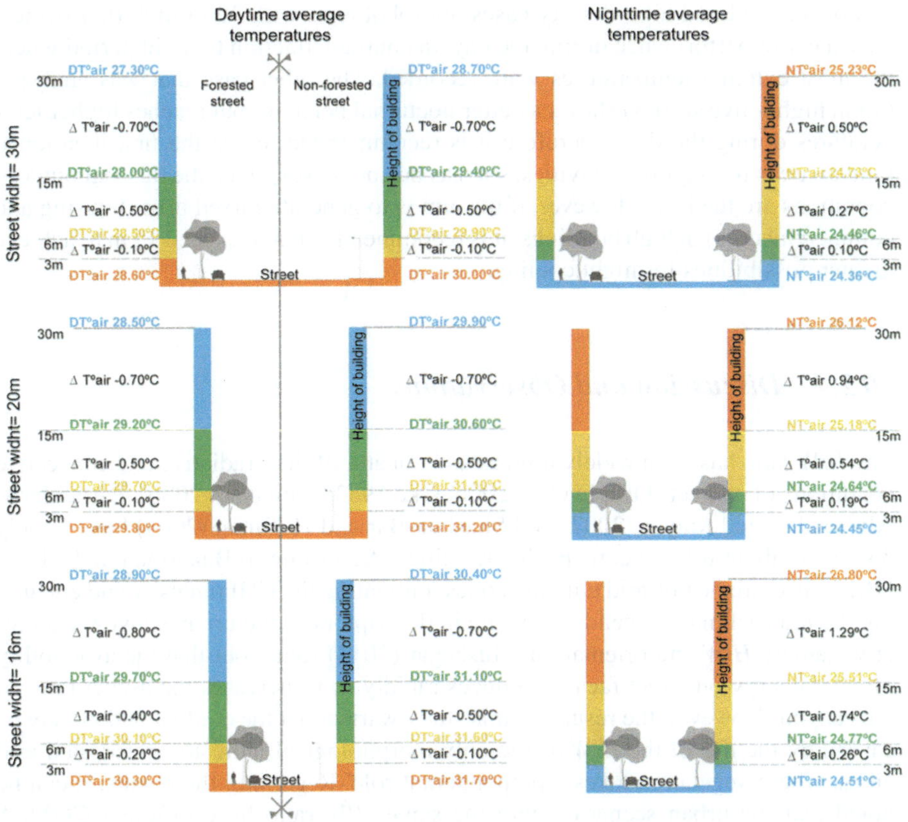

Fig. 20.3 Urban cross sections of forested and non-forested canyons and their thermal behavior. Urban canyon width and height are presented as rows, and the time period as columns. In addition, temperature values and thermal differences for each combination are displayed. The color scale represents the range of temperature for the urban configurations: light blue is the coolest one and orange is the warmest one. DTair stands for daytime air temperature, NTair stands for nighttime air temperature, and Δ*T*air stands for air temperature difference between configurations

combinations. Radiative cooling strategies are possible due to the broad sky view. This means that for high-density configurations, in 16 m street widths, the possibility of vision is partially blocked by buildings and trees. In this width, thermal differences reach 2.3 °C between 3 and 30 m in edification height. This fact allows us to deduce that cooling possibilities are greater in lower and wider urban canyons compared to more higher and narrow ones, in this particular period.

The results aim to identify the best urban configurations in terms of thermal behavior. From a point of view of urban growth planning, this study recommends buildings of 6 m high (two stories) with forested streets for residential areas of the city—low density—as this is the combination with the best performance in both periods. Therefore, the optimum street width is 30 m for both day and night, resulting in an *H*/*W* = 0.20 ratio.

For forested and high-density cases, morphologies with 30 m high (ten stories) have a better performance during the day, in contradistinction to night period where the most extreme temperatures were recorded. The other case that was analyzed (15 m high—five stories) shows a better nocturnal behavior but reaches higher temperatures during the day. Therefore it is recommended to use the first mentioned kind of areas for daytime activities, such as business uses, since the thermal outdoor conditions are the best. However, if the aim is to generate mixed uses (housing and commercial uses) in high buildings, it is recommended not to exceed 15 m to achieve greater possibilities for night cooling.

20.2.4 Discussion and Observations

The *H*/*W* ratio has been widely used as an indicator of the irradiative and convective conditions of a street (Offerle et al. 2006; Oke 1982; Ratti et al. 2006). Particularly, Ali-Toudert and Mayer (2007) and Shashua-Bar and Hoffman (2004) have largely discussed about urban geometry in arid cities. According to Bakarman and Chang (2015), in extreme hot-arid climatic zones, mitigating the UHI intensity and enhancing the microclimate depend on two main descriptors of the urban canyon's geometry, namely *H*/*W* and orientation. Shishegar (2013) found out that the availability of solar energy on street facades reduces rapidly as it increases the aspect ratio of the canyon. However, the results of this study warn about the confusion that may be caused by the use of the *H*/*W* as the only morphological indicator in forested arid cities, where vegetation takes a preponderant role. Regarding the results, it can be noted that the urban scenarios with the same *H*/*W* ratio have different COMFA responses (e.g., forested scenarios with an $H/W = 0.40$ have ΔCOMFA = 46% and non-forested equal to 2%). Another important result is that scenarios with better conditions of outdoor thermal comfort during the day have the highest nighttime air temperatures. These high night temperatures generate an increase of the energy consumption. For instance, Papanastasiou et al. (2013) studied the impact of UHI on energy consumption in Volos, Greece. It was found that UHI intensity reaches 3.1 °C during summer nights, with almost twice cooling load in the city, in relation to the suburb. For this reason, it is important to arrive to a compromise to develop a solution in terms of urban planning. Giving proper zoning to the city can reduce urban warming, which would also improve the thermal habitability and avoid overheating. In terms of zoning, the residential areas (low density) must have better nighttime air temperature (lower energy consumption). Outdoor thermal comfort can be improved with more green spaces near these areas (e.g., forested squares and parks for recreational activities). In the studied city, low-density areas are defined as those buildings with no more than two floors (6 m height) inserted at any street width (16, 20, or 30 m) and forested with any tree species. The mixed-use areas (high density) should have better outdoor comfort. This criterion is based on the idea that the activities in these areas mostly occur during daytime. In order to achieve this, the previously developed design tools predict that narrow streets (16 or

20 m width) with buildings of 12–24 m height, forested with a second magnitude tree, have the best thermal comfort. For 30 m width forested streets, only 24 m height buildings achieve acceptable comfort conditions. However, it is important to remember that high concentrations of buildings and impervious surfaces increase irradiative heating, intensifying the urban warming at night (Ruiz et al. 2017).

20.3 Use of Urban Grid Forms as a Strategy to Reduce Heat Island Effect in Arid Cities

In order to identify urban grid forms (UGs) that contribute to the reduction of the UHI in Mendoza, Argentina, Sosa et al. (2017) assessed the microclimatic conditions of ten urban canyons (UCs) during summertime. These urban canyons were located in different urban grid forms: multi-azimuthal, cul-de-sac, rectangular, and reticular grid. The microclimatic data monitored the morphological features of neighborhood grid: block area, street area, built-up area and compactness ratio, and structure of urban canyons: length, width, *H/W*, street surface, azimuth, built-up volume, SVF, and quantity of trees were statistically analyzed. The results suggest that the minimum air temperature is related to the effects of the neighborhood grid and the urban canyon configuration combined. However, the maximum and average air temperatures are only related to the urban canyon configuration. Considering the importance of mitigating the urban heat island in Mendoza city and knowing that the rate of increase in the number of extreme heat events in the most sprawling metropolitan regions is more than twice the rate of the observed increase in the most compact metropolitan (Stone et al. 2010), the authors concluded that to achieve a sustainable planning of urban built environments in arid zones, the form of the urban neighborhood grids and the structure of the urban canyons must be combined. This combination aims to improve the thermal comfort and to diminish the urban heat island and the energy consumption of buildings. Within this framework, the research aims to evaluate the impact that different social housing neighborhood designs have on outdoor air temperatures and energy consumption for cooling during summertime. The final goal is to improve the energy efficiency at neighborhood scale, by providing planning and design strategies for social housing settlements in oasis cities with arid climates.

20.3.1 Methodology

The research methodology consists of the following procedures: (1) study site characterization and selection process, (2) microclimate monitoring, (3) CFD simulations, and (4) energy consumption estimation.

The MMA is characterized by low building densities constituted mostly by single-family attached houses. The selected neighborhoods for microclimatic monitoring have different layouts and block orientations but maintain certain analog features that allow the microclimatic behavior contrast. See in Fig. 20.4 the typology distribution of urban grids in the MMA, and in Fig. 20.5 the grid location and features assessed.

For this study, urban neighborhoods with similar house types were chosen to prevent the incidence of the built-up characteristics (see Fig. 20.6). The three neighborhoods are forested and have equal low-density social housing typologies (i.e., second magnitude street trees, 16 and 20 m street widths, 3 m house height, albedos of 0.2 in roofs, 0.3 in facades, 0.5 in streets). Moreover, according to the LCZ classification the three study sites correspond to the open-low-rise class with scattered tree subclass (LCZ 6b). In the urban canyons selected the microclimatic data were obtained from measurement campaign. A detailed description and discussion of the methodology procedure can be found in the study of Sosa et al. (2017, 2018).

For the purpose of the study, ENVI-met software was used. ENVI-met is a 3D grid-based computational fluid dynamics (CFD) tool for the simulation of surface-plant-air interactions based on the fundamental laws of fluid dynamics and

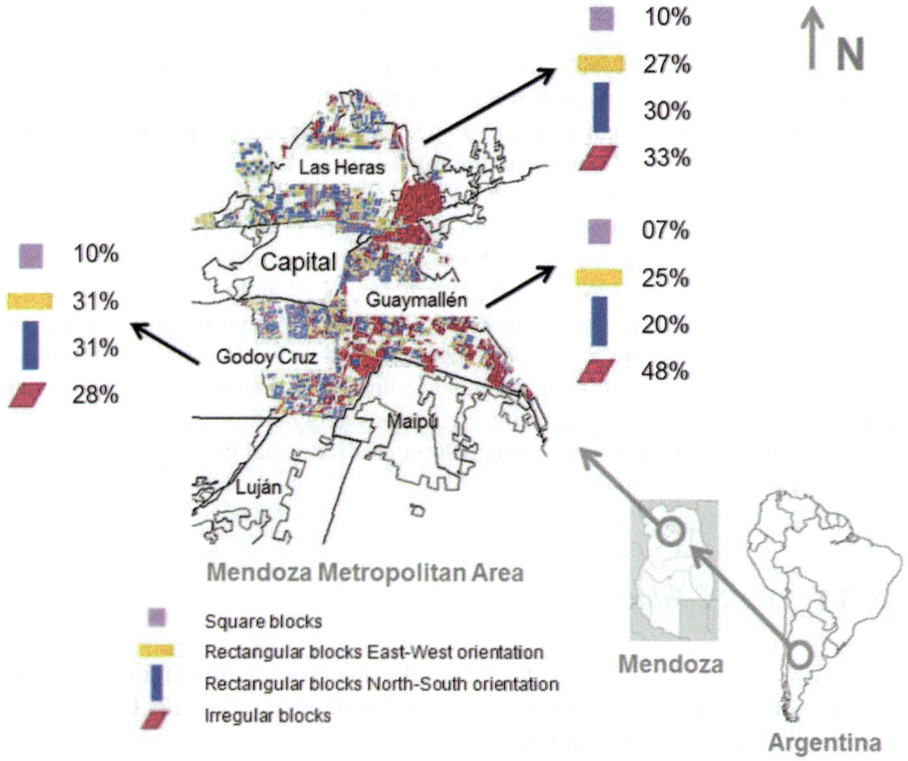

Fig. 20.4 Type and quantity of urban blocks in each analyzed district of the MMA

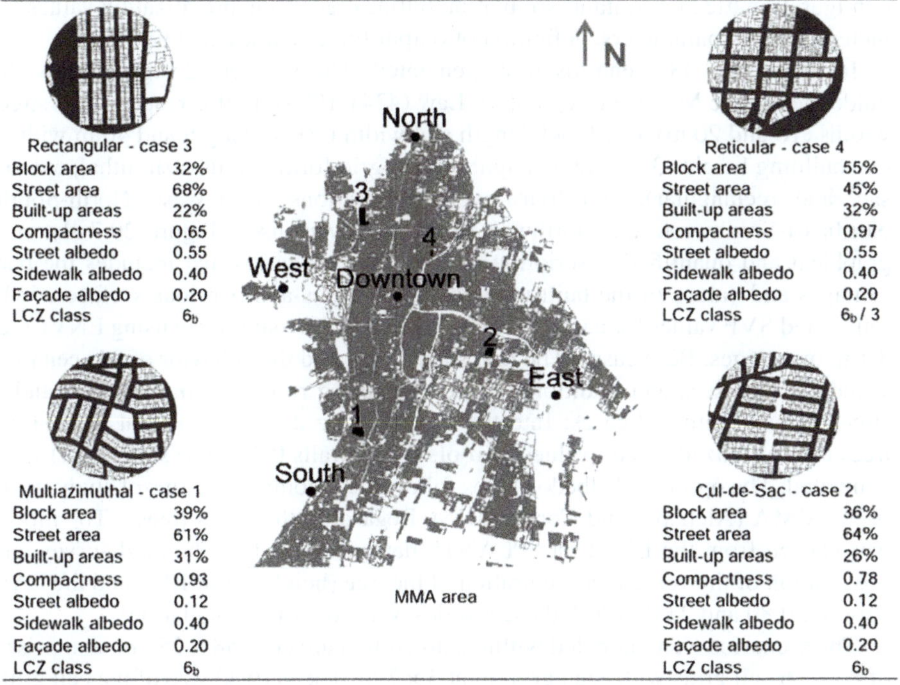

Fig. 20.5 Indicator values of urban grids considered and their location within the MMA

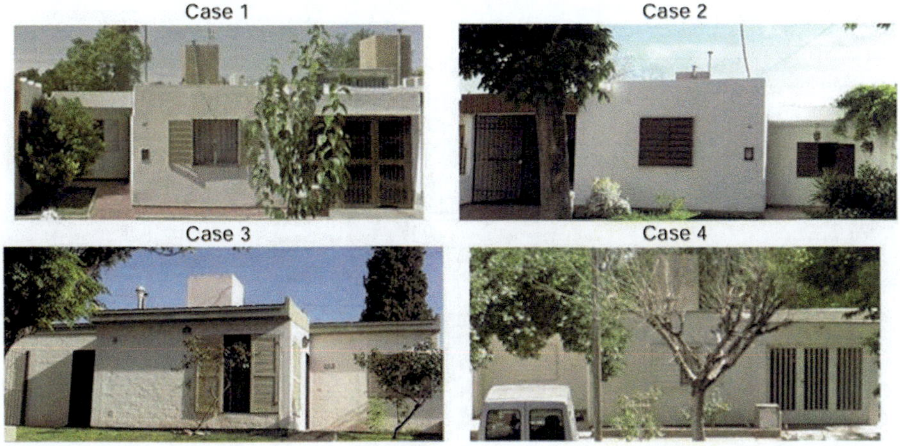

Fig. 20.6 Types of houses in each urban neighborhood

thermodynamics. The use of this software is widely validated by the scientific community (Sun et al. 2017; Roth and Lim 2017; Qaid et al. 2016). ENVI-met required two main input files: (1) an area input file, where the building layout, vegetation, soil type, receptors, and project location parameters are defined, and (2) a configuration file that contains simulation settings regarding initialization values for meteorological parameters, definition of output folder names, and timings.

In this study, 48 scenarios were generated. The scenario designs follow the guidelines of the Mendoza Regulatory Law (4341/1978). In this respect, two street widths (16 and 20 m), one block length and width (100 m length and 40 m width), one building height (3 m), three neighborhood grid forms (multi-azimuthal, cul-de-sac, and rectangular), and four layout orientations (East-West, North-South, Northeast-Southwest, and Northwest-Southeast) were used. Figure 20.7 shows a graphical and quantified description of the urban scenarios that includes the percentages and values of the built-up areas, impervious and pervious surfaces, *H*/*W* ratios, and SVF values for each case. The scenarios were simulated using ENVI-met 3.1 in two stages. Base cases: The first stage simulated the behavior of 24 scenarios without street trees and traditional albedo values (α roofs 0.2, α walls 0.3, and α streets 0.5). Optimized cases: Into the second stage 24 scenarios that have street trees and optimized albedo values (α roofs 0.7, α walls 0.3, and α streets 0.5) were simulated. The optimized albedo values follow the recommendations of study made in the MMA (Alchapar and Correa 2015). Regarding the street trees, "Tb" model was chosen from the ENVI-met PLANTS database based on the field survey and observation of the vertical configuration of the tree (height set in 10 m and the LAD between 0.80 and 2.00 m^2/m^3; these profiles were input for calculation).

The scenarios were modeled within a horizontal area of 258 × 258 m, comprised of 86 × 86 cell grids with each grid set at 3 × 3 m. The vertical size of the cell grids was 3 m for each with a total number of 30 (i.e., the spatial height of each model is 90 m) (Morakinyo et al. 2017; Kong et al. 2016).

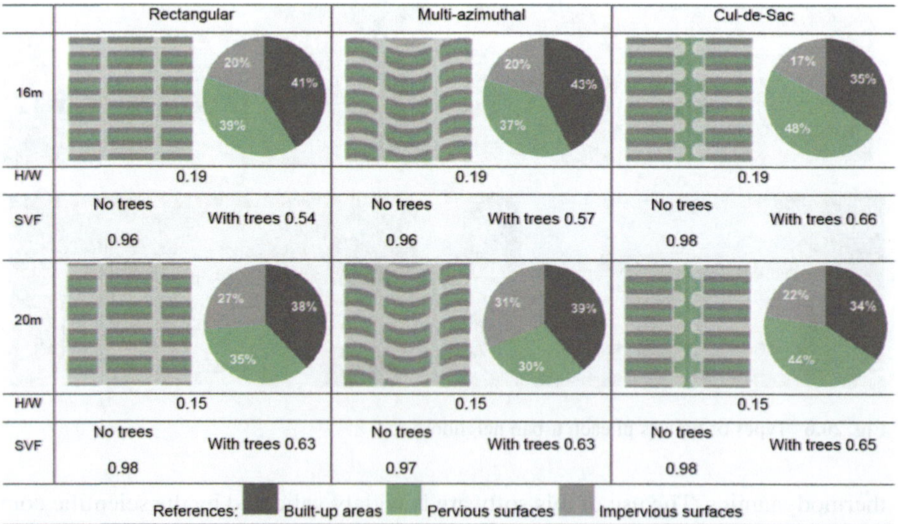

Fig. 20.7 Urban scenario characteristics

The accuracy of the ENVI-met model was assessed comparing the series of air temperature measurements, with the corresponding simulated values. In addition, the SVF values were considered regarding the adjustment of the real study case values and the ones generated by the ENVI-met model. The statistical evaluation of the model was performed using a set of recommended indices that describe the difference between observation and prediction. The indices are (1) the root mean square error, systematic and unsystematic (RMSE, RSMEs, RSMEu); (2) the mean absolute error (MAE); and (3) the mean bias error (MBE). The models can be considered as a reliable tool for microclimate simulation, taking into account statistical index values accepted in previous ENVI-met evaluation studies (Chow and Brazel 2012; Tsoka et al. 2017). The adjustment values of the three models are shown in Fig. 20.8. Furthermore, the maximum, minimum, and average air temperatures and the SVF values are presented and contrasted.

Based on the study aims, we estimated the demand of auxiliary energy (Qaux) for cooling the houses in the 48 scenarios analyzed during summertime. The energy balance equations used give a first approximation of the impact that the outdoor microclimate has on the indoor energy consumption; see Eq. (4) (Sosa et al. 2017). For this diagnosis phase (predesign stage) PREDISE freeware was used. It was developed by the INENCO research group, the Science Department at the National University of Salta, Argentina (Hernández, 2002); (download link: http://www.unsa.edu.ar/alejo/predise/).

PREDISE estimated the average indoor temperature and the energy fluxes that the total volume of a building exchanges with the environment through all its exposed surfaces in a steady state. PREDISE estimated the Qaux from the following energy balance equation:

$$\text{CTU} * \left(T^{\circ}\text{ext} - T^{\circ}\text{ int}\right) - Q\text{solar} - Q\text{gen} = Q\text{aux} \tag{20.4}$$

where CTU is the heat load unit, $T°$int is the indoor average temperature set in 25 °C as it is the temperature suggested by the Argentinean Energy Department, $T°$ext is the exterior average temperature obtained from the ENVI-met simulations, Qsolar is the direct solar gains throughout the windows, and Qgen is the internal heat gains. PREDISE requires a geographical, meteorological, building, and internal heat gain database to function. Geographical data identifies the study site (MMA), meteorological data were obtained from the ENVI-met simulation results, building data includes the material characteristics of all the exposed surfaces (roof, walls, windows, and doors), and the data of internal heat gains reflects the average consumption of a four-person household (one man, one woman, two children) with their basic electrical and natural gas appliances.

Rectangular

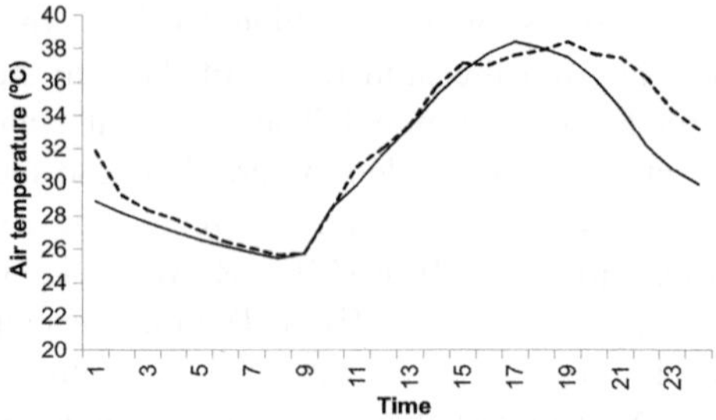

T° maximum (°C)	38.4	38.4*
T° minimum (°C)	25.7	25.5*
T° average (°C)	32.3	31.3*
SVF	0.69	0.62*
MAE	0.5	
RSME	1.7	
RSMEs	1.0	
RSMEu	2.4	
MBE	-1.0	
R^2	0.91	

Multi-azimuthal

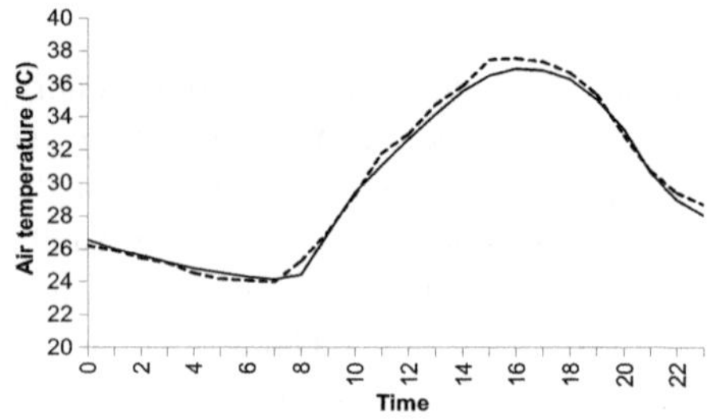

T° maximum (°C)	36.9	37.5*
T° minimum (°C)	24.1	24.0*
T° average (°C)	29.9	30.1*
SVF	0.61	0.61*
MAE	0.6	
RSME	0.4	
RSMEs	0.3	
RSMEu	0.7	
MBE	-0.2	
R2	0.99	

Cul-de-Sac

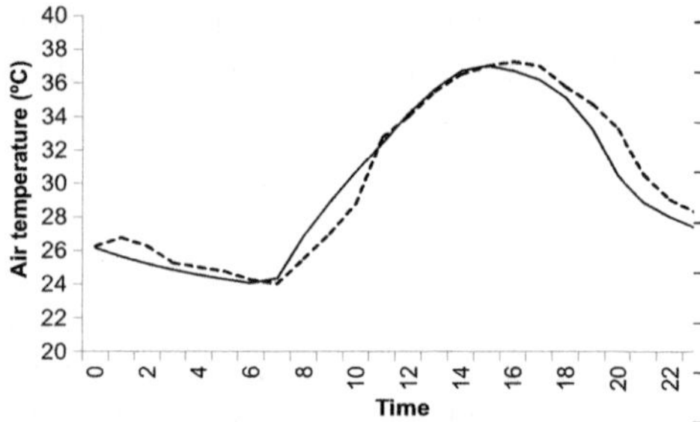

T° maximum (°C)	37.3	37.0*
T° minimum (°C)	24.0	24.1*
T° average (°C)	30.2	29.9*
SVF	0.52	0.53*
MAE	0.5	
RSME	1.1	
RSMEs	0.3	
RSMEu	0.2	
MBE	-0.3	
R2	0.95	

Fig. 20.8 Adjustments of models. *Monitored values

20.3.2 Results

The results of the CFD simulations of the 48 scenarios (24 base cases and 24 optimized cases) are presented in Fig. 20.9. The figure is color coded from light blue (cool) to red (warm) regarding the air temperature values and the scenarios are named as "opti." for the optimized cases and "base" for the base cases.

After analyzing the thermal behaviors, it can be noticed that:

Scenarios			N-S			NO-SE			E-O			NE-SO		
			Opti.	Base	Δ	Opti.	Base	Δ	Opti.	Base	Δ	Opti.	Base	Δ
Rectangular	*16 m*	*T° maximum (°C)*	37,5	42,0	-4,5	37,2	40,7	-3,5	37,7	42,2	-4,5	37,8	42,5	-4,7
		T° minimum (°C)	25,1	27,6	-2,4	25,8	28,1	-2,3	25,3	27,8	-2,5	25,8	27,8	-2,1
		T° average (°C)	30,6	33,9	-3,3	30,7	33,5	-2,9	31,0	34,4	-3,4	30,9	34,3	-3,4
	20 m	*T° maximum (°C)*	38,1	42,1	-4.0	37,7	40,9	-3,1	38,3	42,3	-3,9	38,5	42,6	-4,1
		T° minimum (°C)	25,5	27,6	-2,1	26,2	28,1	-1,9	25,7	27,9	-2,2	26,0	27,9	-1,9
		T° average (°C)	31,1	34,0	-3.0	31,1	33,6	-2,5	31,5	34,4	-2,9	31,4	34,4	-3.0
Multi-azimuthal	*16 m*	*T° maximum (°C)*	37,7	41,5	-3,8	37,5	40,6	-3,1	37,7	41,8	-4.0	38,0	42,1	-4,1
		T° minimum (°C)	25,2	27,6	-2,3	25,8	28,0	-2,2	25,4	27,8	-2,3	25,9	27,7	-1,8
		T° average (°C)	30,8	33,7	-2,9	30,9	33,5	-2,6	31,0	34,1	-3,1	31,1	34,1	-3.0
	20 m	*T° maximum (°C)*	38,0	41,7	-3,7	37,8	40,8	-2,9	38,1	41,9	-3,8	38,4	42,3	-3,9
		T° minimum (°C)	25,6	27,8	-2,2	26,2	28,1	-1,9	25,8	28,0	-2,2	26,1	27,9	-1,9
		T° average (°C)	31,1	33,9	-2,8	31,3	33,6	-2,4	31,4	34,3	-2,9	31,5	34,3	-2,8
Cul-de-Sac	*16 m*	*T° maximum (°C)*	38,3	41,9	-3,7	38,0	40,4	-2,4	38,3	41,8	-3,5	38,8	42,2	-3,4
		T° minimum (°C)	24,8	27,2	-2,4	26,1	27,9	-1,9	25,1	27,4	-2,3	25,5	27,5	-1,9
		T° average (°C)	30,7	33,7	-3.0	31,3	33,4	-2,1	31,1	34,1	-3.0	31,5	34,1	-2,6
	20 m	*T° maximum (°C)*	38,3	41,8	-3,5	38,0	40,4	-2,4	38,4	42,0	-3,6	38,8	42,4	-3,6
		T° minimum (°C)	25,3	27,4	-2,1	26,3	28,0	-1,7	25,5	27,6	-2,1	26,0	27,7	-1,6
		T° average (°C)	30,9	33,8	-2,9	31,4	33,4	-2.0	31,3	34,2	-2,9	31,6	34,3	-2,6

Fig. 20.9 Maximum, minimum, and average air temperature values of the 48 scenarios

- The optimized cases are the coolest in all orientations and street widths. The maximum differences compared with the base cases reach 4.7 °C in the maximum, 2.5 °C in the minimum, and 3.4 °C in the average air temperatures.
- The NO-SE and N-S are the coolest grid orientations. In the NO-SE the maximum air temperatures are the coolest, in the N-S the minimum air temperatures are the coolest, and in both orientations the average air temperatures are the coolest.
- The 16 m street width is the coolest in any street orientation.
- The rectangular neighborhood grid is the coolest; 63% of the scenarios have maximum temperatures below 38 °C, 50% minimum temperatures below 26 °C, and 31% average temperatures lower than 31 °C.
- Regarding the multi-azimuthal grid, 56% of the scenarios have maximum temperatures below 38 °C, 44% minimum temperatures below 26 °C, and 25% average temperatures below 31 °C.
- Finally, in the cul-de-sac grid, 13% of the scenarios have maximum temperatures below 38 °C, 38% minimum temperatures below 26 °C, and 6% average temperatures below 31 °C.

The influence of the grid form on the energy consumption for cooling the houses was quantified. Figure 20.10 shows the differences of energy consumption, net and percentage, between the optimized and base case scenarios. The table is divided according to three urban characteristics: neighborhood grid, street width, and orientation. *Q*aux values were estimated using PREDISE to maintain the indoor temperature at 25 °C.

Regardless of the optimization strategies applied—albedo modification and urban forestry—the results of the rectangular grid are the following: The energy saving varies between 32% and 25%. The most efficient scenario is the 16 m street width and N-S orientation (179 kWh–2.2 kWh/m^2 for an 80 m^2 single-story attached house). The scenario with the greatest possibility of optimization in terms of its base condition is the 20 m street width, oriented E-O (Δ89 kWh of energy saving). By analyzing the values of the multi-azimuthal grid we can see that the energy saving varies between 30% and 27%. The most efficient scenario is the 16 m street width and N-S orientation (184 kWh–2.3 kWh/m^2 for an 80 m^2 single-story attached house). The scenario with the greatest possibility of optimization in terms of its base condition is the 16 m street width, oriented E-O (Δ81 kWh of energy saving). By analyzing the values of the cul-de-sac grid we can see that the energy saving varies between 30% and 21%. The most efficient scenario is the 16 m street width and N-S orientation (182 kWh–2.2 kWh/m^2 for an 80 m^2 single-story attached house). The scenario with the greatest possibility of optimization in terms of its base condition is the 16 m street width, oriented N-S and E-O (Δ78 kWh of energy saving).

Figure 20.11 shows and compares the energy effectiveness of the optimized scenarios versus the base cases for each urban grid, street width, and orientation. The figure reveals that the grid orientation has greater impact on the base cases (i.e., more difference of *Q*aux between orientations). Regarding the orientation, the N-S grids are the most effective for the optimized cases and the NO-SE for the base cases.

Street width	Orientation	Rectangular grid				Multi-azimuthal grid				Cul-de-Sac grid			
		Optimized (kWh)	Base (kWh)	Δ (kWh)	%	Optimized (kWh)	Base (kWh)	Δ (kWh)	%	Optimized (kWh)	Base (kWh)	Δ (kWh)	%
16m	N-S	179	265	-86	32	184	260	-76	29	182	260	-78	30
	NO-SE	182	255	-73	29	187	255	-68	27	197	252	-55	22
	E-O	189	278	-89	32	189	270	-81	30	192	270	-78	29
	NE-SO	187	275	-88	32	192	270	-78	29	202	270	-68	25
20m	N-S	192	268	-76	28	192	265	-73	28	187	262	-75	29
	NO-SE	192	257	-65	25	192	265	-73	28	200	252	-52	21
	E-O	202	278	-76	27	200	275	-75	27	197	273	-76	28
	NE-SO	200	278	-78	28	202	275	-73	27	205	275	-70	25

Fig. 20.10 Electricity consumption for cooling the housing in both urban scenarios

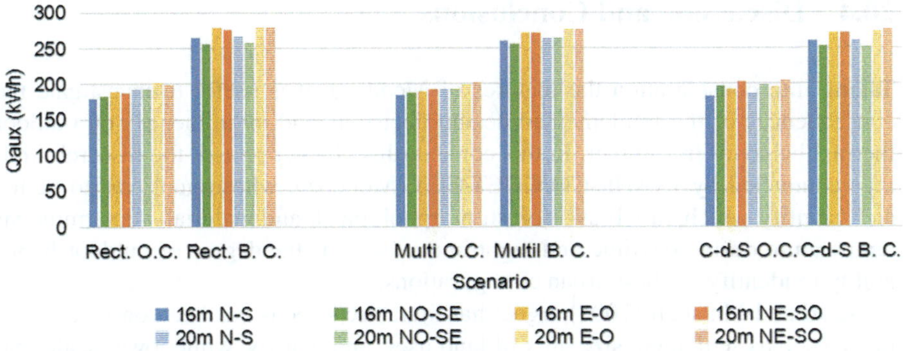

Fig. 20.11 Electricity consumption for cooling the housing in both urban scenarios

20.3.3 Design Strategies at Neighborhood Scale

In order to improve the outdoor thermal behavior of low-density residential neighborhoods, a series of recommendations and strategies were presented, taking into account the results obtained. The urban design variables addressed were the following:

- Built-up areas: The higher the built-up areas, the lower the maximum air temperatures. The multi-azimuthal grid form is recommended as it is the one that has more built-up areas (43%) when comparing the same urban area to the other two grid forms.
- Impervious surfaces: The lower the impervious surfaces, the lower the air temperatures. The cul-de-sac grid form is recommended as it is the one that has lower impervious surfaces (17% versus 20%) when compared to the other two grid forms.
- Layout orientation: The N-S and NO-SE street orientations have the lowest air temperatures. The N-S and E-O orientations behave similarly in terms of minimum air temperatures, regardless of the neighborhood grid.
- SVF: For higher sky view factor values, the minimum air temperatures are lower. The recommended value of this indicator is in the order of 0.50, because this allows us to control the incoming solar radiation and to improve the night release.
- Albedos: Optimized albedo values (α roofs 0.7, α walls 0.3, and α streets 0.5) are more effective to reduce maximum air temperatures. This strategy is recommended to use on horizontal surfaces (roofs, pavements, and sidewalks) in order to obtain a greater impact on outdoor thermal behavior.
- Trees: Street trees planted at regular distances with a homogeneous crown development make it easier to regulate solar exposure during the day and allow cooling during the night. Second magnitude species are recommended for low-density urban areas.

20.4 Discussion and Conclusions

Taking into consideration the features of Mendoza, it is useful to investigate the consequences of the implemented design decisions and its influences on outdoor thermal behavior in summer. In this context, this chapter revises the research findings conducted by INAHE-CONICET-CCT, Mendoza, whose goal was to define and quantify which variables—forest, morphological, and material—determine air temperature during daytime and nighttime as well as the degree of outdoor habitability to identify the best urban configurations.

Stead and Marshall (2001) argue that literature's focus has been on macro features, such as settlement size, mix of land uses, and density, while lower scale features, such as neighborhood design, may also be relevant but have been overlooked. Ko and Radke (2014) have demonstrated that urban forms have a statistically significant impact on energy saving for building cooling. They consider that an optimized community layout and vegetation planting will have a significant impact on macroscale environments in the long term. In the MMA the incorporation of street trees and optimized albedo values reaches an average air temperature reduction of 3.4 °C. For this reason, the incorporation of suitable envelope materials and street trees into the neighborhood design stage is imperious.

The optimization of a single building orientation is estimated to have minimal reductions on annual energy use and costs (Silva et al. 2017). However, for a whole community or urban area, important savings may be achieved by improving building orientation (Hemsath 2016). In this respect, this study suggests two layout orientations (NO-SE and N-S) that demonstrate to be more energy efficient in summertime for the Southern Hemisphere.

Urban form and land-use patterns significantly influence a city's thermal behavior and energy consumption. Regarding the thermal behavior, results show that with an adequate layout orientation, tree selection, and an improvement in the material albedo, a traditional social house built without any bioclimatic design strategy would consume at least 21% less auxiliary energy to achieve an indoor comfort temperature (25 °C) during summertime. Moreover, the study detected that with every 1 °C average air temperature increase the auxiliary energy consumption will increase 26 kWh (0.3 kWh/m^2). Likewise, within the three urban grids tested in the "optimized scenarios" there are energy consumption differences between 1% and 8% when comparing orientations and street widths (i.e., the 16 m street width orientated NO-SE in the rectangular grids consumes 8% less energy than the same scenario in the cul-de-sac grid). The observation above highlights the importance that design and planning decisions have to build social housing settlements in oasis cities with arid climates that are more responsible towards the environment and more efficient in terms of energy use.

References

Alchapar, N., & Correa, E. (2015). Comparison the performance of different facade materials for reducing building cooling needs. In P. Torgal (Ed.), *Eco-efficient materials for mitigating building cooling needs* (pp. 155–194). Cambridge: Woodhead Publishing.

Ali-Toudert, F., & Mayer, H. (2007). Effects of asymmetry, galleries, overhanging façades and vegetation on thermal comfort in urban street canyons. *Solar Energy, 81*(6), 742–754. https://doi.org/10.1016/j.solener.2006.10.007.

Bakarman, M. A., & Chang, J. D. (2015). The influence of height/width ratio on urban heat island in hot-arid climates. *Procedia Engineering, 118*, 101–108, ISSN 1877-7058. https://doi.org/10.1016/j.proeng.2015.08.408.

Bruse, M. (2004). ENVI-met website. Retrieved from http://www.envi-met.com/

Cantón, M. A., de Rosa, C., & Kasperidus, H. (2003). Sustentabilidad del bosque urbano en el área metropolitana de la ciudad de Mendoza: Análisis y diagnóstico de la condición de las arboledas. *Avances en Energías Renovables y Medio Ambiente, 7*(1), 29–34.

Chow, W. T., & Brazel, A. J. (2012). Assessing xeriscaping as a sustainable heat island mitigation approach for a desert city. *Building and Environment, 47*, 170–181. https://doi.org/10.1016/j.buildenv.2011.07.027.

Correa, E., de Rosa, C., & Lesino, G. (2008). Urban heat island effect on heating and cooling degree day's distribution in Mendoza's metropolitan area. Environmental costs. In *Proceedings of the 1st International Conference on Solar Heating, Cooling and Buildings (EUROSUN 2008)* (Vol. 2, pp. 951–958). Red Hook, NY: Curran Associates, Inc..

Correa, E., Ruiz, M. A., & Cantón, M. A. (2010). Urban forest structure and thermal comfort in oasis cities of arid zones. *Ambiente construído, 10*, 119. https://doi.org/10.1590/S1678-86212010000400009.

Correa, E., Ruiz, M. A., Cantón, M. A., & Lesino, G. (2012). Thermal comfort in forested urban canyons of low building density. An assessment for the city of Mendoza, Argentina. *Building and Environment, 58*, 219–230. https://doi.org/10.1016/j.buildenv.2012.06.007.

Gaitani, N., Mihalakakou, G., & Santamouris, M. (2007). On the use of bioclimatic architecture principles in order to improve thermal comfort conditions in outdoor spaces. *Building and Environment, 42*, 317–324. https://doi.org/10.1016/j.buildenv.2005.08.018.

González Loyarte, M. M., Menenti, M., & Diblasi, M. A. (2009). Mapa bioclimatico para las travesías de Mendoza (Argentina) basado en la fenología foliar. *Revista de la Facultad de Ciencias Agrarias, 41*(1), 105–122. Retrieved from http://bdigital.uncu.edu.ar/3121.

Hemsath, T. L. (2016). Housing orientation's effect on energy use in suburban developments. *Energy and Buildings, 122*, 98–106. https://doi.org/10.1016/j.enbuild.2016.04.018.

Hernández, A. (2002). PREDISE. Un novedoso y práctico programa de evaluación térmica de edificios. *Avances en energías renovables y medio ambiente, 6*(2), 61–66. Retrieved from www.unsa.edu.ar/alejo/predise/#link1.

IEA, International Energy Agency. (2017). *WEO-2017 special report: Energy access outlook*. Retrieved from www.iea.org/publications/freepublications/publication/WEO2017SpecialReport_EnergyAccessOutlook.pdf

Jusuf, S. K., Wong, N. H., & Hagen, E. (2007). The influence of land use on the urban heat island in Singapore. *Habitat Internacional, 31*(2), 232–242. https://doi.org/10.1016/j.habitatint.2007.02.006.

Ko, Y., & Radke, J. D. (2014). The effect of urban form and residential cooling energy use in Sacramento, California. *Environment and Planning B: Planning and Design, 41*, 573–593. https://doi.org/10.1068/b12038p.

Kong, F., Sun, C., Liu, F., Yin, H., Jiang, F., Pu, Y., Cavan, G., Skelhorn, C., Middel, A., & Dronova, I. (2016). Energy saving potential of fragmented green spaces due to their temperature regulating ecosystem services in the summer. *Applied Energy, 183*, 1428–1440. https://doi.org/10.1016/j.apenergy.2016.09.070.

Kottek, M., Grieser, J., Beck, C., Rudolf, B., & Rubel, F. (2006). World map of the Koppen-Geiger climate classification updated. *Meteorologische Zeitschrift, 15*(3), 259–263. https://doi.org/10.1127/0941-2948/2006/0130.

Lee, S., & Lee, B. (2014). The influence of urban form on GHG emissions in the U.S. household sector. *Energy Policy, 68*, 534–549. https://doi.org/10.1016/j.enpol.2014.01.024.

Li, C., Song, Y., & Kaza, N. (2018). Urban form and household electricity consumption: A multilevel study. *Energy and Buildings, 158*, 181–193. https://doi.org/10.1016/j.enbuild.2017.10.007.

Lin, T. (2009). Thermal perception, adaptation and attendance in a public square in hot and humid regions. *Journal of Building and Environment, 44*(10), 2017–2026. https://doi.org/10.1016/j.buildenv.2009.02.004.

McPherson, E. G. (1994). Cooling urban heat islands with sustainable landscapes. In R. H. Platt et al. (Eds.), *The ecological city: Preserving and restoring urban biodiversity* (pp. 151–171). Amherst, MA: University of Massachusetts Press.

Middel, A., Häb, K., Brazel, A. J., Martin, C. A., & Guhathakurta, S. (2014). Impact of urban form and design on mid-afternoon microclimate in Phoenix local climate zones. *Landscape and Urban Planning, 122*, 16–28. https://doi.org/10.1016/j.landurbplan.2013.11.004.

Middel, A., Chhetri, N., & Quay, R. (2015). Urban forestry and cool roofs: Assessment of heat mitigation strategies in Phoenix residential neighborhoods. *Urban Forestry and Urban Greening, 14*(1), 178–186. https://doi.org/10.1016/j.ufug.2014.09.010.

Morakinyo, T., Kong, L., Lau, K., Yuan, C., & Ng, E. (2017). A study on the impact of shadow-cast and tree species on in-canyon and neighborhood's thermal comfort. *Building and Environment, 115*, 1–17. https://doi.org/10.1016/j.buildenv.2017.01.005.

NRDC, Raimi + Associates and the Natural Resources Defense Council. (2011). *A citizen's guide to LEED for neighborhood development: How to tell if development is smart and green*. Retrieved from https://www.nrdc.org/sites/default/files/citizens_guide_LEED-ND.pdf

Offerle, B., Grimmond, S., Fortuniak, K., & Pawlak, W. (2006). Intraurban differences of surface energy fluxes in a central European city. *American Meteorological Society, 45*, 125–136. https://doi.org/10.1175/JAM2319.1.

Oke, T. R. (1982). The energetic basis of the urban heat island. *Quarterly Journal Royal Meteorological Society, 108*(45), 1–24. https://doi.org/10.1002/qj.49710845502.

Papanastasiou, D. K., Fidaros, D., Bartzanas, T., & Kittas, C. (2013). Impact of urban heat island development on buildings' energy consumption. *Fresenius Environmental Bulletin, 22*(7A), 2087–2092.

Qaid, A., Lamit, H. B., Ossen, D. R., & Shahminan, R. N. R. (2016). Urban heat island and thermal comfort conditions at micro-climate scale in a tropical planned city. *Energy Buildings, 133*, 577–595. https://doi.org/10.1016/j.enbuild.2016.10.006.

R Development Core Team (2011) R: A language and environment for statistical computing. R Foundation for Statistical Computing, Vienna. http://www.R-project.org.

Ratti, C., Di Sabatino, S., & Britter, R. (2006). *Theoretical and Applied Climatology, 84*, 77–90. https://doi.org/10.1007/s00704-005-0146-z.

Roth, M., & Lim, V. H. (2017). Evaluation of canopy-layer air and mean radiant temperature simulations by a microclimate model over a tropical residential neighbourhood. *Building and Environment, 112*, 177–189. https://doi.org/10.1016/j.buildenv.2016.11.026.

Ruiz, M. A., & Correa, E. N. (2014). Suitability of different comfort indices for the prediction of thermal conditions in tree-covered outdoor spaces in arid cities. *Theoretical and Applied Climatology, 122*, 69. https://doi.org/10.1007/s00704-014-1279-8.

Ruiz, M. A., & Correa, E. N. (2015). Adaptive model for outdoor thermal comfort assessment in an Oasis city of arid climate. *Building and Environment, 85*, 40–51. https://doi.org/10.1016/j.buildenv.2014.11.018.

Ruiz, M. A., Sosa, M. B., Correa, E., & Cantón, M. A. (2015). Suitable configurations for forested urban canyons to mitigate the UHI in the city of Mendoza, Argentina. *Urban Climate, 14*, 197. https://doi.org/10.1016/j.uclim.2015.05.005.

Ruiz, M. A., Sosa, M. B., Correa, E. N., & Cantón, M. A. (2017). Design tool to improve daytime thermal comfort and nighttime cooling of urban canyons. *Landscape and Urban Planning., 167*, 249–256, ISSN 0169-2046. https://doi.org/10.1016/j.landurbplan.2017.07.002.

Sanusi, R., Johnstone, D., May, P., & Livesley, S. (2016). Street orientation and side of the street greatly influence the microclimatic benefits street trees can provide in summer. *Journal of Environmental Quality, 45*, 167–174. https://doi.org/10.2134/jeq2015.01.0039.

Shashua-Bar, L., & Hoffman, M. E. (2003). Geometry and orientation aspects in passive cooling of canyon streets with trees. *Energy and Buildings, 35*, 61–68. https://doi.org/10.1016/s0378-7788(02)00080-4.

Shashua-Bar, L., & Hoffman, M. E. (2004). Quantitative evaluation of passive cooling of the UCL microclimate in hot regions in summer, case study: Urban streets and courtyards with trees. *Building and Environment, 39*, 1087–1099. https://doi.org/10.1016/j.buildenv.2003.11.007.

Shishegar, N. (2013). Street design and urban microclimate: Analyzing the effects of street geometry and orientation on airflow and solar access in urban canyons. *Journal of Clean Energy Technologies, 1*, 52–56. https://doi.org/10.7763/JOCET.2013.V1.13.

Silva, M., Oliveira, V., & Leal, V. (2017). Urban form and energy demand. *Journal of Planning Literature, 32*, 346. https://doi.org/10.1177/0885412217706900.

Sosa, M. B., Correa, E. N., & Cantón, M. A. (2017). Urban grid forms as a strategy for reducing heat island effects in arid cities. *Sustainable Cities and Society, 32*, 547–556. https://doi.org/10.1016/j.scs.2017.05.003.

Sosa, M. B., Correa, E. N., & Cantón, M. A. (2018). Neighborhood designs for low-density social housing energy efficiency: Case study of an arid city in Argentina. *Energy and Buildings, 168*, 137–146, ISSN: 0378-7788. https://doi.org/10.1016/j.enbuild.2018.03.006.

Stead, D., & Marshall, S. (2001). The relationships between urban form and travel patterns. An international review and evaluation. *European Journal of Transport and Infrastructure Research, 1*(2). https://doi.org/10.18757/ejtir.2001.1.2.3497.

Stone, B., Hess, J., & Frumkin, H. (2010). Urban form and extreme heat events: Are sprawling cities more vulnerable to climate change than compact cities? *Environmental Health Perspectives, 10*, 1425–1428. https://doi.org/10.1289/ehp.0901879.

Sun, S., Xu, X., Lao, Z., Liu, W., Li, Z., Higueras García, E., He, L., & Zhu, J. (2017). Evaluating the impact of urban green space and landscape design parameters on thermal comfort in hot summer by numerical simulation. *Building and Environment, 123*, 277–288. https://doi.org/10.1016/j.buildenv.2017.07.010.

Tsoka, S., Tsikaloudaki, K., & Theodosiou, T. (2017). Urban space's morphology and microclimatic analysis: A study for a typical urban district in the Mediterranean city of Thessaloniki, Greece. *Energy and Buildings, 156*, 96–108. https://doi.org/10.1016/j.enbuild.2017.09.066.

UN-Habitat Working Paper. (2013). The relevance of street patterns and public space in urban areas.

Zhao, C., Fu, G., Liu, X., & Fu, F. (2011). Urban planning indicators, morphology and climate indicators: A case study for a north-south transect of Beijing, China. *Building and Environment, 46*, 1174–1183. https://doi.org/10.1016/j.buildenv.2010.12.009.

Chapter 21
Human Biometeorological Models: Existing and Future Reflections for Lisbon

André Santos Nouri and Andreas Matzarakis

21.1 Introduction

When approaching the case of Lisbon, it has already been recognised that additional action is required to adapt to future climatic conditions that are already being witnessed within its urban environment (Andrade 2003; Lopes 2003; Alcoforado and Matzarakis 2010; Lopes et al. 2013; Alcoforado et al. 2014). Furthermore, and as previously identified by earlier studies (e.g. Alcoforado and Vieira 2004; Alcoforado et al. 2009a), the scientific community has recognised a growing need for studies that concretely focus upon Mediterranean climates with hot-dry summers and local approaches to thermo-physiological thresholds. Including in the Lisbon case, this weakness frequently relays to a lack of local scale design guidelines and precedents which could otherwise inform practices for urban planning and design during periods of accentuated urban thermal stimuli (Nouri 2018; Nouri et al. 2018b). Within this chapter, and based upon the application of human biometeorological models within Lisbon, its existing and future thermo-physiological conditions are examined in the context of maturing bottom-up approaches to guide local thermal adaptation efforts. Divided into two predominant sections, the chapter deliberates upon (1) the existing and future thermal conditions within Lisbon based upon more encompassing top-down climatic assessments, and subsequently (2) how the application of

A. S. Nouri (✉)
Department of Interior Architecture and Environmental Design, Faculty of Art, Design and Architecture, Bilkent University, Bilkent, Turkey
e-mail: andre.nouri@bilkent.edu.tr

A. Matzarakis
Research Centre Human Biometeorology, German Meteorological Service, Freiburg, Germany

Environmental Meteorology, Faculty of Environment and Natural Resources, Albert-Ludwigs-University, Freiburg, Germany
e-mail: andreas.matzarakis@dwd.de

M. Palme, A. Salvati (eds.), *Urban Microclimate Modelling for Comfort and Energy Studies*, https://doi.org/10.1007/978-3-030-65421-4_21

biometeorological models and thermo-physiological indices can shed better understandings and assessments of local wholesome thermal comfort conditions in an era moreover prone to climate change impacts. The resulting perspective, in part, suggests a paradigm shift whereby innovative microclimatic investigations can be utilised to combat both climatic uncertainty and downscaling/transposing of climatic know-how into scales that directly influence the human thermoregulatory system. More specifically, implied by this approach are new forms of thinking that consider both extreme weather effects and the less, but constant, effects of meteorological parameters such as radiation fluxes upon the long-term thermal well-being and safety of within Lisbon's public realm.

21.2 Lisbon's Existing and Future Thermal Risk Factors

Determined in the study conducted by Peel et al. (2007), Lisbon witnesses a *K*öppen-*G*eiger (KG) climate classification of '*Csa*', representing a Mediterranean climate with hot and dry summers. Moreover, and according to *G*lobal *C*irculation *M*odel (GCM) output studies conducted by Rubel and Kottek (2010), it is estimated that, by 2100, the entirety of continental Portugal shall witness the same KG classification currently witnessed in Lisbon. As a result, this in turn implies that major northern cities such as Porto shall transit from a classification of '*Csb*' (representing a Mediterranean climate with warm summers) to a classification of '*Csa*'.

Focused specifically upon the '*Csa*' classification, Fig. 21.1a demonstrates the variation of this climate typology between 2001–2025 and 2076–2100. Both are constructed upon temperature and precipitation scenarios extrapolated from the previous *S*pecial *R*eport on *E*missions *S*cenarios (SRES) A1FI scenario, formerly disseminated/utilised by the *I*ntergovernmental *P*anel on *C*limate *C*hange (IPCC 2000). Presently, it is worth noting that based upon intergovernmental sociopolitical consensus during the fifth assessment report, such scenarios were subsequently replaced by the *R*epresentative *C*oncentration *P*athways (RCP) (IPCC 2013). Comparatively, such a scenario typology would now roughly equate to the current $RCP_{8.5}$ scenario, similarly associated to a high-emission future, yielding a temperature increase ranging between 4.0 and 6.1 °C by 2100 (Rogelj et al. 2012). Nevertheless the results presented by Rubel and Kottek (2010) extracted through GCM are still very pertinent in obtaining general top-down understandings of climatic transitional versatility between current and future conditions. Similar methodologies were also later utilised by the authors to investigate very different, yet interrelated, effects of KG variation until the end of the century within the European context (Brugger and Rubel 2013).

In the case of Western Europe, it is possible to verify that this specific KG subclassification shall occupy most of north-western France, leaving the edge of African continent, extending all the way to the southern edge of the UK by 2100. Consequently northern Portuguese cities will also be susceptible to higher temperatures during the summer that are concomitant with those currently found in cities

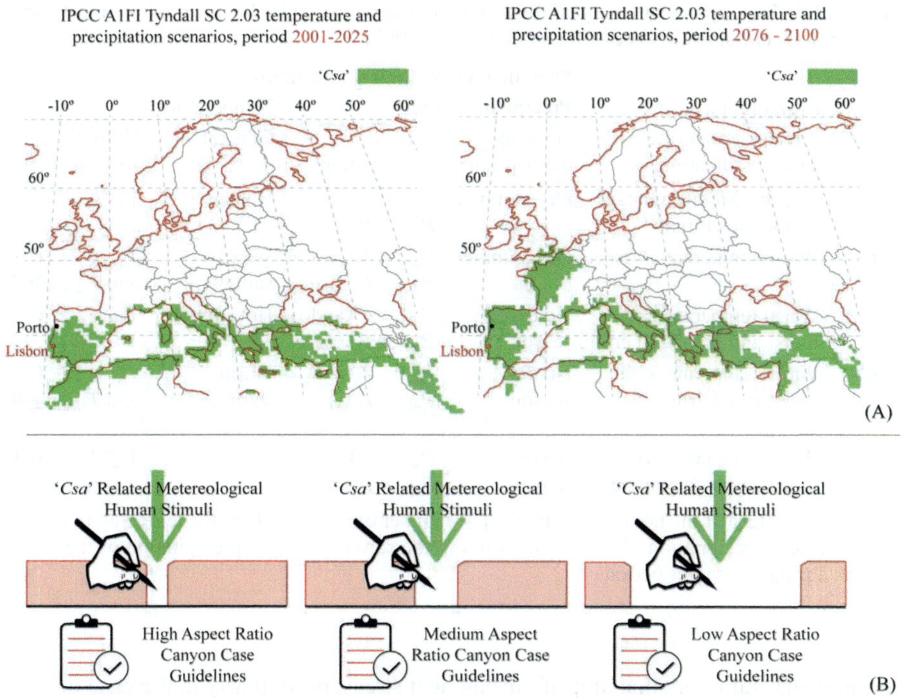

Fig. 21.1 (**a**) Modifications of the '*Csa*' KG sub-classification through GCM outputs between the periods 2001–2025 and 2076–2100. (Source: Adapted from Rubel and Kottek 2010). (**b**) Informing local bottom-up adaptation guidelines based on '*Csa*'-related thermal stimulus

such as Lisbon. It is important to note that the '*Csa*' sub-classification is not the only KG classification that is expected to radically change by 2100. As revealed in detail by Peel et al. (2007), within the Eurasian area, sub-classifications such as '*Cfa*', '*Cfb*' and '*Cfc*' (explained in Table 21.1) will expand to the north-east and replace regions that are currently associated to 'Cold' classifications. These variations are particularly tangible within (1) northern Europe and Scandinavia—including regions of Finland, Sweden and Norway—and (2) Eurasia—including Romania, Ukraine, Belarus, Lithuania and a large portion of western Russia. Finally, it is also key to recognise that beyond the expansion of the 'Temperate' sub-classifications, existing countries with such climates shall also partly transition to 'Arid' subclasses (e.g. eastern Spain, southern Italy, eastern Greece, Cyprus, eastern Romania and a large section of central Turkey). Within the international arena, subsequent studies have also identified the implications of KG on human thermal comfort in numerous countries and specific regions, including China (Yang and Matzarakis 2016) and Australia (Djamila and Yong 2016).

As illustrated in Table 21.1, the differentiation between the 'Temperate' classifications can be divided into two dominant climatic factors: precipitation/aridity and temperature. In the four exemplified sub-classifications, both '*Csa*' and '*Csb*'

Table 21.1 Descriptive summary of the different 'Temperate' sub-classifications of the *K*öppen-*G*eiger (KG) classification according to Peel et al. (2007)

KG	Common label of selected temperate (*C*) sub-classifications	Specific meteorological thresholds			
		Precipitation/aridity		Temperature	
		General description	Climatic specification	General description	Climatic specification
'*Csa*'	Hot-dry Mediterranean/ temperate climate	Dry summer	$P_{sdry} < 40$ and $P_{sdry} < P_{wwet/3}$	Hot summer	$T_{hot} \geq 22$ °C
'*Csb*'	Warm temperate climate	Dry summer	$P_{sdry} < 40$ and $P_{sdry} < P_{wwet/3}$	Warm summer	$T_{hot} \leq 21$ °C and $T_{mon10} \geq 4$
'*Cfa*'	Warm temperate with moist and hot summers	No dry season	$P_{sdry} < 40$ and $P_{sdry} < P_{wwet/3}$	Hot summer	$T_{hot} \geq 22$ °C
'*Cfb*'	Warm temperate with moist and warm summers	No dry season	$P_{sdry} > 40$ and $P_{sdry} > P_{wwet/3}$	Warm summer	$T_{hot} \leq 21$ °C and $T_{mon10} \geq 4$
'*Cfc*'	Warm temperate with moist and cold summers	No dry season	$P_{sdry} > 40$ and $P_{sdry} > P_{wwet/3}$	Cold summer	$1 \leq T_{mon10} < 4$

T_{hot}—temperature of the hottest month, T_{mon10}—number of months where the temperature is above 10 °C, P_{sdry}—precipitation of the driest month in the summer, P_{wwet}—precipitation of the wettest month during the winter period

witness dry summers that amplify urban heat stress, particularly in the case of '*Csa*' which is additionally prone to hot summers (thus colloquially referred to as a hot-dry Mediterranean summer climate). On the other hand, the remaining '*Cfa*', '*Cfb*' and '*Cfc*' climates observe no dry season, which can be beneficial to avoid aspects such as urban droughts and dry spells in hotter sub-classifications. Nevertheless, and particularly in the case of '*Cfa*', the presence of high atmospheric humidity and hot summers also creates a series of obstacles when addressing local thermal comfort thresholds (Ishii et al. 2009; Farnham et al. 2011; Nouri and Matzarakis 2019).

Returning to the case of Lisbon, while it is projected that its sub-classification shall not shift during the twenty-first century, this does not imply that climatic conditions shall remain the same. More specifically, through the use of *R*egional *C*limate *M*odels (RCMs) it was identified by Miranda (2006) that Lisbon could potentially further witness a (1) total of 50 annual days where diurnal T_a surpasses that of 35 °C (i.e. constituting a *V*ery *H*ot *D*ay (VHD)); (2) doubling in duration of days that consecutively surpass 35 °C (i.e. implying a potential span of 20 uninterrupted days); (3) radical upsurge of 'tropical nights', increasing from 20 to a potential 80 annual nights where minimum T_a surpasses 20 °C; and lastly (4) potential escalation of 'summer days', where T_a surpasses 25 °C up to 180 days (representing half of an entire calendar year). It is additionally worth mentioning that these projections were constructed upon the previous $SRES_{A2}$ scenario, equating to '*A very heterogeneous world with an emphasis on family values and local traditions*' (IPCC 2000). Consequently, these projections are not based upon the previous SRES worst-case scenario, suggesting that these aggravations of Lisbon's urban climate could be even direr within the $SRES_{A1FI}$ scenario.

Adjacently to these studies, it has also been made clear by the scientific community that in light of extreme heat events, such as the heatwave in 2003, Western European countries such as Portugal require additional measures to warn, cope and prevent the recurrence of such events (Kovats and Ebi 2006; Matzarakis 2016). As an example of such thermal risk factors upon human morbidity and mortality during such extreme events, Nogueira et al. (2005) identified that in the district of Lisbon during the heatwave of 2003, between the 29th of July and the 13th of August, there (1) were 15 days with a maximum T_a above 32 °C; (2) was a noteworthy consecutive run of 10 days with maximum T_a surpassing 32 °C; and lastly (3) was an estimated mortality rate increase of 37.7% (equating to 1316 excess deaths) in comparison to what would be expected under typical conditions. Such extreme heat events are representative of current conditions, which have, in addition, been argued to be already presenting thermal aggravations associated to climate change impacts in Lisbon (Alcoforado et al. 2005, 2014, 2015; Alcoforado and Matzarakis 2010; Nouri et al. 2018d). These vital studies have thus far, respectively, embraced that increases in heatwaves are 'very likely' throughout the twenty-first century (IPCC 2007), whereby the fifth assessment report moreover stipulated that the influences of climate change upon heatwaves shall be more significant than the impacts upon global average temperatures (IPCC 2013).

21.3 Application of Human Biometeorological Models

Within the aforementioned study conducted by Alcoforado et al. (2015), the importance of using biometeorological models within Lisbon to determine local thermo-physiological indices such as the *P*hysiologically *E*quivalent *T*emperature (PET) (Mayer and Höppe 1987; Höppe 1999) was highlighted. Such a use also extended into the effective subsequent translation into urban interdisciplinary planning efforts and guidelines.

21.3.1 Overview of Thermal Indices and Stress Thresholds

So far within the scientific community, and out of hundreds of others, this thermo-physiological index has been identified as one of the most commonly utilised steady-state models in thermal comfort studies (Pantavou et al. 2014; Potchter et al. 2018; Charalampopoulos 2019; Charalampopoulos and Nouri 2019; Staiger et al. 2019). The justification for its greater use in comparison to other indices can moreover be likely associated to its (1) feasibility in being calibrated on easily obtainable microclimatic variables and (2) base measuring unit being (°C), thus simplifying its comprehension by non-climatic experts (including urban planners and architects) when approaching microclimatic aspects and risk factors.

Based upon the *M*unich *E*nergy-balance *M*odel for *I*ndividuals (MEMI) (Höppe 1984, 1993), the PET index is defined as the air temperature at which, in typical indoor setting, the human energy budget is maintained by the skin temperature, core temperature and perspiration rate, which are equivalent to those under the conditions to be assessed (VDI 1998, Höppe 1999). The composition of the MEMI model is established upon the energy balance equation for the human body, as shown in summary in Eq. (21.1):

$$M + W + R + C + E_D + E_{RE} + E_{Sw} + S = 0 \tag{21.1}$$

where M is the metabolic rate (i.e. internal energy production), W is the physical work output level, R is the net radiation of the body, C is the convective heat flow, E_D is the latent heat flow to the evaporated water diffusing through skin, E_{RE} is the sum of heat flows for heating and humidifying the inspired air, E_{Sw} is the heat flow due to the evaporation of sweat and S is the storage heat flow for thermoregulating body mass—all units of heat flow are in Watt (Höppe 1999).

By referring the earlier studies conducted by Fanger (1972) concerning the constitution of the *P*redicted *M*ean *V*ote (PMV) index to aid mechanical interior air-conditioning efforts to attain thermally comfortable interior conditions, Matzarakis and Mayer (1996) translated the ranges of PMV into grades of *P*hysiological *S*tress (PS) based upon PET values (Table 21.2).

Utilised extensively by the scientific community to investigate local human thermo-physiological conditions within different urban environments and climatic contexts (e.g. Alcoforado et al. 2009b; Matzarakis and Endler 2010; Gómez et al. 2013; de Abreu-Harbich et al. 2015; Algeciras and Matzarakis 2015; Martinelli et al. 2015; Algeciras et al. 2016, 2017; Charalampopoulos et al. 2016; Matzarakis et al. 2016; Lin et al. 2017; Nouri and Costa 2017; Nouri et al. 2018c; Fröhlich et al. 2019), the human biometeorological model RayMan Pro© (Matzarakis and Rutz

Table 21.2 Ranges of the *P*hysiologically *E*quivalent *T*emperature (PET) for different grades of *P*hysiological *S*tress (PS) on human beings, configured upon an internal heat production of 80 W, and heat transfer resistance of clothing of 0.9 Clo (according to Matzarakis and Mayer 1997). Source: Adapted from Matzarakis et al. (1999)

PET range (°C)	PS grade
<4	Extreme cold stress
4–8	Strong cold stress
8–13	Moderate cold stress
13–18	Slight cold stress
18–23	No thermal stress
23–29	Slight heat stress
29–35	Moderate heat stress
35–41	Strong heat stress
>41	Extreme heat stress

2006; Matzarakis et al. 2010) can be utilised to calculate PET and subsequent PS grades. Contiguously, it is worth noting that RayMan can also be utilised to calculate other thermal 'energy balance stress model' (Freitas and Grigorieva 2015) indices, including the *P*erceived *T*emperature (PT) (Tinz and Jendrizky 2003), the *S*tandard *E*ffective *T*emperature (SET*) (Gonzalez et al. 1974; Gagge et al. 1986) and the *U*niversal *T*hermal *C*limate *I*ndex (UTCI) (Jendritzky et al. 2002; Jendritzky et al. 2012).

In accumulation, the RayMan model has moreover been utilised within adjacent fields related to that of urban human thermal comfort studies. Furthermore, it is believed that this interdisciplinary application of biometeorological models and use of thermo-physiological indices within more medical contexts will naturally grow throughout the unravelling of the twenty-first century. The amalgamated message in this growth is the resourceful facilitation in bridging encompassing climatic factors specifically, and directly, with the consequential reactions of the human thermoregulatory system. More concretely, in the recent study undertaken by Hanefeld et al. (2019), the impacts of extreme climatic events upon German emergency medical services were investigated. Within the study, both PT and PET, although not directly processed through RayMan, were calculated from the same embedded MEMI equation (Eq. 21.1). The utilisation of such thermo-physiological indices aided the determination that cold spells resulted in an increased rate of emergency calls concerning cardiovascular afflictions. It was additionally acknowledged that such distress calls were not only observable during the particularly cold day, but moreover, during the next three subsequent days as well. Such a result enforces that cumulative thermal exposure carries a significant interdisciplinary role in longer term human thermoregulatory processes/reactions as well (Tsuzuki et al. 2004; Nouri and Matzarakis 2019). In alignment with this perspective, and still within the medical setting, the earlier study undertaken by Nastos and Matzarakis (2008) utilised RayMan to calculate PET to assess the relationship between hot conditions and events of sleep disorder admissions within the psychiatric emergency unit in Eginition Hospital, Athens. Within the research, it was acknowledged that during extended periods of higher diurnal/nocturnal thermal stimuli, there were substantial increases in sleep disorder-related admissions, which did not placate to the accentuated thermal conditions over time.

Returning to the possible variable outputs of the RayMan model, the *M*ean *R*adiant *T*emperature (MRT) (referred to as *R* within Eq. 21.1) can also be determined. This variable is a complex parameter to calculate yet crucial for human thermal comfort evaluations given its inherent assimilation within the human energy model (Matzarakis et al. 2007).

Within the large amount of studies that have utilised RayMan to determine human thermo-physiological risk factors, many have also revealed important lessons for the case of Lisbon. These lessons reinforce the need for such biometeorological models within Lisbon to continue to properly monitor, address and adapt to both current and future thermo-physiological risk factors associated to climates with a KG '*Csa*' sub-classification (Alcoforado et al. 2015; Nouri 2018).

21.3.2 Impacts on Thermal Indices as a Result of Climate Change

At the outset, and adjacently suggesting the inherent risks of depending solely on climatic evaluations that do not consider non-temperature variables, the application of thermo-physiological indices to approach future climatic conditions has already raised alarming results. In the particular case of Lisbon, while it has been demonstrated that its KG sub-classification shall not be modified by 2100, this does not imply that Lisbon will not witness extreme changes in thermo-physiological stress levels.

Within the study conducted by Matzarakis and Amelung (2008), through the use of synoptic global radiation (G_{rad}) estimations were obtained from monthly fractions (later translated into cloud cover (octas)) attained from the Hadley Centre's HadCM3 model (one of the predominant models utilised by the IPCC's earlier third assessment report in 2001). Once obtained, MRT and PET for various time slices (including 2070–2099) were calculated with the RayMan model based upon the input parameters of T_a, *R*elative *H*umidity (RH), wind speed (V) and octas. The method for calculating PET from the historical data set was identical. Such data was retrieved from the CRU CL 1.0 data set for the 1961–1990 (CNTRL) time slice facilitated by the Climatic Research Unit, University of East Anglia.

The results from the study identified drastic changes of PET within a $SRES_{A1F}$ scenario by the end of the century. As demonstrated in Fig. 21.2, areas such as the Iberian Peninsula and northern Turkey, in comparison to CNTRL data set, presented potential increases by up to 15 °C. Correspondingly, such results equate to numerous interconnected conclusions.

Firstly, there is a very strong possibility that the concrete thermo-physiological impacts associated to climate change are being dangerously underestimated. Thus far, the overlooking of fundamental non-temperature variables such as radiation fluxes, and use of simpler indices (elaborated from a combination of both T_a and RH), does not give a wholesome outlook into the biometeorological implications of climate change by 2100. In concurrence with Matzarakis and Amelung (2008), later studies (e.g. Nouri et al. 2018a, b) have also discussed the risks of international bodies, including the IPCC, describing the effect of weather and climate on humans while excluding crucial meteorological components, including V and G_{rad}. While it is conceded that variables such as K_i and L_i are generally not available in climatic records, paramount results demonstrated in Fig. 21.2 indicate that such data can be both obtained and utilised through biometeorological models such as RayMan to obtain such synoptic estimations. These methodologies/models raise new opportunities to further develop the direct cause-and-effect understanding of the climatic system upon the human body. These respective prospects are particularly tangent given the identified potential of PET increasing by 15 °C, a value that is dramatically higher than any singularly constructed T_a projection for the end of the century.

Secondly, and returning to the climatic variations for the time period as illustrated in Fig. 21.1a, given the fundamental constitution of the KG classification

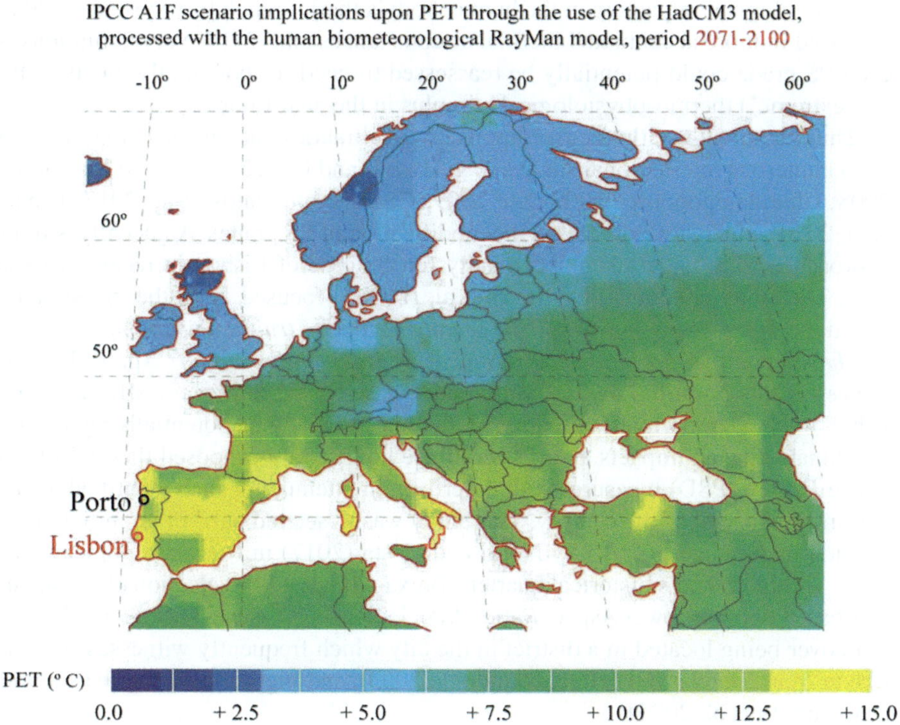

Fig. 21.2 Expected increases in *P*hysiologically *E*quivalent *T*emperature (PET) values for the end-of-century time slice in comparison to CNTRL values. (Source: Adapted from Matzarakis and Amelung 2008)

system (whereby the first letter determines predominant typology of the class, the second accounts for precipitation/aridity and the third accounts for temperature (Table 21.1)) one is presented with similar constraints as those previously mentioned. As the KG classification focuses categorically upon T_a and RH, the projections undertaken by Rubel and Kottek (2010), while representative of essential top-down studies of existing/future climate conditions, fall short in communicating biometeorological evaluations. In the case of Lisbon, while its KG sub-classification will not alter by the end of the century, as demonstrated in Fig. 21.2, it is specifically one of the areas which are projected to witness alarming increases in general thermo-physiological stimulus by the end of the century.

Thirdly, and more generally, the potential for such increases in PET (and consequential PS classifications shown in Table 21.1) raises the natural endorsement and/or requirement to reconsider the existing breadth of the PS levels. More specifically, in the event of a PET amplification of such proportions, new PS grades would be required to plot such PET values that surpass the 41 °C (i.e. extreme heat stress PS grade). The respective extension would indorse further imperative study into the already existing PS grades as determined by Matzarakis et al. (1999), whereby (1) the physiological impact upon the human body's thermoregulatory system could be reviewed in more accentuated thermal circumstances and (2) the very definitions of each PS grade could potentially be reassessed to heed for potentially much higher (or 'extreme') thermo-physiological stimulus in the near future.

Thus far, albeit mostly focused upon current climatic contexts, there have already been numerous international studies (e.g. Hwang and Lin 2007; Lin and Matzarakis 2008; Charalampopoulos 2009; Lin 2009; Charalampopoulos et al. 2017; Binarti et al. 2020) which have discussed the modification of PS grades. Adjacently, similar approaches were undertaken specifically for the case of Lisbon in recent years as well. The study conducted by Nouri et al. (2018d) focused upon the investigative question: '*If the* $SRES_{A1FI}$ *(or* $RCP_{8.5}$*) scenario came to fruition, what impacts could this have upon the local biometeorological environment in Rossio?*' (p. 9). In order to answer this question, the RayMan model was utilised to process MRT and PET values to (1) identify microclimatic risk factors that could be potentially aggravated by climate change impacts and (2) scrutinise how certain proposed thermal *P*ublic *S*pace *D*esign (PSD) measures would perform in attenuating such identified aggravations by the end of the century. This study was a 'second part' of an initial investigation undertaken in 2015 by Nouri and Costa (2017) in Rossio, a popular civic space within Lisbon's historical quarter, 'Baixa-Chiado'. The selection of the square was based upon its lower *A*spect *R*atio (AR), increasing its susceptibility to G_{rad}, and moreover being located in a district in the city which frequently witnesses stronger *U*rban *H*eat *I*sland (UHI) effects (Andrade 2003), and higher T_a during the summer (Alcoforado et al. 2005).

Within the initial study, and through the use of portable meteorological equipment, five different microclimatic in situ measurements during the summer of 2015 were undertaken on-site within six pre-designated locations, identified as *P*oints *of I*nterests (PoIs) (based upon initial *C*omputational *F*luid *D*ynamic (CFD) calculations for *V* estimations, and virtual shadow behaviour studies for G_{rad} estimations). The undertaken measurements were *V*, T_a, RH, G_{rad} and surface temperature (T_{surf}) which were then processed through RayMan, along with vectorised fisheye-lens photography to obtain each of the PoI's sky view factor (SVF). Resultantly, average PET values were obtained within six different areas of the square (four lateral and two central) throughout the month of July. Once obtained, it was possible to assess actual PS risk factors, which, in combination with numerous psychological assessments undertaken in the square, permitted the consideration of how such thermal stimuli could be respectively reduced during the summer period. As abstractly demonstrated in Fig. 21.3, such a bottom-up approach enabled in situ human thermo-physiological conditions to be evaluated within the square of Rossio based upon its own urban characteristics, amenities and overall morphological layout.

Summarising the results obtained by Nouri and Costa (2017), within the four hottest PoIs, it was calculated that through thermal sensitive PSD measures, the incorporation of (1) nylon canopies between tree crowns within the north-eastern sidewalk of the square could reduce PET by up to 9.9 °C; (2) six shading trees (*Tipuana tipu*) with a central water/misting feature within the centre of the square could reduce PET by up to 16.6 °C; (3) personifiable acrylic based shelter canopies

Fig. 21.3 Graphical representation of in situ thermo-physiological variations between different regions of the square, namely beneath a lateral tree crown and within the centre of the square (Rossio). (Source: Adapted from Nouri and Costa 2017)

could reduce PET by up to 12.3 °C; and lastly (4) misting systems within the tree crowns of the south-western sidewalk alone could render reductions in PET by up to 4.7 °C. As a result of the disclosed reduction in PET, the associated PS levels decreased substantially, with a maximum reduction of three complete grades within the centre of the square (from 'extreme heat stress' down to 'moderate heat stress'). Such overall results demonstrated the promising potentiality for PSD to address existing thermo-physiological risk factors within Lisbon's hot-dry Mediterranean summers.

Within the subsequent segment of the research, the respective potential impacts of an $SRES_{A1FI}/RCP_{8.5}$ climatic scenario on such existing biometeorological conditions (both with and without the proposed PSD interventions) were estimated within a pilot analysis. Although the study was based upon the projected increases in PET as projected by Matzarakis and Amelung (2008) for the 2071–2100 time slice, instead of the +15 °C identified for zones such as Lisbon, the study was constructed upon a more modest PET oscillation of +10 °C. Subsequently, to plot the new results of the study, the existing PS grades were extended.

As indicated in the study, based on the increment of roughly 5 °C per PS threshold within the original grading system, three new classifications beyond the 'extreme heat stress' were added (Table 21.3). Naturally and as previously suggested, such an expansion requires new study into the precise implications of such dramatic thresholds upon the human biometeorological system. Once such studies are underway,

Table 21.3 Grade extension of physiological stress (PS) on human beings to accompany the projected increased physiologically equivalent temperature (PET) value of 41 °C (see Table 21.2). Source: Adapted from Matzarakis et al. (1999) and Nouri et al. (2018d)

PET range (°C)	PS grade	Grade extension
<4	Extreme cold stress	Existing grades
4–8	Strong cold stress	
8–13	Moderate cold stress	
13–18	Slight cold stress	
18–23	No thermal stress	
23–29	Slight heat stress	
29–35	Moderate heat stress	
35–41	Strong heat stress	
41–46	Extreme heat stress 1	
46–51	'Extreme heat stress 2'	New grades
51–56	'Extreme heat stress 3'	
>56	'Extreme heat stress 4'	

Table 21.4 Description of the assessed physiologically equivalent temperature (PET) data sets before and after interventions for existing and future $SRES_{A1FI}/RCP_{8.5}$ scenario. Source: Adapted from Nouri et al. (2018d)

Existing scenarios	
$\overline{PET}_{D3}$	Showing existing PET without PSD interventions
$\overline{PET\Delta}_{D3}$	Representing existing PET with PSD interventions
Projected scenarios (2100)	
$\overline{PET}_{A1FI/8.5}$	Representing projected PET without PSD interventions in a scenario with no policy changes to reduce global emissions
$\overline{PET\Delta}_{A1FI/8.5}$	Representing projected PET with PSD interventions in a scenario with no policy changes to reduce global emissions

D3: Average values obtained from the site visit with the hottest conditions (*day 3*)

then the entire terminology of the PS labelling system can be potentially adjusted to account for the increased susceptibility to augmented PET values which shall more frequently (and severely) surpass the 41 °C threshold.

Concentrating upon the average PET results obtained between 15:00 and 16:00 (representing the hottest hour during the day), numerous PET data sets were elaborated and four of them are summarised in Table 21.4.

In addition, and based upon a constructed grid network to determine the dimensional extents of the interventions, some of the PSD measures introduced in the PoIs were extended throughout the entire square, including the (1) extension of nylon canopies throughout the entire eastern sidewalk and (2) insertion of misting systems within all of the tree crowns within the entire western sidewalk. In summary, the results of the study revealed that in accordance with the outcomes obtained by Nouri and Costa (2017):

For existing scenarios:

- The $\overline{\text{PET}\Delta_{D3}}$ data set revealed reductions of PET conducive of PS reductions by three levels within the centre of the square. In addition, the $\overline{\text{PET}\Delta_{D3}}$ data set also revealed reductions in PS grades by two levels on the eastern sidewalk. On the other hand, the lack of PSD measures led to the $\overline{\text{PET}_{D3}}$ data set revealing PS grades of 'extreme heat stress 2' as a result of PET values reaching 48.6 °C.

For future scenarios:

- The $\overline{\text{PET}_{A1FI/8.5}}$ data set revealed dramatic increases of PS around the entire square of Rossio. Given the modest augment of PET values by 10 °C (and not 15 °C as determined by Matzarakis and Amelung (2008)), in the case of $\overline{\text{PET}_{A1FI/8.5}}$ data set, even with the presence of tree shades, wind currents and shadows cast by the building facades, extreme conditions were identified throughout the whole square. More specifically, most of the square presented a PS of either 'extreme heat stress 3' or 'extreme heat stress 4' due to synoptic PET values exceeding those of 51 and 56 °C, respectively.
- The $\overline{\text{PET}\Delta_{A1FI/8.5}}$ data set presented comparatively more attenuated PS thresholds due to the proposed thermal PSD interventions. More precisely, within the (1) eastern sidewalk, the nylon canopies decreased PS by three levels, and (2) the measures within the centre of the square revealed reductions by four PS levels, from 'extreme heat stress 4' to 'strong heat stress'.

Although such results were based upon synoptic future projections undertaken through the use of the RayMan model, it demonstrated a method to explore bottom-up methods that considered imperative meteorological factors (including radiation fluxes) beneath the urban canopy layer. Moreover, it enabled the resulting impacts directly upon the human thermoregulatory system. As concluded by the study, thermal PSD was able to effectively attenuate existing thermo-physiological risk factors during the summer, and moreover presented significant reductions to extreme PS levels by the end of the century. Just as significantly, in the event of a worse-case scenario by 2100, such thermal PSD measures alone were identified to be insufficient to maintain a thermally comfortable and safe public realm within hot-dry Mediterranean summers. Based upon the same adaptation of the PS grade system, the study by Nouri et al. (2018a) examined the disparity between singular climatic variables and thermo-physiological indices for current and future thermal comfort thresholds for hot-dry Mediterranean summers. Unlike in the previous studies where RayMan was processed exclusively from in situ measurements, recordings obtained from the *W*orld *M*eteorological *O*rganization (WMO) weather station located in Lisbon (with the WMO index N°08535) were utilised to process PET values. Of the assessed years within the study, the results pertaining to the 2016 data set shall be subsequently discussed. Between the hours of 09:00 and 18:00, hourly recordings of octas, T_a, RH and V were retrieved. With regard to the last variable, and in order to determine V relatable to the gravity centre of the human body as defined by Kuttler (2000), V outputs were adapted to a height of 1.1 m (henceforth $V_{1.1}$). Within Fig. 21.4, and calculated through RayMan, the PET/PS results

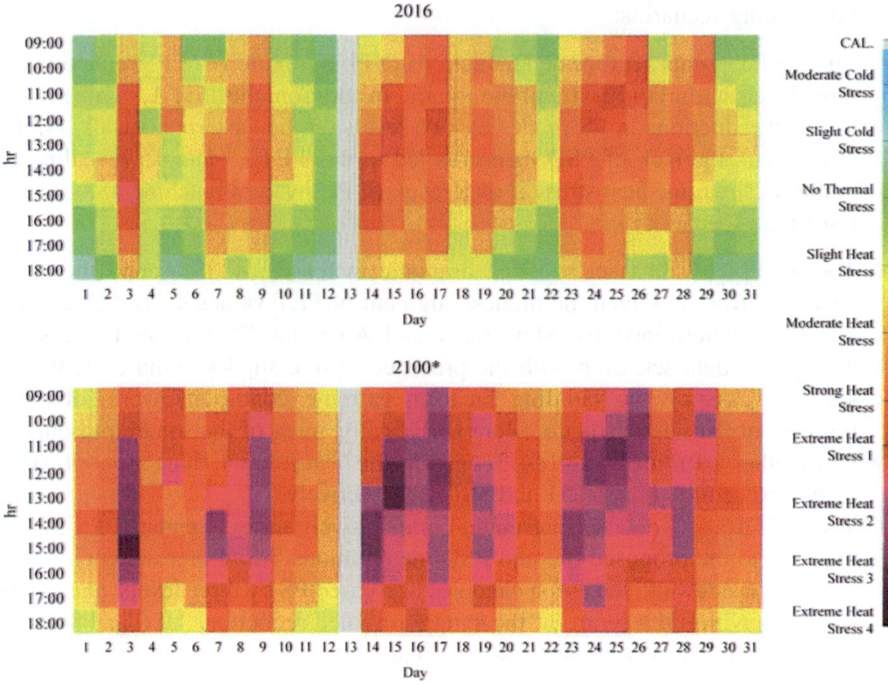

Fig. 21.4 Identification of hourly/diurnal variations of *P*hysiological *S*tress (PS) as a result of *P*hysiologically *E*quivalent *T*emperature (PET) values between 09:00 and 18:00 for 2016, and their projected aggravations in light of a $SRES_{A1FI}/RCP_{8.5}$ scenario by the end of the century. (Source: Adapted from Nouri et al. 2018a)

pertaining to the diurnal/hourly oscillations during July 2016, and their respective synoptic aggravations for 2100, are illustrated.

To facilitate the representation of the PS/PET variations, once the results were processed through RayMan, the *C*limate *T*ourism/Transfer *I*nformation *S*cheme (CTIS) model was utilised to construct the diagrams in Fig. 21.4. Such a method was based upon the calibration of the desired thresholds to enable the fluctuations of PS to be colour coded based upon their lower and upper limits, hence facilitating the reading/interpretation of the data. An additional classicisation of 'CAL' was created to account for days (such as the 13th) in which the station was turned off due to calibration and/or maintenance purposes. Within this study, the same extension of existing PS grades was undertaken to better plot the higher PET values (Table 21.3). Adjacently, the same synoptic projected increment of 10 °C was applied to the 2016 data set to consider how the existing bioclimatic conditions could be amplified by 2100. It is important to note, once again, that such data is based upon a WMO station data, meaning that unlike in the study of Nouri and Costa (2017) and Nouri et al. (2018d), the configured SVF was set to 1.00 (or 100%) in order to reproduce the total exposure to solar radiation. During this phase of the study, the intention behind this was to focus specifically upon the existing-projected thermo-physiological variables themselves, and not the influences of urban canyons upon these results.

In summary, and beyond showing the influences of extreme heat events in July 2016, including the thermo-physiological impacts of the heatwaves throughout the 7th–9th, the 14th–17th and 23rd–28th of July 2016, it was also possible to identify that PS almost never presented any type of diurnal cold stress, even during the mornings or evening period. On the other hand, and as exemplified by the VHD on the 3rd of July, even in the 2016 data set, there were hours of the day (e.g. 15:00) which reached the PS grade of 'extreme heat stress 2' resultant of PET values reaching 47.3 °C. Subsequently, these results are already indicative of the already existing risk factors for thermal comfort upon Lisbon's public realm. Within the 2100 projection, it was demonstrated that the lowest projected PS grade was of 'slight heat stress' (even in the early morning and late afternoon hours). During the rest of the day, and particularly during the corresponding extreme heat events, PS oscillated between the four extreme heat stress grades between 09:00 and 18:00. Such outcomes indicate, once again, the potentially acute repercussions associated to a worst-case scenario of climate change upon urban biometeorological conditions by the end of the century.

21.3.3 Approaching Generic Canyon Thermal Attenuation Priorities

As previously represented in Fig. 21.1b, urban climatic conditions vary significantly beneath the urban canopy layer. One of the predominant influences upon such conditions is the variation of ARs which can render different very thermal conditions specifically within different symmetrical (Ali-Toudert and Mayer 2006; Herrmann and Matzarakis 2012; Ketterer and Matzarakis 2014; Algeciras et al. 2016) and asymmetrical urban canyons (Qaid and Ossen 2015; Algeciras et al. 2017).

Considering the case of Lisbon, and again focused within its downtown district, the study undertaken by Nouri et al. (2017) especially focused upon how different ARs render different thermal comfort conditions, including within the different regions of the canyons themselves. Such an assessment was undertaken also through the use of the SkyHelios model (Matzarakis and Matuschek 2011), which in succinct terms trails similar processing algorithms to those embedded within RayMan, with the additional feature of also considering the spatial dimension of the input parameters. Resultantly, five different *H*eight-to-*W*idth (*H*/*W*) ARs (varying among very high—*H*/*W* of 2.00, high—*H*/*W* of 1.00, medium—*H*/*W* of 0.50, low and very low—*H*/*W* of 0.17) were constructed within the SkyHelios. Once configured, it was possible to calculate SVF at specific locations within each of the constructed canyons. All were configured with a *N*orth-to-*S*outh *O*rientation (NSO) and a *W*est-to-*E*ast *O*rientation (WEO), with three analytical points within the canyon (two laterals, and one central) to further identify fluctuations of radiation. Within the RayMan model, in addition to the climatic parameters (i.e. $V_{1.1}$, T_a and RH) retrieved from the WMO meteorological station (index N°08535) the latter part of the study also utilised hourly recordings of in situ G_{rad}. Although RayMan is capable to calculate such radiation fluxes, this supplementary parameter was utilised to replace the introduced octas values obtained from the WMO station. Resultantly, the study was constructed upon a mixture of both meteorological station and in situ recordings to evaluate in further detail the thermo-physiological conditions within various morphological canyons.

One output of the study was the identification of *T*hermal *A*ttenuation *P*riorities (TAP) for PSD given the identified existing thermo-physiological risk factors in each of the default canyons in Lisbon's downtown district. Such an output was elaborated through the use of the newly updated PET index for Western Europe developed by Chen and Matzarakis (2017) called the *m*odified *PET* (mPET) index, also calculated in RayMan. In summary, the main alterations of mPET are the integrated thermoregulation model (altered from a single double-node body model to a multiple-segment model) and the updated clothing model, relaying to a more accurate analysis of the human bio-heat transfer mechanism. Although recent, the index has already proven by numerous studies to be more efficient at determining PS, particularly during periods of higher thermal stimuli (Lin et al. 2018; Nouri et al. 2018c). In addition, and based upon the research outputs from Charalampopoulos et al. (2016), the PET load (PETL) and the cumulative diurnal PETL (cPETL) were also utilised, whereby (1) PETL denotes the outcome difference from optimum conditions, allowing for the retrieval of a value which specifically denotes the amount of thermal strain upon 'optimum conditions', and (2) cPETL denotes the cumulative sum of the PETL specifically during a fixed amount of time. Respectively, by adapting such equations to incorporate mPET, it became possible to determine hourly PS load based upon the divergent morphological conditions and time of day (Eq. 21.2):

$$\mathrm{mPETL} = \mathrm{mPET_h} - \mathrm{BC} \tag{21.2}$$

where $\mathrm{mPET_h}$ is the average hourly mPET value, and BC (*B*ackground *C*onditions) in the study was set to denote the maximum PET for the PS grade of 'no thermal stress' (i.e. of 23 °C). Average mPET values were based upon hourly measurements retrieved for July 2016.

To construct the TAP results shown in Fig. 21.5, five different priority levels were determined based on the amount of variation from BC. The increment between each level of priority was based upon ±4 °C, allowing thus the mPET results to be evenly distributed, and moreover relatable to the grades of PS as described in Tables 21.2 and 21.3.

In brief, the TAP results within one of Lisbon's most vulnerable districts to thermo-physiological risk factors indicated that there is a generally high priority to attenuate PS levels. Particularly in medium to very low ARs, after 12:00, mPETL values almost always led to very high TAP during the afternoon within both NSO and WEO canyons. It is important to note that while very high ARs revealed lower TAP levels, there were periods of the day that still revealed high TAP levels, namely

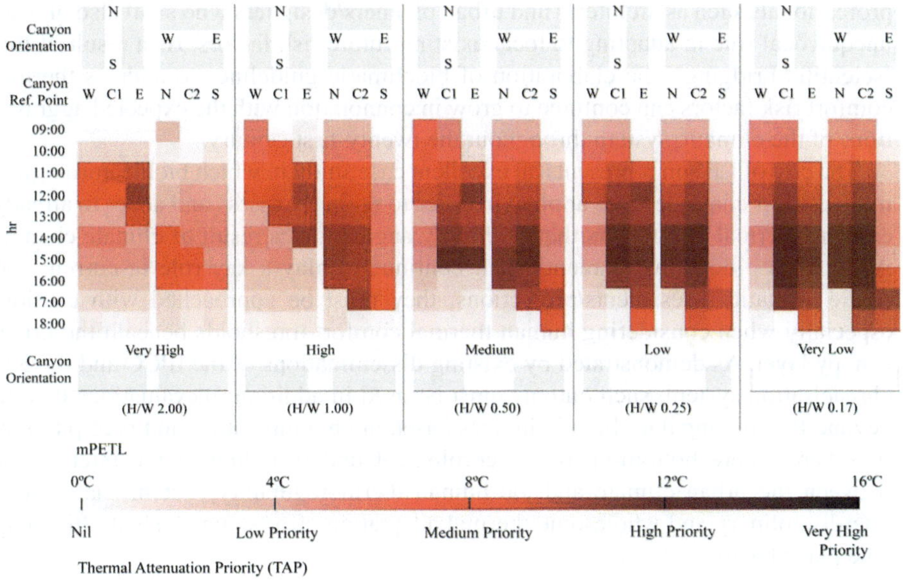

Fig. 21.5 *T*hermal *A*ttenuation *P*riority (TAP) for different morphological canyons through the identification of *m*odified *P*hysiologically *E*quivalent *T*emperature *L*oad (mPETL) between 09:00 and 18:00 based upon hourly averages obtained for July 2016. (Source: Adapted from Nouri et al. 2017)

the eastern regions of the NSO canyons between 11:00 and 13:00. Such outcomes moreover suggest the significance of going beyond the general AR, and explore how specific locations within the same canyon may present completely divergent thermo-physiological conditions.

21.4 Final Remarks

Within this chapter, and with a particular focus upon the Lisbon case, it has been discussed how the use of human biometeorological models plays a fundamental role in determining evaluations of both current and future thermo-physiological risk factors. It is expected that these models, such as RayMan and SkyHelios, shall continue to grow both in analytical capacity and in weight within the intrinsically interdisciplinary nature of adapting to existing/future thermal risk factors in consolidated urban environments.

Adjacently, in addition to the use of such models, there is also the quintessential requirement to ensure the communication of such information to non-climatic experts. Such interdisciplinary bridging can be aided by the use of models such as the CTIS model that can aid the ease of communication and comprehension of essential biometeorological outputs for non-climatic experts. These experts include professionals such as architects and urban planners/designers who shall also play an unequivocal role in adapting to local existing/future risk factors. As a result of this 'scientific bridging', the elaboration of bioclimatic guidelines to address thermal comfort risk factors can continue to grow in connotation with the expected aggravations of the climatic system throughout the twenty-first century.

The city of Lisbon represents an excellent case study in which biometeorological models must continue to be applied in order to properly assess and adapt to already existing thermal risk factors that shall only augment as a result of climate change. Although top-down assessments shall continue to play a key role in current and future climatic assessments/projections, they must be approached with caution, especially when considering human thermal comfort thresholds beneath the urban canopy layer. As demonstrated by existing disseminations of the IPCC and the KG classicisation system, such outputs must be used in addition to evaluations that go beyond that of singular climatic indices based upon temperature and precipitation. It is here where bottom-up biometeorological understanding of the interactions between the urban climate and the human thermoregulatory system shall aid in interdisciplinary and wholesome thermal adaptation efforts, which shall ultimately take place locally.

References

Alcoforado, M. J., & Matzarakis, A. (2010). Planning with urban climate in different climatic zones. *Geographicalia, 57*, 5–39.

Alcoforado, M. J., & Vieira, H. (2004). Urban climate in Portuguese management plans (in Portuguese with abstract in English). *Sociedade e Território, 37*, 101–116.

Alcoforado, M. J., Lopes, A., Andrade, H., & Vasconcelos, J. (2005). *Orientações climáticas para o ordenamento em Lisboa (relatório 4)*. Lisboa: Centro de Estudos Geográficos da Universidade de Lisboa.

Alcoforado, M.-J., Andrade, H., Lopes, A., & Vasconcelos, J. (2009a). Application of climatic guidelines to urban planning—The example of Lisbon (Portugal). *Landscape and Urban Planning, 90*, 56–65.

Alcoforado, M. J., Andrade, H., & Oliveira, S. (2009b). Alterações climáticas e desenvolvimento urbano Pol'tica de Cidades—4. V. Higgs. Lisboa: Direcção-Geral do Ordenamento do Território e Desenvolvimento Urbano.

Alcoforado, M. J., Lopes, A., Alves, E., & Canário, P. (2014). Lisbon heat island; statistical study. *Finisterra, 98*, 61–80.

Alcoforado, M. J., Lopes, A., & Andrade, H. (2015). Urban climatic map studies in Portugal. In E. Ng & C. Ren (Eds.), *The urban climatic map: A methodology for sustainable urban planning*. London: Routledge.

Algeciras, J. A. R., & Matzarakis, A. (2015). Quantification of thermal bioclimate for the management of urban design in Mediterranean climate of Barcelona, Spain. *International Journal of Biometeorology, 8*, 1261–1270.

Algeciras, J. A. R., Consuegra, L. G., & Matzarakis, A. (2016). Spatial-temporal study on the effect of urban street configurations on human thermal comfort in the world heritage city of Camagüey-Cuba. *Building and Environment, 101*, 85–101.

Algeciras, J. A. R., Tablada, A., & Matzarakis, A. (2017). Effect of asymmetrical street canyons on pedestrian thermal comfort in warm-humid climate of Cuba. In *Theoretical and applied climatology* (pp. 1–17). Wien: Springer.

Ali-Toudert, F., & Mayer, H. (2006). Numerical study on the effects of aspect ratio and orientation of an urban street canyon on outdoor thermal comfort in hot and dry climate. *Building and Environment, 41*, 94–108.

Andrade, H. (2003). *Bioclima humano e Temperatura do ar em Lisboa*. PhD, Universidade de Lisboa.

Binarti, F., Koerniawan, M., Triyadi, S., Utami, S., & Matzarakis, A. (2020). A review of outdoor thermal comfort indices and neutral ranges for hot-humid regions. *Urban Climate, 31*.

Brugger, K., & Rubel, F. (2013). Characterizing the species composition of European Culicoides vectors by means of the Köppen-Geiger climate classification. *Journal of Parasites and Vectors, 6*, 1–6.

Charalampopoulos, I. (2009). *Development of a methodology and applications for the bioclimatic conditions' evaluation of open space with diverse configurations* (in Greek). PhD, Agricultural University of Athens.

Charalampopoulos, I. (2019). A comparative sensitivity analysis of human thermal comfort indices with generalized additive models. In *Theoretical and applied climatology* (pp. 1–18). Wien: Springer.

Charalampopoulos, I., & Nouri, A. S. (2019). Investigating the behaviour of human thermal indices under divergent atmospheric conditions: A sensitivity analysis approach. *Atmosphere, 10*, 580.

Charalampopoulos, I., Tsiros, I., Chronopoulou-Sereli, A., & Matzarakis, A. (2016). A methodology for the evaluation of the human-bioclimatic performance of open spaces. In *Theoretical and applied climatology* (pp. 1–10). Wien: Springer.

Charalampopoulos, I., Nastos, P., & Didaskalou, E. (2017). Human thermal conditions and north Europeans' web searching behaviour (Google Trends) on Mediterranean touristic destinations. *Urban Science, 1*, 8.

Chen, Y.-C., & Matzarakis, A. (2017). Modified physiologically equivalent temperature—Basics and applications for western European climate. In *Theoretical and applied climatology* (pp. 1–15). Wien: Springer.

de Abreu-Harbich, L. V., Labaki, L., & Matzarakis, A. (2015). Effect of tree planting design and tree species on human thermal comfort in the tropics. *Landscape and Urban Planning, 138*, 99–109.

Djamila, H., & Yong, T. (2016). A study of Köppen-Geiger system for comfort temperature prediction in Melbourne city. *Sustainable Cities and Society, 27*, 42–48.

Fanger, P. (1972). *Thermal comfort: Analysis and applications in environmental engineering*. New York: McGraw-Hill Book Company.

Farnham, C., Nakao, M., Nishioka, M., Nabeshima, M., & Mizuno, T. (2011). Study of mist-cooling for semi-enclosed spaces in Osaka, Japan. *Urban Environmental Pollution, 4*, 228–238.

Freitas, C., & Grigorieva, E. (2015). A comprehensive catalogue and classification of human thermal climate indices. *International Journal of Biometeorology, 59*, 1–12.

Fröhlich, D., Gangwisch, M., & Matzarakis, A. (2019). Effect of radiation and wind on thermal comfort in urban environments—Applications of the RayMan and SkyHelios Model. *Urban Climate, 27*, 1–7.

Gagge, A., Fobelets, P., & Bergland, L. (1986). A standard predictive index of human response to thermal environment. *ASHRAE Transactions, 92*, 709–731.

Gómez, F., Cueva, A., Valcuende, M., & Matzarakis, A. (2013). Research in ecological design to enhance comfort in open spaces of a city (Valencia, Spain). Utility of the physiological equivalent temperature (PET). *Journal of Ecological Engineering, 57*, 27–39.

Gonzalez, R., Nishi, Y., & Gagge, A. (1974). Experimental evaluation of standard effective temperature: A new biometeorological index of man's thermal discomfort. *International Journal of Biometeorology, 18*, 1–15.

Hanefeld, C., Klaaßen-Mielke, R., Miebach, J., Muthers, S., Haschemi, A., Trampisch, H., Kloppe, C., Matzarakis, A., Krogias, C., & Schroeder, C. (2019). Einfluss von Wetterextrema auf Einsatzzahlen im Notarztdienst. *Medizinische Klinik—Intensivmedizin und Notfallmedizin* (pp. 1–7).

Herrmann, J., & Matzarakis, A. (2012). Mean radiant temperature in idealised urban canyons—Examples from Freiburg, Germany. *International Journal of Biometeorology, 56*, 199–203.

Höppe, P. (1984). *The energy balance in humans (Original title—Die Energiebilanz des menschen)*. Munich: Universitat Munchen, Meteorologisches Institut.

Höppe, P. (1993). Heat balance modelling. *Experientia, 49*, 741–746.

Höppe, P. (1999). The physiological equivalent temperature—A universal index for the biometeorological assessment of the thermal environment. *International Journal of Biometeorology, 43*, 71–75.

Hwang, L., & Lin, T. (2007). Thermal comfort requirements for occupants of semi-outdoor and outdoor environments in hot-humid regions. *Journal of Architectural Science Review, 50*(4), 60–67.

IPCC. (2000). *IPCC special report emission scenarios, a special report of working group III of the Intergovernmental Panel on Climate Change* (p. 27). Cambridge: IPCC.

IPCC. (2007). *Synthesis report—An assessment of the Intergovernmental Panel on Climate Change—Adopted section by section at IPCC Plenary XXVII (Valencia, Spain, 12–17 November 2007)* (p. 104). Geneva: IPCC.

IPCC. (2013). *Climate change 2013: The physical science basis. Working group contribution to the IPCC 5th assessment report. Changes to the underlying scientific/technical assessment.* Cambridge: IPCC.

Ishii, T., Tsujimoto, M., Yoon, G., & Okumiya, M. (2009). Cooling system with water mist sprayers for mitigation of heat-island. In *The Seventh International Conference on Urban Climate*. Yokohama: ICUC.

Jendritzky, G., Maarouf, A., Fiala, D., & Staiger, H. (2002). An update on the development of a Universal Thermal Climate Index. In *15th Conf. Biomet. Aerobiol. and 16th ICB02* (pp. 129–133). Kansas City: AMS.

Jendritzky, G., de Dear, R., & Havenith, G. (2012). UTCI—Why another thermal index? *International Journal of Biometeorology, 56*(3), 421–428.

Ketterer, C., & Matzarakis, A. (2014). Human-biometeorological assessment of heat stress reduction by replanning measures in Stuttgart, Germany. *Landscape and Urban Planning, 122*, 78–88.

Kovats, R., & Ebi, K. (2006). Heatwaves and public health in Europe. *Journal of Public Health, 16*(6), 592–599.

Kuttler, W. (2000). Stadtklima. In R. Guderian (Ed.), *Handbuch der Umweltveränderungen und Ökotoxologie* (Atmosphäre) (Vol. 1B, pp. 420–470). New York: Springer-Verlag.

Lin, T.-P. (2009). Thermal perception, adaptation and attendance in a public square in hot and humid regions. *Building and Environment, 44*, 2017–2026.

Lin, T.-P., & Matzarakis, A. (2008). Tourism climate and thermal comfort in Sun Moon Lake, Taiwan. *International Journal of Biometeorology, 52*, 281–290.

Lin, T.-P., Chen, Y.-C., & Matzarakis, A. (2017). Urban thermal stress climatic mapping: Combination of long-term climate data and thermal stress risk evaluation. *Journal of Sustainable Cities and Society, 34*, 12–21.

Lin, T.-P., Yang, S.-R., Chen, Y.-C., & Matzarakis, A. (2018). The potential of a modified physiologically equivalent temperature (mPET) based on local thermal comfort perception in hot and humid regions. In *Theoretical and applied climatology*. Wien: Springer.

Lopes, A. (2003). *Modificações no Clima da Lisboa Como Consequência do Crescimento Urbano*. Doctoral Thesis, University of Lisbon.

Lopes, A., Alves, E., Alcoforado, M. J., & Machete, R. (2013). Lisbon Urban heat island updated: New highlights about the relationships between thermal patterns and wind regimes. *Journal of Advances in Meteorology, 2013*, 1–11.

Martinelli, L., Lin, T., & Matzarakis, A. (2015). Assessment of the influence of daily shading pattern on human thermal comfort and attendance in Rome during summer period. *Building and Environment, 92*, 30–38.

Matzarakis, A. (2016). The heat health warning system of DWD—Concept and lessons learned. In T. Karacostas, A. Bais, & P. Nastos (Eds.), *Perspectives on atmospheric sciences* (Springer atmospheric sciences). Cham: Springer.

Matzarakis, A., & Amelung, B. (2008). Physiological equivalent temperature as indicator for impacts of climate change on thermal comfort of humans. In M. C. Thomson, R. Garcia-Herrera, & M. Beniston (Eds.), *Seasonal forecasts, climatic change and human health* (Advances in global research) (Vol. 30, pp. 161–172). Berlin: Springer.

Matzarakis, A., & Endler, C. (2010). Climate change and thermal bioclimate in cities: Impacts and options for adaptation in Freiburg, Germany. *International Journal of Biometeorology, 54*, 479–483.

Matzarakis, A., & Matuschek, O. (2011). Sky view factor as a parameter in applied climatology rapid estimation by the SkyHelios model. *Meteorologische Zeitschrift, 20*(1), 39–45.

Matzarakis, A., & Mayer, H. (1996). Another kind of environmental stress: Thermal stress. WHO Collaborating Centre for air quality management and air pollution control. *Newsletters, 18*, 7–10.

Matzarakis, A., & Mayer, H. (1997). Heat stress in Greece. *International Journal of Biometeorology, 41*, 34–39.

Matzarakis, A., & Rutz, F. (2006). In A. Matzarakis (Ed.), *RayMan Pro—Modelling of mean radiant temperature in urban structures calculation of thermal indices. Version 2.1 Pro*. Freiburg im Breisgau: Meteorological Institute, University of Freiburg.

Matzarakis, A., Mayer, H., & Iziomon, G. M. (1999). Applications of a universal thermal index: Physiological equivalent temperature. *International Journal of Biometeorology, 42*(2), 76–84.

Matzarakis, A., Rutz, F., & Mayer, H. (2007). Modelling radiation fluxes in simple and complex environments—Application of the RayMan model. *International Journal of Biometeorology, 51*, 323–334.

Matzarakis, A., Rutz, F., & Mayer, H. (2010). Modelling radiation fluxes in simple and complex environments—Basics of the RayMan model. *International Journal of Biometeorology, 54*, 131–139.

Matzarakis, A., Fröhlich, D., & Gangwisch, M. (2016). Effect of radiation and wind on thermal comfort in urban environments—Application of the RayMan and SkyHelios model. In *4th International Conference on Countermeasures to Urban Heat Island*. Singapore: National University of Singapore.

Mayer, H., & Höppe, P. (1987). Thermal comfort of man in different urban environments. *Theoretical and Applied Climatology, 38*, 43–49.

Miranda, P. (2006). The climate in Portugal during the XX and XXI centuries climate change in Portugal. In F. Santos & P. Miranda (Eds.), *Scenarios, impacts and adaptation measures (SIAM)* (pp. 46–113). Lisbon: Gradiva.

Nastos, P., & Matzarakis, A. (2008). Human-biometeorological effects on sleep in Athens, Greece: A preliminary evaluation. *Indoor and Built Environment, 17*, 535–542.

Nogueira, P., Falcão, J., Contreiras, M., Paixão, E., Brandão, J., & Batista, I. (2005). Mortality in Portugal associated with the heat wave of August 2003: Early estimation of effect, using a rapid method. *Eurosurveillance, 10*(7), 5–6.

Nouri, A. S. (2018). *Addressing urban outdoor thermal comfort thresholds through public space design—A bottom-up interdisciplinary research approach for thermal sensitive urban design in an era of climate change: The Lisbon case*. PhD, University of Lisbon.

Nouri, A. S., & Costa, J. P. (2017). Addressing thermophysiological thresholds and psychological aspects during hot and dry Mediterranean summers through public space design: The case of Rossio. *Building and Environment, 118*, 67–90.

Nouri, A. S., & Matzarakis, A. (2019). The maturing interdisciplinary relationship between human biometeorological aspects and local adaptation processes: An encompassing overview. *Climate, 7*, 134.
Nouri, A. S., Costa, J. P., & Matzarakis, A. (2017). Examining default urban-aspect-ratios and sky-view-factors to identify priorities for thermal-sensitive public space design in hot-summer Mediterranean climates: The Lisbon case. *Building and Environment, 126*, 442–456.
Nouri, A. S., Charalampopoulos, I., & Matzarakis, A. (2018a). Beyond singular climatic variables—Identifying the dynamics of wholesome thermo-physiological factors for existing/future human thermal comfort during hot dry Mediterranean summers. *International Journal of Environmental Research and Public Health, 15*, 2362.
Nouri, A. S., Costa, J. P., Santamouris, M., & Matzarakis, A. (2018b). Approaches to outdoor thermal comfort thresholds through public space design: A review. *Atmosphere, 9*(3), 108.
Nouri, A. S., Fröhlich, D., Silva, M. M., & Matzarakis, A. (2018c). The impact of Tipuana tipu species on local human thermal comfort thresholds in different urban canyon cases in Mediterranean climates: Lisbon, Portugal. *Atmosphere, 9*(12), 2–28.
Nouri, A. S., Lopes, A., Costa, J. P., & Matzarakis, A. (2018d). Confronting potential future augmentations of the physiologically equivalent temperature through public space design: The case of Rossio, Lisbon. *Sustainable Cities and Society, 37*, 7–25.
Pantavou, K., Santamouris, M., Asimakopoulos, D., & Theoharatos, G. (2014). Empirical calibration of thermal indices in an urban outdoor Mediterranean environment. *Building and Environment, 80*, 283–292.
Peel, M., Finlayson, B., & McMahon, T. (2007). Updated world map of the Köppen-Geiger climate classification. *Journal of Hydrology and Earth System Sciences, 11*, 1633–1644.
Potchter, O., Cohen, P., Lin, T., & Matzarakis, A. (2018). Outdoor human thermal perception in various climates: A comprehensive review of approaches, methods and quantification. *Science of the Total Environment, 631–632*, 390–406.
Qaid, A., & Ossen, D. (2015). Effect of asymmetrical street aspect ratios on microclimates in hot, humid regions. *International Journal of Biometeorology, 59*, 657–677.
Rogelj, J., Meinshausen, M., & Knutti, R. (2012). Global warming under old and new scenarios using IPCC climate sensitivity range estimates. *Nature Climate Change: pp., 2*, 248–253.
Rubel, F., & Kottek, M. (2010). Observed and projected climate shifts 1901-2100 depicted by world maps of the Köppen-Geiger climate classification. *Journal of Meteorologische Zeitschrift, 19*(2), 135–141.
Staiger, H., Laschewski, G., & Matzarakis, A. (2019). Selection of appropriate thermal indices for applications in human biometeorological studies. *Atmosphere, 10*, 18.
Tinz, B., & Jendrizky, G. (2003). Europa- und Weltkarten der gefühlten Temperatur. In F. Chmielewski & T. Foken (Eds.), *Beiträge zur Klima- und Meeresforschung* (pp. 111–123). Berlin und Bayreuth: Eigenverlag.
Tsuzuki, K., Okamoto-Mizuno, K., & Mizuno, K. (2004). Effects of humid heat exposure on sleep, thermoregulation, melatonin, and microclimate. *Journal of Thermal Biology, 29*, 31–36.
VDI. (1998). *Part I: Environmental meteorology, methods for the human-biometeorological evaluation of climate and air quality for the urban and regional planning at regional level. Part I: Climate. VDI/DIN-Handbuch Reinhaltung der Luft. VDI 3797* (p. 29). Düsseldorf: Verein Deutscher Ingenieure.
Yang, S., & Matzarakis, A. (2016). Implementation of human thermal comfort information in Köppen-Geiger climate classification—The example of China. *International Journal of Biometeorology, 60*(11), 2–6.

Chapter 22
Impact of Local Urban Climate on Building Energy Performance: Case Studies in Mendoza, Argentina

Carolina Ganem Karlen, Julieta Balter, and Noelia Liliana Alchapar

22.1 Introduction

Globally, the significant increases in populations in urban areas over the recent decades due to rapid urbanization have induced the formation of the local climate change. One of the major factors in this change is the urban heat island (UHI) effect, which describes the excess warmth of the urban atmosphere in comparison to the rural areas (Levermore et al. 2017). Around 3.5 billion people live in urban areas all over the world and by 2050 more than two-thirds of the urban population will live in cities (DOE 2017). And around two-thirds of global primary energy demand is attributed to urban areas, inducing 71% of global direct energy-related GHG emissions (IEA 2014). According to the fifth assessment report on climate change from the Intergovernmental Panel on Climate Change (IPCC 2014), there is no doubt that anthropogenic greenhouse gas emissions (GHG) are responsible for the current climate change. Rapid urbanization increases demand for energy and consequently GHG emissions in cities.

The combination of the projected population and economic growth together with climate change results in placing greater stress on vital resources in the future if there is a continuation of the business-as-usual scenario. The energy sector in urban areas could thus play an important role in tackling climate change and in decreasing the carbon/energy footprint of urban areas. To support climate action plans and achieve more sustainable and resilient cities, understanding and managing urban energy use is essential (UN 2017). However, most of the time, the climatic challenges (e.g., comfort, energy demand, energy systems) are assessed individually while they are likely to be interrelated and require a holistic understanding of the ecosystem and human activities and the built environment (its form and materials) (Moonen et al. 2012). To achieve cities' energy and climate goals, it is necessary to

C. Ganem Karlen (✉) · J. Balter · N. L. Alchapar
INAHE, CCT-CONICET, Mendoza, Argentina
e-mail: cganem@mendoza-conicet.gov.ar

M. Palme, A. Salvati (eds.), *Urban Microclimate Modelling for Comfort and Energy Studies*, https://doi.org/10.1007/978-3-030-65421-4_22

reduce energy use in buildings through energy conservation and efficiency improvements. Computational tools can model performance of buildings at the urban scale to provide quantitative insights for stakeholders and inform their decision-making on urban energy planning, as well as building energy retrofits at scale, to achieve efficiency, sustainability, and resilience of urban buildings (Hong et al. 2020).

In recent years, multiple software tools have been developed for the assessment of urban climate, building energy demand, thermal comfort, and energy systems. Building energy simulation (BES) programs are capable of modelling building energy performance in detail in a dynamic model (building energy models (BEM)). Whole building energy simulation (BES) models play a significant role in the design and optimization of buildings. Simulation models may be used to compare the cost-effectiveness of energy conservation measures (ECMs) in the design stage as well as assess various performance optimization measures during the operational stage. Nevertheless, these tools often address only one or two of these urban planning aspects (Mauree et al. 2019). It is clear that building thermal performance and its energy consumption are affected by the energy exchange processes taking place between the outer skin or envelope of the building and the surrounding environment. It is a dynamic system in which there are continuous changes in a daily and seasonal range. Quantity and quality of the exposed envelope as well as albedo, vegetation, and urban geometry are significant factors in determining the impact of urban microclimates on energy building consumption.

Many previous studies have documented significant impacts of the urban microclimate on the thermal loads, and thus building energy performance. For instance, the effects of UHI, a term raised by Manley (1958), might lead to changes to building energy demand (e.g., a decrease in heating but an increase in cooling) depending on the city, type of building, or meteorological conditions, which yield a wide range of impacts on energy consumption (Davies et al. 2008; Xu et al. 2012; van Hooff et al. 2016). An example is that the weather data collected at airports are commonly used for building energy simulations and these data do not take into account real temperature distribution in cities. A recent modelling study based on a building located in the center of Rome, Italy, indicated that the energy consumption of cooling would be underestimated by 35–50% if the climatic effect of the heat island is not considered (Ciancio et al. 2018; Zinzi et al. 2018). This report is consistent with the results found by Correa (2006) for the city of Mendoza, Argentina, located in a similar climate.

The existing methods and tools have limitations in representing a realistic urban energy model and supporting energy performance evaluation at urban or neighborhood scales. There is a lack of an integrated approach for modelling and analyzing different components of urban energy use. In simulation-based methods, oversimplification of the urban context, urban microclimate, and inter-building effects are the major limitations (Abbasabadi and Ashayeri 2019). The urban context is often simplified or neglected in building energy models (BEMs) due to the difficulties of taking accurately into account all the heat fluxes emanating from the environment. Oversimplifying the urban context can impact the accuracy of BEM predictions. Nevertheless, several approaches can be used to allow for the impact of the urban environment on the dynamic behavior of a building, its heating and cooling demands, and thermal comfort (Lauzet et al. 2019). Co-simulation between models is necessary to understand the impact of systems interacting with each other in real time (Hong et al. 2020).

With the aim of finding a solution, in the scientific field there are some experiences regarding the integration of two different scales in dynamic simulation programs: for the building scale by using EnergyPlus or TRNSYS (Transient Systems Simulation) and for the neighborhood scale by using ENVI-met. Yang et al. (2012) developed a simulation methodology to integrate ENVI-met and EnergyPlus programs, based on the mapping of the surfaces of a building. The work uses the Building Controls Virtual Test Bed (BCTVB) software to develop a coupling module in order to transfer the simulation results between the two programs. Morakinyo et al. (2016) worked on validating the urban model in ENVI-met with in situ measurements and then specific external meteorological data was used to create the object of "a design day" and apply it as a boundary condition for each building in the simulation with EnergyPlus. Pastore et al. (2017) performed a simulation with ENVI-met of a neighborhood in order to obtain climatic boundary conditions (without on-site measurements). The climatic variables were used to simulate the neighborhood on a smaller scale with different scenarios. Finally, the microclimatic output variables were used to simulate with EnergyPlus temperatures and indoor comfort conditions.

Moreover, Kuo-Tsang and Yi-Jhen (2017) used the two simulators mentioned above based on data from meteorological bases of a typical year to simulate the microclimatic conditions of urban canyons for the warmest climatic conditions of the year. Lassandro and di Turi (2019) evaluated the possibility of integration of EnergyPlus and ENVI-met by comparing external surface temperature values in three different cities obtained by in situ thermograph measurements. On a local scale, Balter et al. (2018) evaluated the predictive potential of ENVI-met as a tool for building thermal analysis in an arid climate context. The high degree of adjustment—R^2 above 0.94—of the indoor air temperature monitored and adjusted with EnergyPlus versus that used with the microclimatic data calculated with the urban simulator ENVI-met supports the reliability of the predictive results of the integration method of both software.

There are also examples of integration of ENVI-met and TRNSYS software. Schwed and Sheng (2017) applied an algorithm to translate climatic data (EPW) into annual databases for specific locations and microclimates through simulation with ENVI-met. This information was used to simulate thermal conditions and cooling demands in buildings through the TRNSYS program. Perini et al. (2017) incorporated CFD-based simulation tool ENVI-met and TRNSYS (Transient Systems Simulation) by means of Grasshopper. The results confirmed the potentialities of combining both software for the calculation of urban features affecting urban microclimate (urban form, vegetation, canyon proportion) incrementing the simulation accuracy in terms of outdoor thermal comfort, especially during night.

Due to the complexity of the built environment and prevalence of large numbers of independent interacting variables, it is difficult to achieve an accurate representation of real-world building operation. Even though existing buildings and their microclimates can be monitored in situ, this practice is very useful but time and resource consuming. Only some punctual cases can be evaluated thoroughly, and it is impossible to measure buildings that are still in project. Therefore, by reconciling model outputs with measured data, we can achieve more accurate and reliable results in simulations (Ganem Karlen 2006; Coakley et al. 2014). Regarding all of the expressed above, it is clear that to provide reliable estimates microclimate models need to be parameterized based on empirically obtained data.

This chapter deals with the comparison of microclimatic information obtained with different methods:

1. Local weather stations average climate data from the past 20 years. As averaged, they prevent visualization of increasing temperatures over time due to climate change.
2. On-site microclimatic measurements: Very expensive and time consuming but very accurate.
3. Results of ENVI-met software: They allow a reduction in time and costs with respect to in situ measurements. Prior validation is necessary.

A case study in a high-density area in the city of Mendoza, Argentina, is presented in which year-round in situ measurements of temperature, humidity, radiation, and air movement were taken in two different scales: within the streets in a neighborhood and outside and inside a building. The micro-urban scale and the building scale were covered. A specific weather file was created for each scale, to be integrated in simulation software ENVI-met and EnergyPlus, respectively. Models were calibrated with the real data, to be run again with the information provided by local weather stations. Also, as a third term of comparison, the simulation workflow moves from the micro-urban- to a building-scale assessment by linking the ENVI-met software microclimatic results to the building energy simulation program EnergyPlus.

Results obtained with the three alternatives—(a) with the local weather stations' average climate input, (b) with the on-site microclimatic measurements, and (c) with ENVI-met software—are compared in order to assess each case reliability in assessing the impact of local urban climate on building energy performance.

22.1.1 Application Case: Mendoza, Argentina

The city of Mendoza is located at the foot of the Andes Mountains, in central western Argentina (32° 40′ South Latitude, 68° 51′ West Longitude, and 750 m above sea level). Although Mendoza is located in a semiarid continental climate with low percentages of atmospheric relative humidity, high heliophany, and annual precipitation of 218 mm—BWk according to Köppen classification (Kottek et al. 2006)—it does not follow a compact urban model. Its urban model is defined by its wide and tree-lined streets that form green tunnels. The checkered frame contains the buildings while the main strategy for minimizing the sun exposure is the vegetal frame. Mendoza's urban structure intermingles three types of meshes that overlap in space: a water network, the characteristic Spanish regular orthogonal grid, and, finally, a green mesh that arises due to the interaction of the first two. See climatic data for the city of Mendoza, Argentina, in Table 22.1.

The analysis was carried out within a grid of 6 × 6 ha, which totals 36 blocks in Mendoza Metropolitan Area (MMA). This area, mainly for residential use, was selected because of its high building density with more than 800 inhabitants/ha. It includes the five main squares of the city: Independence, Chile, San Martin, Spain, and Italy. The building height ranges from 3 to 57 m, with a higher percentage of buildings between 1 and 3 levels.

The distribution of surfaces in the evaluated sector corresponds to 44% of roofs, 28% of vehicular roads, 10% of squares, and 18% of private uncovered surfaces (courtyards, gardens). With respect to the material configuration, there is a generalized use of the calcareous pedestrian pavement in different colors: yellow (31%), red (21%), black (11%), and gray cement (22%), among others, with an average albedo of 0.3. The opaque surface materials that make up the facades of the urban canyon are predominantly of stone, brick, and/or paintings. The average albedo of vertical materials is 0.2. Most of the roofs are flat (80%) mainly built in reinforced

Table 22.1 Climatic data for the city of Mendoza, Argentina

Annual values	Average maximum temperature	22.6 °C
	Average minimum temperature	11.0 °C
	Mean temperature	15.9 °C
	Global horizontal irradiance	18 MJ/m^2
	Relative humidity	54.7%
	Mean rainfall	218 mm
July (winter)	Average maximum temperature	14.7 °C
	Average minimum temperature	3.4 °C
	Mean temperature	7.8 °C
	Thermal amplitude	11.3 °C
	Global horizontal irradiance	9.9 MJ/m^2
	Mean wind velocity	7.6 km/h
January (summer)	Average maximum temperature	30.1 °C
	Average minimum temperature	18.4 °C
	Mean temperature	25.3 °C
	Thermal amplitude	11.7 °C
	Global horizontal irradiance	25.7 MJ/m^2
	Mean wind velocity	10.8 km/h
Annual heating degree-days (T_b = 18 °C)		1384
Annual cooling degree-days (T_b = 23 °C)		163

Source: SMN (2019)

concrete and silver waterproof membranes (average albedo = 0.3). The remaining 20% is tilt and constructed with red ceramic tiles (average albedo = 0.35) (Alchapar et al. 2014).

The trees planted in alignment in the city of Mendoza correspond 78% to three species: "white mulberry" (*Morus alba*) 38%; "London plane" (*Platanus hispanica*) 21%; and "European ash" (*Fraxinus excelsior)* 19%. The remaining 22% corresponds to "American ash" (*Fraxinus americana*), "*Senegalia visco*" (*Acacia visco*), "white cedar" (*Melia azedarach*), "tipa" (*Tipuana tipu*), "poplar" (*Populus* spp.), and "Acer" (*Acer negundo*) (Martinez et al. 2017).

22.2 Microclimate and Buildings

22.2.1 Design and Validation of the Numerical Model at Urban Scale with ENVI-met

The free access program ENVI-met 3.1, developed by Michael Bruse at the Geography Institute of the University of Mainz, Germany, was used to perform the numerical design. This three-dimensional computational model works on an urban microclimate scale and simulates the interactions between the air and the surface of the urban environment with a typical resolution of 0.5–10 m in space and every 10 s in time. ENVI-met 3.1 is based on the fundamental laws of fluid dynamics and thermodynamics. The model includes the simulation of flows around and between buildings; heat and steam exchange processes of floor and wall surfaces; turbulence; vegetation parameters; bioclimatology; and dispersion of pollutants (Bruse 2006).

The data input for the numerical modelling of the urban area evaluated can be divided into three groups:

- Design of the physical space: The model was made in a 200 × 200 × 30 version. The resolution of the area is 3.5 × 3.5 × 3 m and a mesh of 197 (x) and 197 (y) because the reference surface is 690 × 690 m.
- Climate variables: The ENVI-met 3.1 software requires the entry of undisturbed variables that characterize the simulation boundary conditions, such as (1) wind speed, wind direction (m/s) at 10 m height, and roughness from ground ($z0$) to the reference point; (2) initial atmospheric temperature (K) and specific humidity (gr. water/kg air) at 2500 m high: the data were obtained from Francisco Gabrielli Airport—Station n° 87418, Aero de Mendoza Observatory—in collaboration with the University of Wyoming; and (3) relative humidity (%) at 2 m height, registered with ONSET Weather sensor, type HOBO H08-003-02 (fixed point observed -Po-, for adjustment).

Table 22.2 Input parameters for simulation in ENVI-met

Data		December 19th
Main	Wind velocity at 10 m [m/s]	2
	Wind direction (0:N; 90:E; 180:S; 270:O)	150
	Roughness *z*0 to the reference point	0.1
	Initial atmospheric temperature [K]	295
	Specific humidity at 2500 m [gr. water/kg air]	2.8
	Relative humidity at 2 m [%]	35
Buildings	Interior temperature [K]	295
	Wall thermal transmittance [W/m^2 K]	2
	Roof thermal transmittance [W/m^2 K]	0.7
	Wall albedo	0.2
	Roof albedo	0.3
Ground	Initial temperature—top layer (0–20 cm) [K]	293
	Initial temperature—middle layer (20–50 cm) [K]	293
	Initial temperature—deep layer (below 50 cm) [K]	293
	Relative humidity—top layer (0–20 cm) [%]	50
	Relative humidity—middle layer (20–50 cm) [%]	60
	Relative humidity —deep layer (below 50 cm)	60
Number of meshes		*x*-Meshes: 197
		y-Meshes: 197
		z-Meshes: 23
Size of the mesh (m)		d*x*: 3.5
		d*y*: 3.5
		d*z*: 5
Vegetation		Tree 15 m light: Height 15 m
		LAD1: 0.04; LAD6: 0.150; LAD10: 0.00 (m^2/m^3)
		Tree 15 m very dense: Height 15.0 m
		LAD1: 0.15; LAD6: 2.15; LAD10: 0.00 (m^2/m^3)
Road pavement		Concrete: albedo 0.3
Sky view factor at point A	Building	0.648
	Building + vegetation	0.332

Source: Own elaboration (2019)

- Thermal properties of the urban theoretical model: For the characterization of buildings it is necessary to define internal temperature, thermal transmittance, albedo, conductivity, and specific heat of walls, roofs, and pavements. In order to specify the behavior of the soil, temperature and humidity must be specified for different soil layers. Table 22.2 lists the simulation conditions and the properties used.

22.2.1.1 Monitoring and Calibration

The microclimatic conditions of the area were monitored during the summer period of 2012–2013 by means of the fixed reference point (Po) located within the analyzed road channel and indicated in Fig. 22.1. The urban monitoring point (Po) was selected from a grid of available monitoring points in the city, because it is at a close distance (less than 500 m) from the reference building (Bo). Moreover, the orientation of the road channels where both sensors (Po and Bo) are located is coincident. Both channels have an E-O arrangement. Ruiz et al. (2017) have demonstrated that E-O is the most appropriate orientation for urban channels in the city of Mendoza, allowing full North exposure in building facades (towards the Equator in the South Hemisphere). Furthermore, road channels with E-O orientation represent almost between 55% and 60% of the city's frame.

To calibrate the numerical model, the air temperature curve was contrasted Po, with the air temperature curve of the simulated area with ENVI-met (Ps). December 19th was selected as reference day, as ENVI-met version 3.1 software allows calculating the microclimatic characteristics of a single reference day. See Fig. 22.2.

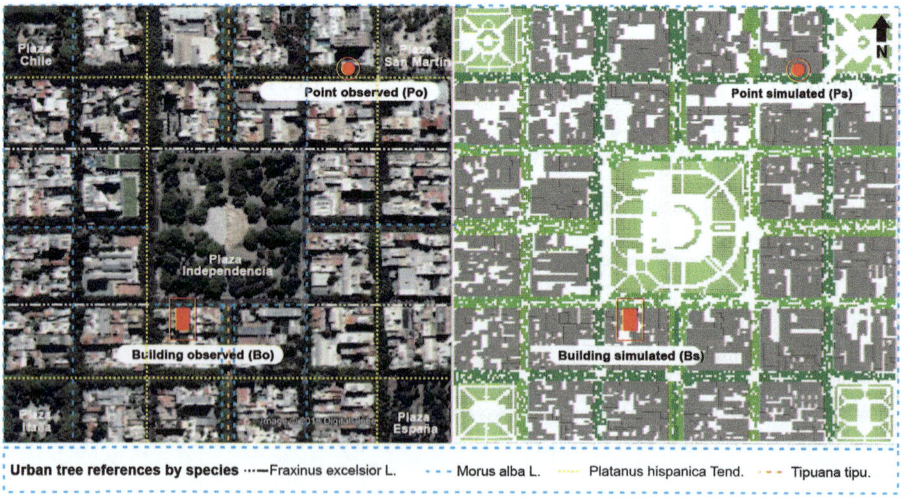

Fig. 22.1 Aerial view of the studied area and numerical model configuration with ENVI-met 3.1. Location of point observed (Po) and simulated (Ps), and building observed (Bo) and simulated (Bs). Description of urban trees. (Source: Own elaboration 2019)

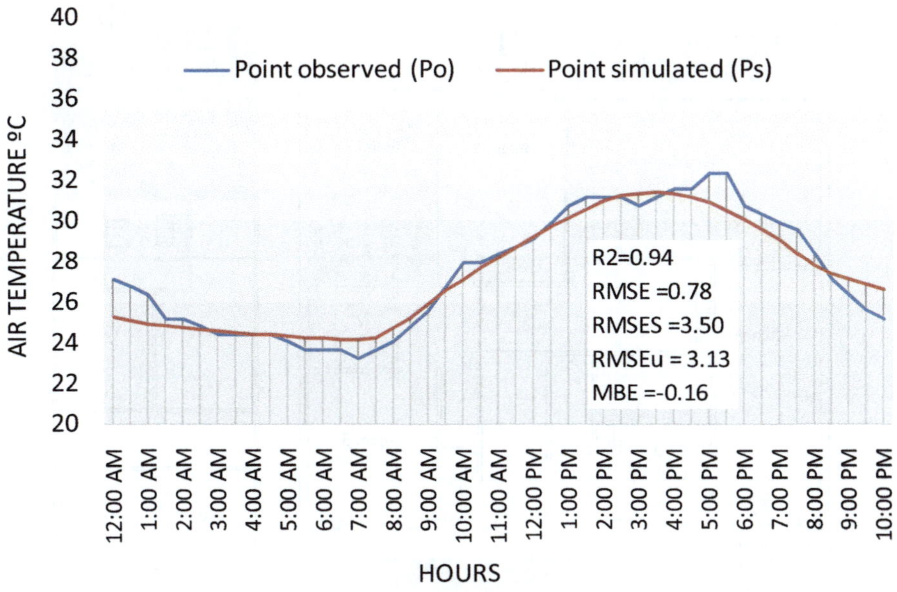

Fig. 22.2 Simulated (Ps) and observed (Po) point adjustment chart at a pedestrian scale (2.5 m) in the area for December 19th. (Source: Own elaboration 2019)

22.2.2 Design and Validation of the Numerical Model at Building Scale with EnergyPlus

The selected case is a north-facing building corresponding to the Tower typology. This typology corresponds to one of the three morphological classifications existing in Mendoza according to the building regulations at the time of its construction (Balter et al. 2013).

The monitored housing unit, located on the fifth level, has an area of 122 m^2 and location is frontal, that is, oriented to the public road and therefore to the urban forestry. Likewise, for its selection it was considered pertinent, since the building is implanted in front of a square that ensures the absence of shadows by the surroundings.

As for materiality, it is a mostly massive building, with 73% opaque materials in its envelope and 10% in the exposed reinforced concrete envelope (structure). The exterior walls are of 0.30 m hollow ceramic brick with plaster and paint without insulation and the interior divisions are of the same material of 0.10 m thickness. The glasses are simple 4 mm (K = 5.8 W/m^2 °C, solar factor = 0.87). As sun protection elements, the building has 1.20 m deep balconies and sliding shutters with white wooden lattices. See Fig. 22.3.

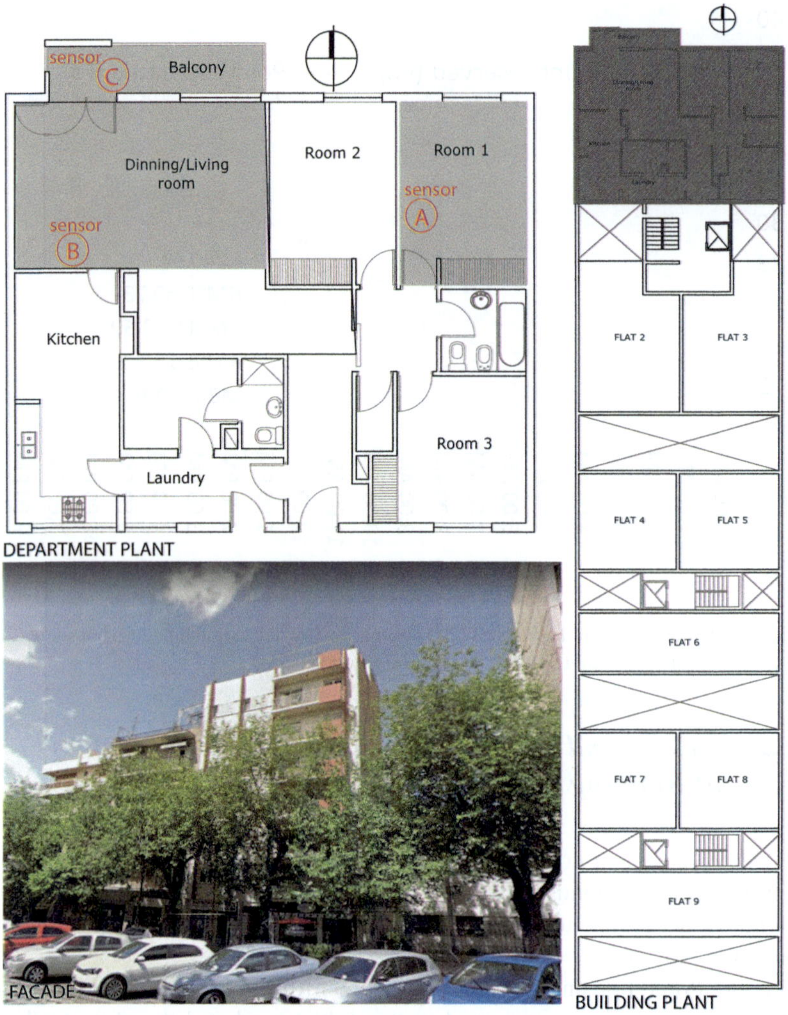

Fig. 22.3 Image and plan of the selected building. Plan of the monitored housing unit in three different heights: first floor, fifth floor, and eighth floor (upper level). (Source: Own elaboration 2019)

22.2.2.1 Monitoring and Calibration

On-site audits were carried out in a period from December 14th to January 10th. Micro-acquirers of HOBO U12 data of temperature and relative humidity were used and recording intervals were set every 15 min synchronously in all instruments, criteria adopted according to the recommendations of Longobardi and Hancock (2000). The sensors were installed in different environments: two inside (living room and bedroom) and one outside (balcony) protected from solar radiation. They were located at an average height of 2 m, following the recommendations of Kolher and Hassler (2002), and at a sufficient distance from the mass of the walls in order to avoid distortions in the data (Oke 2004).

The measurements made were used to validate the dynamic simulation model using the EnergyPlus program, version 8.8. This free program was developed by the Lawrence Berkeley National Laboratory (LBNL) and is currently the official simulation software of the US Department of Energy. Table 22.3 shows the description of the opaque materiality.

The validation was carried out in the fifth-level housing unit due, on the one hand, to the need to isolate the heating or cooling contributions inside the spaces: this unit was unoccupied, with windows and curtains closed, during the period of measurement; therefore, mechanical air-conditioning means were not used. This situation was observed in the audited thermal behavior and was corroborated in the interviews conducted with the users. The simulation was scheduled 10 days before the selected date as it is important that the physical model enters into regime in advance.

The climate file (EPW) used for the validation of the model was made with the temperature and relative humidity data monitored outside the building, and with the radiation data measured at the Mendoza Scientific and Technological Centre, which is located within a radius of 2.6 km, distance that is appropriate for the validity of

Table 22.3 Properties of the materials input for the EnergyPlus model

Layers	Roughness	Thickness (m)	Conductivity (W/m °C)	Density (kg/m³)	Specific heat (J/kg °C)
Exposed envelope (structure)					
Concrete	Rough	0.12	1.7	2400	800
Exposed envelope (exterior walls)					
Exterior revetment	Very rough	0.025	0.93	1900	1000
Hollow ceramic brick	Rough	0.3	0.41	1200	600
Interior revetment	Very rough	0.025	0.93	1900	1000
Interior walls					
Exterior revetment	Very rough	0.025	0.93	1900	1000
Hollow ceramic brick	Rough	0.1	0.41	1200	600
Interior revetment	Very rough	0.025	0.93	1900	1000
Floors/ceilings					
Plaster	Smooth	0.025	0.48	741.3	836.3
Concrete	Rough	0.12	1.7	2400	800
Concrete mortar	Medium roughness	0.1	1.63	2400	800
Wooden floor	Smooth	0.025	0.11	500	2800

the data according to what is indicated for the solar radiation records of the Solarimetric Network of the Argentine Republic (Grossi Gallegos et al. 1983). Figure 22.4 shows the curves of the adjustment made in the living room, for the period from December 19th to December 21st.

22.2.3 Simulation of Energy Consumption and Temperatures for Houses Above and Below the Tree Canopy, According to Different External Microclimatic Data

The energy consumption simulations were carried out by programming 24 °C thermostats for all areas of the building in three identical housing units located in different relative positions in height within the monitored building. A height limit to define whether the housing unit is located above or below the tree canopy was defined based on the types of trees: 12 m from street level for "white mulberry" (*Morus alba*) which is considered as from the fourth story (12 m) of the building. The following indices were determined (Balter 2015):

- Housing below the tree canopy: up to and including the fourth story (ground floor +3), corresponds to a height up to 12 m
- Housing above the tree canopy: starting from the fifth story (ground floor +4), corresponds to a height greater than 12 m

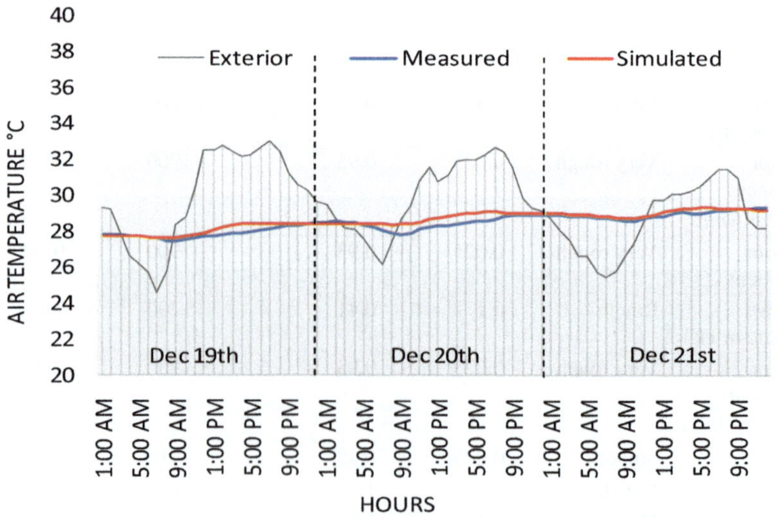

Fig. 22.4 Adjustment of the measurements in the EnergyPlus model for housing located above the tree canopy. (Source: Own elaboration 2019)

Three cases were defined to predict buildings' heating and cooling needs and energy consumption through the dynamic simulation method:

- *First floor* (under the tree canopy)
- *Fifth floor* (above the tree canopy) with adiabatic ground and roof because the housing unit is surrounded by other units under and above
- *Eighth and upper floor* (above the tree canopy) where the exposed envelope is extender because of the exposure of the roof to the microclimatic conditions

Predicting buildings' heating and cooling needs through dynamic simulation methods requires the input of hourly weather data, so as to represent the typical meteorological characteristics of a specific location. Hence, the so-called typical weather years (TWY), mainly deduced from multi-year records of meteorological stations outside the urban centers, cannot account for the complex interactions between solar radiation, wind speed, and high urban densities which lead to the formation of the urban heat island effect and to higher ambient air temperatures.

To take these facts into account, five different types of files (EPW extension) were formed for the entry of climatic data with the required data: global radiation on horizontal surface, diffuse radiation on horizontal surface, direct normal radiation to the beam, external dry bulb temperature, and external relative humidity, atmospheric pressure, and wind speed and direction. The data required for the conformation of the climate file (EPW) were repeated for the period of 10 days prior to the selected analysis date so that the model enters into regime with the calculated climatic variables. We worked with a total of five climate data files, according to the following configurations (see Fig. 22.5 for a location of the different points on the map):

1. Statistical data (Sd point): average climatic databases for a period of 14 years (2003–2017) from Francisco Gabrielli Airport—Station n° 87418, Mendoza Observatory, outside the city 8.6 km apart (OB Org 2019). In this case we worked with a single climate file, without differentiating the situation above and below the tree canopy.
2. Monitored building data (Bo point):

   2A. Temperature and relative humidity data obtained from measurements taken outside the monitored apartments above the tree canopy. Solar radiation data were those measured at the Mendoza Scientific and Technological Centre (CCT), 2.6 km apart.

   2B. Temperature and relative humidity data obtained from measurements taken outside the monitored apartments below the tree canopy. In order to contemplate the situation under the tree canopy, the incident radiation under the urban grove was affected with a local permeability factor. According to Cantón et al. (2003) global radiation permeability corresponding to the existing urban forestry in this case study (*Morus alba*) is 31.4% at midday in summer.

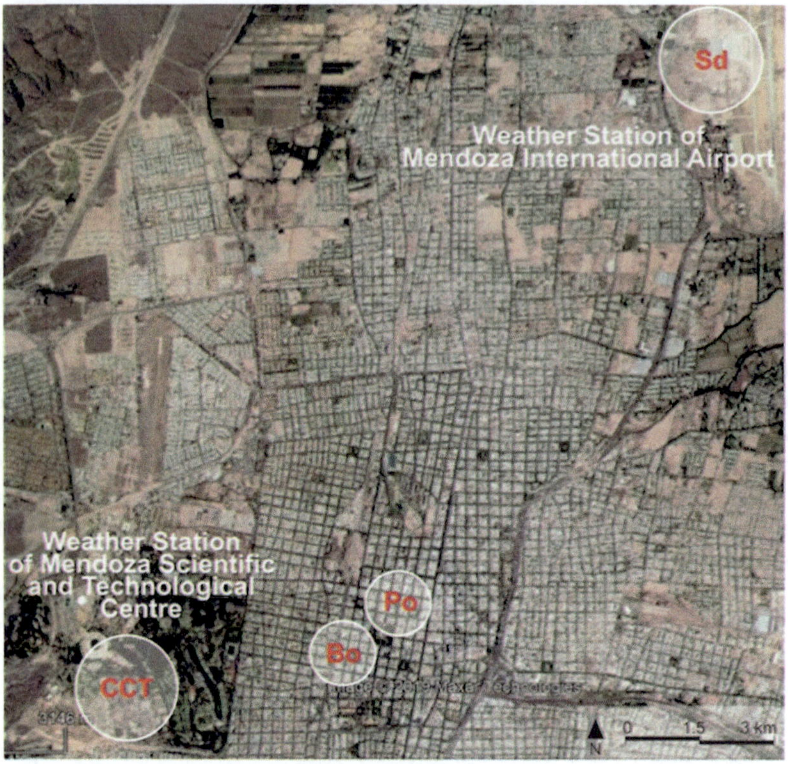

Sd point	To Airport whether station	8.6 Km
CCT point	To Scientific and to Technological Centre whether station	2.6 Km
Po Point	To Observed fixed point	0.7 Km

Fig. 22.5 Distance from building observed (Bo point). (Source: Own elaboration 2019)

3. Data calculated by ENVI-met (Po point): Two climate files were formed, based on the data obtained from the ENVI-met model for monitor points located in the road channel where the building is located, 0.7 km apart, simulated for December 19:

    3A. Above the tree canopy: for a height of 12.5 m
    3B. Below the tree canopy: for a height of 2.5 m

Figure 22.6 compares global radiation curves for each climatic model. Model 1 (statistical data) presents an averaged curve with a maximum value at midday of 1050 W/m^2. Model 2 (monitored building data) is with a maximum value at midday of 1067 W/m^2 (2A), which is reduced to 704 W/m^2 by the effect of the urban forestry in this case study (2B). Model 3 (data calculated by ENVI-met) is with values at midday of 1047 and 715 W/m^2 above and under the tree canopy, respectively (3A and 3B).

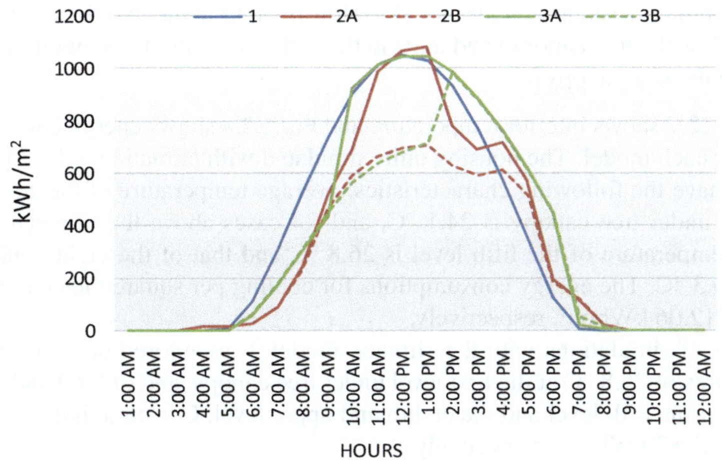

REFERENCES	1.Statistical data	2.Monitored data	3. Data calculated by ENVI-met
Above tree canopy	1	2A	3A
Below tree canopy		2B	3B

Fig. 22.6 Global radiation for the five climatic models. (Source: Own elaboration 2019)

The three models above the tree canopy (1, 2A, and 2B) follow similar curves coherent with the particularities of each methodology. In curve 2A the climatic particularities of the day within some hours were partially cloudy. These particularities also reflect on curve 2B. As for ENVI-met results, on curve 3B as per the simulation motor calculated from 2:00 p.m., the tree canopy no longer shadows the evaluated point, as values rise to couple with 3A curve (above the tree canopy).

22.3 Impact of Local Urban Climate on Building Energy Performance

Five simulation sequences were run in the validated BSM (EnergyPlus) with the established microclimatic models:

- One simulation sequence for model 1 (this model does not distinguish the relative position of the housing unit above or under the tree canopy): Units in the first floor, fifth floor, and eighth-upper floor
- Two simulation sequences for model 2: Housing unit in the first floor with model 2B (below the tree canopy) and units in the fifth and eighth floors with model 2A (above the tree canopy)

– Two simulation sequences for model 3: Housing unit in the first floor with model 3B (below the tree canopy) and units in the fifth and eighth floors with model 3A (above the tree canopy)

Figure 22.7 shows interior temperature and Fig. 22.8 shows energy consumption results for each model. The housing units simulated with climatic model 1 (statistical data) have the following characteristics: average temperature of the case of the first floor under tree canopy is 24.1 °C, and for cases above the tree canopy, the average temperature of the fifth level is 26.8 °C and that of the eighth and upper level is 30.3 °C. The energy consumptions for cooling per square meter were 0.16, 1.75, and 12.06 kWh/m^2, respectively.

Results of simulations with the climatic model 2 (monitored data) are average temperatures of 25.8 °C at the first level under tree canopy, and of 29.9 and 31.1 °C for cases of the fifth level and the eighth and upper level. Consumptions were 0.64, 4.28, and 25.87 kWh/m^2, respectively.

The resulting values simulated with climatic model 3 (ENVI-met) are average temperatures of 26.9 °C in the case of the first floor under tree canopy, 32.2 °C in the fifth level, and 36.9 °C in the eighth and upper level above the tree canopy. The consumptions in these cases were 0.89, 4.38, and 23.96 kWh/m^2, respectively.

Evaluation of the housing unit performance with climatic model 1 results in the lowest indoor temperatures and energy consumption. These are also the farther apart from in situ data obtained through monitoring the existing housing units. The underestimation of energy consumption presents differences of more than 50% of monitored values that double the simulated results. This is explained because of the averaged climatic information that flattens the curve and uncovers the climate change effect. Also the heat island effect is not taken into account as climatic data was taken outside the city. Moreover, without a differentiation of the relative height of the housing unit, particularities of the microclimate of the city of Mendoza are also omitted.

On the other hand, the results given with the ENVI-met model (3A and 3B) are very close to those of the monitored models (2A and 2B) with differences of ±3.5%. This study reveals the capabilities and advantages of working with this tool for the generation of microclimatic data, which when integrated with EnergyPlus presents a less expensive and fast alternative to in situ monitoring.

In conclusion, theoretical models of urban microclimatic simulation are an adequate and necessary tool to be able to not only diagnose the thermo-environmental behavior of an outdoor area, but also predict the building behavior inserted in a given existing urban configuration or proposed scenario.

The results of the study demonstrate the reliability of the proposed method. A combination of ENVI-met and EnergyPlus increases the simulation accuracy in terms of interior temperatures and energy consumption. The coupling of the urban and building dynamic prediction software can be performed all year round following the same methodology.

If local in situ temperatures are not available for the assessment, ENVI-met offers a valid option to climatic 10/20 years' averaged local data. The average climatic data is lower than real on-site measurements on the one hand, because it is

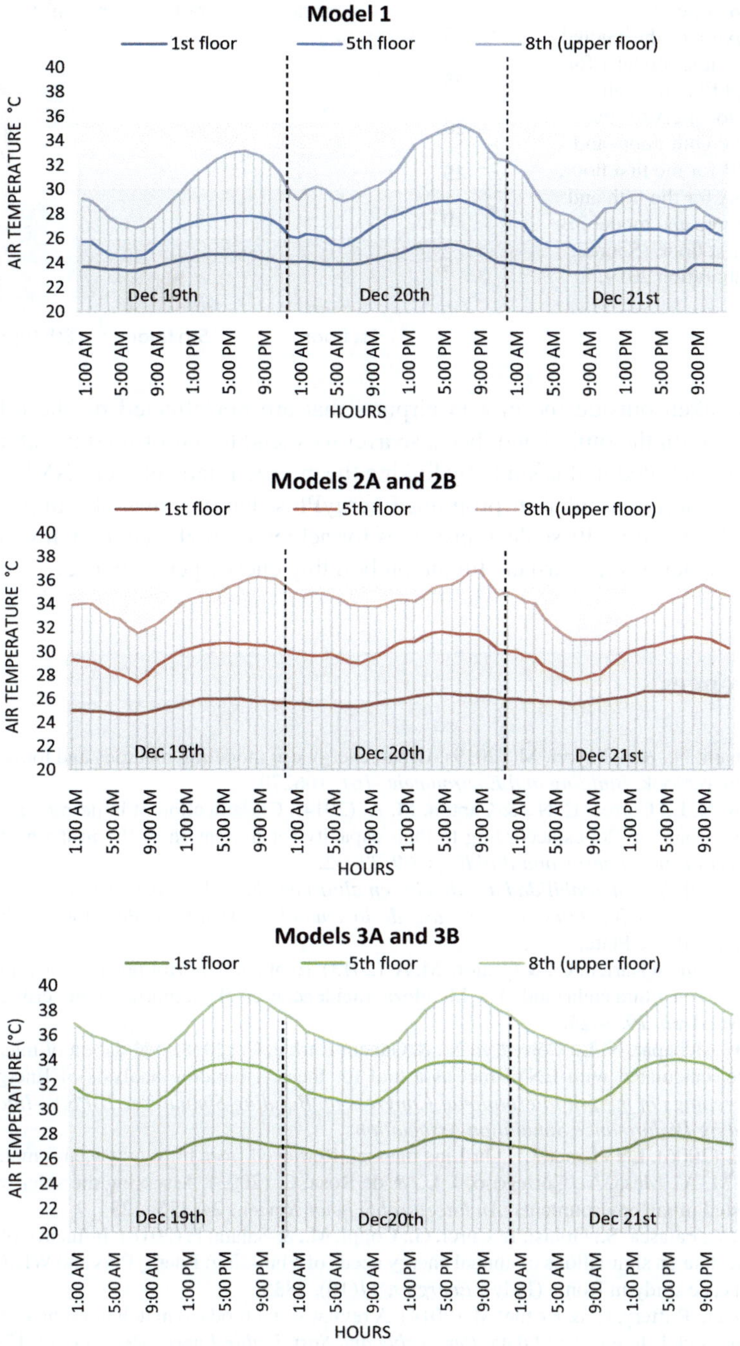

Fig. 22.7 Interior temperatures of the studied housing units. Model 1 for the first, fifth, and eighth floors. Model 2A for the fifth and eighth floors and model 2B for the first floor. Model 3A for the fifth and eighth floors and model 3B for the first floor. (Source: Own elaboration 2019)

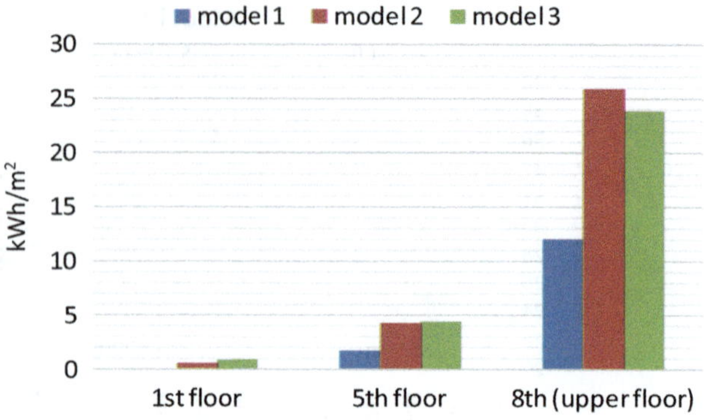

Fig. 22.8 Energy consumption of the studied housing units. Model 1 for the first, fifth, and eighth floors. Model 2A for the fifth and eighth floors and model 2B for the first floor. Model 3A for the fifth and eighth floors and model 3B for the first floor. (Source: Own elaboration 2019)

usually taken outside the city in airports that are not affected by the urban heat island, and on the other hand, because averages tend to uniform temperatures.

Data generated in this study by linking the microclimatic model ENVI-met to the building energy simulation program EnergyPlus demonstrated the importance of more robust and multi-scale approaches for achieving high-accuracy investigations on the impact of local urban climate on building energy performance.

References

Abbasabadi, N., & Ashayeri, M. (2019). Urban energy use modeling methods and tools: A review and an outlook. *Building and Environment, 161*, 106270.

Alchapar, N. L., Correa, E. N., & Cantón, M. A. (2014). Classification of building materials used in the urban envelopes according to their capacity for mitigation of the urban heat island in semiarid zones. *Energy and Buildings, 69*, 22–32.

Balter, J. (2015). *Sustentabilidad de edificios en altura en Ciudades-Oasis. Características arquitectónicas y ambientales para el caso de la ciudad de Mendoza*. PhD thesis. Universidad Nacional de La Plata.

Balter, J., Ganem Karlen, C., & Cantón, M. A. (2013). Evolución morfológica y materialización en edificios en altura en la ciudad de Mendoza. Incidencias en el comportamiento térmico interior. *Revista Área, 19*, 8–25.

Balter, J., Alchapar, N. L., Correa, E. N., & Ganem Karlen, C. (2018). Validation of microclimatic model calculated with ENVI-met as a tool for thermal building analysis of EnergyPlus. In *Proceedings of the 5th International Congress on Building Simulation. IBPSA—International Building Performance Simulation Association*.

Bruse, E. (2006). ENVI-met 3.1. On-Line manual. Retrieved from http://www.envi-met.com

Cantón, M. A., Mesa, A., Cortegoso, J. L., & de Rosa, C. (2003). Assessing the solar resource in forested urban environments. *Architectural Science Review, 24*, 115–123.

Ciancio, V., Falasca, S., Golasi, I., Curci, G., Coppi, M., & Salata, F. (2018). Influence of input climatic data on simulations of annual energy needs of a building: EnergyPlus and WRF modeling for a case study in Rome (Italy). *Energies, 11*(10), 2835.

Coakley, D., Raftery, P., & Keane, M. (2014). A review of methods to match building energy simulation models to measured data. *Renewable and Sustainable Energy Reviews, 37*, 123–141.

Correa, E. N. (2006). *Isla de Calor Urbana. El caso del área metropolitana de Mendoza*. PhD thesis. Facultad de Ciencias Exactas. Área Energías Renovables. Universidad Nacional de Salta.

Davies, M., Steadman, P., & Oreszczyn, T. (2008). Strategies for the modification of the urban climate and the consequent impact on building energy use. *Energy Policy, 36*(12), 4548–4551.

DOE—Department of Energy US. United Nations Population Division. (2017). *World population prospects: The 2017 revision, key findings and advance tables.*

Ganem Karlen, C. (2006). *Rehabilitación ambiental de la envolvente de viviendas. El caso de Mendoza.* PhD thesis. Escuela Técnica Superior de Arquitectura de Barcelona. Universidad Politécnica de Cataluña. Barcelona Tech.

Grossi Gallegos, H., Nollman, U., Lopardo, R., & Atienza, G. (1983). Evaluación preliminar del recurso solar en Argentina. *Actas de 8° ASADES, 1*, 179–194.

Hong, T., Chen, Y., Luo, X., Lou, N., & Lee, S. H. (2020). Ten questions on urban building energy modelling. *Building and Environment, 168*, 106508.

IEA—International Energy Agency. (2014). World energy outlook 2014.

IPCC—International Panel on Climate Change. (2014). *Climate change 2014: Synthesis report. Contribution of working groups I, II and III to the fifth assessment report of the Intergovernmental Panel on Climate Change.* Geneva: IPCC.

Kolher, N., & Hassler, U. (2002). The building stock as a research object. *Building Research & Information, 30*, 226–236.

Kottek, M., Grieser, J., Beck, C., Rudolf, B., & Rubel, F. (2006). World map of the Köppen-Geiger climate classification update. Sustainable United Nations. *Meteorologische Zeitschrift, 15*, 259–253.

Kuo-Tsang, H., & Yi-Jhen, L. (2017). Impact of street canyon typology on building's peak cooling energy demand. *Energy and Buildings, 154*, 448–464.

Lassandro, P., & di Turi, S. (2019). Multi-criteria and multiscale assessment of building envelope responsibility to rising heat waves. *Sustainable Cities and Society, 51*, 101755.

Lauzet, N., Rodler, A., Musy, M., Azamde, M. H., Guernouti, S., Mauree, D., & Colinart, T. (2019). How building energy models take the local climate into account in an urban context—A review. *Renewable and Sustainable Energy Reviews, 116*, 109390.

Levermore, G., Parkinson, J., Lee, K., Laycock, P., & Lindley, S. (2017). The increasing trend of the urban heat island intensity. *Urban Climate, 24*, 360–368. https://doi.org/10.1016/j.uclim.2017.02.004.

Longobardi, M., & Hancock, M. (2000). Field trip strategies. In *Proceedings of TIA, Oxford.*

Manley, G. (1958). On the frequency of snowfall in metropolitan England. *Quarterly Journal of the Royal Meteorological Society, 84*(359), 70–72.

Martinez, C., Ruiz, M., & Atencio, L. (2017). Proyecto integral de forestación y reforestación urbana para Mendoza. Sustenta-bilidad ambiental del bosque urbano para ciudades de zonas áridas de Argentina—Etapa 1 UNICIPIO—Provincia de Mendoza. *Instituto de Ambiente, Hábitat y Energía INAHE – CCT CONICET Mendoza, 1*, 96.

Mauree, D., Naboni, E., Coccolo, S., Perera, A. T. D., Nik, V. M., & Scartezzini, J. L. (2019). A review of assessment methods for the urban environment and its energy sustainability to guarantee climate adaptation of future cities. *Renewable and Sustainable Energy Reviews, 112*, 733–746.

Moonen, P., Defraeye, T., Dorer, V., Blocken, B., & Carmeliet, J. (2012). Urban physics: Effect of the micro-climate on comfort, health and energy demand. *Frontiers of Architectural Research, 1*(3), 197–228.

Morakinyo, T. E., Dahanayake, K. W. D., Adegun, O. B., & Balogun, A. A. (2016). Modelling the effect of tree shading on summer indoor and outdoor thermal condition of two similar buildings in a Nigerian university. *Energy and Buildings, 130*, 721–732.

OB Org—One Building Organization. (2019). *Repository of free climate data for building performance simulation.* Retrieved from http://climate.onebuilding.org/

Oke, T. R. (2004). *Initial guidance to obtain representative meteorological observations at urban sites. Iom Report.* Geneva: World Meteorological Organization.

Pastore, L., Corrao, R., & Heiselberg, P. K. (2017). The effects of vegetation on indoor thermal comfort: The application of a multi-scale simulation methodology on a residential neighborhood renovation case study. *Energy and Buildings, 146*, 1–11.

Perini, K., Ottelé, M., Giulini, S., Magliocco, A., & Roccotiello, E. (2017). Quantification of fine dust deposition on different plant species in a vertical greening system. *Ecological Engineering, 100*, 268–276.

Ruiz, M. A., Sosa, M. B., Correa, E. N., & Cantón, M. A. (2017). Design tool to improve daytime thermal comfort and nighttime cooling of urban canyon. *Landscape and Urban Planning, 167*, 249–256.

Schwed, D., & Sheng, M. (2017). Assessment of the annual energy demand for cooling of buildings in their urban context in 26 cities in China. *Procedia Engineering, 198*, 305–312.

SMN. Servicio Meteorológico Nacional Argentino [National Weather Service of Argentina] (2019). Fuerza Aérea Argentina. Ministerio de Defensa.

UN. United Nations (2017). *United Nations framework convention on climate change, initiatives in the area of human settlements and adaptation*. Retrieved from https://unfccc.int/news/rapidurbanization-increases-climate-risk-for-billions-of-people

van Hooff, T., Blocken, B., Timmermans, H. J. P., & Hensen, J. L. M. (2016). Analysis of the predicted effect of passive climate adaptation measures on energy demand for cooling and heating in a residential building. *Energy, 94*, 811–820.

Xu, T., Sathaye, J., Akbari, H., Garg, V., & Tetali, V. (2012). Quantifying the direct benefits of cool roofs in an urban setting: Reduced cooling energy use and lowered greenhouse gas emissions. *Building and Environment, 48*, 1–6.

Yang, X., Zhao, L., Bruse, M., & Meng, Q. (2012). An integrated simulation method for building energy performance assessment in urban environments. *Energy and Buildings, 54*, 243–251.

Zinzi, M., Carnielo, E., & Mattonia, B. (2018). On the relation between urban climate and energy performance of buildings. A three-years experience in Rome, Italy. *Applied Energy, 221*, 148–160.

Chapter 23
Green Infrastructure to Reduce the Energy Demand of Cities

Riccardo Privitera, Gianpiero Evola, Daniele La Rosa, and Vincenzo Costanzo

23.1 Introduction

23.1.1 Green Infrastructure and Urban Climate Regulation

A green infrastructure (GI) is a network of different ecological systems—both natural and artificial—that can be implemented at different spatial scales and are able to provide a wide set of services (Ahern 2007; Hansen and Pauleit 2014), contributing to the well-being of humans. Such services include CO_2 sequestration, reduction of air pollutants and noise, regulation of microclimate and heat island effect, oxygen production, water run-off regulation, water filtering, pollination, support to biodiversity and provision of cultural and recreational values (Pappalardo et al. 2017).

In cities, GI includes many ecosystems: urban parks and woodland, lakes, streams and rivers, different types of urban greenery, agricultural areas and other more artificial urban features such as green or vertical roofs (Wild et al. 2017). These ecosystems thus play a fundamental role in health, well-being and social safety (Kim and Miller 2019).

Among the services provided by GI, climate regulation is of utmost importance in cities, where the microclimatic benefits of urban vegetation can contribute to the mitigation of the urban heat island effect. Vegetation contributes to regulating the urban temperature through three main actions: shading the built environment,

R. Privitera (✉) · D. La Rosa
Department of Civil Engineering and Architecture, University of Catania, Catania, Italy
e-mail: riccardo.privitera@darc.unict.it; dlarosa@darc.unict.it

G. Evola · V. Costanzo
Department of Electric, Electronic and Computer Engineering, University of Catania, Catania, Italy
e-mail: gevola@unict.it; vincenzo.costanzo@unict.it

M. Palme, A. Salvati (eds.), *Urban Microclimate Modelling for Comfort and Energy Studies*, https://doi.org/10.1007/978-3-030-65421-4_23

modifying the airflow around it and directly lowering the outdoor air temperature through evapotranspiration processes (Hwang et al. 2017).

Climate regulation has relevant positive impacts on the energy demand of buildings, as demonstrated by a growing body of research and experimental measurements (Simpson and McPherson 1998; Konarska et al. 2015) and confirmed in different climate conditions and type of buildings (Laband and Sophocleus 2009; Palme et al. 2017). Shade effect by trees reduces the amount of solar energy a building absorbs and therefore reduces their cooling energy use. Vegetation also cools the air around buildings and this has an indirect effect on the need of energy for cooling inside the buildings. When these two processes act at city or district level, the positive cooling process induced by vegetation can generate relevant electrical energy savings (Simpson 1998; Wang et al. 2019) with performances that can be further increased by the evapotranspiration effect (Hsieh et al. 2018).

A GI is also defined as a large-scale planning framework that takes the environment as its core, delineating hubs, links and spots, and setting aside areas of core environmental function ahead of transformation (Benedict and McMahon 2006). It can increase the overall liveability of cities and regenerate deprived districts through urban regeneration solutions making use of different forms of greenery, improving the quality of life for citizens, reducing urban violence and decreasing social tensions through an increased social cohesion (Fan et al. 2017).

The objective of planning, managing and increasing GI is therefore of growing interest for public administrations that aim at sustainable development. Many approaches and urban policies for GI have been proposed (Lennon 2015), but a much limited number have been officially launched, tested and ultimately implemented (Di Marino et al. 2019). However, a consolidated translation of GI in planning processes is still complicated (Matthews et al. 2015).

Despite their importance in providing the above-mentioned ecosystem services, the implementation of GI has to face the lack of public open spaces to be set as new green areas. This may be even more problematic when suitable open spaces are located within private residential compounds or belong to private landowners, therefore hampering the economic feasibility of the new GI (Privitera and La Rosa 2018).

Different approaches and models have been developed for evaluating the potentiality of trees in cooling energy reduction, but limited research has focused on the different relations between buildings and trees in the urban environment (Farhadi et al. 2019; Wang et al. 2019), especially when private property must be considered to understand the feasibility of planning and design alternatives for GI. To this end, this chapter has a threefold objective. Firstly, it analyses how urban vegetation, and specifically urban trees, can generate a cooling effect and consequent energy saving in the urban environment, with particular reference to the residential stock. Secondly, it explores the relation between urban trees and buildings, identifying morphological constraints and characteristics of the built environment that must be considered when planning and designing new forms of greenery. Finally, it argues about the opportunities and limitations of tree planting for planning a new public GI.

23.1.2 The Effect of Trees on Energy Demand: Basic Principles

Trees in urban areas can help reduce buildings' energy needs through three main actions (Fig. 23.1): (1) shading buildings' surfaces and other man-made constructions, (2) modifying the airflow around buildings and (3) lowering the outdoor air temperature through evapotranspiration processes (Hwang et al. 2017).

As far as the *shading effect* is concerned, trees block direct sunlight and thus reduce the solar heat gains through the opaque and the glazed envelope. This effect is normally beneficial in the summer season, especially in buildings with low albedo and poor insulation; however, opposite results are expected in winter when solar gains are useful to reduce the heating load. Therefore, the choice of deciduous species is an effective and largely documented option, since the loss of leaves during autumn and winter allows for sunlight penetration and absorption by the building shell (Hwang et al. 2017; Calcerano and Martinelli 2016; Ko 2018; Nikoofard et al. 2011; Liu and Harris 2008). In fact, although trunks and branches still cast shadows in the winter season, their impact on the heating needs is found to be negligible, i.e. less than 0.5% according to Nikoofard et al. (2011). Wide consensus is also achieved upon the choice of the best orientation for tree planting: in the Northern Hemisphere, priority should be given to west orientation, followed by east orientation, whereas south orientation for trees has lower effects in summer and hinders solar gains in winter, thus justifying it only in mild and hot climates (Calcerano and Martinelli 2016; Hsieh et al. 2018; Ko 2018; Simpson and McPherson 1996). Local exceptions may occur: for example, coastal areas may frequently experience fog in the morning, which reduces the cooling benefits from trees planted along the east-facing walls (McPherson and Simpson 2003). North orientation is not considered because only diffuse radiation is received by the surfaces facing this direction.

The trees also generate a *sheltering effect*, which modifies wind patterns and significantly reduces the wind speed (*windbreak action*); this contributes to lowering the building energy needs in different ways, especially in winter. First, air pressure

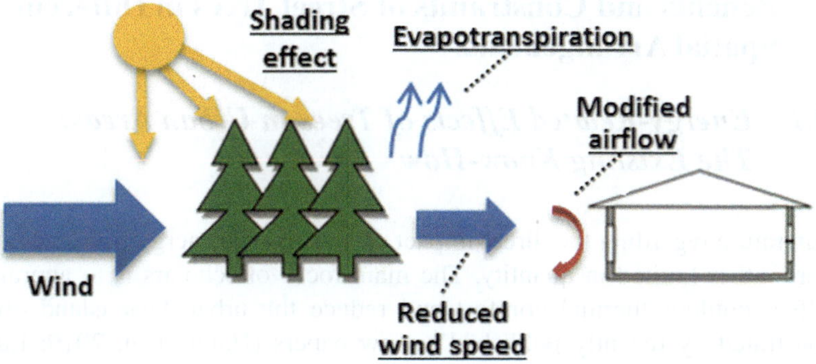

Fig. 23.1 Overview of the effects of trees on a building energy balance. (Modified after Liu and Harris 2008)

at the facades is reduced, thus diminishing the rate of adventitious air (cold in winter and hot in summer) that infiltrates through air leakages in the envelope (Liu and Harris 2008). Secondly, lower wind speeds result in a reduced intensity for the convective heat transfer with the outdoors, thereby lowering conductive heat losses. Finally, the outdoor air temperature decreases as well due to the reduced wind speed around the fabric; indeed, during daytime the ground surface—which is typically hotter than air—would release less heat by convection. The opposite occurs at night-time because of the low ground surface temperatures.

Evapotranspiration processes taking place in summer months reduce ambient air temperatures due to their endothermic nature (Berry et al. 2013). The physical processes behind such mechanisms are two, and depend on the soil and foliage layers. In fact, under the action of air temperature and relative humidity, as well as of wind and solar radiation, the soil releases water vapour to the air by evaporation. Further, plants absorb water from the soil through the roots and carry it in liquid form to the leaves where the photosynthesis process takes place. The stomata, which are tiny pores located in the leaves and in some stems, release water vapour to the air as a result of this endothermic mechanism. The combination of evaporation and plant transpiration results in the evapotranspiration process, which locally increases air humidity and reduces the dry bulb air temperature (Morakinyo et al. 2014): as a result, a reduction in the sensible cooling needs and an increase in the latent cooling needs of buildings can be observed (Calcerano and Martinelli 2016). However, according to some researches the inclusion of evapotranspiration effects generated from trees planted in urban areas does not significantly modify the predictive capability of simulation tools aimed to calculate the cooling needs of neighbouring buildings (Lun et al. 2011).

Finally, one more possible consequence of planting trees close to a building consists of the alteration of the long-wave heat exchange between the building and its surroundings. However, this effect is negligible if compared to the previously discussed processes (Simpson 2002).

23.2 Benefits and Constraints of Street Trees in Different Spatial Arrangements

23.2.1 Energy-Related Effects of Trees in Urban Areas: The Existing Know-How

The literature regarding the direct impact of trees on the energy needs of buildings appears rather limited in quantity. The main focus of scholars is to appraise how GI affect outdoor thermal comfort and reduce the urban heat island effect, as demonstrated by recently published review papers (Hami et al. 2019; Lai et al. 2019). Within this niche, in this section the reviewed papers have been categorised according to the processes outlined in the previous section. Particular attention is paid to discuss how trees have been modelled and how different design solutions, such as the number of trees to be planted, their species and arrangement close to the building, can impact the energy demand for space heating and cooling.

23.2.1.1 The Trees and Their Shading Effect

A noticeable simulation exercise on the appraisal of shading effects on cooling energy needs has been carried out by Calcerano and Martinelli (2016). In this work, dynamic energy simulations with the EnergyPlus-based Ladybug/Honeybee tools in Grasshopper environment have been coupled with genetic algorithms to determine the optimal arrangement of trees around a fictitious stand-alone building located in the Mediterranean climate of Rome (Italy) in order to maximise energy savings in the cooling season. The simulations considered both a single and a two-storey house, with 17 possible locations of idealised trees at a distance of 3 m from the outer walls, not overlapping with each other. The total height of the trees is 8 m, the trunk height is 2 m and the crown diameter is 6 m; no reference is made to any specific species. The complex three-dimensional shape of the trees is simplified and simulated by intersecting two octagons orthogonally.

The results of this exercise revealed that arranging the trees along east and west walls is a favourable option since solar gains are reduced by 50–70% in summer. The predicted energy savings thus range from around 11%, when only one tree is placed on the centred east position, up to around 45% when five trees are arranged around the building. Interestingly, it is found that the relative incidence of the number of trees on the energy consumption decreases as their number increases, the first two trees being sufficient to introduce a 20% reduction.

Hwang et al. (2017) used remote sensing to identify the existing placement of trees purposely planted around homes in three US cities with different climates (Minneapolis, Charlotte and Orlando, namely). After a sampling process that allowed to select representative parcels for each city, high-resolution aerial pictures and LiDAR data were used to extract information concerning houses and parcel orientation, tree stature, tree form and placement in relation with the houses. Then, these variables were used as an input for energy simulations in EnergyPlus to appraise the shading effect on the annual energy demand of a fixed-size house according to the existing configuration and to possible alternative arrangements. Both deciduous and evergreen trees have been modelled in three dimensions assuming an ellipsoid crown shape with two different sizes: small (height: 7.3 m, crown diameter: 7.6 m) and large (height: 15.2 m, crown diameter: 10.6 m). The possible locations of the trees are along eight directions (four cardinal and four inter-cardinal) and at three different distances from the building (5, 10 and 15 m, respectively). The study also considers different values of the crown opacity, corresponding to a shade coefficient as high as 75% during shoulder and summer seasons and 25% during winter for deciduous trees, and a fixed 80% value for the evergreen species. The outcomes of the simulations revealed that the optimal tree arrangement along the west and east walls would introduce an average additional energy savings of 22 ± 2.4 kWh per parcel annually with respect to the existing arrangement of the trees.

Other studies (Akbari and Taha 1992) found that a 30% increase in the vegetative cover of an urban neighbourhood in Toronto, with about three trees per house, would increase the annual heating energy consumption by up to 1% while decreasing the cooling energy consumption by up to 30%.

Another study conducted in Sacramento, California (Simpson and McPherson 1998), found that adding to each property an average of three deciduous trees would reduce the annual and peak cooling energy use by 22% and 7.1%, respectively, while also increasing the heating energy use by 5.9%. In this study, trees were located according to local tree-siting guidelines, and their distance from the houses had to be included between 4.5 and 10.5 m. Akbari et al. (1997) demonstrated that by shading two real houses from the south and the west sides with eight tall trees (about 6 m high) and eight short trees (about 2.4 m high), the measured cooling energy savings could range between 26% and 47%.

A different approach has been employed by Zhao et al. (2017), who looked for the optimal tree arrangement that maximises the shading of walls, windows and doors while minimising that at rooftop, but without calculating the direct energy savings. Here the authors presented a three-dimensional spatial optimisation model applied to residential premises located in the area of Tempe in Arizona (USA) by integrating geographic information system (GIS) data with spatial optimisation methods. Trees are modelled in a very detailed way using the three-dimensional tree plug-in available for Google SketchUp tool to create a fictitious thornless mature mesquite tree of 7 m height, 6 m crown diameter and 3 m trunk height, which is the most common species in desert Tempe area. An example of the outputs of this investigation is reported in Fig. 23.2 where the shading hours for a south-facing wall of a representative residential building on a summer day are plotted in a grey-colour scale considering only the hours with the highest insolation values, i.e. from 9:30 a.m. to 3:30 p.m. The main outcome is given by the cumulated shaded area, which can approach 40% in summer: this is expected to induce significant cooling energy savings.

In the framework of a wider investigation on the role played by external shading objects in the energy consumption for space heating and cooling of typical

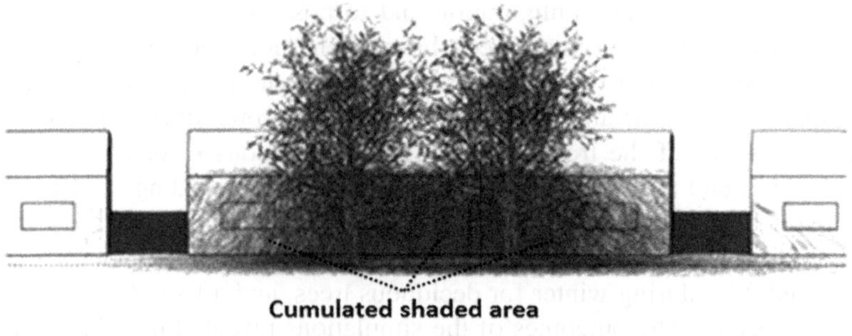

Fig. 23.2 Shading hours from trees on a summer day. (Modified after Zhao et al. 2017)

two-storey detached houses in Canada, Nikoofard et al. (2011) ran dynamic energy simulations with the ESP-r software for four major cities representative of the main national climate zones (Atlantic, Central, Prairies and Pacific, namely). Trees were modelled as simple fully opaque obstructions, with a trunk height of 2 m, a crown diameter of 4 m and a crown height of 6 m. In accordance with the findings of the other works discussed so far, cooling energy demand decreases more if trees are placed along the west and east walls. Such reduction falls within 10% and 20% with only a single tree placed at 4.7 m from the shaded house, but it can rise up to 30% if trees are planted at 2.3 m from the walls. Conversely, heating energy needs increase more if trees are planted along the south wall; however, this effect does not exceed 2% when a single tree is planted, and is negligible in case of deciduous trees.

Balter et al. (2016) demonstrated that, in high-rise buildings located in urban context, there is a remarkable difference in terms of thermal comfort between the apartments above and below the treetops, due to their shading effect. In lightweight buildings, the apartments above the treetops showed indoor temperature values exceeding by up to 1.5 °C those measured in the equivalent apartments placed at the lower floors. However, this effect is less remarkable when the building has a massive envelope: in this case, the indoor temperature difference between top and bottom floors does not exceed 0.7 °C. The results refer to rooms facing north: please consider that the study was conducted in the city of Mendoza (Argentina), in the Southern Hemisphere.

Finally, it is worth mentioning that a few works made an attempt to integrate urban microscale effects with single-building ones through experimental measurements and simulations carried out with ENVI-met software. In their research, Wu and Chen (2017) studied the dry bulb outdoor air temperature variation caused in summer by different vegetation patterns in a high-rise residential quarter of Beijing (China). The investigation was based on simulations validated through on-site measurements from a portable meteorological station, while the incident short-wave radiation at land surfaces was assessed through a purposely built module in GIS environment. The temperature variation was studied from the pedestrian level to the above-roof level according to the relative layers' height as reported in Fig. 23.3 (the top levels are indirectly affected by trees, while the two bottom layers are directly affected).

Based on a clustering analysis concerning the amount of short-wave radiation at land surfaces, low-, medium- and high-intensity solar radiation classes were first defined. For each class, the same amount of additional trees with respect to the existing configuration was added (10% more) to appraise the cooling energy savings potential of vegetation. The results showed a reduction in air temperatures at the pedestrian layer (C2), with an average predicted value of 0.22 °C in the areas with the highest radiation intensity. However, the reduction in the sensible heat released from the ground surface to the air varied according to the solar radiation classes. This means that planting new trees in areas that are already well shaded gives only a modest contribution.

A further step in this direction has been made by Aboelata and Sodoudi (2019) who ran ENVI-met simulations for a summer day considering a $250 \times 250\ m^2$ area

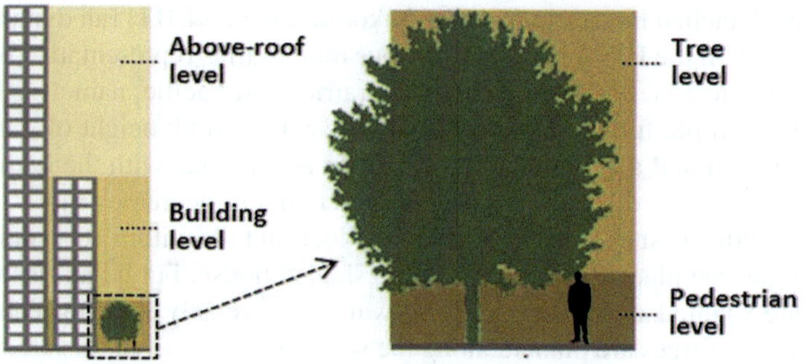

Fig. 23.3 Different atmospheric layers. (Modified after Wu and Chen 2017)

in downtown Cairo (Egypt) comparing the existing scenario with three design scenarios involving planting 30% more trees, 50% more trees and a combination of 30% more trees and 70% more grass, respectively. The height of the simulated trees is 11 m for *Ficus microcarpa* and 10 m for *Delonix regia*; the crown diameter is 11 m and 5 m, respectively. *Ficus microcarpa* was placed in large areas, main streets and parking lots, while *Delonix regia* was used in narrow streets; in any case, trees were arranged so as to keep the same shade density everywhere in the study area.

Then, some of the outputs from the urban-scale simulations—namely the outdoor dry bulb air temperature and the wind speed—were used to modify the weather file and to run the energy simulations of a selected ten-storey commercial building in DesignBuilder, thus accounting for some of the urban microclimate features together with the impact of trees on the sensible heat transfer. The best scenario for energy saving purposes involves planting 50% more trees, with an estimated 4% energy savings for cooling with respect to the existing scenario.

23.2.1.2 The Trees and Their Windbreak Effect

The sheltering effect of trees, which slow down the wind close to the buildings, thus reducing the convective heat losses and the infiltration rates, is particularly interesting in windy, cold and frequently overcast sites.

Liu and Harris (2008) considered the effect of deciduous shelterbelt trees on wind in Scotland while overlooking their shading effect. Firstly, they conducted a field experiment to derive a correlation between the convective coefficient and the wind speed for buildings protected by shelterbelt trees, based on measurements made on a single-storey building surrounded by trees on its east, south and west sides in Edinburgh. Then, the authors simulated a two-storey, middle-sized open-plan office building with ESP-r, in order to calculate its heating energy use. The arrangement of trees in simulations was established according to the following best practice rules:

- The ideal arrangement of shelterbelt trees is perpendicular to the prevailing wind.

- Shelterbelt trees should have a medium porosity (about 40%) so as to provide satisfactory wind speed reduction over a long distance.
- Shrubs should be planted at the basis of the trees, to avoid any vertical gaps occurring in the shelterbelt.
- The suggested distance between the shelterbelt trees and the protected building is five times the height of the trees.
- Although tall trees provide better sheltering effect, excessively tall trees cause insufficient daylight availability for the sheltered building. Therefore, the suggested height of the trees is 1.2 times the height of the protected building.
- The shelterbelt trees should protect the entire length of the building.

In the simulations, two different weather data sets were considered, namely the original Edinburgh's hourly weather data and a modified version taking the sheltering effect of trees into account, that is to say containing reduced values for the wind speed.

As shown in Fig. 23.4 for a typical winter day, the shelterbelt trees reduce by more than a half the external convective coefficient, with significant reduction of the hourly convective heat losses through the envelope; the predicted savings in the annual heating needs are 18.1%.

23.2.1.3 The Trees and Their Evapotranspiration Effect

Some studies investigated the effect of trees in urban areas by focusing on the reduction in the dry bulb temperature due to evapotranspiration, and by quantifying the consequent variation in the cooling load of buildings.

As an example, Shahidan et al. (2012) addressed this topic with reference to the tropical climate of Malaysia through validated numerical simulations performed with ENVI-met 3.1. Starting from the actual arrangement of the trees along a 4200 m long × 100 m wide boulevard in Putrajaya, they simulated several possible scenarios with different number of trees and different canopy density, measured by the leaf area index (from LAI = 0.9 to LAI = 9.7). The results of this study show that

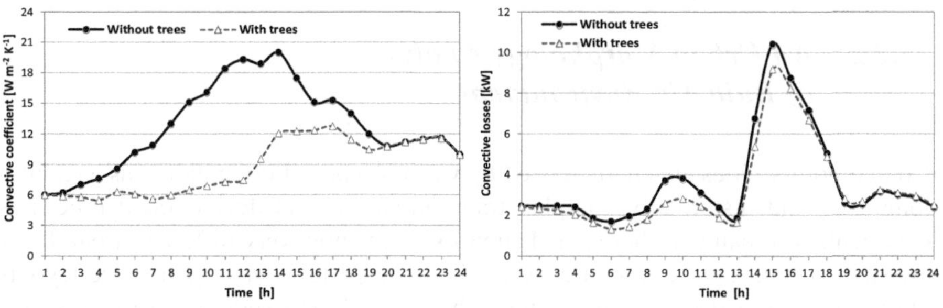

Fig. 23.4 Hourly variations in the external convective coefficient and the convective heat losses for a simulated building over a 24-h period. (Modified after Liu and Harris 2008)

doubling the number of trees would generate a mean reduction in the maximum outdoor air temperature by 0.9 °C for low-density foliage and 1.2 °C for high-density foliage. This results in a considerable decrease in the cooling load for six representative buildings, ranging from 7.4% to 21.8% in case of low-density foliage and from 12.1% to 29.4% in case of high-density foliage.

Furthermore, Hsieh et al. (2018) considered both tree shading and evapotranspiration for a building in Nanjing Forestry University, China. Based on 2-year experimental measurements taken in two different positions of the campus, they created different weather files to simulate the outdoor conditions close or far from the trees (*Platanus acerifolia* and *Metasequoia glyptostroboides*, with an average height of 20 m). Then, they were also able to correlate the outdoor air temperature to the evapotranspiration rate, which made it possible to consider the effect of other species with a different canopy density.

Based on these preliminary investigations, the authors considered four different scenarios:

- Scenario 1 simulates an exposed building without any surrounding trees; accordingly, simulations were run with the weather data taken far from the trees.
- Scenario 2 is the actual situation of the building, simulated by using actual climatic data measured close to the trees; the trees are placed at 6 m from the walls.
- Scenario 3 simulates a condition of larger shading area due to a lower distance between the trees and the building (3 m).
- Scenario 4 considers a different species of trees, having higher transpiration rate, but placed at 6 m from the walls, as in Scenario 2; the higher transpiration rate results in a lower outdoor air temperature.

The results of the simulations show that the absence of trees (Scenario 1) would increase the cooling load of the building by approximately 10.3% if compared to the actual situation (Scenario 2). In Scenario 3, a further reduction by 4.9% can be achieved. Finally, in Scenario 4 the use of species with higher transpiration rate results in a lower cooling demand than that for Scenario 2 (−2%), thus confirming the role of evapotranspiration.

23.2.2 An Urban Morphology Analysis of Built-Up Environment

Contemporary cities are characterised by various types of urban fabric as a result of long-term and historically complex transformations and development processes. City centres, urban peripheries and suburbs often show very different urban forms typified by built-up areas with different building patterns and open-space layouts. During the latest decades, increasing urban growth and sprawl processes shaped new settlements that even demolished the traditional and consolidated urban morphology types.

Urban morphology involves relationship among the primary elements of urban fabric such as plot, street, constructed space and open space (Levy 1999). While these features can be measured by parameters such as built-up density, building coverage ratio and percentage of open spaces, urban morphology is not only a simple aggregation of geometry factors. Firstly, specific land uses can allocate different functions and activities (residential, trading, manufacturing, services) which call for different layouts of urbanisation. Land covers can also variably occur and shape the green equipment of urban fabric with different types of greenery and other surfaces such as trees, shrubs and lawns, bare soil, and impervious surfaces. Moreover, landownership assets can strongly characterise the morphology types in that the proportion of public and private open spaces affects the spatial distribution of functions and activities and the related accessibility. Urban morphology also takes into account the interaction among a set of parameters that include constructional, functional and building regulation constraints concerning a range of different scales as well (Tsirigoti and Bikas 2017). At urban and district scales, morphology of urban fabrics is primarily resulting from public road network layouts enclosing built-up areas. Most common urban morphology types are grids, continuous linear/curvilinear settlements alongside the road, loops and cul-de-sac configurations, and hybrids of these types (Vernez Moudon 1992) (Fig. 23.5).

At block and sub-block levels, urban morphology types can be identified by looking at the different spatial arrangements in terms of relationship among buildings, streets and public and private open spaces (Vernez Moudon 1997). The type of buildings can vary from single and terraced houses to multistorey apartments

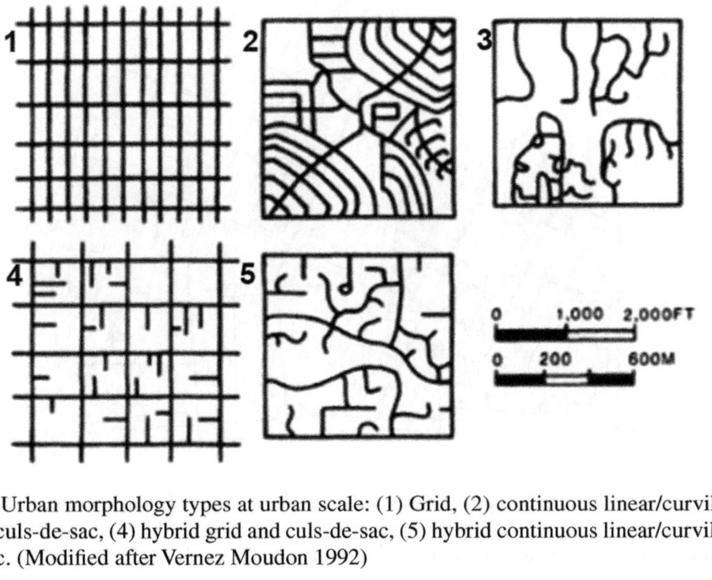

Fig. 23.5 Urban morphology types at urban scale: (1) Grid, (2) continuous linear/curvilinear, (3) loops and culs-de-sac, (4) hybrid grid and culs-de-sac, (5) hybrid continuous linear/curvilinear and culs-de-sac. (Modified after Vernez Moudon 1992)

and towers (with different number of storey and building height); streets can vary in terms of width and amount and type of lanes (cars, public transit, bikes, pedestrian sidewalks); open spaces can be characterised by public plots (squares, parking lots, urban gardens, parks) and private ones such as interior courtyards or front/rear/side setback yards. Different combinations of these features can produce different morphological layouts (Fig. 23.6). Moreover, within these different types, open spaces can be characterised by different size, shape, and land-use and land cover features.

Across the different urban morphology types, particularly in dense urban fabrics, the amount of existing trees and greenery within public and private open spaces is often limited as well as the opportunity to increase its extent and size. Indeed, a potential tree planting strategy can be affected by the complexity of existing morphology types, which show different patterns with different opportunities for green equipment enhancement.

Following up the objectives of shading buildings' surfaces, modifying the airflow around buildings and lowering the outdoor air temperature, street trees should be planted somewhere within the open spaces alongside the buildings. However, the actual availability of space for planting new trees could be strongly limited by the constraints posed by different morphological features. In urban fabrics characterised

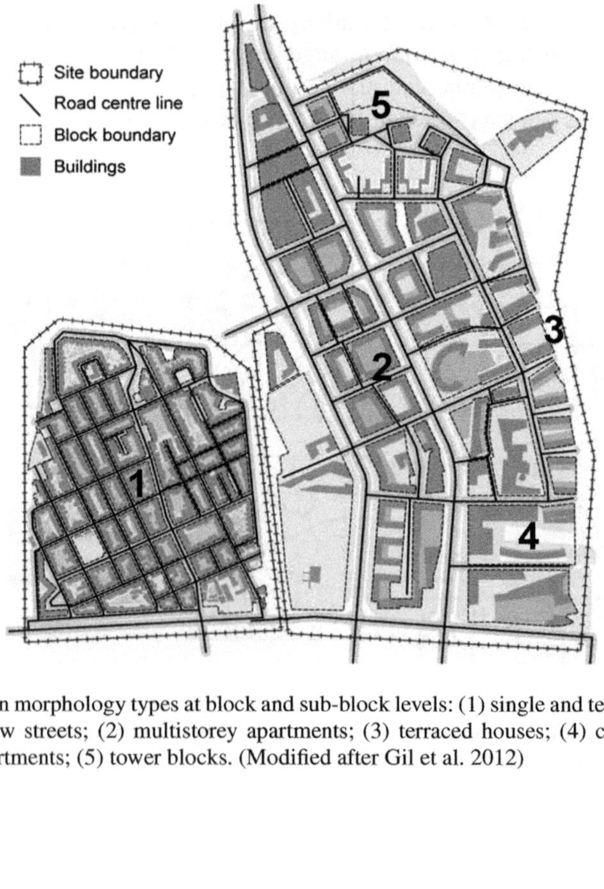

Fig. 23.6 Urban morphology types at block and sub-block levels: (1) single and terraced houses alongside narrow streets; (2) multistorey apartments; (3) terraced houses; (4) compounds of multistorey apartments; (5) tower blocks. (Modified after Gil et al. 2012)

by grid with terraced houses, public streets could be set up with narrow sidewalks, but in this case there would not be enough space for planting tree lines in front of the street-facing buildings. In multistorey apartments, usually arranged as compounds, open spaces are set up as private domestic gardens and shared open setback yards (walkways, common gardens, playgrounds, parking plots). In such a spatial arrangement, new trees could be bedded out in places too much faraway or even too much close to the buildings (Privitera et al. 2018). This would make it difficult to shade buildings alongside their total height and width front. Finally, in tower building morphology type, specific features of street trees (such as height, canopy and transparency) could not be suitable for adequately shading building facades.

In these respects, three main parameters should be taken into account as morphological key constraints when implementing a tree line planting strategy:

- Tree line 'distance' (distance between buildings and line of trees planted in front of them)
- Tree line 'length' (length of the line of trees planted in front of buildings)
- Tree 'species' (height, canopy width and transparency)

These three parameters can be selected as a basis for analysing the geometry of open space arrangement and its relationship with surrounding buildings, in order to check the suitability of the space available for tree planting (Fig. 23.7).

The actual availability of open space for tree planting should be checked in the proximity of buildings along all the facades. A limited range of distances could arise between the suitable tree line location and the facing building facade (Fig. 23.7a). The minimum distance could be too high, and the maximum distance too close, for shading the entire facade or a part of it. The length of tree line could be limited by the tiny availability of space and, as a consequence, only partially shade the facade (Fig. 23.7b); geometric shape of open space could restrict the bedding out of large canopy trees while smaller trees could not be enough for adequately shading the building and even obstruct the facade (Fig. 23.7c). Moreover, the final availability of space for tree planting could be further constrained and reduced by current uses and covers of the land plot.

The above-mentioned geometric parameters, and the other aspects such as land use, cover and ownership, are real constraints that can considerably reduce the potential tree planting effects and consequently reduce the benefits from tree shading. In this respect, a set of analyses such as landownership, land use and land cover analyses could support the detecting of public and private property assets and contribute to providing information about the quality of the open spaces through identifying residential and non-residential uses and the impervious, permeable and vegetated surfaces (Privitera et al. 2018). This would allow to build a more comprehensive strategy for planting new trees and more effectively achieve the objectives of reducing energy demand of buildings through cooling effect of street trees.

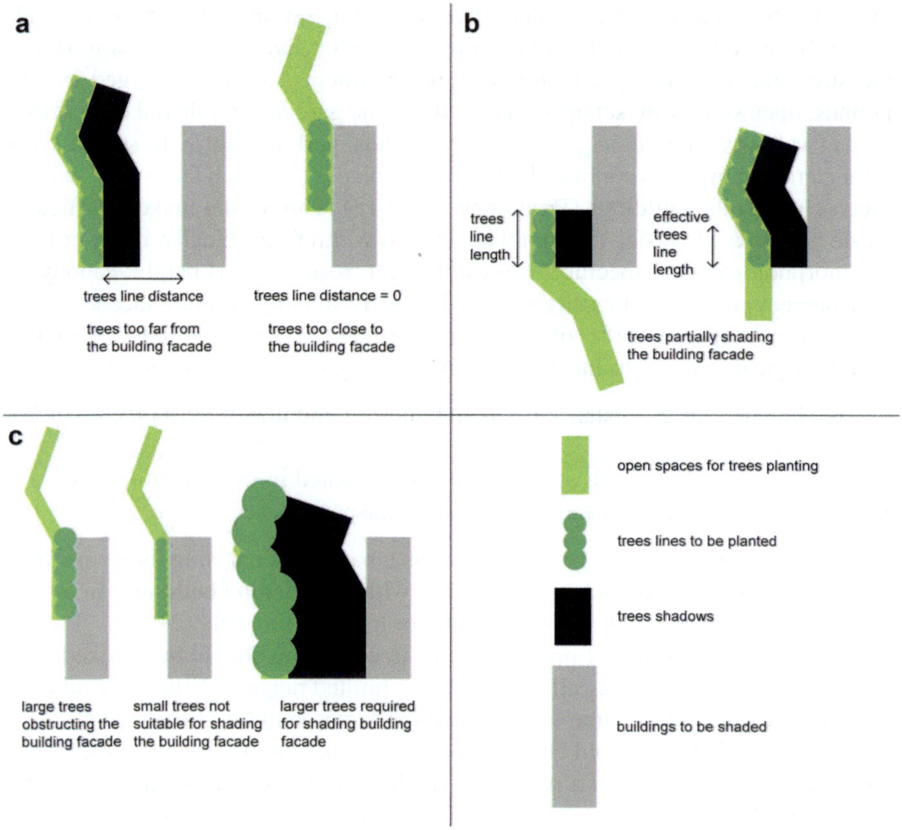

Fig. 23.7 Three parameters as morphological key constraints: Tree line distance (**a**), tree line length (**b**), tree species (**c**). (Authors' conceptualisation)

23.3 Implications for Urban Planning

A strategy of planting street trees for reducing the energy needs of buildings would eventually provide urban fabric with supplementary amount of greenery. This new potential vegetation represents a fundamental source for providing several ecosystem services as well as an important opportunity for planning a GI by designing and implementing public green spaces alongside buildings. The basic approach of just planting tree lines around buildings should move toward a wider strategy aimed at designing larger and accessible public green spaces equipped with pathways, bike lanes, playgrounds and benches to be enjoyed by local residents. Nevertheless, the complexity of urban morphology could produce poor-quality green spaces with fragmented and scattered patches that would entail difficulties in implementing a new public GI as a well-connected and accessible network.

Despite its recognised importance, the implementation of GI, especially in dense urban contexts, has to face several difficulties and limitations. Particularly, distinguishing public and private open space ownership is basically the main point to be taken into account in urban planning practices for exploring the economic viability of a GI development. Within private assets, implementation must deal with land acquisition of those private plots where GI could be designed. Effectively, land acquisition is directly linked to the economic feasibility of the green space implementation, as direct public acquisitions of land are often economically unsustainable for local administration and face resistance from private landowners (Kelly 1993; Bengston et al. 2004). Public intervention reflects a traditional approach characterised by a strong role of public actors who act through an articulated set of urban design projects and relevant state funding for covering investment and maintenance costs. However, the recent issues related to climate change risks imply a deep change in leading urban transformations, which cannot be addressed solely by the public actor through the application of traditional tools. Operating on a public-owned land implies less negotiation with several landowners, provided that there is a consensus on implementing these strategies at the political level, whereas action on private land has to be mainly based on market conveniences. Accordingly, compensation and incentive approaches for managing transformation of urban fabrics in private land parcels are the tools that can be used, such as Carbon Offset Fund and Transfer of Development Rights, respectively. The Carbon Offset Fund approach allows to reduce the net carbon emissions of new development, retrofitting or regeneration projects by collecting investment from developers and channelling it into a fund that is invested in a range of projects that deliver carbon savings (O'Rourke 2010). On the other hand, implementation of new public open spaces has often to deal with land acquisition of private open space plots that is economically unsustainable for local administration which faces resistance from landowners. The issue of economic feasibility for managing the provisioning of accessible public open spaces could be addressed through the Transfer of Development Rights and the purchase of development rights or conservation easements (Brabec and Smith 2002). This approach defines an area to be protected from development and one where development will be allowed to occur. Landowners can transfer the rights from the area to be protected (sending area) to the area to be developed (receiving area). As a consequence, the parcel from which the development rights are being transferred can no longer be developed, or can be developed only in a limited way. As a result, landowners are compensated for regulatory restrictions that reduce the property values (Bengston et al. 2004).

Nevertheless, these economic advantages for private landowners (to be added to the energy savings due to the shading effect of trees) could not be compensated by other relevant limitations derived by the potential reduction of their private open spaces. Decreasing of usability level of private open spaces after tree planting could arise in terms of loss of parking plots, which could be quite relevant in that urban fabrics with no public parking area available.

After tree planting, other significant disservices for private landowners could occur such as view blocking, increasing greenery maintenance costs, cleaning costs and storm drain obstruction due to the falling leaves and bird droppings.

23.4 Conclusions

The importance of urban vegetation in providing ecosystem services is widely acknowledged and efforts are increasingly being made to enhance their quality and quantity in cities. This is even more important for cities where greenery is lacking or where the access to green spaces is either low or unevenly distributed.

This chapter has analysed the complex relation between vegetation and built environment, by discussing the cooling potential of urban vegetation, the morphological relation between buildings and trees and the opportunities and limitations for planning a new public GI.

The review of existing research highlighted some general limitations in the current approaches to measure the shading effect of vegetation, due to the limited number of empirical field studies, the small sample size and the lack of validation for the performed simulations. However, some conclusions were derived, which can provide practical indications for planning and design.

Looking at the effectiveness in cooling the buildings, vegetation shows a significant impact especially on stand-alone buildings, whereas in dense urban areas with limited solar access its role is minor. Ideally, trees planted for summer shade should shelter western and eastern windows and walls, and be tall enough not to block views or breezes. The shading effect tends to decrease as the building-to-tree distance increases: this implies that, where possible, new trees should be planted so that, when reaching their maturity stage, the edge of the canopy is close to the building wall (Simpson 1998). Energy savings do not scale linearly with distance, and there is often a maximum impact on energy reduction for a distance between 4.6 and 15 m (Simpson 2002).

Although trees that are too far from buildings do not provide direct shading, they can still induce modifications to the local microclimate in terms of windbreak effect, whose magnitude in terms of cooling demand reduction is around 25–50% of what is ensured by tree shading on a west wall. If evergreen trees planted in proximity to a building may reduce solar gains in winter, thus potentially increasing the costs for space heating, this effect is usually offset by the reduced wind speed, which limits the convective heat losses. Moreover, changes in the convective heat transfer coefficient at the surface of buildings, originated by shelterbelt trees, have a relatively low impact for highly insulated facades, while for glazed facades they become significant. Finally, changes in the infiltration rates, due to the reduced wind speed ensured by shelterbelt trees, are almost negligible in modern airtight buildings, but cannot be neglected in old non-airtight buildings.

The complex relation among different land covers in the built environment requires a better understanding of the morphological features and the different property assets to develop feasible and effective policies aimed at implementing GI while reducing the energy demand of cities. To this end, a comprehensive planning policy could be based on an assessment of the economic viability of public investment and maintenance costs and related public benefits (such as ecosystem services) as well as private costs (loss of private/shared open spaces, parking plots, view blocking, cleaning costs and storm drain obstruction due to the falling leaves and bird droppings) and benefits (building energy savings, gained development credits, etc.). Such a strategy should be aimed at identifying different scenarios and alternatives of a GI that could better balance public and private costs and benefits. However, education and outreach effort should be made to raise the awareness of land developers and private owners on the shading potential of trees in reducing energy consumption of buildings, so as to ease the acceptability of any planning interventions of tree planting in residential contexts.

References

Aboelata, A., & Sodoudi, S. (2019). Evaluating urban vegetation scenarios to mitigate urban heat island and reduce buildings' energy in dense built-up areas in Cairo. *Building and Environment, 166*, 106407. https://doi.org/10.1016/j.buildenv.2019.106407.

Ahern, J. (2007). Green infrastructure for cities: The spatial dimension. In V. Novotny (Ed.), *Cities of the future: Towards integrated sustainable water and landscape management* (pp. 267–283). London: IWA Publishing.

Akbari, H., & Taha, H. (1992). The impact of trees and white surfaces on residential heating and cooling energy use in four Canadian cities. *Energy, 17*, 141–149. https://doi.org/10.1016/0360-5442(92)90063-6.

Akbari, H., Kurn, D. M., Bretz, S. E., & Hanford, J. W. (1997). Peak power and cooling energy savings of shade trees. *Energy and Buildings, 25*, 139–148. https://doi.org/10.1016/S0378-7788(96)01003-1.

Balter, J., Ganem, C., & Discoli, C. (2016). On high-rise residential buildings in an oasis-city: Thermal and energy assessment of different envelope materiality above and below tree canopy. *Energy and Buildings, 113*, 61–73.

Benedict, M. A., & McMahon, E. (2006). *Green infrastructure: Linking landscapes and communities*. Washington, DC: Island Press.

Bengston, D. N., Fletcher, J. O., & Nelson, K. C. (2004). Public policies for managing urban growth and protecting open space: Policy instruments and lessons learned in the United States. *Landscape and Urban Planning, 69*, 271–286.

Berry, R., Livesley, S. J., & Aye, L. (2013). Tree canopy shade impacts on solar irradiance received by building walls and their surface temperature. *Building and Environment, 69*, 91–100. https://doi.org/10.1016/j.buildenv.2013.07.009.

Brabec, E., & Smith, C. (2002). Agricultural land fragmentation: The spatial effects of three land protection strategies in the eastern United States. *Landscape and Urban Planning, 58*, 255–268.

Calcerano, F., & Martinelli, L. (2016). Numerical optimisation through dynamic simulation of the position of trees around a stand-alone building to reduce cooling energy consumption. *Energy and Buildings, 112*, 234–243. https://doi.org/10.1016/j.enbuild.2015.12.023.

Di Marino, M., Tiitu, M., Lapintie, K., Viinikka, A., & Kopperoinen, L. (2019). Integrating green infrastructure and ecosystem services in land use planning. Results from two Finnish case studies. *Land Use Policy, 82*, 643–656. https://doi.org/10.1016/j.landusepol.2019.01.007.

Fan, P., Ouyang, Z., Basnou, C., Pino, J., Park, H., & Chen, J. (2017). Nature-based solutions for urban landscapes under post-industrialization and globalization: Barcelona versus Shanghai. *Environmental Research, 156*, 272–283.

Farhadi, H., Faizi, M., & Sanaieian, H. (2019). Mitigating the urban heat island in a residential area in Tehran: Investigating the role of vegetation, materials, and orientation of buildings. *Sustainable Cities and Society, 46*, 101448.

Gil, J., Beirão, J. N., Montenegro, N., & Duarte, J. P. (2012). On the discovery of urban typologies: Data mining the many dimensions of urban form. *Urban Morphology., 16*(1), 27–40.

Hami, A., Abdi, B., Zarehaghi, D., & Maulan, S. B. (2019). Assessing the thermal comfort effects of green spaces: A systematic review of methods, parameters, and plants' attributes. *Sustainable Cities and Society, 49*, 101634. https://doi.org/10.1016/j.scs.2019.101634.

Hansen, R., & Pauleit, S. (2014). From multifunctionality to multiple ecosystem services? A conceptual framework for multifunctionality in green infrastructure planning for urban areas. *Ambio, 43*, 516–529.

Hsieh, C. M., Li, J. J., Zhang, L., & Schwegler, B. (2018). Effects of tree shading and transpiration on building cooling energy use. *Energy and Buildings, 159*, 382–397. https://doi.org/10.1016/j.enbuild.2017.10.045.

Hwang, W. H., Wiseman, P. E., & Thomas, V. A. (2017). Enhancing the energy conservation benefits of shade trees in dense residential developments using an alternative tree placement strategy. *Landscape and Urban Planning, 158*, 62–74. https://doi.org/10.1016/j.landurbplan.2016.09.022.

Kelly, E. D. (1993). *Managing community growth: Policies, techniques, and impacts* (p. 264). Westport, CT: Praeger.

Kim, G., & Miller, P. A. (2019). The impact of green infrastructure on human health and well-being: The example of the Huckleberry Trail and the Heritage Community Park and Natural Area in Blacksburg, Virginia. *Sustainable Cities and Society, 48*, 101562.

Ko, Y. (2018). Trees and vegetation for residential energy conservation: A critical review for evidence-based urban greening in North America. *Urban Forestry and Urban Greening, 34*, 318–335. https://doi.org/10.1016/j.ufug.2018.07.021.

Konarska, J., Uddling, J., Holmer, B., Lutz, M., Lindberg, F., Pleijel, H., & Thorsson, S. (2015). Transpiration of urban trees and its cooling effect in a high latitude city. *International Journal of Biometereology, 60*(1), 159–172.

Laband, D., & Sophocleus, V. (2009). An experimental analysis of the impact of tree shade on electricity consumption. *Arboriculture and Urban Forestry, 35*, 197–202.

Lai, D., Liu, W., Gan, T., Liu, K., & Chen, Q. (2019). A review of mitigating strategies to improve the thermal environment and thermal comfort in urban outdoor spaces. *Science of the Total Environment, 661*, 337–353. https://doi.org/10.1016/j.scitotenv.2019.01.062.

Lennon, M. (2015). Green infrastructure and planning policy, a critical assessment. *Local Environment, 20*(8), 957–980.

Levy, A. (1999). Urban morphology and the problem of the modern urban fabric: Some questions for research. *Urban Morphology, 3*(2), 79–85.

Liu, Y., & Harris, D. J. (2008). Effects of shelterbelt trees on reducing heating-energy consumption of office buildings in Scotland. *Applied Energy, 85*(2–3), 115–127. https://doi.org/10.1016/j.apenergy.2007.06.008.

Lun, I., Mochida, A., & Ooka, R. (2011). Progress in numerical modelling for urban thermal environment studies. *Advances in Building Energy Research, 3*(1), 147–188. https://doi.org/10.3763/aber.2009.0306.

Matthews, T., Lo, A. Y., & Byrne, J. A. (2015). Reconceptualizing green infrastructure for climate change adaptation: Barriers to adoption and drivers for uptake by spatial planners. *Landscape and Urban Planning, 138*, 155–163. https://doi.org/10.1016/j.landurbplan.2015.02.010.

McPherson, E. G., & Simpson, J. R. (2003). Potential energy savings in buildings by an urban tree planting programme in California. *Urban Forestry and Urban Greening, 2*, 73–86. https://doi.org/10.1078/1618-8667-00025.

Morakinyo, T. E., Balogun, A. A., & Adegun, O. B. (2014). Comparing the effect of trees on thermal conditions of two typical urban buildings. *Urban Climate, 3*, 76–93. https://doi.org/10.1016/j.uclim.2013.04.002.

Nikoofard, S., Ugursal, V. I., & Beausoleil-Morrison, I. (2011). Effect of external shading on household energy requirement for heating and cooling in Canada. *Energy and Buildings, 43*(7), 1627–1635. https://doi.org/10.1016/j.enbuild.2011.03.003.

O'Rourke, T (2010). *Scoping report: Feasibility of a carbon offset mechanism for Cambridgeshire for Cambridgeshire horizons—Final report* (Technical report), Cambridge.

Palme, M., Inostroza, L., Villacreses, G., Lobato, A., & Carrasco, C. (2017). From urban climate to energy consumption. Enhancing building performance simulation by considering the urban heat island effect. *Energy and Buildings, 145*, 107–120.

Pappalardo, V., La Rosa, D., La Greca, P., & Campisano, A. (2017). The potential of GI application in urban runoff control for land use management: A preliminary evaluation from a southern Italy case study. *Ecosystem Services, 26*(Part B), 345–354. https://doi.org/10.1016/j.ecoser.2017.04.015.

Privitera, R., & La Rosa, D. (2018). Reducing seismic vulnerability and energy demand of cities through green infrastructure. *Sustainability, 10*(8), 2591. https://doi.org/10.3390/su10082591.

Privitera, R., Palermo, V., Martinico, F., Fichera, A., & La Rosa, D. (2018). Towards lower carbon cities: Urban morphology contribution in climate change adaptation strategies. *European Planning Studies, 26*, 812–837.

Shahidan, M. F., Jones, P. J., Gwilliam, J., & Salleh, E. (2012). An evaluation of outdoor and building environment cooling achieved through combination modification of trees with ground materials. *Building and Environment, 58*, 245–257. https://doi.org/10.1016/j.buildenv.2012.07.012.

Simpson, J. R. (1998). Urban forest impacts on regional cooling and heating energy use: Sacramento County case study. *Journal of Arboriculture, 24*(4), 201–214.

Simpson, J. R. (2002). Improved estimates of tree-shade effects on residential energy use. *Energy and Buildings, 34*, 1067–1076. https://doi.org/10.1016/S0378-7788(02)00028-2.

Simpson, J. R., & McPherson, E. G. (1996). Potential of tree shade for reducing residential energy use in California. *Arboriculture and Urban Forestry, 22*, 10–18.

Simpson, J. R., & McPherson, E. G. (1998). Simulation of tree shade impacts on residential energy use for space conditioning in Sacramento. *Atmospheric Environment, 32*, 69–74. https://doi.org/10.1016/S1352-2310(97)00181-7.

Tsirigoti, D., & Bikas, D. (2017). A cross scale analysis of the relationship between energy efficiency and urban morphology in the Greek city context. *Procedia Environmental Sciences, 38*, 682–687.

Vernez Moudon, A. (1992). The evolution of the twentieth-century residential forms: An American case study. In J. W. R. Whitehead & P. J. Larkham (Eds.), *Urban landscapes-international perspectives*. London-New York: Routledge.

Vernez Moudon, A. (1997). Urban morphology as an emerging interdisciplinary field. *Urban Morphology, 1*, 3–10.

Wang, Y., Ni, Z., Chen, S., & Xia, B. (2019). Microclimate regulation and energy saving potential from different urban green infrastructures in a subtropical city. *Journal of Cleaner Production, 226*, 913–927. https://doi.org/10.1016/j.jclepro.2019.04.114.

Wild, T., Henneberry, J. M., & Gill, L. (2017). Comprehending the multiple 'values' of green infrastructure-valuing nature-based solutions for urban water management, from multiple perspectives. *Environmental Research, 158*, 179–187.

Wu, Z., & Chen, L. (2017). Optimizing the spatial arrangement of trees in residential neighborhoods for better cooling effects: Integrating modeling with in-situ measurements. *Landscape and Urban Planning, 167*(July), 463–472. https://doi.org/10.1016/j.landurbplan.2017.07.015.

Zhao, Q., Wentz, E. A., & Murray, A. T. (2017). Tree shade coverage optimization in an urban residential environment. *Building and Environment, 115*, 269–280. https://doi.org/10.1016/j.buildenv.2017.01.036.

Chapter 24
Cool Materials for Passive Cooling in Buildings

Claudia Fabiani and Anna Laura Pisello

24.1 Introduction

At the current urbanization stage, over 60% of the world population will live in cities by 2030, and more than 68% by 2050 (European Commission 2018). This remarkable trend deserves proper attention, especially since urban sprawl is one of the major causes of harmful environmental issues with serious repercussions on human health (Vardoulakis et al. 2015). Urban overheating, among them, is regarded as one of the most important (Chapman et al. 2017; Susan 2007). Its detrimental escalation has led to the definition of a unique phenomenon: the urban heat island (UHI), which consists of a local temperature increment affecting metropolitan areas with respect to their rural surroundings (Kolokotroni and Giridharan 2008), with the maximum temperatures experienced within the densest city blocks (Xu et al. 2018). Further intensifying the risk is the synergistic interaction between UHIs and regional heat waves (Li and Bou-Zeid 2013), which was shown to produce repercussions that might exceed a simple superposition of the two hazards (Zhao et al. 2018). Beyond the well-known environmental and health-related impacts, UHIs are also involved in the exacerbation of building cooling loads (Santamouris 2014) and in the consequent increase in peak electricity demand (Ihara et al. 2008).

As a result, in recent years, several research contributions have been developing and investigating UHI mitigation techniques with the aim of tackling local and global overheating. In particular, using of cool materials was found to be a highly effective solution for reducing both surface and air temperature peaks in the urban environment during summer (Li et al. 2014). In this chapter, we firstly present and

C. Fabiani · A. L. Pisello (✉)
CIRIAF—Interuniversity Research Center, University of Perugia, Perugia, PG, Italy

Department of Engineering, University of Perugia, Perugia, PG, Italy
e-mail: anna.pisello@unipg.it

M. Palme, A. Salvati (eds.), *Urban Microclimate Modelling for Comfort and Energy Studies*, https://doi.org/10.1007/978-3-030-65421-4_24

classify these materials, together with the methods that are generally used for their characterization, with special focus on their numerical assessment via several computational methods. At a later stage we discuss the impact of cool materials on building energy consumption and we examine their most common shortcomings. Finally, the most promising future trends and the potential advances in this technology are presented.

24.2 Basics of Cool Material Physical Behaviour

24.2.1 Generalities

Cool materials represent a cost-effective passive technique that contributes to achieving lower cooling energy loads by reducing surface and air temperatures in the built environment, particularly in the building fabric. By definition cool materials are characterized by (1) high solar reflectance (SR) and (2) high infrared emittance (ε).

The former property estimates the ability of a surface to reflect the incoming solar radiation. It is measured on a scale from 0 to 1, and its definition is carried out considering the hemispherical radiation and integrating over the complete solar spectrum. Infrared emittance, on the other hand, measures the capability of a surface to release heat upon absorbing it from the surrounding environment. It is also measured on a scale from 0 to 1 and quantifies how well a selected surface radiates energy compared to a black body maintained at the same temperature. Solar reflectance and infrared emittance are combined in a secondary indicator largely used in building engineering: the solar reflectance index (SRI) (ASTM E1980-11 2019). The SRI, which also takes into account convective cooling effects, is defined from 0, i.e. a standard black surface (SR = 5%, ε = 90%), to 1, i.e. standard white surface (SR = 80%, ε = 90%).

Surfaces combining high solar reflectance and high infrared emittance can reflect most of the incoming solar radiation and, at the same time, re-emit a large portion of the absorbed one, albeit small. Therefore, they will remain cooler and maintain a lower surface temperature if compared to similar surfaces with lower SR and ε values. This unique behaviour can be particularly favourable in building applications, where the amount of heat transferred through the building envelop is directly proportional to the indoor/outdoor surface temperature difference. Thereon, the use of cool materials in buildings allows to reduce the amount of heat transferred from the outdoor to the indoor, above all in case of extreme solar heat gains.

24.2.2 Thermal Performance of Cool Materials

As previously said, cool surfaces show lower surface temperatures compared to non-cool solutions. In fact, the high solar reflectance and high thermal emittance allow them to absorb less energy from the sun, and release more ambient heat. The general law representing all the main heat-exchanging phenomena, apart from the latent contributions, taking place on a flat surface is described in the following (see Fig. 24.1):

$$(1-\mathrm{SR})I = \varepsilon\sigma\left(T_{\mathrm{s}}^4 - T_{\mathrm{sky}}^4\right) + h_{\mathrm{c}}\left(T_{\mathrm{s}} - T_{\mathrm{air}}\right) - \lambda\frac{dT}{dx} \tag{24.1}$$

where SR is the solar reflectance of the surface (–); I is the insolation (W/m^2); ε is the thermal emittance (–); σ is the Stefan–Boltzmann constant ($\sigma = 56.685 \times 10^{-8}$ W/m^2 K^4); T_s, T_{sky} and T_{air} are the surface, sky and air temperatures (K); h_c is the convection coefficient (W/m^2 K); λ is the thermal conductivity of the surface (W/m K); and dT/dx is the temperature gradient (in the x direction).

As it can be seen, the first term, which represents the net radiation absorbed by the surface, is directly proportional to the solar reflectance, while the term with the highest dependence on the temperature difference on the right-hand side of Eq. (24.1) is directly proportional to the thermal emittance. These two terms play the major role in the final thermal balance of a highly insulated flat surface and so they are the main factors influencing the thermal performance of a roof or more generally of a surface during the day and the night, respectively. Whatever the orientation of a surface, sky view factors (SVFs) should always be considered and used to split the amount of insolation to be considered. In radiative heat transfer, a view factor $F_{A\rightarrow B}$ is generally defined as the proportion of the radiation which leaves surface A that strikes surface B. Similarly, the sky view factor is a measure of the openness of a surface: a SVF equal to 1 means an unobstructed while a SVF of 0 means a completely obstructed view of the sky. For a given built form, the sky view factor at ground level decreases built density and height.

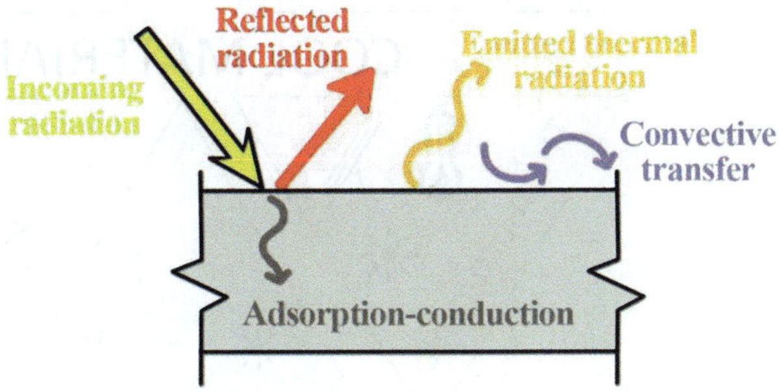

Fig. 24.1 Scheme of the energy balance of a flat surface exposed to solar radiation

24.2.3 *Classification of Cool Materials*

Cool materials are typically divided based on their application, or their visible reflectance. In the first case, we talk about cool materials for roofs, and cool materials for roads and paving solutions (Santamouris et al. 2011). In the second, we define white or light-coloured cool materials, and coloured cool materials (Synnefa and Santamouris 2016). Material science, however, has opened the way to a huge number of innovative applications that are nowadays difficult to classify following such schematizations. All this considered, in this chapter we present a new classification encompassing all the solutions explored by the present research panorama on cool materials. As shown in Fig. 24.2, six different groups of materials can roughly be defined and presented: high-reflective, cool-coloured, retroreflective, photoluminescent, thermochromic and recycled/reused materials.

The knowledge that light-coloured materials allow to maintain cooler surface temperatures is deeply rooted in the past; however, from a scientific point of view, one of the first contributions to analyse such an effect was the study by Givoni and Hoffiman (1968) comparing white and grey-coloured building facades under unventilated conditions. This work paved the way for developing the first group of cool materials, i.e. the high-reflective materials.

Inspired by vernacular architectures in Mediterranean and other hot climates, said materials traditionally focused the efforts on improving the overall visible reflectance of a surface. Since 1968, several research teams all around the world developed different kinds of high-reflective solutions for the built environment. The

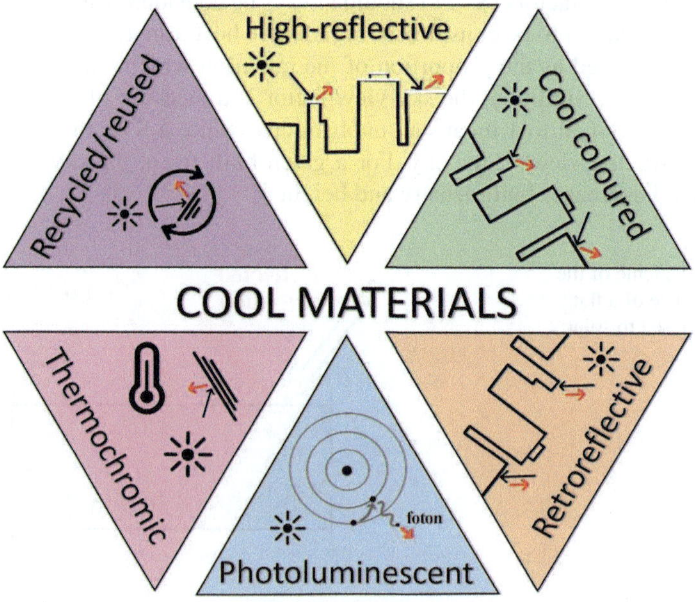

Fig. 24.2 Classification of cool materials, according to the latest research developments

main applications regarded high-reflective roofs (Akbari et al. 1998; Akridge 1998; Pisello and Cotana 2014), paints and coatings for walls (Revel et al. 2014; Zinzi 2016), paving (Doulos et al. 2004; Santamouris et al. 2012), shading devices (Pisello 2015), elastomeric membranes (Fabiani et al. 2020a; Pisello et al. 2016a, b, 2017) and natural cool materials (Castaldo et al. 2015; Pisello et al. 2014). Most of these solutions are based on two white pigments with high reflectance in the whole visible and near-infrared (Vis–NIR) range, i.e. rutile (the most common polymorph of titanium dioxide) and calcite (the most stable polymorph of calcium carbonate) (Pisello et al. 2016c).

The use of white building components is often associated to important limitations in terms of architectural integration, long-term deterioration and glare penalty. All this considered, several scientists struggled for developing cool-coloured alternatives to integrate in more acceptable applications for the built environment. Cool-coloured materials can highly reflect in the near-infrared part of the solar spectrum while maintaining the same visual aspect of their conventional coloured counterparts. A general, initial investigation about these dyes was carried out by Levinson et al. (2005) in their study identifying different pigments capable to absorb less than 10% of the solar energy in the NIR. Several researches followed, producing and investigating the thermal energy performance of coatings for roofs (Levinson et al. 2007; Synnefa et al. 2007), shading devices (Zinzi et al. 2012) and paving systems (Synnefa et al. 2011; Xie et al. 2019). The use of cool-coloured solutions is particularly important for the energy retrofit of historic buildings. In this view, Rosso et al. (2017a, b), among others, focused their attention on the integration of cool-coloured pigments in cement pastes and concretes, in the attempt of identifying the perfect compromise between innovation and heritage for tackling surface overheating.

Both white and coloured cool materials are produced and investigated as surfaces following the Lambertian reflection law. As a consequence, each of these solutions shows an isotropic solar reflectance value, although variable within the solar spectrum. Retroreflective (RR) materials, on the other hand, are a class of cool applications equipped with the skill of reflecting the incident solar radiation directly back to its source, and only on that direction (Fig. 24.3) (Hernández-Pérez et al. 2017; Qin et al. 2016). Opposed to the previously described diffuse solutions, retroreflective materials are particularly effective when applied on vertical walls, in densely built urban areas where sharp incident angles between the solar radiation and the surface allow to maximize their beneficial effect (Rossi et al. 2015a, 2016).

An additional class of cool materials, i.e. the photoluminescent one, makes use of photoluminescence together with simple reflection to improve its resilience to surface overheating (Berdahl et al. 2016; Levinson et al. 2017). Photoluminescence is one of the many forms of luminescence in nature and refers to the physical phenomenon by which an electron emits energy in the form of light as a consequence of a relaxation process. The extra energy radiated in this way was originally transferred to the electron by a photon, causing its migration to a higher energy level in the atom. Therefore, the integration of photoluminescent materials in urban components could produce a surface with excellent effective solar absorption capacity, given its potential to reflect and dissipate heat by means of both common thermal

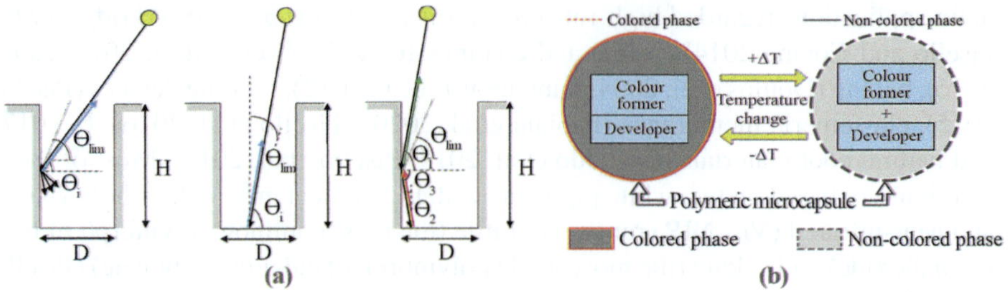

Fig. 24.3 Cooling mechanism of (**a**) retroreflective (RR) and (**b**) thermochromic materials. (Rossi et al. 2015b; Fabiani et al. 2020b, elaboration with permission from Elsevier Ltd.)

emittance and unique photoluminescent phenomenon. In this context, Berdahl et al. (2016) focused their attention on the unique deep red and near-infrared efficient emissivity of ruby (aluminium oxide doped with chromium) and proposed it as a robust fluorescent solution for cooling applications. They also suggested ND-doped YAG, cadmium pigments and their alloys as remarkable fluorescence examples. Over and above, the introduction of phosphorescent materials, with their capability to re-emit light even up to 104 s later, could represent a promising solution, particularly in road applications, given their potential to be integrated in existing artificial lighting systems for electricity consumption reduction.

All types of cool materials presented above share a detrimental and yet inescapable setback, that is, increased heating energy demand during winter. Building on this, a further class of cool materials was introduced and is lately being investigated by the scientific community: thermochromics, i.e. advanced materials capable of reversibly tuning their colour from darker to lighter tones as a function of the local temperature boundary conditions (Fabiani et al. 2019; Garshasbi and Santamouris 2019). Thermochromic solutions have been mostly integrated in advanced coatings (Fabiani et al. 2020b; Yiping et al. 2002), mortars (Perez et al. 2018) and roof (Hu and Yu 2019) applications. They show an improved thermal energy performance that allows to obtain better indoor thermal comfort conditions throughout the year, although they also show non-negligible photodegradation occurrences (Karlessi et al. 2009).

The last kind of cool materials is composed of all those applications aiming to reduce the amount of embodied energy stored in their production process. The group of recycled/reused materials would probably be better defined as a cluster, rather than a class, since it arranges together all those solutions that, although integrating different cooling mechanisms, are born under an eco-friendly approach. As an example, Zinzi and Fasano (2009) used milk and vinegar to produce a cool blend with impressive cooling properties. Ferrari et al. (2013) manufactured a high-albedo engobe by using recycled glass and alumina on a ceramic support, while Libbra et al. (2011) explored the cooling potential of cool-coloured tile coverings as a substitution of conventional roof tiles in historical centres with propitious results.

24.3 Experimental Investigation of Cool Materials

In this section, references, standards and experimental procedures for measuring the main properties of a cool material are firstly dealt with, particularly focusing on recent research developments in this field. As previously discussed, several experimental in-lab and in-field studies have abundantly demonstrated the potential of high-albedo solutions in buildings in terms of energy saving for cooling, indoor comfort conditions and surface thermal characteristics. For this reason, many governmental authorities and technical organizations have supported the elaboration of new standards, i.e. ASHRAE, ASTM and ISO standards, for encouraging the implementation of highly reflective applications in the built environment (Akbari et al. 2001). The progress achieved in these years was partly determined by the work of important clusters such as the Heat Island Group from the Lawrence Berkeley National Laboratory, the Cool Roof Rating Council (CRRC) and the European Cool Roofs Council (EU-CRC), linking public and private institutions in order to control material quality and provide reliable measurement procedures. The following subsections will (1) describe the most important experimental procedures for quantifying solar reflectance and thermal emittance of cool materials (according to specific reference standard) and (2) present the main results of advanced numerical studies assessing the potential of cool solution at large scale.

24.3.1 Solar Reflectance Analysis

Solar reflectance values can be measured by using (1) a spectrophotometer with an integrating sphere, (2) a reflectometer and (3) a pyranometer.

SR measurements carried out using a spectrophotometer with an integrating sphere follow the ASTM E903-12 standard (2012). The use of the integrating sphere allows to obtain the spectral near-normal hemispherical reflectance in the range of 250–2500 nm. The final SR values are then calculated through a weighted average value with respect to the standard solar spectral irradiance (ASTM G173-03 2012). The main components of a spectrophotometer are the light source (covering the range of 250–2500 nm), the diffraction grating and mirrors, and the integrating sphere, which is internally coated with a highly reflective finish. Such an experimental equipment allows to measure solar reflectance and transmittance of a planar sample with respect to a reference (standard white of known reflectance) as a function of the wavelength (λ). Additionally, it can also be used to calculate the solar absorbance of the investigated surface.

The measurement procedure to be followed when using a reflectometer is reported in ASTM C1549-16 (2016). This instrument is used for field measurements and it can measure the reflectance by varying the air mass setup. The calibration procedure of a reflectometer is carried out using a blackbody cavity (reference for null reflectance), and a surface of known reflectance, previously characterized through solar spectrophotometer according to ASTM E903-12. This measurement method is particularly suitable for characterizing reflectance properties of opaque and homogeneous materials through fast and relatively simple field measurements (Fig. 24.4).

The last kind of in-field measurement procedure for measuring the reflectance of a surface makes use of a pyranometer and follows ASTM E1918-16 standard (2016). A pyranometer can be used to measure clear sky global solar reflectance of a horizontal or low-sloped surface when the sun angle to the normal surface is lower than 45° (during the central hours of the day). This measuring technique is potentially compromised by the presence of shadows, and background and instrumental errors. Additionally, it requires accurate instruments, able to detect global solar radiation in the range of 300–2500 nm, and a minimum surface of 4 m². In order to reduce differential effects in convective heat exchanges and optimize the precision of the albedo measurements, a double-pyranometer apparatus, i.e. albedometer, is normally used. In this case, the upward-oriented pyranometer measures the radiation arriving to the sample, while the downward orientation captures the reflected radiation. The albedo is then calculated as the ratio between reflected and incoming global solar radiation values.

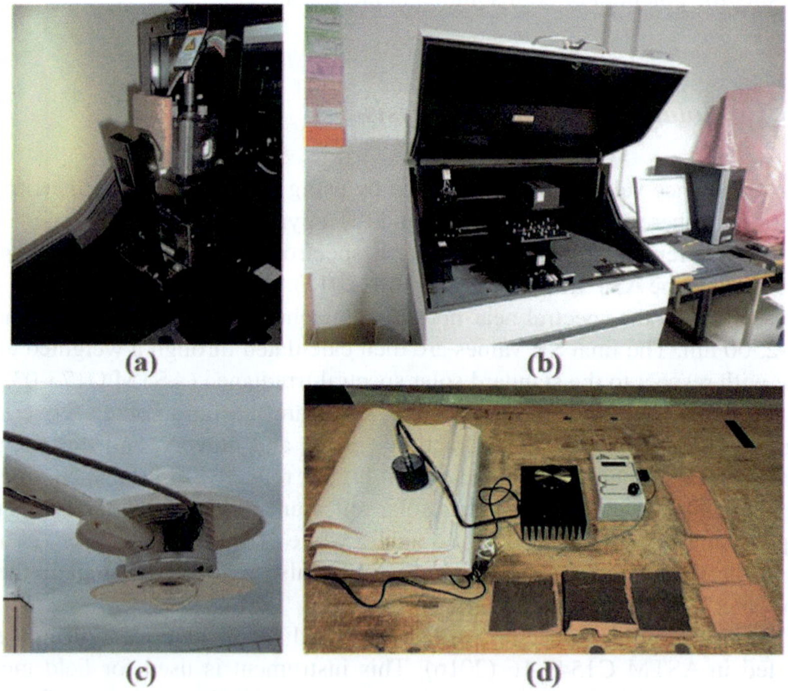

Fig. 24.4 (**a**, **b**) Spectrophotometer with an integrating sphere for in-lab SR measurements, (**c**) albedometer for in-field SR measurements and (**d**) portable emissometer for in-lab thermal emittance measurements

24.3.2 *Thermal Emittance Analysis*

The main reference standards which can be used to measure the thermal emittance of a surface are the ASTM C1371-15 (2015) and the ASTM E408-13 (2019) standard. The former makes use of portable emissometers, and the latter describes the inspection meter technique.

The first experimental technique makes use of a portable differential thermopile emissometer and it allows to measure hemispherical emittance of a surface or a coating. When using a portable emissometer, it is important to firstly operate a simple yet essential calibration procedure using samples with known high-emittance and known low-emittance values. Additionally, the sample, which has to have a minimum surface of 15 cm^2, needs to be positioned over a heat sink to prevent excessive temperature increase during the course of the measurement.

The second procedure generally used to measure thermal emittance is the inspection meter technique and it consists of two different test methods (A and B), characterized by different limitations and accuracy. Both these methods are based on thermal reflectance measurement and only estimate emittance as a calculated value. All this considered, although their results are not strongly affected by temperature drifts, they often generate erroneous estimations of the thermal emittance and, for this reason, the use of portable emissometers is preferred in the common practice.

24.4 Computational Investigation of Cool Materials

In parallel to the most common experimental investigation techniques, the analysis of the UHI mitigation potential of highly reflecting cool materials was also carried out through numerical simulations, as a reliable tool to explore a variety of configurations and boundary conditions with little experimental effort. This field of investigation was particularly boosted by the remarkable advances in computational resources and the ever-greater need of high spatial modelling accuracy in microclimate studies of the past decades. Among others, surface energy balance or budget (SEB) models, computational fluid dynamics (CFD) and building energy performance simulations (BEPS) are widely acknowledged tools, particularly appreciated by the scientific community. In this section, we present the main characteristics of these models, with special focus on cool materials' applications. We present the main contributions in the current research background and discuss the main findings derived from their application.

24.4.1 Cool Materials in Surface Energy Balance (SEB) Models

Computational methods for urban microclimate analysis originally involved the "simplified approach" of surface energy balance or surface energy budget (SEB) models. Based on the law of energy conservation for a specific control volume, these models follow the principle of surface energy balance proposed by Oke (1982), which allows to write Eq. (24.2) for each facet of the control volume:

$$R_N = H + \mathrm{LE} + G \tag{24.2}$$

where $R_N = S^{\downarrow} + L^{\downarrow} + S^{\uparrow} + L^{\uparrow}$ is the net radiation (S is the short wave and L is the long wave, while the downward and upward arrows denote the downwelling and upwelling components), H is the sensible heat flux, LE is the latent heat flux and G is the conductive heat flux, respectively. This equation cannot realistically be solved for every point on the urban surface and, therefore, a certain level of approximation is required. Four basic approaches have been used to simplify this problem (Fig. 24.5):

- *Slab models*, treating the urban geometry as a flat surface with a large roughness length and small albedo (Best 2005)
- *Volumetric averaging models*, considering the energy balance of a volume incorporating buildings, air and underlying substrate (Grimmond et al. 1991)

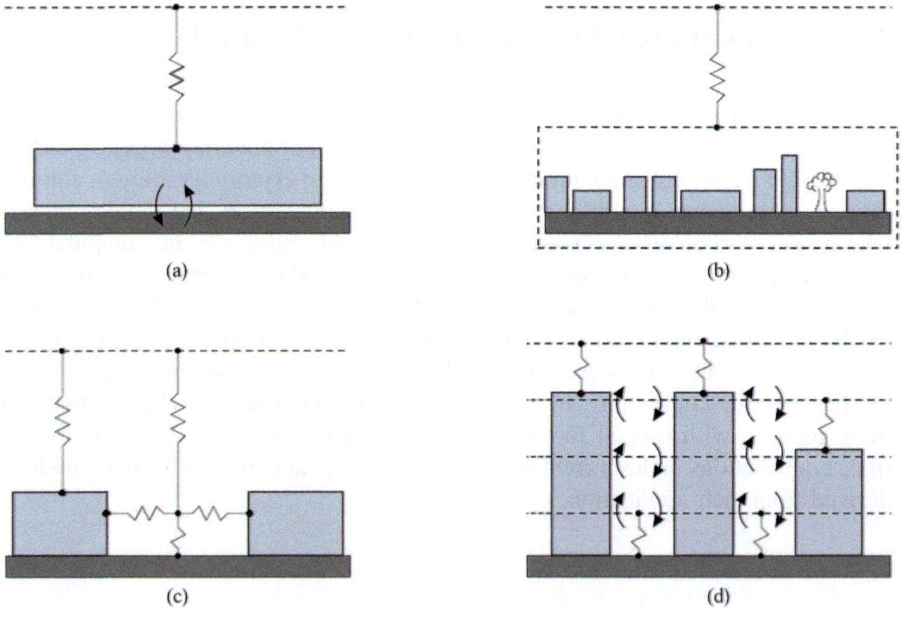

Fig. 24.5 Schematic of the four basic approaches in modelling the urban energy balance. (**a**) Slab models; (**b**) volumetric averaging models; (**c**) single-layer models; (**d**) multilayer models

- *Single-layer urban street canyon or urban canopy models*, considering the different building facets in a two-dimensional, symmetrical street canyon with infinite length (Masson 2000)
- *Multilayer urban street canyon or urban canopy models*, considering the different building facets in a three-dimensional geometry (Martilli et al. 2002)

Volume-averaging models (VAMs) and single-layer urban canopy models (S-UCMs), in particular, have been widely used in the scientific community. They provide the basis for numerical microclimate analyses, yet they give no information about wind velocity fields and they cannot properly model flow patterns that seriously affect sensible heat flux in urban surfaces (Toparlar et al. 2017). The main difference between volume-averaging and urban canopy models lies in the way the urban surface is discretized and its energy balance calculated. VAMs indeed only allow to consider a volume average representative of all the surfaces and the air enclosed within it, while UCMs distinguish among different types of urban facets (e.g. building roofs, walls and ground); therefore, they allow to have a more detailed representation of local energy exchanges in urban areas (Masson 2000; Kusaka et al. 2001; Grimmond and Oke 2002), and they are particularly useful for quantifying the local effect of advanced materials for urban surfaces. By using this unique feature, for example, Fabiani et al. (2019) were able to quantify the urban heat island mitigation potential of a high-reflective roof and of an equivalent thermochromic-based application, and compare it to that of a traditional dark solution. Results showed that the thermochromic produces enhanced short-wave solar reflection in summer conditions while also reducing the reflected solar fraction in winter. This is due to the adaptive nature of the thermochromics, which allows to obtain a 5.6% reduction in the cooling energy needs in the face of almost negligible increases in the heating demand. In the same way, Manni et al. (2019) developed an innovative algorithm using a Monte Carlo-based routine for conducting full ray-tracing solar analyses aimed at evaluating the influence of retroreflective materials on the energy balance of the urban canopy.

Despite their exceptional capability to quantify local changes in the urban energy balance, UCMs are most largely used in combination with mesoscale atmospheric models for correctly parameterizing the interactions between urban surfaces and bordering atmosphere (Ramamurthy et al. 2014). In his work about the effectiveness of different UHI mitigation strategies, for example, Zhang et al. (2017) used the weather research and forecasting (WRF) model coupled with a physically based single-layer UCM to predict both local and regional effect of cool roofs. He found that the integration of cool materials seriously affects surface energy partition, and the sensible flux of the urban skin, in particular, was found to be strictly correlated to local variations in solar reflectance. The regional effect could be found in both surface and near-surface air temperatures when the roofs were covered with high-reflectance materials (SR = 0.7) or, alternatively, when the fraction of green roofs reached 50% (Fig. 24.6). In a similar way, Li et al. (2014) quantified cool and green roof mitigation effect at the city scale by coupling WRF with the Princeton urban canopy model (PUCM). Based on their results, covering 30% of the roof areas with

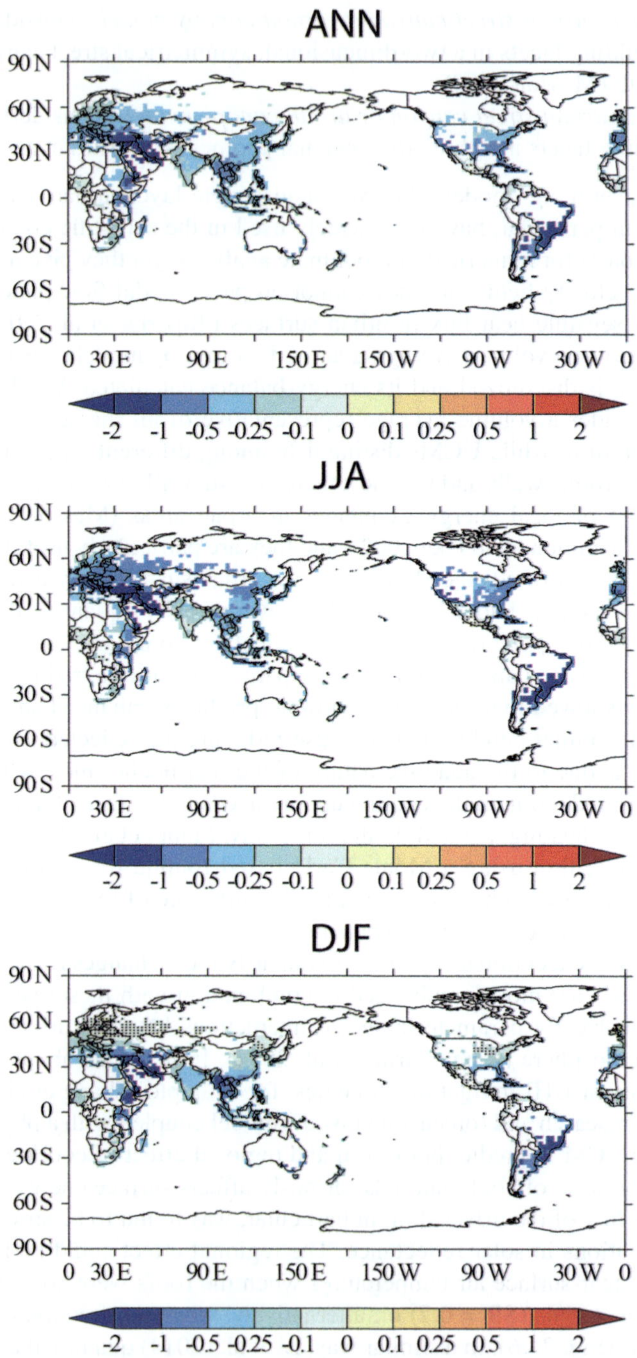

Fig. 24.6 Changes in urban heat islands (urban minus rural surface air temperature) from increasing roof albedo, averaged annually (ANN), in summer (JJA) and in winter (DJF). Dotted areas are where differences are not statistically significant at 95% confidence interval. Each panel shows differences between COOL and DARK cases (Zhang et al. 2016)

cool roofs with an albedo of 0.7 could produce a 1 °C near-surface air temperature reduction in the Baltimore-Washington metropolitan area. An additional reduction of 0.14 °C could be obtained if the cool roof fraction were to be raised to 50%, and the albedo value changed from 0.7 to 0.9. Zhang et al. (2018) combined urban canopy and mesoscale atmospheric models to quantify and compare the effectiveness of cool roofs and walls with SR of 0.5 and 0.9 as UHI mitigation strategies. They found that the average air temperature reductions induced by both these applications is proportional to the increase in albedo.

24.4.2 Cool Materials in Computational Fluid Dynamic (CFD) Models

Computational fluid dynamic (CFD) models analyse and solve problems that involve fluid flows. CFD is based on the Navier-Stokes equations, describing the correlation among velocity, pressure, temperature and density of a moving fluid. In contrast with simple energy balance models, CFD couples explicitly velocity and temperature fields, possibly considering humidity and pollution fields as well, and it can resolve the flow field at much finer scales (Toparlar et al. 2017). CFD simulations, however, require a much more detailed representation of the urban geometry, together with a detailed definition of specific boundary conditions for all relevant flow variables, and adequate computational resources (Blocken 2015). CFD models allow the study of the urban microclimate at different scales of investigation, ranging from the meteorological mesoscale to the building scale and the indoor environment (Fig. 24.7).

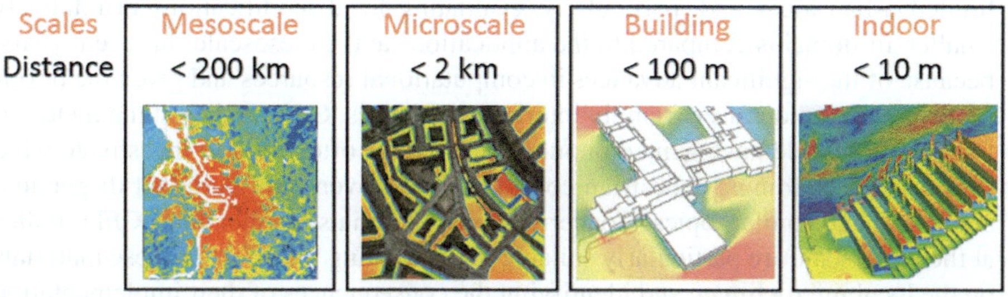

Fig. 24.7 Schematic representation of the spatial scales in climate modelling, with typical horizontal dimensions (Toparlar et al. 2017)

24.4.2.1 Mesoscale Computational Fluid Dynamic

Meteorological mesoscale simulations refer to climatic studies investigating atmospheric developing over horizontal distances of several hundred kilometres, numerical weather prediction (NWP) models (Lynch 2008). These models date back to the 1970s and were originally applied for 2D computational domains, where urban areas were defined as localized heat sources (Bornstein 1975). Yet, the necessity to further increase the representativeness of these models led to including specific 3D representations of the built environment, initially in the form of basic forcing functions for topography, building heights, roughness variations and temperature perturbations (Vukovich Dunn and Crissman 1976; Saitoh et al. 1996), and finally, as fully developed surface energy balance models (Li et al. 2014; Ramamurthy et al. 2014). Effectively coupling SEBs and NWPs allowed to quantify the mesoscale effect of different kinds of cool materials as previously described in Sect. 24.4.1 for the specific case of UCMs. Zhou and Shepherd (2010), for example, modelled the area of Atlanta, USA, with the Weather Research and Forecasting (WRF)-NOAH Land Surface Model (LSM) and concluded that tripling city's albedo could effectively attenuate UHI intensity. Similarly, Zhang et al. (2016) used the Community System Model (version 1.2.0) at the urban, continental and global scales. They reported a statistically significant annual and global mean temperature decrease due to large-scale implementation of cool materials, which registered everywhere, with the exception of some regions in Africa and Mexico.

24.4.2.2 Microscale Computational Fluid Dynamic

CFD simulations at the meteorological microscale consider horizontal distances up to a few kilometres. They are useful for accurately modelling specific buildings or limited urban areas and, particularly, evaluating local modifications in relatively smaller air domains compared to the applications at the mesoscale. In recent years, because of the significant advances in computational resources and given the development of evermore user-friendly interfaces and tools, CFD studies at the meteorological microscale have gained popularity. They are often used to investigate wind flow around buildings, pedestrian comfort, wind-driven rain, pollutant dispersion, snow drift and similar topics. Concerning cool materials' applications, CFD studies at the microscale are particularly suitable for evaluating the effect of these materials on the local microclimate and identifying the consequences of their implementation in a selected urban context. Most of these studies solve RANS equations using the acknowledged k-ε model.

Georgakis et al. (2014), for example, used Ansys CFX to evaluate the effect of different coatings with albedo of 0.3, 0.6 and 0.9 on a deep street urban canyon. They found that the integration of these coatings on the surfaces of the canyon could reduce local surface temperatures by 7–8 °C at the ground level, and by 2–3 °C on the walls of the overlooking buildings. Finally, they also concluded that the ambient air temperature inside the canyon may decrease up to 1 °C. A different package of the Ansys suite, i.e. Ansys Fluent, was used by Herbert et al. (1998) to evaluate the effect of cool materials' implementation in Los Angeles. The authors designed an imaginary canyon with an above-canyon wind speed of 3 ms^{-1} and ran two simulations: the former taking into account common building materials, and the latter assuming that all the surfaces of the canyon were painted white. Results showed lower overall air temperatures and within-canyon air temperatures during the overall extent of the simulation for the "white" city configuration. In particular, both these temperatures were reduced by approximately 1 °C during the peak hours of the day.

Another acknowledged CFD tool, i.e. Star CCM+, was used by Botham-Myint et al. (2015) for measuring the thermal perturbation or "thermal footprint" caused by a white roof in different urban geometries. They found that introducing the cool material on the roof of a short building surrounded by taller constructions could improve the effectiveness of this solution. The same tool was used by Yang et al. (2017) for quantifying the UHI mitigation effect of a PCM cool roof system applied on a residential and on a commercial city block. Results showed a greater effect in the commercial zone (characterized by a larger roof area), where the urban canopy layer registered a 7.8 °C temperature reduction in summer, and a reduction of 11.3 °C in winter (Fig. 24.8). Concerning the residential zone, in this case the temperatures were decreased by 6.4 °C in summer and 10.5 °C in winter, respectively (Fig. 24.9).

Several open-source CFD codes also exist and were used for evaluating the effect of cool applications at the microclimate scale. Qu et al. (2012), for example, used the Code_Saturne (Archambeau et al. 2004) for simulating the effect of materials with increasing albedo values in a specific building block. They found that increasing the solar reflectance of all the urban surface case study from 0.1 to 0.6 drastically reduces the net radiation term in surface energy balance of the urban skin. Consequently, the potential temperature close to the cool walls and roofs may decrease by 2 °C. ENVI-met is another free tool which has frequently been used for simulating the thermal performance of cool materials in the built environment. Pigliautile et al. (2020), for instance, used ENVI-met for assessing the UHI mitigation potential of evapotranspiration and high-reflectance surfaces in a full-scale experimental set-up composed of more than 20 continuously monitored cubicles in a Mediterranean continental climate. Based on their results, they state that the high-albedo solution has the highest potential in mitigating summer overheating, while the introduction of greenery could be more effective in packed configurations with low-albedo envelopes. The same tool coupled with an energy performance model was used by Cardinali et al. (2020) for assessing the impact of outdoor microclimate improvements, including cool solutions, on building energy performance (Fig. 24.10).

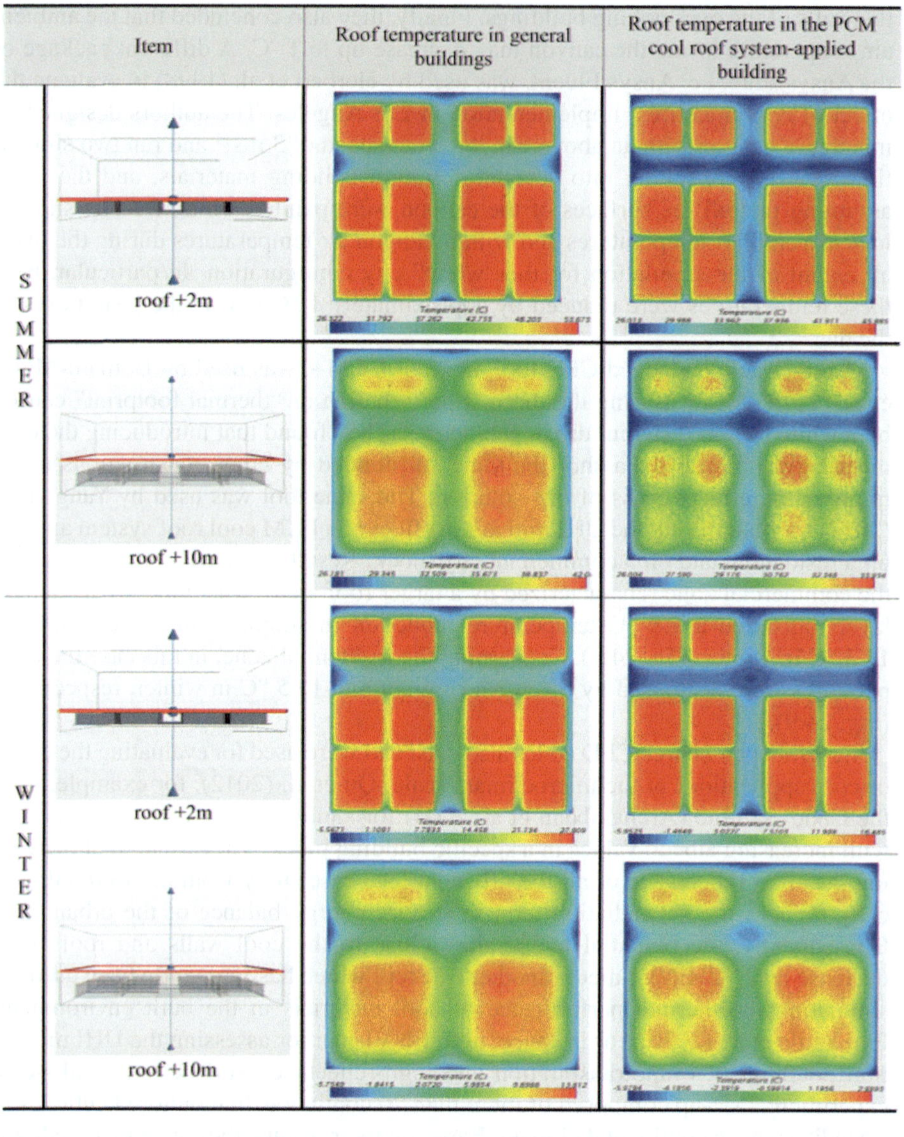

Fig. 24.8 Temperature distribution of commercial area (horizontal section). (Yang et al. 2017, with permission from Elsevier Ltd.)

24.4.2.3 Building-Scale Computational Fluid Dynamic

Computational fluid dynamic can also be used for producing a detailed analysis of the local microclimate around one or a few buildings. In this case we talk about building-scale CFD and we generally consider distances below 100 m on the horizontal surface. This application makes use of the same tools and platforms used in

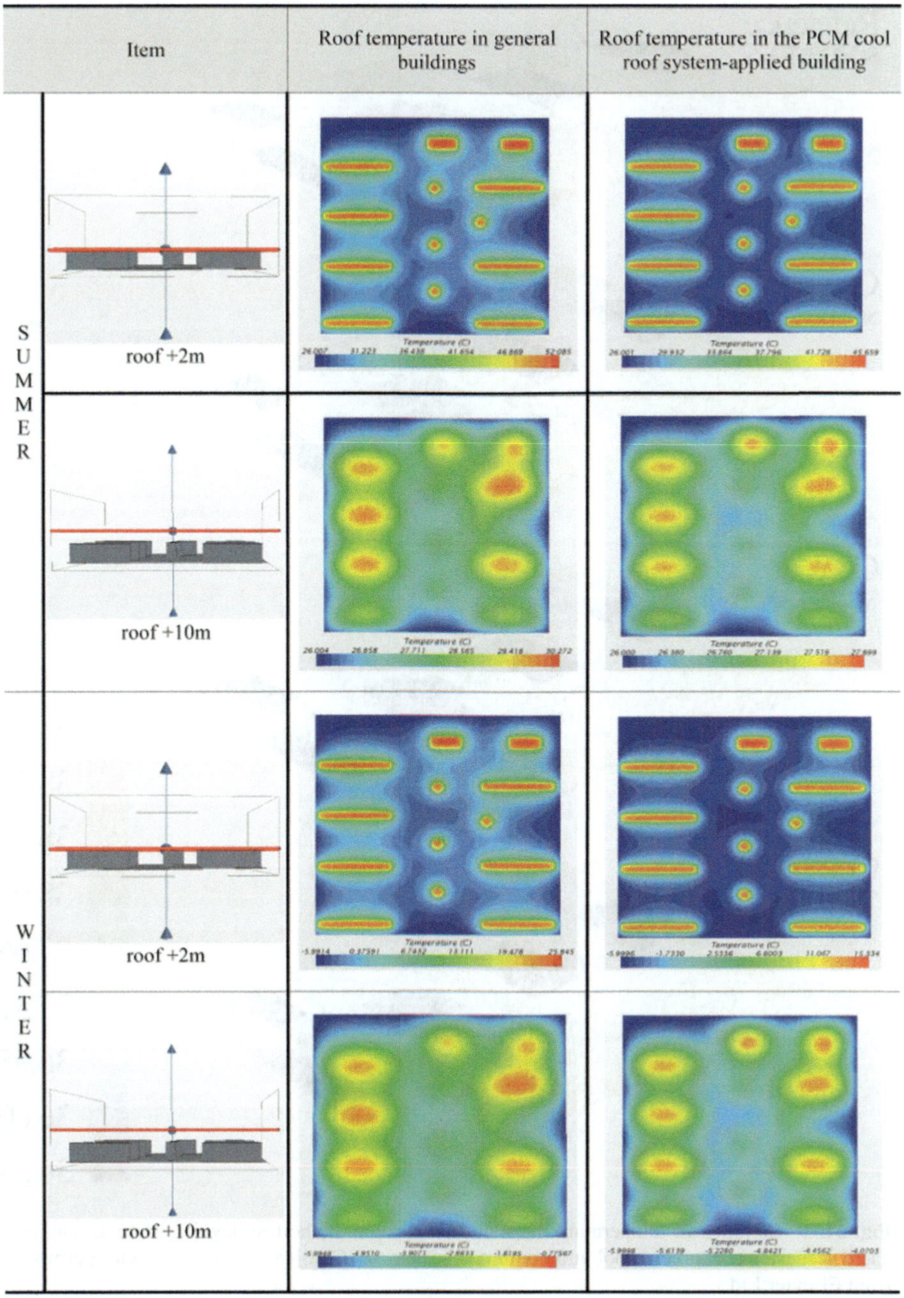

Fig. 24.9 Temperature distribution of residential area (horizontal section). (Yang et al. 2017, with permission from Elsevier Ltd.)

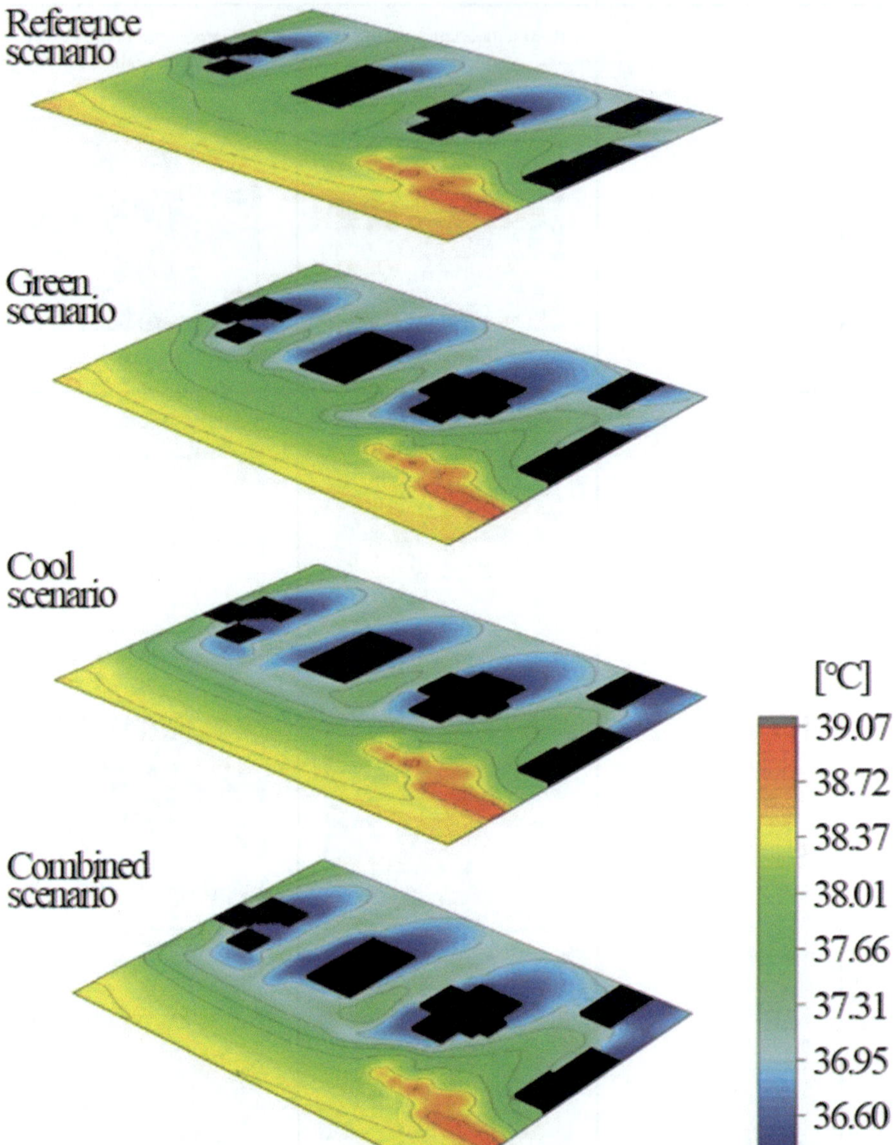

Fig. 24.10 Summer air temperature maps at 3:00 p.m. on the hottest day for reference and mitigated scenarios (i.e. green, cool and combined scenario). (Cardinali et al. 2020, with permission from Elsevier Ltd.)

the microclimate scale, yet more detailed input data are needed in terms of both geometry and microclimate parameters. The improved precision in the input and local boundary conditions allows to obtain detailed outputs from the model and, consequently, to more carefully describe the evolution of local flow field. This kind of models is often used for investigating pollutant dispersion and local temperature gradients. Haghighat and Mirzaei (2011), for example, used Fluent software for studying the effect of non-uniform wall surface temperature distribution on the pollution dispersion and flow pattern within a short street canyon. In particular, they investigated the capability of active and passive techniques (e.g. cool materials) to reduce the pollutant concentration level in the canyon. Ghaffarianhoseini et al. (2015), for instance, used ENVI-met for simulating the thermal performance of courtyards in the hot and humid climate of Kuala Lumpur, Malaysia. Based on the obtained results, they specify guidelines for optimizing the design of courtyards and enhancing their thermal performance. In particular, they found that although the introduction of cool materials generally reduces local air temperatures in the simulated environment, they also generally produce considerably higher mean radiant temperatures, and consequently produce higher discomfort conditions. As a matter of fact, results from their simulations show that the predicted mean vote (PMV) value of the courtyard rises from 4 to 5, and finally 6, with increasing albedo from 0.3 to 0.55, and 0.93, respectively (Fig. 24.11).

24.4.2.4 Computational Fluid Dynamic in the Indoor Environment

Given their complex nature, and the generally high computation time of most CFD analyses, these models were originally mostly used to perform heat transfer analysis in closed cavities, generally representing portions of buildings with varying external coating reflectance. The use of such an advanced technique allowed to further corroborate the positive results obtained in dynamic simulations with a more careful investigation of the indoor thermal comfort in 3D models focused on airflows and local discomfort conditions. Revel et al. (2014), for example, used Star CCM+ code for conducting a transient (24-h long) conjugate heat transfer study for investigating the effect of cool facade tiles and roof membrane on the external and internal surface temperature of the simulated building, as well as on the indoor air temperature. They found a 3.0 and 1.5 °C reduction in terms of external and internal surface temperature, respectively (Fig. 24.12). In terms of indoor temperature, reductions of peak values of 0.9 and 0.7 °C at the centre of the air volume have been found when using the cool facade and roof. In a similar way, Pisello et al. (2016d) developed and validated a two-dimensional CFD analysis of an attic room in a residential building, with the aim of quantifying the effect of a cool roof solution within the indoor environment of the thermal zone adjacent to the roof, and analysing the attic local thermal comfort conditions. They found up to 2.8 and 1.5 °C air temperature reduction in the cool configuration, in summer and winter conditions, respectively. Additionally, they also reported the capability of the cool application to

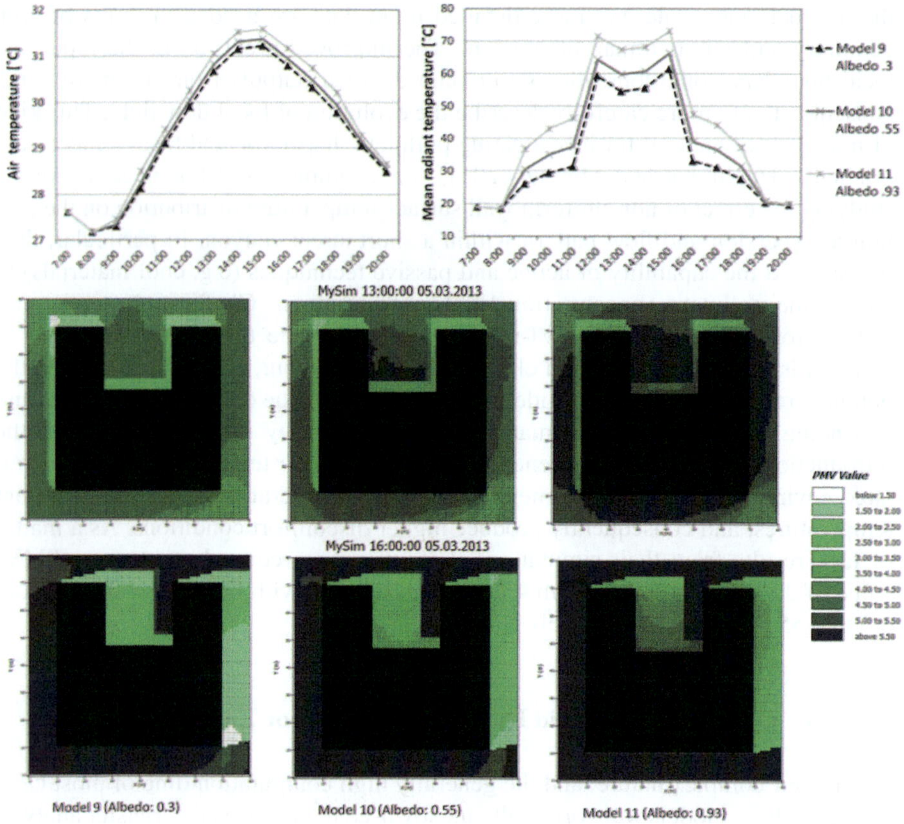

Fig. 24.11 Hourly ambient air temperature (top left), mean radiant temperature (top right) and simulated distributions of PMV in courtyard models with albedo values of 0.3, 0.55 and 0.93 at 13:00 and 16:00, at 2 m height (bottom). (Ghaffarianhoseini et al. 2015, elaborated with permission from Elsevier Ltd.)

significantly reduce the intense thermal stratification caused by summer overheating, which seriously affects local comfort conditions in the attic (Fig. 24.13).

24.4.3 Cool Materials in Building Energy Performance Simulations (BEPS)

An initial approach with numerical simulations of cool components for the built environment mainly aimed at quantifying their possible impact on the final energy requirements for cooling and heating in residential buildings. All this considered, different researchers used dynamic thermal energy simulation tools to investigate the introduction of cool materials in different cities around the world and estimate possible winter penalties with varying climate boundary conditions.

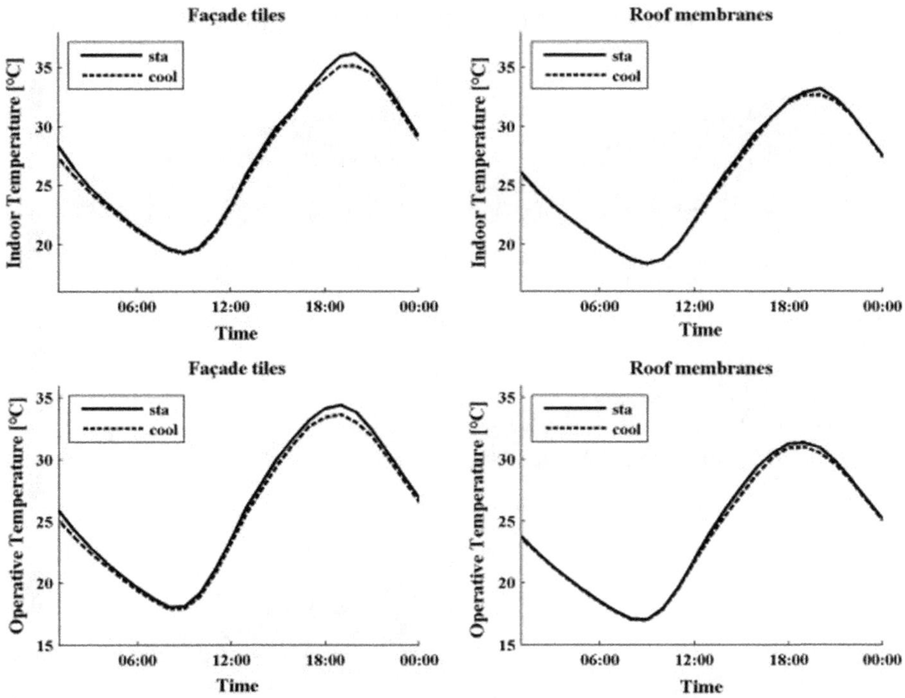

Fig. 24.12 Daily profiles of indoor and operative temperature at the centre of the indoor air volume for the standard and cool models. (Revel et al. 2014, with permission from Elsevier Ltd.)

Piselli et al. (2017), for instance, minimized building annual energy requirements for air conditioning by optimizing the roof coating solar reflectance capability under different boundary conditions and climate contexts. Based on their study, they concluded that the optimum roof solar reflectance to reduce the annual building HVAC energy consumption is mainly affected by the climate context. The same authors used a similar optimization approach for defining the most promising roof configuration in terms of solar reflectance of the external coating and thermal insulation layer thickness in the case of a standard small office building under varying climate conditions. They concluded that the cool roof was able to optimize the annual HVAC energy consumption of the case study building in almost all considered climate conditions, except the coldest zones (Fig. 24.14) (Piselli et al. 2019).

The optimum combination of cool coatings and insulation thickness was also the main topic of the work by Saafi and Daouas (2018). In this case however, the authors took into account two typical Tunisian roof structures for residential buildings, and three types of thermal insulation materials. They carried out a life cycle cost analysis considering the energy consumption derived from a dynamic simulation tool, and concluded that in the selected climate conditions, the cooling energy loads are more sensitive to the variations of solar reflectivity compared to the heating loads and the effect of reflectivity is always more pronounced for low-roof insulation

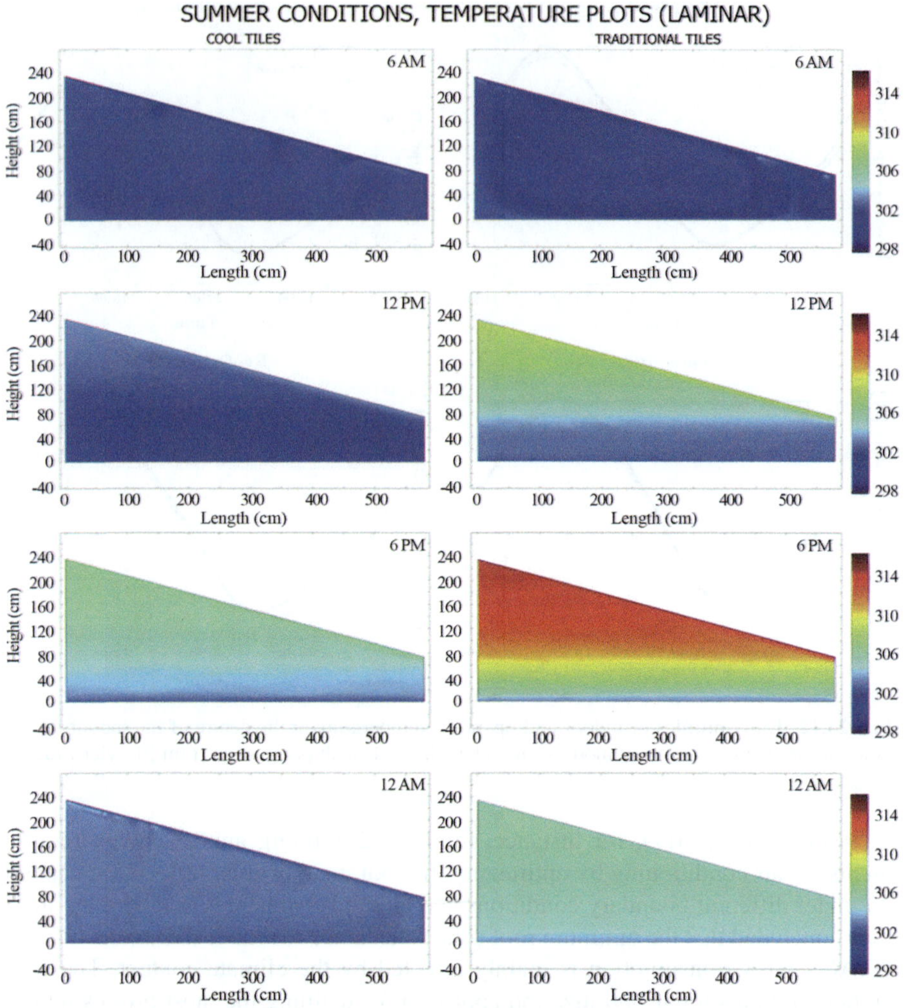

Fig. 24.13 Temperature distribution for cool and traditional tiles at different time steps in summer conditions. (Pisello et al. 2016d, with permission from Elsevier Ltd.)

levels. The effect of cool materials on the annual energy consumption of a case study building and the related costs was also investigated by Jo et al. (2010). In this case, however, the thermal energy investigation concerned the implementation of cool roof coatings (with an albedo value equal to 0.72, against the original value of 0.3) on a single-storey office building equipped with a data processing centre in Phoenix, Arizona. Given the necessity of maintaining a constant and relatively lower temperature profile 24 h, the introduction of the cool solution allowed to obtain 3.16% energy savings per year, corresponding to about 22,000$/year.

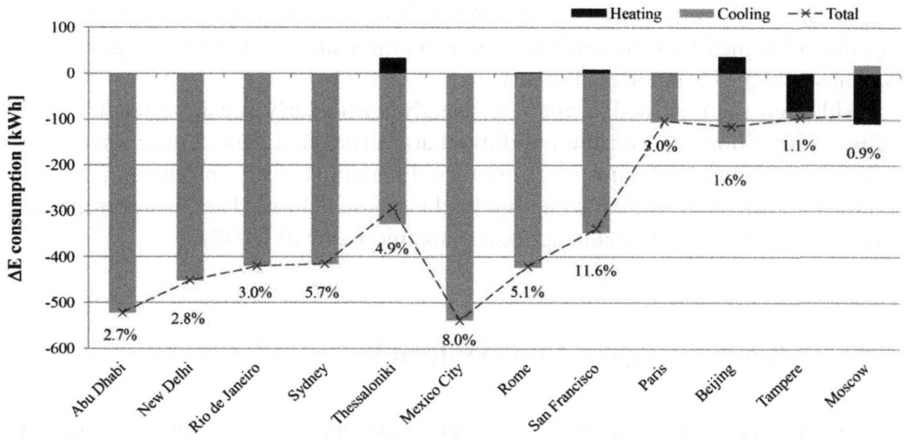

Fig. 24.14 Difference of building HVAC energy consumption between the optimum and the "standard roof" configuration in each climate zone, reporting the annual and the separate contributions for heating and cooling (Piselli et al. 2019)

Building energy simulations are particularly useful for identifying the possible energy savings associated to the implementation of innovative building components equipped with cool coatings based on advanced material technologies. Nie et al. (2020), for example, used this kind of analysis for quantifying the potential energy saving derived from the implementation of an optical coating integrating glass bubbles within a polymer film which allows a selective solar reflectivity increase from 0.06 to 0.92 while maintaining the mid-infrared emissivity above 0.85. They estimated that the applications of the polymer coating on common buildings could guarantee a potential annual energy savings of 2–12 MJ/m^2 and CO_2 emission savings of about 0.3–1.5 kg/m^2. In a similar study, Park and Krarti (2016) estimated that by introducing a variable reflectivity roof coating, characterized by a solar reflectance of 0.55 in summer and 0.3 in winter, it could be possible to reduce building cooling loads without increasing the heating loads in the winter period.

All the above simulations were carried out using a widely acknowledged software called EnergyPlus, which, given its dynamic nature, ease of use and free availability, probably represents the most important software for building energy performance simulations in scientific applications. Yet other tools exist and are widely used by scientists all over the world. Revel et al. (2014), for example, used ESP-r for carrying out dynamic energy-saving calculations and defining the amount of energy that could be saved by replacing standard facade tiles with their equivalent cool alternatives on a standard multistorey building. They found that a maximum of 1.1 kWh/m^2 could be saved in Rome, while this value could increase up to 1.7 kWh/m^2 in Palermo. Wang et al. (2008), on the other hand, used EDSL TAS for investigating the thermal energy performance of a retail shed with a pitched roof equipped with air conditioning for cooling and heating purpose in different locations around the world. The case study building was equipped with 12 different coatings characterized by an albedo value ranging from 0.05 to 0.65. As expected, their results confirmed that the high-reflective coatings are associated to higher energy savings (25–38%) in hotter climates.

Although the obtained results are usually particularly sensitive to the specific climate conditions and on the building characteristics, a general reference pattern can be extrapolated. The introduction of cool materials always produces a reduction on the cooling load, while the heating load increase is limited and is generally compensated by the cooling load decrease (Santamouris et al. 2008).

24.5 Shortcomings of Cool Materials

As previously mentioned, high-reflectance components for buildings and pavements are becoming increasingly important as a passive strategy to save energy for cooling and improve indoor and outdoor thermal comfort conditions. The benefits derived from their use arise from their capability to reduce surface overheating, and it can be accounted for at building, city and even global scales. In this context, however, all cool materials share two inescapable shortcomings: winter penalty and reduction of their performance over time due to ageing processes. The former problem is gradually being solved by resorting to adaptive materials that can tailor their behaviour in response to local boundary conditions. The latter represents an unsolved issue that needs to be tackled seriously, since ageing phenomena such as soiling and weathering can progressively modify the surface properties of the investigated solutions. Therefore, solar reflectance properties of such materials should be specified before and after environmental exposure, in a way to exhaustively describe their thermal optic performance during the course of their life cycle. In this section we give a brief analysis of the main environmental agents affecting cool materials' durability and we describe the main techniques used for studying this phenomenon.

24.5.1 *The Role of Weathering*

Weathering is a detrimental phenomenon due to the effect of all those environmental agents that can influence the physical behaviour and durability of a coating or a surface exposed to light, heat and moisture (Santamouris et al. 2011). The science of weathering is a complex topic that should be studied as a part of the materials' production, measurement, test design, exposure and evaluation as a whole.

The first possible source of degradation is the interaction with light. Shorter and consequently higher energy wavelengths are easily absorbed by most polymeric materials. The interaction with solar radiation, in particular, characterized by huge variations in intensity and spectral distribution, due to different local boundary conditions (as time of the year, location and atmospheric effects) can easily induce destructive reactions and colour changes on a surface.

A second harmful interaction concerns intense temperature fluctuations that can directly lead to a disruptive degradation process. The effect depends on the mechanical stresses produced on a matric by the differential thermal expansion. The effect depends, of course, on the heat source and the specific properties of the sample such as reflectance, absorbance, emissivity and conductivity.

Moisture can touch and penetrate within a surface inducing both chemical and physical stresses in different ways (rain, snow and ice are the most important ones). When ice is formed the associated volume expansion is responsible for increased internal physical stress. The de-icing process, in particular, can cause abrupt contractions and these cyclic dimensional changes can seriously deteriorate the mechanical strength of the material. At a later stage, acid water from the rain often carries atmospheric pollutants that might create corrosive or chemically adverse phenomena. Wind pressure, finally, can cause structural/mechanical modifications in both horizontal and vertical building components, inducing cracks and other damaging effects.

24.5.2 The Role of Soiling

Soiling mechanisms consist of the deposition of atmospheric particulate over a surface and can be coupled or not with the presence or growth of microorganisms (Sleiman et al. 2014).

Deposition of atmospheric particulate matter is the dominant source of soiling agents accumulating on exposed surfaces in the built environment. The combination of these pollutants, particularly black carbon and organic matter emitted during combustion processes and traffic, with oxygen and water vapour is responsible for the formation of oxides and hydroxides. These last molecules can easily corrode materials and damage external surfaces, affecting their solar reflectance capability.

Biological contaminants consist of fungi and other organic agents that proliferate in environments filled with moisture. These microorganisms or microplants deposit over a surface and radically affect its optical properties. All this considered, nowadays periodic removal and restoration tasks are scheduled for most advanced high-performance building applications making use of cool materials.

24.5.3 Quantifying Progressive Degradation of Cool Materials

The estimation of the long-term performance of cool surfaces can be tested in three different ways: natural weathering exposition, accelerated weathering tests and artificial soiling experiments.

In natural weathering exposition, cool components for the building envelope or cool paving solutions are exposed to real weathering agents such as the sun, the rain and the wind. This kind of degradation analysis makes use of natural forcing and is carried out by means of specific test racks, with appropriate distance above ground and to the surrounding area. Other important parameters in this kind of analyses are the sunlight angle, period of exposure and length of the test, generally determined by the purpose of the experiment and the parameters to be measured. In any case, natural weathering procedures are usually long-term experimental campaigns focused on material durability. Of course, during these campaigns a huge set of environmental and geometric characteristics are controlled and monitored, and the solar reflectance and emittance properties are quantified at least before and after the exposure period.

Accelerated weathering tests are in-lab tests carried out using weathering test chambers equipped with a xenon lamp, in order to simulate severe outdoor exposure conditions. This kind of tests is used to anticipate and predict the damaging of coatings and membranes, under the effect of accelerated degradation conditions. No unique correlation exists between outdoor natural and accelerated weathering procedures, so this degradation analysis should always be designed for comparison purposes and analysed with care.

Advanced test chambers and apparatus are also used in artificial soiling experiments. In this case, however, not only the most severe outdoor climatic conditions but also atmospheric pollutants are reproduced and integrated in specifically designed experimental campaigns. In this kind of tests, microbial growth reproduction is particularly important being a major agent of roof soiling in humid climates.

24.6 Future Trends

This chapter provided an in-depth investigation of one of the most interesting kind of materials for building applications, i.e. cool materials. As described above, cool materials are an environmentally friendly and relatively cost-effective solution aimed at reducing building energy needs and improving local and eventually global environmental conditions. Recent research findings showed the importance of developing new technologies capable of reducing the major shortcomings of these relatively simple components: winter penalty and ageing.

Winter penalty can be drastically reduced by using adaptive cool solutions, that is, an innovative kind of materials capable of tailoring their thermo-optic properties in response to local boundary conditions. Future trends in this field will focus on the development and use of nanoscale thermochromics based on quantum dots and plasmonic and photonic structures, among others. These advanced solutions could successfully be used to obtain innovative structures that could highly reflect solar radiation in the UV, visible and near-infrared range while behaving as a black body within the atmospheric window, i.e. between 8 and 13 μm, allowing the so-called radiative cooling phenomenon.

Concerning the issue of ageing, as already discussed, long-term durability is nowadays the real weakness of cool applications that cannot guarantee adequate performance throughout the lifespan of a building, but are susceptible to non-negligible changes over time due to the action of external agents and to the exposure to ambient and adverse weather conditions. For this reason, future trends in research should also focus on the improvement of the thermal energy performance of such solutions in terms of extreme deterioration over time.

References

Akbari, H., Gartland, L., & Konopacki, S. (1998). Measured energy savings of light colored roofs: Results from three California demonstration sites.

Akbari, H., Pomerantz, M., & Taha, H. (2001). Cool surfaces and shade trees to reduce energy use and improve air quality in urban areas. *Solar Energy, 70*(3), 295–310. https://doi.org/10.1016/S0038-092X(00)00089-X.

Akridge, J. M. (1998). High-albedo roof coatings-impact on energy consumption. In *ASHRAE Transactions.*

Archambeau, F., Méchitoua, N., & Sakiz, M. (2004). Code Saturne: A finite volume code for the computation of turbulent incompressible flows—Industrial applications. *International Journal on Finite Volumes, 1*(1). Retrieved from http://www.latp.univ-mrs.fr/IJFV/spip.php?article3, https://hal.archives-ouvertes.fr/hal-01115371

ASTM C1371-15. (2015). *Standard test method for determination of emittance of materials near room temperature using portable emissometers*. West Conshohocken, PA: American Society for Testing Materials. https://doi.org/10.1520/C1371-15.

ASTM C1549-16. (2016). *Standard test method for determination of solar reflectance near ambient temperature using a portable solar reflectometer*. West Conshohocken, PA: American Society for Testing Materials. https://doi.org/10.1520/C1549-16.

ASTM E1918-16. (2016). *Standard test method for measuring solar reflectance of horizontal and low-sloped surfaces in the field*. West Conshohocken, PA: American Society for Testing Materials. https://doi.org/10.1520/E1918-16.

ASTM E1980-11. (2019). *Standard practice for calculating solar reflectance index of horizontal and low-sloped opaque surfaces*. West Conshohocken, PA: American Society for Testing Materials. https://doi.org/10.1520/E1980-11.2.

ASTM E408-13. (2019). *Standard test methods for total normal emittance of surfaces using inspection-meter techniques*. West Conshohocken, PA: American Society for Testing Materials. https://doi.org/10.1520/E0408-13R19.

ASTM E903-12. (2012). *Standard test method for solar absorptance, reflectance, and transmittance of materials using integrating spheres*. West Conshohocken, PA: American Society for Testing Materials.

ASTM G173-03. (2012). *Standard tables for reference solar spectral irradiances: Direct normal and hemispherical on 37° tilted surface*. West Conshohocken, PA: American Society for Testing Materials. https://doi.org/10.1520/G0173-03R12.

Berdahl, P., Chen, S. S., Destaillats, H., Kirchstetter, T. W., Levinson, R. M., & Zalich, M. A. (2016). Fluorescent cooling of objects exposed to sunlight—The ruby example. *Solar Energy Materials and Solar Cells, 157*, 312–317. https://doi.org/10.1016/j.solmat.2016.05.058.

Best, M. J. (2005). Representing urban areas within operational numerical weather prediction models. *Boundary-Layer Meteorology, 114*(1), 91–109. https://doi.org/10.1007/s10546-004-4834-5.

Blocken, B. (2015). Computational fluid dynamics for urban physics: Importance, scales, possibilities, limitations and ten tips and tricks towards accurate and reliable simulations. *Building and Environment, 91*, 219–245. https://doi.org/10.1016/j.buildenv.2015.02.015.

Bornstein, R. D. (1975). The two-dimensional URBMET urban boundary layer model. *Journal of Applied Meteorology, 14*(8), 1459–1477. https://doi.org/10.1175/1520-0450(1975)014<1459:TTDUUB>2.0.CO;2.

Botham-Myint, D., Recktenwald, G. W., & Sailor, D. J. (2015). Thermal footprint effect of rooftop urban cooling strategies. *Urban Climate, 14*(2015), 268–277. https://doi.org/10.1016/j.uclim.2015.07.005.

Cardinali, M., Pisello, A. L., Piselli, C., Pigliautile, I., & Cotana, F. (2020). Microclimate mitigation for enhancing energy and environmental performance of Near Zero Energy Settlements in Italy. *Sustainable Cities and Society, 53*, 101964. https://doi.org/10.1016/j.scs.2019.101964.

Castaldo, V. L., Coccia, V., Cotana, F., Pignatta, G., Pisello, A. L., & Rossi, F. (2015). Thermal-energy analysis of natural "cool" stone aggregates as passive cooling and global warming mitigation technique. *Urban Climate, 14*, 301–314. https://doi.org/10.1016/j.uclim.2015.05.006.

Chapman, S., Watson, J. E. M., Salazar, A., Thatcher, M., & McAlpine, C. A. (2017). The impact of urbanization and climate change on urban temperatures: A systematic review. *Landscape Ecology, 32*(10), 1921–1935. https://doi.org/10.1007/s10980-017-0561-4.

Doulos, L., Santamouris, M., & Livada, I. (2004). Passive cooling of outdoor urban spaces. The role of materials. *Solar Energy, 77*(2), 231–249. https://doi.org/10.1016/j.solener.2004.04.005.

European Commission. (2018). *Continuing urbanisation—Knowledge for policy*. Retrieved from https://ec.europa.eu/knowledge4policy/foresight/topic/continuing-urbanisation_en

Fabiani, C., Pisello, A. L., Bou-Zeid, E., Yang, J., & Cotana, F. (2019). Adaptive measures for mitigating urban heat islands: The potential of thermochromic materials to control roofing energy balance. *Applied Energy, 247*, 155–170. https://doi.org/10.1016/j.apenergy.2019.04.020.

Fabiani, C., Piselli, C., & Pisello, A. L. (2020a). Thermo-optic durability of cool roof membranes: Effect of shape stabilized phase change material inclusion on building energy efficiency. *Energy and Buildings, 207*, 109592. https://doi.org/10.1016/j.enbuild.2019.109592.

Fabiani, C., Castaldo, V. L., & Pisello, A. L. (2020b). Thermochromic materials for indoor thermal comfort improvement: Finite difference modeling and validation in a real case-study building. *Applied Energy, 262*, 114147.

Ferrari, C., Libbra, A., Muscio, A., & Siligardi, C. (2013). Design of ceramic tiles with high solar reflectance through the development of a functional engobe. *Ceramics International, 39*(8), 9583–9590. https://doi.org/10.1016/j.ceramint.2013.05.077.

Garshasbi, S., & Santamouris, M. (2019). Using advanced thermochromic technologies in the built environment: Recent development and potential to decrease the energy consumption and fight urban overheating. *Solar Energy Materials and Solar Cells, 191*, 21–32. https://doi.org/10.1016/j.solmat.2018.10.023.

Georgakis, C., Zoras, S., & Santamouris, M. (2014). Studying the effect of "cool" coatings in street urban canyons and its potential as a heat island mitigation technique. *Sustainable Cities and Society, 13*, 20–31. https://doi.org/10.1016/j.scs.2014.04.002.

Ghaffarianhoseini, A., Berardi, U., & Ghaffarianhoseini, A. (2015). Thermal performance characteristics of unshaded courtyards in hot and humid climates. *Building and Environment, 87*, 154–168. https://doi.org/10.1016/j.buildenv.2015.02.001.

Givoni, B., & Hoffiman, M. E. (1968). *Effect of building materials on internal temperatures*. Research report. Haifa: Building Research Station, Technion.

Grimmond, C. S. B., & Oke, T. R. (2002). Turbulent heat fluxes in urban areas: Observations and a local-scale urban meteorological parameterization scheme (LUMPS). *Journal of Applied Meteorology, 41*(7), 792–810. https://doi.org/10.1175/1520-0450(2002)041<0792:THFIUA>2.0.CO;2.

Grimmond, C. S. B., Cleugh, H. A., & Oke, T. R. (1991). An objective urban heat storage model and its comparison with other schemes. *Atmospheric Environment. Part B. Urban Atmosphere, 25*(3), 311–326. https://doi.org/10.1016/0957-1272(91)90003-W.

Haghighat, F., & Mirzaei, P. A. (2011). Impact of non-uniform urban surface temperature on pollution dispersion in urban areas. *Building Simulation, 4*(3), 227–244. https://doi.org/10.1007/s12273-011-0035-6.

Herbert, J. M., Johnson, G. T., & Arnfield, A. J. (1998). Modelling the thermal climate in city canyons. *Environmental Modelling and Software, 13*(3–4), 267–277. https://doi.org/10.1016/S1364-8152(98)00027-9.

Hernández-Pérez, I., Xamán, J., Macías-Melo, E. V., & Aguilar-Castro, K. M. (2017). Reflective materials for cost-effective energy-efficient retrofitting of roofs. In *Cost-effective energy efficient building retrofitting* (pp. 119–139). Duxford: Woodhead Publishing. https://doi.org/10.1016/B978-0-08-101128-7.00004-6.

Hu, J., & Yu, X. B. (2019). Adaptive thermochromic roof system: Assessment of performance under different climates. *Energy and Buildings, 192*, 1–14. https://doi.org/10.1016/j.enbuild.2019.02.040.

Ihara, T., Kikegawa, Y., Asahi, K., Genchi, Y., & Kondo, H. (2008). Changes in year-round air temperature and annual energy consumption in office building areas by urban heat-island countermeasures and energy-saving measures. *Applied Energy, 85*(1), 12–25. https://doi.org/10.1016/j.apenergy.2007.06.012.

Jo, J. H., Carlson, J. D., Golden, J. S., & Bryan, H. (2010). An integrated empirical and modeling methodology for analyzing solar reflective roof technologies on commercial buildings. *Building and Environment, 45*(2), 453–460. https://doi.org/10.1016/j.buildenv.2009.07.001.

Karlessi, T., Santamouris, M., Apostolakis, K., Synnefa, A., & Livada, I. (2009). Development and testing of thermochromic coatings for buildings and urban structures. *Solar Energy, 83*(4), 538–551. https://doi.org/10.1016/j.solener.2008.10.005.

Kolokotroni, M., & Giridharan, R. (2008). Urban heat island intensity in London: An investigation of the impact of physical characteristics on changes in outdoor air temperature during summer. *Solar Energy, 82*(11), 986–998. https://doi.org/10.1016/j.solener.2008.05.004.

Kusaka, H., Kondo, H., Kikegawa, Y., & Kimura, F. (2001). A simple single-layer urban canopy model for atmospheric models: Comparison with multi-layer and slab models. *Boundary-Layer Meteorology, 101*(3), 329–358. https://doi.org/10.1023/A:1019207923078.

Levinson, R., Berdahl, P., & Akbari, H. (2005). Solar spectral optical properties of pigments—Part II: Survey of common colorants. *Solar Energy Materials and Solar Cells, 89*(4), 351–389. https://doi.org/10.1016/j.solmat.2004.11.013.

Levinson, R., Akbari, H., & Reilly, J. C. (2007). Cooler tile-roofed buildings with near-infrared-reflective non-white coatings. *Building and Environment, 42*(7), 2591–2605. https://doi.org/10.1016/j.buildenv.2006.06.005.

Levinson, R., Chen, S., Ferrari, C., Berdahl, P., & Slack, J. (2017). Methods and instrumentation to measure the effective solar reflectance of fluorescent cool surfaces. *Energy and Buildings, 152*, 752–765. https://doi.org/10.1016/j.enbuild.2016.11.007.

Li, D., & Bou-Zeid, E. (2013). Synergistic interactions between urban heat islands and heat waves: The impact in cities is larger than the sum of its parts. *Journal of Applied Meteorology and Climatology, 52*(9), 2051–2064. https://doi.org/10.1175/JAMC-D-13-02.1.

Li, D., Bou-Zeid, E., & Oppenheimer, M. (2014). The effectiveness of cool and green roofs as urban heat island mitigation strategies. *Environmental Research Letters, 9*(5). https://doi.org/10.1088/1748-9326/9/5/055002.

Libbra, A., Tarozzi, L., Muscio, A., & Corticelli, M. A. (2011). Spectral response data for development of cool coloured tile coverings. *Optics and Laser Technology, 43*(2), 394–400. https://doi.org/10.1016/j.optlastec.2009.07.001.

Lynch, P. (2008). The origins of computer weather prediction and climate modeling. *Journal of Computational Physics, 227*(7), 3431–3444. https://doi.org/10.1016/j.jcp.2007.02.034.

Manni, M., Lobaccaro, G., Goia, F., Nicolini, A., & Rossi, F. (2019). Exploiting selective angular properties of retro-reflective coatings to mitigate solar irradiation within the urban canyon. *Solar Energy, 189*, 74–85. https://doi.org/10.1016/j.solener.2019.07.045.

Martilli, A., Clappier, A., & Rotach, M. W. (2002). An urban surface exchange parameterisation for mesoscale models. *Boundary-Layer Meteorology, 104*(2), 261–304. https://doi.org/10.1023/A:1016099921195.

Masson, V. (2000). A physically-based scheme for the urban energy budget in atmospheric models. *Boundary-Layer Meteorology, 94*(3), 357–397. https://doi.org/10.1023/A:1002463829265.

Nie, X., Yoo, Y., Hewakuruppu, H., Sullivan, J., Krishna, A., & Lee, J. (2020). Cool white polymer coatings based on glass bubbles for buildings. *Scientific Reports, 10*(1). https://doi.org/10.1038/s41598-020-63027-2.

Oke, T. R. (1982). The energetic basis of the urban heat island. *Quarterly Journal of the Royal Meteorological Society, 108*(455), 1–24. https://doi.org/10.1002/qj.49710845502.

Park, B., & Krarti, M. (2016). Energy performance analysis of variable reflectivity envelope systems for commercial buildings. *Energy and Buildings, 124*, 88–98. https://doi.org/10.1016/j.enbuild.2016.04.070.

Perez, G., Allegro, V. R., Corroto, M., Pons, A., & Guerrero, A. (2018). Smart reversible thermochromic mortar for improvement of energy efficiency in buildings. *Construction and Building Materials, 186*, 884–891. https://doi.org/10.1016/j.conbuildmat.2018.07.246.

Pigliautile, I., Chàfer, M., Pisello, A. L., Pérez, G., & Cabeza, L. F. (2020). Inter-building assessment of urban heat island mitigation strategies: Field tests and numerical modelling in a simplified-geometry experimental set-up. *Renewable Energy, 147*, 1663–1675. https://doi.org/10.1016/j.renene.2019.09.082.

Piselli, C., Saffari, M., de Gracia, A., Pisello, A. L., Cotana, F., & Cabeza, L. F. (2017). Optimization of roof solar reflectance under different climate conditions, occupancy, building configuration and energy systems. *Energy and Buildings, 151*, 81–97. https://doi.org/10.1016/j.enbuild.2017.06.045.

Piselli, C., Pisello, A. L., Saffari, M., de Gracia, A., Cotana, F., & Cabeza, L. F. (2019). Cool roof impact on building energy need: The role of thermal insulation with varying climate conditions. *Energies, 12*(17). https://doi.org/10.3390/en12173354.

Pisello, A. L. (2015). Experimental analysis of cool traditional solar shading systems for residential buildings. *Energies, 8*(3), 2197–2210. https://doi.org/10.3390/en8032197.

Pisello, A. L., & Cotana, F. (2014). The thermal effect of an innovative cool roof on residential buildings in Italy: Results from two years of continuous monitoring. *Energy and Buildings, 69*, 154–164. https://doi.org/10.1016/j.enbuild.2013.10.031.

Pisello, A. L., Pignatta, G., Castaldo, V. L., & Cotana, F. (2014). Experimental analysis of natural gravel covering as cool roofing and cool pavement. *Sustainability (Switzerland), 6*(8), 4706–4722. https://doi.org/10.3390/su6084706.

Pisello, A. L., Castaldo, V. L., Pignatta, G., Cotana, F., & Santamouris, M. (2016a). Experimental in-lab and in-field analysis of waterproof membranes for cool roof application and urban heat island mitigation. *Energy and Buildings, 114*, 180–190. https://doi.org/10.1016/j.enbuild.2015.05.026.

Pisello, A. L., Fortunati, E., Mattioli, S., Cabeza, L. F., Barreneche, C., Kenny, J. M., & Cotana, F. (2016b). Innovative cool roofing membrane with integrated phase change materials: Experimental characterization of morphological, thermal and optic-energy behavior. *Energy and Buildings, 112*, 40–48. https://doi.org/10.1016/j.enbuild.2015.11.061.

Pisello, A. L., Castaldo, V. L., Fabiani, C., & Cotana, F. (2016c). Investigation on the effect of innovative cool tiles on local indoor thermal conditions: Finite element modeling and continuous monitoring. *Building and Environment, 97*, 55–68. https://doi.org/10.1016/j.buildenv.2015.11.038.

Pisello, A.L., Paolini, R., Diamanti, M. V., Fortunati, E., Castaldo, V., & Torre, L. (2016d). Nanotech-based cool materials for building energy efficiency. In F. Pacheco Torgal, C. Buratti, S. Kalaiselvam, C.-G. Granqvist, & V. Ivanov (Eds.), *Nano and biotech based materials for energy building efficiency*. (pp. 245–278), Springer International Publishing. https://doi.org/10.1007/978-3-319-27505-5_9.

Pisello, A. L., Castaldo, V. L., Piselli, C., Fabiani, C., & Cotana, F. (2017). Thermal performance of coupled cool roof and cool façade: Experimental monitoring and analytical optimization procedure. *Energy and Buildings, 157*, 35–52. https://doi.org/10.1016/j.enbuild.2017.04.054.

Qin, Y., Liang, J., Tan, K., & Li, F. (2016). A side by side comparison of the cooling effect of building blocks with retro-reflective and diffuse-reflective walls. *Solar Energy, 133*, 172–179. https://doi.org/10.1016/j.solener.2016.03.067.

Qu, Y., Milliez, M., Musson-Genon, L., & Carissimo, B. (2012). Numerical study of the thermal effects of buildings on low-speed airflow taking into account 3D atmospheric radiation in urban canopy. *Journal of Wind Engineering and Industrial Aerodynamics, 104–106*, 474–483. https://doi.org/10.1016/j.jweia.2012.03.008.

Ramamurthy, P., Bou-Zeid, E., Smith, J. A., Wang, Z., Baeck, M. L., Saliendra, N. Z., et al. (2014). Influence of subfacet heterogeneity and material properties on the urban surface energy budget. *Journal of Applied Meteorology and Climatology, 53*(9), 2114–2129. https://doi.org/10.1175/JAMC-D-13-0286.1.

Revel, G. M., Martarelli, M., Emiliani, M., Celotti, L., Nadalini, R., De Ferrari, A., Hermanns, S., & Beckers, E. (2014). Cool products for building envelope–Part II: Experimental and numerical evaluation of thermal performances. *Solar Energy, 105*, 780–791. https://doi.org/10.1016/j.solener.2014.02.035.

Rossi, F., Morini, E., Castellani, B., Nicolini, A., Bonamente, E., Anderini, E., & Cotana, F. (2015a). Beneficial effects of retroreflective materials in urban canyons: Results from seasonal monitoring campaign. *Journal of Physics: Conference Series, 655*(1). https://doi.org/10.1088/1742-6596/655/1/012012.

Rossi, F., Castellani, B., Presciutti, A., Morini, E., Filipponi, M., Nicolini, A., & Santamouris, M. (2015b). Retroreflective façades for urban heat island mitigation: Experimental investigation and energy evaluations. *Applied Energy, 145*, 8–20. https://doi.org/10.1016/j.apenergy.2015.01.129.

Rossi, F., Castellani, B., Presciutti, A., Morini, E., Anderini, E., Filipponi, M., & Nicolini, A. (2016). Experimental evaluation of urban heat island mitigation potential of retro-reflective pavement in urban canyons. *Energy and Buildings, 126*, 340–352. https://doi.org/10.1016/j.enbuild.2016.05.036.

Rosso, F., Pisello, A. L., Castaldo, V. L., Fabiani, C., Cotana, F., Ferrero, M., & Jin, W. (2017a). New cool concrete for building envelopes and urban paving: Optics-energy and thermal assessment in dynamic conditions. *Energy and Buildings, 151*, 381–392. https://doi.org/10.1016/j.enbuild.2017.06.051.

Rosso, F., Pisello, A. L., Castaldo, V. L., Ferrero, M., & Cotana, F. (2017b). On innovative cool-colored materials for building envelopes: Balancing the architectural appearance and the thermal-energy performance in historical districts. *Sustainability (Switzerland), 9*(12). https://doi.org/10.3390/su9122319.

Saafi, K., & Daouas, N. (2018). A life-cycle cost analysis for an optimum combination of cool coating and thermal insulation of residential building roofs in Tunisia. *Energy, 152*, 925–938. https://doi.org/10.1016/j.energy.2018.04.010.

Saitoh, T. S., Shimada, T., & Hoshi, H. (1996). Modeling and simulation of the Tokyo urban heat island. *Atmospheric Environment, 30*(20), 3431–3442. https://doi.org/10.1016/1352-2310(95)00489-0.

Santamouris, M. (2014). Cooling the cities—A review of reflective and green roof mitigation technologies to fight heat island and improve comfort in urban environments. *Solar Energy, 103*, 682–703. https://doi.org/10.1016/j.solener.2012.07.003.

Santamouris, M., Synnefa, A., Kolokotsa, D., Dimitriou, V., & Apostolakis, K. (2008). Passive cooling of the built environment – use of innovative reflective materials to fight heat islands and decrease cooling needs. *International Journal of Low-Carbon Technologies, 3*(2), 71–82. https://doi.org/10.1093/ijlct/3.2.71

Santamouris, M., Synnefa, A., & Karlessi, T. (2011). Using advanced cool materials in the urban built environment to mitigate heat islands and improve thermal comfort conditions. *Solar Energy, 85*(12), 3085–3102. https://doi.org/10.1016/j.solener.2010.12.023.

Santamouris, M., Gaitani, N., Spanou, A., Saliari, M., Giannopoulou, K., Vasilakopoulou, K., & Kardomateas, T. (2012). Using cool paving materials to improve microclimate of urban areas—Design realization and results of the flisvos project. *Building and Environment, 53*, 128–136. https://doi.org/10.1016/j.buildenv.2012.01.022.

Sleiman, M., Kirchstetter, T. W., Berdahl, P., Gilbert, H. E., Quelen, S., Marlot, L., et al. (2014). Soiling of building envelope surfaces and its effect on solar reflectance—Part II: Development of an accelerated aging method for roofing materials. *Solar Energy Materials and Solar Cells, 122*, 271–281. https://doi.org/10.1016/j.solmat.2013.11.028.

Susan, C. (2007). Urbanization and global environmental change: Local effects of urban warming. *Geographical Journal, 60*, 83–87. https://doi.org/10.1111/j.1475-4959.2007.232.

Synnefa, A., & Santamouris, M. (2016). Mitigating the urban heat with cool materials for the buildings' fabric. *Urban Climate Mitigation Techniques*, 67–92. https://doi.org/10.4324/9781315765839.

Synnefa, A., Santamouris, M., & Akbari, H. (2007). Estimating the effect of using cool coatings on energy loads and thermal comfort in residential buildings in various climatic conditions. *Energy and Buildings, 39*(11), 1167–1174. https://doi.org/10.1016/j.enbuild.2007.01.004.

Synnefa, A., Karlessi, T., Gaitani, N., Santamouris, M., Assimakopoulos, D. N., & Papakatsikas, C. (2011). Experimental testing of cool colored thin layer asphalt and estimation of its potential to improve the urban microclimate. *Building and Environment, 46*(1), 38–44. https://doi.org/10.1016/j.buildenv.2010.06.014.

Toparlar, Y., Blocken, B., Maiheu, B., & van Heijst, G. J. F. (2017). A review on the CFD analysis of urban microclimate. *Renewable and Sustainable Energy Reviews, 80*, 1613–1640. https://doi.org/10.1016/j.rser.2017.05.248.

Vardoulakis, S., Dimitroulopoulou, C., Thornes, J., Lai, K. M., Taylor, J., Myers, I., et al. (2015). Impact of climate change on the domestic indoor environment and associated health risks in the UK. *Environment International, 85*, 299–313. https://doi.org/10.1016/j.envint.2015.09.010.

Vukovich Dunn, F. M. J. W., & Crissman, B. W. (1976). A theoretical study of the St. Louis Heat Island: The wind and temperature distribution. *Journal of Applied Meteorology, 15*(5), 417–440. https://doi.org/10.1175/1520-0450(1976)0152.0.CO;2.

Wang, X., Kendrick, C., Ogden, R., & Maxted, J. (2008). Dynamic thermal simulation of a retail shed with solar reflective coatings. *Applied Thermal Engineering, 28*(8–9), 1066–1073. https://doi.org/10.1016/j.applthermaleng.2007.06.011.

Xie, N., Li, H., Abdelhady, A., & Harvey, J. (2019). Laboratorial investigation on optical and thermal properties of cool pavement nano-coatings for urban heat island mitigation. *Building and Environment, 147*, 231–240. https://doi.org/10.1016/j.buildenv.2018.10.017.

Xu, X., González, J. E., Shen, S., Miao, S., & Dou, J. (2018). Impacts of urbanization and air pollution on building energy demands—Beijing case study. *Applied Energy, 225*, 98–109. https://doi.org/10.1016/j.apenergy.2018.04.120.

Yang, Y. K., Kang, I. S., Chung, M. H., Kim, S. M., & Park, J. C. (2017). Effect of PCM cool roof system on the reduction in urban heat island phenomenon. *Building and Environment, 122*, 411–421. https://doi.org/10.1016/j.buildenv.2017.06.015.

Yiping, M., Xiong, Z., Beirong, Z., & Keru, W. (2002). Research on reversible effects and mechanism between the energy-absorbing and energy-reflecting states of chameleon-type building coatings. *Solar Energy, 72*(6), 511–520. Retrieved from http://cat.inist.fr/?aModele=afficheN&cpsidt=13841821.

Zhang, J., Zhang, K., Liu, J., & Ban-Weiss, G. (2016). Revisiting the climate impacts of cool roofs around the globe using an Earth system model. *Environmental Research Letters, 11*(8). https://doi.org/10.1088/1748-9326/11/8/084014.

Zhang, N., Chen, Y., Luo, L., & Wang, Y. (2017). Effectiveness of different urban heat island mitigation methods and their regional impacts. *Journal of Hydrometeorology, 18*(11), 2991–3012. https://doi.org/10.1175/JHM-D-17-0049.1.

Zhang, J., Mohegh, A., Li, Y., Levinson, R., & Ban-Weiss, G. (2018). Systematic comparison of the influence of cool wall versus cool roof adoption on urban climate in the Los Angeles basin. *Environmental Science and Technology, 52*(19), 11188–11197. https://doi.org/10.1021/acs.est.8b00732.

Zhao, L., Oppenheimer, M., Zhu, Q., Baldwin, J. W., Ebi, K. L., Bou-Zeid, E., et al. (2018). Interactions between urban heat islands and heat waves. *Environmental Research Letters, 13*(3). https://doi.org/10.1088/1748-9326/aa9f73.

Zhou, Y., & Marshall Shepherd, J. (2010) Atlanta's urban heat island under extreme heat conditions and potential mitigation strategies. *Natural Hazards, 52*(3), 639–668. https://doi.org/10.1007/s11069-009-9406-z

Zinzi, M. (2016). Exploring the potentialities of cool facades to improve the thermal response of Mediterranean residential buildings. *Solar Energy, 135*, 386–397. https://doi.org/10.1016/j.solener.2016.06.021.

Zinzi, M., & Fasano, G. (2009). Properties and performance of advanced reflective paints to reduce the cooling loads in buildings and mitigate the heat island effect in urban areas. *International Journal of Sustainable Energy, 28*(1–3), 123–139. https://doi.org/10.1080/14786450802453314.

Zinzi, M., Carnielo, E., & Agnoli, S. (2012). Characterization and assessment of cool coloured solar protection devices for Mediterranean residential buildings application. *Energy and Buildings, 50*, 111–119. https://doi.org/10.1016/j.enbuild.2012.03.031.

Index

M. Palme, A. Salvati (eds.), *Urban Microclimate Modelling for Comfort and Energy Studies*, https://doi.org/10.1007/978-3-030-65421-4

B

C

U

9783030654207